Biomedical Engineering Design

Biomedical Engineering Design

Joseph Tranquillo, PhD
Associate Provost for Transformative Teaching and Learning
Professor of Biomedical Engineering
Bucknell University
Lewisburg, Pennsylvania

Jay Goldberg, PhD, PE
Professor of Practice in Biomedical Engineering
Department of Biomedical Engineering
Marquette University and the Medical College of Wisconsin
Milwaukee, Wisconsin

Robert Allen, PhD, PE
Research Assistant Professor of Obstetrics & Gynecology and Women's Health
Albert Einstein School of Medicine
Bronx, New York;
Retired Faculty
Department of Biomedical Engineering
Johns Hopkins University
Baltimore, Maryland

ACADEMIC PRESS
An imprint of Elsevier

ELSEVIER

Library of Congress Cataloging-in-Publication Data
A catalog record for this book is available from the Library of Congress

British Library Cataloguing-in-Publication Data
A catalogue record for this book is available from the British Library

For information on all Academic Press publications visit our website at
https://www.elsevier.com/books-and-journals

ISBN: 978-0-12-816444-0

Publisher: Katey Birtcher
Acquisitions Editor: Steve Merken
Editorial Project Manager: Beth LoGiudice
Publishing Services Manager: Shereen Jameel
Senior Project Manager: Manikandan Chandrasekaran
Designer: Ryan Cook

Printed in the United States of America

Last digit is the print number: 9 8 7 6 5 4 3 2

We dedicate this book to our families, without whose support this book would not have been possible. Time spent away from them and family activities were often necessary, and their support during these times is greatly appreciated. During the writing of this book, three of our children were married, three grandchildren were born, and two of our parents, Elaine Goldberg and Tony Tranquillo, passed away. We hope to honor their memories through this labor of love.

Preface

We, the coauthors of this book, followed different pathways into medical device design education. Two of us took the traditional research and teaching path, and the other took a path through industry prior to joining academia. Once we discovered the joys of teaching students about the challenges and rewards of the medical device design process, we chose to make it a focus of our careers. Over the past several decades, through our collective experience of more than 65 years directing design programs at Bucknell University, Marquette University, and Johns Hopkins University, respectively, we have taught a few thousand students about the design process. Although we have witnessed many changes in pedagogy and design education over the last 30 years, one thing remained constant—our desire to help engineering students develop into professionals who have a positive impact on the world. For this reason, teaching biomedical engineering design became a lifelong passion for each of us.

Our objective in writing this book is to share our knowledge and experience from academic and industry practice with students and other design educators. In teaching design, we each implement a hands-on, team-based, experiential learning approach, through open-ended problems. Our exercises aim to have students create deliverables that apply to their project and are similar to those required by medical device companies. We often remind our students that design is an iterative and nonlinear process, and some steps may be revisited multiple times. We reflect these practices throughout the text.

While helping students navigate a wide range of design projects, we observed two common themes that influenced the way we wrote and organized this book. First, an academic design process includes most of the same phases as that of an industry design project, but due to time and resources constraints, some phases cannot be included. We felt that students needed a text that would address the unique requirements of an academic design project (senior capstone design projects and underclass design projects) as well as the knowledge, skills, and mindsets required to create value in an industry setting. Second, we observed that there were several areas of nontechnical expertise needed for successful new product development that engineers in industry would be expected to have. Creating an awareness of these knowledge areas is beneficial to all students, particularly those who pursue a career in the medical device industry.

We hope this text will inspire you to become as passionate about Biomedical Engineering design as we are.

Joseph Tranquillo
Jay Goldberg
Robert Allen

To Instructors

Teaching ancillaries for this book, including instructor guide, and image bank, are available online to qualified instructors. Visit https://inspectioncopy.elsevier.com/book/details/9780128164440 for more information and to register for access.

Acknowledgments

"It takes a village…" to write a book, and many people contributed to this book in different ways. Throughout the years, we have taught thousands of students about design and often learned as much from them as they learned from us. We thank them for trusting us to prepare them for successful careers in biomedical engineering and related professions and for sharing their experiences with us.

We appreciate the advice and support of our colleagues in our respective departments, alumni, reviewers of the early manuscripts, and those who shared their perspectives and advice through the many breakout boxes included in the text.

Specifically, we thank members of our village—BME-IDEA, ASEE BED, and Capstone Design Conference communities—for their opinions, advice, and encouragement during the writing of this book.

The original inspiration for this book came from a Kern Entrepreneurial Engineering Network (KEEN) workshop that focused on methods for instilling an entrepreneurial mindset within our biomedical engineering students. Their support, hospitality, and generosity made it possible for us to collaborate in person on several occasions.

Our publishers provided helpful guidance during this process. Our illustrator, Dave Rank, collaborated with us to produce many of the figures in the book.

We thank God for the opportunity to collaborate on this book and to give back to the biomedical engineering design community.

Finally, we lovingly thank our spouses—Lisa, Susie, and Edith—for their patience, advice, and moral support during this project.

Reviewers

John D. DesJardins, PhD
Robert B. and Susan B. Hambright Leadership Professor
Department of Bioengineering
Director, Laboratory of Orthopaedic Design and Engineering
Clemson University
Clemson, South Carolina

Robin Hissam, PhD
Director of Undergraduate Education
Teaching Associate Professor
Department of Chemical and Biomedical Engineering
Benjamin M. Statler College of Engineering and Mineral Resources
West Virginia University
Morgantown, West Virginia

James Keszenheimer, PhD, MBA
Professor of Engineering Practice
Department of Biomedical Engineering
University of Akron
Akron, Ohio

Hisham Mohamed, PhD
Senior Lecturer, Biomedical Engineering
Rensselaer Polytechnic Institute
Troy, New York

D. Patrick O'Neal, PhD
Chief Technology Officer and Vice President
Wavegate Corporation
Dallas, Texas;
Associate Professor of Biomedical Engineering
College of Engineering and Science
Louisiana Tech University
Ruston, Louisiana

Charles W. Peak, PhD
Instructional Assistant Professor
Department of Biomedical Engineering
The College of Engineering
Texas A&M University
College Station, Texas

Steven S. Saliterman, MD
Adjunct Professor
Department of Biomedical Engineering
College of Science and Engineering
University of Minnesota
Minneapolis, Minnesota

Mary Staehle, PhD
Associate Professor and Undergraduate Coordinator
Department of Biomedical Engineering
Henry M. Rowen College of Engineering
Rowen University
Glassboro, New Jersey

Anandhi Upendran, PhD
Director of Biomedical Innovation
School of Medicine
University of Missouri—Columbia
Columbia, Missouri

Wei Yin, PhD
Associate Professor
Department of Biomedical Engineering;
Associate Dean for Diversity and Outreach (CEAS)
College of Engineering and Applied Sciences
Stony Brook University
Stony Brook, New York

Contents

Introduction

To design is much more than simply to assemble, to order, or even to edit: it is to add value and meaning, to illuminate, to simplify, to clarify, to modify, to dignify, to dramatize, to persuade, and perhaps even to amuse. To design is to transform prose into poetry.
– Paul Rand, Graphic Designer

Chapter outline

Biomedical Engineering Design. https://doi.org/10.1016/B978-0-12-816444-0.00001-8

1.1 Introduction

How does an idea for a new healthcare technology make it out into the world, where it can have an impact? There are many pathways to making a difference; some are well worn and others are emerging. The aim of this text is to help you discover one of the most important pathways: the process of designing a medical device. In this introductory chapter, we explain where device design fits into the overall goal of improving health and well-being and define terms and processes that will guide your journey through the text.

Medical device design is situated at the intersection of the Healthcare Ecosystem, Industry, and Engineering Design, as shown in Figure 1.1. Each of these areas is dynamically changing, has its own established norms and practices, and is composed of many players with their own unique needs and motivations. They overlap and inform one another, as advances in one area can lead to innovations in the others. Having a holistic view of how these three areas intersect will enable you to be an impactful designer who can effectively transform an idea into a medical device.

This text is written for students engaged in a medical device design project. For that reason, we have made several pedagogical choices. First, we have written in an engaging style to keep your interest. When possible, we address you (or your team) directly, as if we are having a conversation. Second, our primary goal is to help you navigate a typical academic design project, whether it is a college club, introductory or capstone design course, or graduate program. The processes, methods, and terminology introduced, however, are used in industry and we provide you with many examples of industry products and projects. This approach is meant to keep you focused on learning the design process while preparing you for careers in the medical device industry. Third, while the focus is on medical device design, woven into the text are the many nontechnical activities that are necessary to commercialize a medical device. Finally, a text is a repository of knowledge, yet design is a set of activities. Although this text

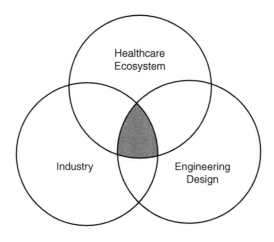

FIGURE 1.1

Medical Device Design at the intersection between the Healthcare Ecosystem, Industry, and Engineering Design.

can serve as a guide, you must do more than simply read and study to fully understand engineering design. Engaging in a unique design project will add additional context and richness to the material presented; experience is the best way to learn the design process.

The first three sections of this introductory chapter are a brief overview of the healthcare ecosystem, industry, and engineering design. The second half of this chapter turns to the unique aspects of medical device design and clarifies the differences and similarities between design in academic and industry settings.

1.2 The Healthcare Ecosystem

Health care accounts for more than 10% of the total expenditures of most developed countries, and has been hovering between 17% and 18% since 2009 in the United States. The **healthcare ecosystem** is composed of many components including the following:

- patients, family members, caregivers, and other healthcare providers
- technological advancements
- basic and clinical research
- the flow of payments and products
- the economics of insurance reimbursement
- governmental and international organizations
- regulatory policies and laws
- medical and political advocacy and nonprofit groups
- healthcare professionals and societies
- healthcare standards and best practices
- broad public perceptions and assumptions

As you read this text and navigate your design process, you will be asked to consider all of these important elements of the healthcare ecosystem.

1.2.1 The Dynamics of the Healthcare Ecosystem

As in any complex system, there are many variables that equilibrate competing interests, not unlike the homeostasis of a biological organism. It is this delicate balance that enables the system to adapt over time through changes in processes, policies, financial incentives (or penalties), marketing campaigns, new devices, and public perception. Many of these dimensions are inherited from the wider sociopolitical environment and therefore vary from culture to culture. Determining the value of an innovation to the healthcare ecosystem therefore requires consideration of several factors; it is often very challenging to predict the holistic value of an innovation. Like a biological organism, the healthcare ecosystem will often oppose an innovation—what one group considers a good idea is often not shared by others. For example, radiologists may oppose automatic interpretation of MRI scans, not because the interpretation is incorrect, but because it encroaches upon their specialty as currently practiced. Successful medical innovators understand the wider dynamics at play in the healthcare ecosystem and will anticipate the many barriers to commercialization and adoption.

1.2.2 Stakeholders, Users, Customers, and Shareholders

As value creation is at the center of all innovation, it is critical to ask *who* will benefit from the value created. Several terms are used throughout this text. A **stakeholder** is anyone who cares about or has influence over the creation or adoption of a product. They may be individual patients, healthcare providers, medical device companies, insurance companies, governmental agencies or professional organizations, as well as society at large. There are also specific types of stakeholders. A **user** benefits (receives value) from the adoption of a product. In the healthcare ecosystem, users often include patients, healthcare providers, and healthcare facilities. A **customer** is willing to pay for a product. Customers can also be users, such as when someone purchases a digital thermometer for personal use. In a hospital, a purchasing agent (the customer) often has the authority to purchase a device that will be used by a physician (one type of user) in a procedure on a patient (a second type of user and beneficiary of the technology). A **shareholder** stands to gain or lose financially from decisions that impact any aspect of the healthcare ecosystem. As pointed out earlier, there are many stakeholders in the healthcare ecosystem, each with their own concerns and opinions that conflict with one another at times.

1.3 Industry

Industry is a term that refers collectively to all of the companies that fall into a particular category of value creators by selling products or services to customers. A **company** is a legal entity that transforms ideas into products and services, often scaling them so that many users benefit from the innovation. For example, Zimmer Biomet is a medical device company with thousands of products that are sold around the world. Between the entire medical device industry and individual companies are *segments*, composed of several companies that sell similar products in the same markets; Zimmer Biomet, Stryker, DePuy Synthes, and Smith & Nephew all work broadly in the orthopedics segments. Many larger companies, such as Johnson & Johnson or Medtronic, work in several segments. The medical device industry spans many segments including orthopedics, imaging, wound care, dialysis, dental applications, medical education simulators, and many others.

1.3.1 Industry Innovation

To stay competitive in a changing world, companies must continuously innovate. This is especially true for the medical device industry, where rapid changes are simultaneously occurring in the societal, legal, political, scientific, and medical domains that compose the wider healthcare ecosystem. For these reasons, innovation is a nonlinear and often unpredictable process. To navigate an innovation process, many companies have implemented frameworks that allow them to focus their attention and resources only on new products that show the most promise. In this section, we will provide a brief overview of a general process for industry innovation. Although this section and much of the text will focus on for-profit industry, many of the same processes are also used by nonprofit, philanthropic, and governmental organizations.

A high-level overview of moving an innovation to market is summarized in Figure 1.2. This **stage-gate model** (sometimes also called phase-gate) is used in many companies to develop products by starting with many ideas and then eliminating those that do not meet specific requirements. As a project advances through the stages of this process, it is reviewed at various checkpoints (gates) and if certain

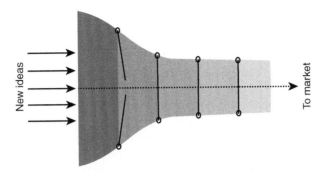

FIGURE 1.2

The Stage-Gate Model of Industry Innovation. During each stage, projects are evaluated against criteria specific to that particular stage, indicated by the changing color (contrast) scheme. To get from one stage to the next, projects must pass through successive gates (only the first gate is shown as open). Only ideas that make it through all gates are released to the market. Companies carefully consider whether a project should move to the next stage.

requirements are not met, it is not allowed to pass through the gate to proceed to the next phase. The first two of these requirements are fulfillment of a customer need and existence of a market opportunity. This involves asking two questions; will a solution solve the problem? And will anyone purchase the product? A third requirement is technical feasibility; could the technology to be developed do what it is supposed to do, and can we make it? A fourth requirement is financial viability; will expected sales revenue over the lifecycle of the product be enough to allow the company to recover its investment in the project and meet its financial goals within a reasonable amount of time? In business parlance, this is referred to as the **return on investment** (ROI). Additional important considerations include legal constraints, regulatory barriers, company values and strategic goals, and societal pressures coming from the wider healthcare ecosystem.

As projects move from one stage to the next, the four requirements discussed above are assessed for each project along with other criteria deemed important to the company. Projects continue to move through the stage-gate process unless they are deemed no longer viable. This can be due to many factors including a lack of a need by customers, changing market conditions, or unreasonable production costs, thereby reducing the ROI for the project. As moving to the next stage requires further expenditure of limited resources, only some projects advance, while most others are terminated or suspended. For this reason, the stage-gate model is often graphically represented as a funnel, whereby many ideas enter but only a few are realized (narrowing of the funnel) and enter the marketplace. The underlying logic behind the stage-gate model is that investments in time, attention, and resources are made early in the process to reduce the number of feasible projects to only those that show real promise.

1.3.2 Multifunctional Groups

In a medical device company, there will typically be several departments that coordinate their actions to move a product through the stage-gate process. Each has its own responsibilities, processes, and core competencies, but all share the goal of introducing a successful new device or technology to the market.

The most common departments within a device company are (in alphabetical order):

- **Finance** estimates the ROI of a new project. They track budgets as a project moves through the stage-gate process.
- **Legal** determines novelty and patentability as well as freedom to operate, apply for copyrights and trademarks, and to anticipate liability concerns. They also manage the patent application process and legal actions involving intellectual property, product liability, and recalls.
- **Manufacturing** scales the production of a completed design. This includes designing, building, and debugging any required tooling or production equipment as well as determining workflows, assembly line layouts, and training production personnel to assemble, package, sterilize and ship new products.
- **Marketing and Sales** identifies and clarifies the unmet needs of customers and obtains feedback on preliminary ideas from users. They generate sales forecasts, create marketing and sales plans, train sales personnel, and work with financial personnel to determine a preliminary ROI for a new project.
- **Operations and Purchasing** finds components that meet specifications and cost requirements, qualifies venders, determines how to manage supply chains, and determines the best ways to move products from the factory to customers.
- **Regulatory Affairs** determines pathways to commercialization in both domestic and international markets and assures compliance with applicable standards and regulations. They are responsible for regulatory submissions, government audits, product recalls, and implementation and mainte-nance of the company's quality system.
- **Research and Development** assesses technical feasibility, converts customer needs into a tech-nical project statement and functional specifications, generates concepts, develops and builds prototypes, and performs functional tests. Medical design engineers are typically part of this department, which is expanded upon in Section 1.3.3 and throughout the text.

Other departments play a wide range of internal roles and often include Human Resources (HR) and Information Technology (IT).

Many of these departments continue to support a product after it has been introduced to the market. Product performance is monitored and evaluated during *post-market surveillance* to ensure that the product is still safe and is continuing to meet customer needs and delivering value to the healthcare ecosystem. A product on the market might reenter the stage-gate process to become a new project to create a new improved version of the product. Upon reentry, the aim is to add value in the form of a new feature, improvement on an existing feature, added safety, cost reduction, or compliance with a new regulation or standard.

No one department can move a potential product through the stage-gate on its own. There is always a great deal of discussion and sharing of information across functional groups within a company. As an engineer, the more you know about the disciplines represented in other departments (i.e., their pro-cesses, goals, language, and perspectives), the more effective and efficient you will be in your role. Throughout the text, we explain the type of information engineers often receive from other functional groups at particular phases of product development. The final chapter explains in more detail the roles that other functional groups play as a product moves through the stage-gate.

1.3.3 Differences Across Companies

Companies tailor their processes to the products they produce, as well as their own unique organizational needs, structure, history, culture, and core competencies. For example, some companies initiate projects by recognizing an unmet market need (referred to as market-pull), while others base a project on a unique core internal technology in anticipation of a market for the new product (referred to as technology-push). As a result, there are variations from company to company in how projects are initiated and managed, timelines, the size and make-up of the teams involved, financial investment, and how success is defined and measured. Every organization navigates innovation a bit differently. Upon being hired, new employees are often taught these specifics and quickly learn the norms and processes of that particular company.

Large companies generally have departments that perform each of the functions in the preceding section. They also likely have many products in various phases of the stage-gate moving along simultaneously. For that reason, decisions regarding particular products are made relative to other products. For example, initiating a product redesign could depend upon company resources already committed to other projects. Startup companies typically have fewer resources than more established companies and do not initially have enough employees to sustain fully functional departments. As a result, employees in startup companies often play multiple roles, or the company may hire outside help (often called outsourcing) to perform critical tasks.

1.3.4 Products, Projects, and Processes

It is important to distinguish between the terms product, project, and process as used in industry and throughout this text. A **product** is a commercialized solution to a problem or response to an opportunity in the marketplace. It is a holistic term that refers to not only the physical object (or service/software), but also the labeling, packaging, servicing, marketing and other aspects that comprise the totality of the user-customer experience. A **project** involves all of the activities and resources (e.g., people, materials, equipment, and money) that are required to make a product a reality. Projects typically have a start and finish date as well as a specific allocation of resources. The product development **process** includes the sequence of activities beginning with the discovery of an opportunity and ending in the production, sales, and delivery of a product. A *project* is executed by following a *process*, resulting in a *product*.

1.3.5 Designs, Devices, Technologies, and Innovations

In creating a medical device, you will encounter four terms: design, device, technology, and innovation. When used as nouns, these terms are related to one another but have distinct meanings in this text. A **design** is a plan for an object or system (often represented in a drawing or prototype) that solves a problem by achieving specific functions. A **device** is a physical realization of the design that performs the specific functions. **Technology** is the application of engineering and scientific knowledge for a practice purpose. Several technologies are often mixed and matched to achieve the functions of the device. An **innovation** is a device that solves a problem in a new and better way and creates value for a specific population, typically by being commercialized. To summarize, a *design* incorporates various *technologies* that, when realized in a *device*, may become an *innovation* when adopted by the healthcare ecosystem.

Derivatives of these terms are also used. For example, to innovate (a verb) is to follow a process that will lead to the commercialization of a new device that provides value to a population. Similarly, a designer (a noun) is someone who follows the design process to create a device.

1.4 Engineering Design

Engineers engage in many critical activities that help move a product through the stage-gate process. The collection of these activities is often referred to as the **engineering design process**. Because each medical device company is different, engineering design processes vary. In fact, a summary published in 2008 identified over 100 unique design processes, many of them linked to particular kinds of products or industries. Rather than present one model in particular, this text follows a general design process, shown in Figure 1.3. Your understanding of each of the boxes, as well as how they are connected together, will emerge as you continue to read and engage in this text and your project.

1.4.1 Important Aspects of Engineering Design

Throughout the text we emphasize many core tenants of good engineering design. These include:

- **Design is purposeful and not tinkering.** Although it is fun to "geek out" by losing yourself in a hobby, engineering design has the end goal of solving a problem that matters.
- **Design is a systematic process that is inherently iterative.** The diagram in Figure 1.3 is a general framework. Each box is often referred to as a **design phase** that has its own output, which will become the input to the next phase. There are many checkpoints, reviews, and tests that provide feedback on progress, and it is not uncommon to revisit previous design decisions and phases based upon new information that could only have been gained by going through the process. A design process is only complete when the design solution has been fully validated and verified and design transfer has occurred.
- **Design draws upon technical knowledge and skills.** Many of the actions of a robust design process require the engineering abilities you have acquired over time. On the other hand, you may need to acquire new skills and knowledge to complete a design process.
- **Design involves wrong turns and failures.** Try to view these events as learning opportunities and ways to further clarify the nature of the problem and solution. Focus on what can be learned through failure and use this to move the process forward.
- **Design is always conducted under constraints.** A constraint is anything that limits the successful completion of a design project. Universal constraints for every project are time, funding, and the number of people available to work on a project. You also inherit constraints from the nature of your problem and the solution you choose to implement. For example, a new surgical tool will inherit constraints including how surgeons will use the instrument, existing patents for similar tools, industry standards, the ability to sterilize the materials, and manufacturability. You should determine the constraints that are unique to your problem as early as possible and then consider them throughout the design process to ensure that you are investing time and resources in solutions that are within the constraints.
- **Design is different for each project.** Even if you have previous design experience, approach each new project with fresh eyes and an open mind.

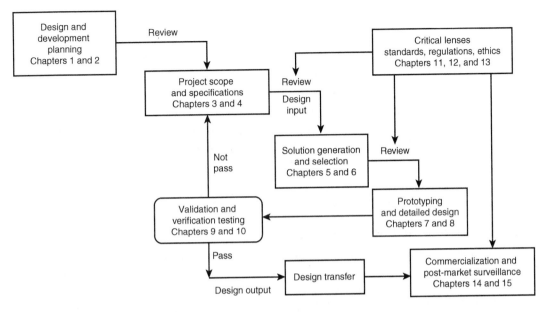

FIGURE 1.3

The design process that will guide you through this text. The model is adapted from the FDA Waterfall Model and the design controls requirements of the ISO 9001 (general quality management systems) and ISO 13485 standards (specific to medical devices). Although not shown graphically, documentation is ubiquitous throughout the design process and is discussed in Section 1.3.2.

- **Design is a social and collaborative enterprise.** Even when an individual or small team engages in a design project, the outcome will be improved by broadly engaging others as advisors, mentors, or consultants. A diverse team that includes people with different backgrounds, experiences, opinions, and perspectives increases the probability of generating and realizing a unique and impactful design solution.
- **Design is a professional activity.** Engineering designers provide a critical service to society and therefore hold themselves to a high standard. We are expected to comply with applicable laws and ethical principles in our decisions and interactions with others. You should strive to meet these same high standards.

1.4.2 Documenting Your Work and Decisions

Documentation is not explicitly shown in Figure 1.3 but it is deeply embedded in every aspect of the design process. The successful navigation of any design project generates a variety of types of work (e.g., written statements, sketches, photographs, prototypes, testing protocols and results, manufacturing instructions), and requires you to make hundreds (or even thousands) of decisions. It is not possible to keep all of this "in your head," especially if you are working as a team. All of your work, whether or not it eventually ends up in your final solution, must be documented. Documentation is part of design control standards, necessary for regulatory approval, and mandated by every company that manufactures products. Documentation is also practical, as it will help you retrace your steps if you make a wrong turn, help you justify your decisions, and demonstrate that you followed a robust design process. Projects are often passed from one team to another in both industry and academia, further emphasizing the need for clear and complete record of the evolution of your design.

Although tempting to view documentation as a distraction from other design activities, documentation will not be overly burdensome if it becomes a habit; it is always easier to record work in the moment than to recreate it after the fact. Chapter 2 introduces documentation as one of the major components of the Design Development and Planning phase. In particular, you will learn to create a repository for your work, called a **Design History File** (DHF). In every subsequent chapter, we offer guidance on how to document your work at that phase of the design process. At all phases of design, a DHF is therefore a complete and up-to-date record of your work and decisions.

1.4.3 The Role of the Design Engineer

Engineers are most often a part of the Research and Development department (often abbreviated as R&D) within a company. Although research and development are often included in the same department, they play different but complimentary roles in the creation of technical products. Companies employ engineers to work on both sides of R&D; it is therefore important to understand the differences. In general, those on the research side generate ideas that could enter into the stage-gate process. Their functions within a company include translating basic science findings into practice, exploring ideas for new-to-the-world products, and identifying needs that stakeholders are not able to articulate. In general, those on the development side are responsible for evaluation design and moving projects through the stage-gate using the design process shown in Figure 1.3.

This text focuses on the role of design engineers who primarily contribute to product development. However, there is rarely a division between research and development. There is always communication between engineers and scientists in the R&D department. How tightly these two groups are coupled depends upon the company. Throughout your career you may find yourself gravitating more toward one side of R&D or the other, or even being asked to allocate time to both as part of your job description.

1.4.4 Design and Engineering Design

The noun *design* has many meanings and modifiers (e.g., engineering, industrial, interior, graphic). When used with no modifier, design is nonspecific but often refers to aesthetic and artistic design such as choosing color schemes, typefaces, graphical representations of data, and, more broadly, how to create a good user experience based upon how our senses perceive stimuli. For example, the late Paul Rand, who is quoted at the beginning of the chapter, was the graphic designer behind the famous logos of IBM, UPS, and ABC. In contrast, engineering design focuses on the creation and integration of functions (often hidden from a user) that achieve a technical goal. These two sides of design often meet in *user interface design* (sometimes called the human-machine interface), which considers how technical functions can be accessed by a user in a way that is safe and effective. As safety and efficacy are critical characteristics of all medical devices, we encourage you to consider your users at every stage of your design project.

1.5 Medical Device Design

Medical device design exists at the intersection of the Healthcare Ecosystem, Industry, and Engineering Design, as shown in Figure 1.1. After briefly introducing each of these areas separately in the preceding sections, this section explores how all three come together. We begin with a high-level overview of the medical device industry, then list unique aspects of working in this area, and conclude with a conceptual framework for navigating your project.

1.5.1 The Medical Device Industry

The medical device design industry is composed of over 27,000 companies based in Europe; 1600 companies in Asia; 6500 companies in the United States; and a rapidly growing number of companies in Africa and South America. These companies and the products they produce make up one of the largest players in the healthcare ecosystem. There is considerable variation across companies in size, products sold, history, processes, and structures. For example, in the US more than 5000 medical device companies have fewer than 50 employees, yet only approximately 30 companies earn 80% of global medical device revenue. Companies such as Johnson & Johnson, Medtronic, and General Electric Healthcare have thousands of employees, manufacture many medical products, and have each been in existence for over 70 years.

1.5.2 The Unique Aspects of Medical Device Design

Designing a medical device includes all of the steps in the general engineering design process. There are, however, unique nontechnical skills, knowledge, and perspectives you will need to possess to successfully navigate a medical device design project. For example, a major difference between general engineering design and medical device design is that medical devices must demonstrate safety and efficacy and must be cleared by a regulatory agency prior to commercialization for use in a medical setting. Unique aspects of medical device design include:

- Fluency with medical and biological terminology required to understand healthcare problems and opportunities;
- Observation and listening skills to identify the needs of multiple stakeholders in the healthcare ecosystem and translate those needs into technical design objectives;
- Lifelong learning habits to keep up with rapid advancements in medical technology and our understanding of biology from the level of molecules to organisms;
- Awareness of changing national and international medical device standards that impact design decisions;
- Familiarity with national and international regulatory bodies in medical device design and how they impact the pathways to market, including animal testing and human clinical trials;
- Consideration of the complexities of healthcare economics, specifically the national and international reimbursement landscape that can act as a barrier to product commercialization;
- Understanding of how a medical product flows through the healthcare ecosystem, including sterilization, package design, the unique environments in which healthcare products are used, and any waste that is generated during the product lifecycle;

- Recognition of the complex ethical dimensions of commercializing medical devices. These considerations include animal and human testing, as well as how the development of healthcare technologies can exacerbate disparities in access to health care.

   As these points are derived from the healthcare ecosystem, you will find that all personnel working in the departments outlined in Section 1.3.2 consider these points at they navigate the stage-gate process.

1.5.3 The Medical Device Design Spiral

Navigating a medical device design project requires you to engage in many technical and nontechnical activities. These are inherited from Figure 1.1, the stage-gate model (Figure 1.2), the general engineering design process (Figure 1.3), and the eight unique aspects of medical device design listed in Section 1.4.2. To help you conceptualize how all of this can come together, Figure 1.4 shows a conceptual framework that moves a medical device project through four quadrants: Administrative, Medical, Technical, and Commercial. This is only a conceptual framework but will help you navigate your medical device design project. The spiral also highlights three important aspects of medical device design.

   A medical design engineer engages in all four quadrants, not only the technical quadrant. This is the first important takeaway from the spiral model. Completing tasks in the medical quadrant ensures that you are solving the right problem and doing so in a way that will actually solve the problem. Completing tasks in the commercial quadrant ensures that your solution has the potential to reach enough users to have the intended impact. Completing tasks in the administrative quadrant ensures that you are getting the input and information you need to keep your project moving forward on schedule. Integrated into each chapter are methods and advice on topics related to all quadrants. For example, you will learn how to run efficient meetings, observe clinical procedures, anticipate and mitigate project risks, search and evaluate relevant medical literature, and explore possible regulatory pathways and reimbursement strategies.

   Every project begins with a great deal of uncertainty and gains clarity over time. This is the second important takeaway from the spiral model. A project begins at or near the origin with very little clarity,

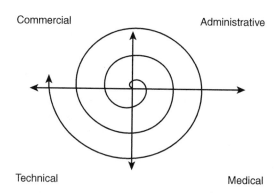

FIGURE 1.4

Spiral Model of a Medical Design Project. Projects typically start at or near the center of one of the quadrants and spiral outward until terminated, put on hold, or commercialized. As actions are taken and new information is gained, the scope of the project and product are further clarified.

then spirals outward, cycling through each quadrant multiple times. Each action you take results in new information about the project (and product) that was not known before. Therefore, each time a quadrant is revisited it is with new information, insights, ideas, and decisions, that moves the design process forward. As you discover new information, project requirements may change prompting you to take actions you may not have anticipated. However, you should not blindly move from one quadrant to the next in a serial manner; Figure 1.4 should not be interpreted as a suggestion to always follow administrative tasks with tasks in the medical quadrant. Rather, you will work in each quadrant simultaneously (in parallel). The spiral is only a conceptual model to remind you to revisit previous decisions as you move your project forward. It is not uncommon to repeat similar tasks (e.g., a patent search) several times throughout a project. Spiraling outward, you will gain confidence that a device is (or is not) moving closer to providing value to the healthcare ecosystem.

Projects that aim to impact human health require a sustained effort, usually through the actions of many people. To coordinate these actions, it is critical to follow an organized process and consider many perspectives. This is the third important takeaway from the spiral model. This text focuses on the design of medical devices; the process followed and perspectives considered mirror that aim. However, the topics, tools, and frameworks, with slight modifications, can be used to commercialize software, translate basic science findings to real-world applications, and design medical technologies for low resource settings.

1.6 The Academic Design Process

The goal of this text is to help you learn how design occurs in medical device companies. Most academic design experiences attempt to emulate this real-world design process, with the intent of helping you develop the skills, knowledge, mindsets, and habits needed to succeed in industry. However, there are differences in how industry engineers engage with the design process and how you will likely engage in medical device design in an academic setting. Because this text is primarily written to support academic design, its organization and content are biased toward navigating an academic design project. In this section we outline the major similarities and differences between design in industry and academia. As you progress throughout the text, we build upon these points.

1.6.1 Similarities Between Academic and Industry Design Processes

Medical device design in both industry and academia has the same goal; to add value to the healthcare ecosystem. Additional similarities between the academic and industry design processes are as follows:

- **The same design activities are executed and phases are followed in the same order.** Figure 1.3 and the points shared in Sections 1.4.1 and 1.5.2 form a common framework. Furthermore, progress within a design process is marked by key milestones reached by creating a variety of deliverables (e.g., written reports, prototypes, oral and poster presentations, test results).
- **The physical output of a design project is similar.** A functional prototype of a product is one goal of industry projects and the ultimate goal of academic design projects.
- **The techniques and tools used are similar.** The various templates, tables, charts, diagrams, and online resources presented in the text in the context of an academic design process are often also used in industry to make and document design decisions.

- **Every project is different.** The design spiral metaphor is a reminder that every project involves an iterative and evolving process of collecting the information needed to make the next design decision. As every problem is different, you never repeat the exact same process.

1.6.2 Differences Between Academic and Industry Design Processes

Most academic design experiences mimic an industry process. However, even industry-sponsored projects are modified to suit the academic environment. For this reason, your experience with design in an academic environment will differ from how engineering design is practiced in industry. Some of these differences are of degree (e.g., the level of emphasis placed on particular topics) whereas others are differences of priority (e.g., emphasizing learning over project outcomes). Additional ways in which industry and design processes differ include:

- **The scope of work and responsibilities is different.** Most companies have all of the departments listed in Section 1.2.2, and therefore specialists in each area contribute to moving a project through the stage-gate process. As part of an academic design experience, you must be more of a generalist and engage in a wider range of design activities to build your holistic awareness of how a medical product becomes a reality. For example, assessing possible regulatory pathways is a common (and important) academic exercise, whereas this task would generally not be asked of an industry engineer.
- **The success of a design project is measured differently.** In industry, a new product that meets or exceeds expected sales is considered to be a success. In academia, the emphasis is placed on learning, with prototype development and testing as a means to that end. The goal of engaging in an academic design process is to prepare you to successfully navigate design projects in the future.
- **Timelines and workloads differ.** In industry, the members of a design team are often dedicated to a single project, whereas in academia, you are typically engaged in other classes and extracurricular activities. In industry, timelines are determined at the beginning of a project and adjusted based upon the project scope and resources. In academia, timelines are usually set by the academic calendar.
- **Available resources vary.** Industry and academia have constructed environments and gathered resources (e.g., equipment, funding, expertise) they can deploy to meet their primary goals. In academia, resources are typically channeled to education and the creation of new knowledge, whereas industry resources are allocated toward the commercialization of specific products.
- **The termination of a project is often triggered in a different way.** In industry, projects are terminated when the product is introduced or when the project fails to pass through a stage of the stage-gate process. In academia, it is rare for a project to be terminated for these reasons and typically ends with the delivery of a functional prototype.

1.7 Organization of this Text

Our aim is to help you navigate your own design process. In this section we provide you with helpful suggestions on how to get the most out of this text.

1.7.1 Organization

This text leads you through the design phases in Figure 1.3. Each of the chapters listed within the boxes provides you with tools, techniques, and resources that will help you advance to the next design phase. Not all of the tools and suggestions may be directly related to your current project; however, they could become critical in navigating future projects. Table 1.1 is a further explanation of the design phases in the diagram, along with the broad topics covered and key deliverables.

In Table 1.1, chapters have been combined together. Although you can read each chapter separately, we suggest reading chapter sets close together (e.g., 5 and 6) because they support a particular phase of the design process. Prior to each grouping of chapters are one-page narratives of that specific design phase's purpose, takeaways, learning objectives, and deliverables that serve as the input to the next design phase. In fact, if you would like a birds-eye view of the text, reading these one-page descriptions is a good next step after completing this chapter.

Table 1.1 An Overview of the Chapters in the Text.

Design Phase	Chapters	Topics	Key Deliverables
Design and Development Planning	1–2	Basic medical device design terminology and processes, teaming and project management	Team trust/culture, project management and documentation system, project risks
Project Scoping and Specifications	3–4	Techniques for identifying and clarifying medical problems, opportunities and design inputs	Project statement, medical need, and technical specifications
Solution Generation and Selection	5–6	Iterative methods of ideation that include generation and selection of solution concepts and design solutions	Record of possible solutions, refinements, and rationale for selecting a "best" design for further development
Prototyping and Detailed Design	7–8	Prototyping methods and considerations for iterating upon a medical device design	Iterations of functional prototypes that are closer and closer to meeting design specifications
Verification and Validation Testing	9–10	Planning, execution, and analysis of verification tests ("did we make the product right?") and validation tests ("did we make the right product?") in living and nonliving systems	Demonstration that design inputs have been met
Critical Lenses	11–13	Standards, regulations, and ethical considerations in the design and commercialization of medical devices	Record of standards, regulations, and ethical considerations that impacted your design process and prototype iterations
Commercialization and Post-Market Surveillance	14–15	Roles industry engineers play in design transfer; late-stage commercialization; post-market surveillance; insights into the roles of the multifunctional groups listed in Section 1.2.2	Plan for the future steps needed to commercialize your current prototype

1.7.2 Emergent Understanding of Design Topics

Throughout the text, some topics come up again and again. One example is engineering standards. We introduced the ISO 13485 and 9001 standards in this chapter in the context of the engineering design process. Standards are also discussed in the context of documenting your decisions and work (Chapter 2), translating customer needs into technical specifications (Chapters 3 and 4), selecting design concepts (Chapter 6), iterating on your design and selecting material (Chapters 7 and 8), and testing (Chapters 9 and 10). In addition, there is an entire chapter dedicated to standards (Chapter 11). By reading the text, a rich understanding of standards will emerge. Several other topics also reappear, such as patents, regulatory considerations, communication, teamwork, project management, and, as already discussed in Section 1.4.2, documentation.

1.7.3 Keeping the End Goal in Mind

The goal of an academic design project is for you to gain experience with the design process. The long-term goal, however, is for you to abstract this singular experience so that it can help you navigate similar experiences in the future. You will almost certainly learn topics not taught to you, such as making difficult decisions, interacting and negotiating with teammates, advisors and mentors, balancing creativity and organization, and communicating with nonengineers. However, these will not be abstract concepts, but rather vivid stories that will be unique to you and this particular project. These experiences will serve you well when you encounter design again and help differentiate you from other engineering designers. We encourage you to pause at times to reflect on what you are learning and capture the details of the stories as they unfold throughout your experience.

We conclude this section with the perspective of a student who worked on two medical device design teams in academia. As you read the Breakout Box 1.1, notice how she leveraged her academic design experience in her professional career. Throughout the text, we include other breakout boxes that provide additional perspectives, design tools, examples of real products, and case studies.

Breakout Box 1.1 Reflections of a Design Team Alumna

During my time as a biomedical engineering undergraduate, I served on two design teams: once as a freshman in the Spring of 2007 and once as a junior for the 2008–09 academic year. As a freshman, the expectations for me on a design team were more focused on being a contributor and gaining hands-on exposure, as opposed to being the driver and a leader on the team. Being part of a team early in my education helped me grow my core engineering knowledge of coding in MATLAB, developing design specs and prototyping in SolidWorks/CAD, building metal prototypes, and testing them in *ex vivo* animal models; I also gained communication experience through delivering elevator pitches and slide deck presentations at various student competitions throughout the country.

As a junior, instead of mainly contributing and following a project plan, the expectations were around execution and critical thinking as well as having an eye ahead on what should be thought about next and what it will take to get there (aka the manager role). It was no longer enough to do the coding, building, and testing, but I was now expected to have stronger communication skills and clearly and succinctly answer the questions around "So what?" and "Yes, and what do we do next?" as well as tell a comprehensive story of what I was trying to achieve to a nontechnical audience. I also started to gain indirect exposure to task delegation as well as the intangible and somewhat "softer" ability to interpret that which is less easy to teach; underlying motivations, interests, and goals to identify appropriate people to excel in the appropriate tasks.

This combination of experience and skill development set the irreplaceable and invaluable foundation I have leveraged throughout my career in biomedical engineering graduate school, fellowship in medical device innovation, business school, and now current career as a medical device consultant. Both within and outside of academia, knowing how to be both a contributor and a leader of teams, while understanding core engineering knowledge as well as "softer" knowledge of communication and organizational design, are necessary skills. These skills have given me a comprehensive view of various engineering and business situations as well as the ability to quickly identify underlying problems, develop novel ideas, efficient solutions, and execution strategies. I am able to bring together unique and diverse people that significantly strengthen our quality of work, and which we subsequently call a team.

Laura Paulsen, BS, MS, MBA
Director of Product Management; Hyalex Orthopedics Inc.

1.8 Closing Thoughts

Engaging in a medical device design project places you at the center of some of the most exciting and promising advancements that will shape our future. Improved health care is recognized as one of the primary ways to address our most pressing global problems, such as poverty, inequity, and lack of sustainable living. Every day, new breakthroughs are being made in the basic sciences about underlying dynamics of human development, disease, and aging, from genes to organisms. The combination of big data, deep learning, and artificial intelligence enables the detection of previously unrecognized patterns that will impact healthcare ethics and governmental policies. At the same time, technology is continuing to advance in areas ranging from materials to miniaturization of devices to patient-specific computer models and new imaging modalities.

As medical device designers, we help bind together these dynamic fields. We design and build tools used by scientists to make discoveries. We identify unmet needs and opportunities to create value in the healthcare ecosystem. We translate medical problems into technical problems where advances in technology can be applied. We move good ideas from creative minds to commercial products that can make a difference in people's lives.

You are embarking on a challenging but fulfilling path, and we wish you well!

Key Points

Each chapter concludes with key points that summarize the main takeaways of the chapter. For this chapter, the key points are:

- Medical device design is at the intersection of three dynamically changing areas—the Healthcare Ecosystem, Industry, and Engineering Design.
- The healthcare ecosystem is composed of many stakeholders, regulations, best practices, and the public at large that together aim to improve human health and well-being.
- The medical device industry uses a stage-gate process across multiple functional groups to design and manufacture products in order to make the greatest impact.
- The engineering design process is an iterative framework for translating an idea into reality. It involves both technical and nontechnical skills and knowledge. Documentation is ubiquitous throughout. The process is purposeful, collaborative, always conducted under constraints, and includes failures and wrong turns.
- There are many similarities between how medical device design is practiced in industry and how it is practiced in academic settings. There are also many differences. This text primarily follows the trajectory of an academic design project.
- Participating in a medical device design project will help you gain skills, knowledge, and perspectives you will use throughout your professional career.

Exercises

Each chapter concludes with a set of exercises that encourage you to explore further and apply the chapter's contents to your design project. As this is the introductory chapter, these exercises are meant to help you better understand the medical device design process.

1. Find one medical technology that you feel has changed the world. Explain how the world is different because of this technology.
2. Identify a medical (engineering) innovator. Explore their biography and the impact they have had on the world.
3. Identify a stakeholder in the healthcare ecosystem (not an engineer) and find out more about them. If possible, arrange to meet with a person who represents that stakeholder. What do they consider the definition of "creating value" when describing a new medical device?
4. Why did you choose to study biomedical engineering (or other STEM discipline)?
5. How do you anticipate using the skills you expected to learn in the design process 10 years from now?
6. Visit the website of a medical device company. What is their mission statement? How do their products support this mission? Do they rely on new product development to drive growth of the company?

References and Resources

Dubberly, H. (2008). *How do you design: A compendium of models*. Technical Report Dubberly Design Office http://www.dubberly.com/wp-content/uploads/2008/06/ddo_designprocess.pdf.

Food and Drug Administration. (1997). *Design control guidance for medical device manufacturers*. https://www.fda.gov/media/116573/download.

Harvard Business School. (1997). *Project management manual*. https://www.designorate.com/stage-gate-new-product-development-process/. https://www.toolshero.com/innovation/stage-gate-process-robert-cooper/.

International Organization for Standardization. (2015). *Quality management systems—requirements*. (ISO standard no. 9001:2015). https://www.iso.org/obp/ui/#iso:std:iso:9001:ed-5:v1:en.

International Organization for Standardization. (2016). *Medical devices—quality management systems—requirements for regulatory purposes*. https://www.iso.org/obp/ui#iso:std:iso:13485:ed-3:v1:en.

The last section in each chapter is a list of references that were used to write the text and resources that can supplement your understanding of aspects of the medical device design process.

Ulrich, K., & Eppinger, S. (2012). *Product design and development* (5th ed., p. 2). McGraw-Hill Irwin.

Yazdi, Y., & Acharya, S. (2013). A new model for graduate education and innovation in medical technology. *Annals of Biomedical Engineering, 41*(9), 1822–1833. https://doi.org/10.1007/s10439-013-0869-4.

DESIGN DEVELOPMENT AND PLANNING

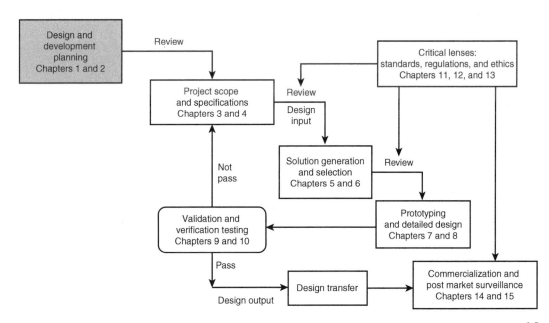

In Chapter 2 you will learn teaming and project management tools and skills that you will use early and often throughout the design process and your professional career. A robust design process is filled with difficult decisions and wrong turns. Effective teamwork and project management are the keys to not only overcoming these inevitable obstacles, but remaining creative, organized, and flexible. Developing good teaming and project management habits now will serve you well no matter where your career takes you.

The first half of Chapter 2 provides guidance on how to build and maintain healthy and trusting relationships within your team as well as with those who will support you. You will learn techniques for fostering open and clear lines of communication, clarifying your team values and goals, as well as detecting and mitigating problems early, before they infect your team.

The second half of Chapter 2 discusses project management tools for planning and tracking your work, documenting decisions and completing tasks, and ensuring that you have the resources you need. You will also learn how to identify, assess, and minimize potential risks to completing your project.

By the end of Chapter 2, you will be able to:

- Establish team values and create a team commitment document.
- Establish team rituals and lines of communication that will help sustain your team through the design process.
- Engage in effective communication within your team, with your mentors and advisors, and with others who will support your design process.
- Create the file structure for a Design History File where you will document your work.
- Conduct effective meetings driven by agendas and action items within your team as well as with mentors, advisors, and other external stakeholders.
- Construct and maintain Gantt Charts and Work Breakdown Structures to organize your work and communicate progress to others.
- Identify potential threats to your project and develop plans to eliminate or minimize the impacts of these threats.

Design Teams and Project Management

2

Alone we can do so little; together we can do so much.
– Helen Keller, deaf-blind author, disability activist, and co-founder of the ACLU

Teamwork is the fuel that allows common people to attain uncommon results.
– Andrew Carnegie, founder of Carnegie Steel Corporation and philanthropist

Fail early, fail often, but always fail forward.
– John Maxwell, author

Chapter outline

Biomedical Engineering Design. https://doi.org/10.1016/B978-0-12-816444-0.00002-X

2.1 Introduction

The goal of every medical device design project is to improve human health by introducing a new or improved device. This is achieved through the *biomedical engineering design process.* In academia, an equally important goal is for you to gain knowledge and practice the skills needed to navigate this design process. In this chapter we present several concepts, tools, techniques, and habits that will help you get off to a good start; you will continue to use these during the design process and throughout your professional careers.

The *biomedical engineering design process* is best achieved through teaming and project management. Even if you are working individually on a current project, you will need to manage your project and you will be working on teams later in your career. As realized long ago by Helen Keller and Andrew Carnegie, working as part of a team allows people to achieve much more than working individually. Teaming in design is necessary when optimizing design decisions and when making decisions that drive the design process forward. Project management provides a broad set of tools to help you plan, organize and coordinate activities, communicate with others outside of your team, overcome obstacles, identify threats and mitigate risks to your project, and document your progress. Project management is a critical aspect of planning and maintaining meaningful relationships within your team as well as with external mentors and advisors. Breakout Box 2.1 describes one biomedical engineering graduate's perception of working on an academic design project and how it helped her in medical school education, residency and fellowship training, and as an attending physician.

In this chapter, we present how academic design teams emulate part of the industry design process. Included are fundamental characteristics of being an effective teammate, the role of a team leader, and the role of peer and self-evaluations. A healthy team that establishes good habits is a prerequisite for performing well in the design process. These include complementary skill sets, trust, agreed-upon work ethic, open and direct communication, shared goals and values, and mutual accountability. Team success is improved with project management, assessing and mitigating risk appropriately, and overcoming common problems on design teams.

Breakout Box 2.1 Reflections of Being Part of a Design Team

Few educational experiences have had a greater impact on my professional growth than participating on a design team. I participated in a design team project as a freshman team member during the 2006 spring semester, and then as a co-team leader on a design team project during the 2007-2008 academic year. In addition to the technical knowledge gained from each experience, I also learned soft skills that still serve me on a daily basis, including building relationships and communicating. Initially, I navigated the dynamics of being a member of a team; then, I learned how to lead a team and leverage each member's individual strengths toward a common goal. In addition to class presentations, our team was also privileged to participate in multiple design and business plan competitions, each of which afforded me lessons in communication and human relationships, which I continue to use today.

I felt fortunate during the medical school admission process that I had the framework and support as part of my college experience to be able to say I worked with a team to identify a clinical need, create a solution, and see it implemented in clinical practice. My experience with my design team influenced my medical education and subsequent clinical training as well. The practice of medicine is a team effort with constantly evolving roles, not unlike on a design team. Along the way, I have enjoyed identifying clinical needs that can become opportunities for innovation. I am so grateful for the opportunity to be part of a design team, acknowledge that this experience helped open many doors in my career, and look forward to participating as a mentor in the future.

Gayathree Murugappan, MD
Stanford University

2.2 Industry Design Teams

As academic teams often try to emulate the commercial product team in some ways, it is helpful to know how teams are formed and function in industry before exploring academic design teams. No one person possesses all the skills and knowledge required to successfully design and commercialize a complex, high-quality product on time and within a budget; therefore, commercial medical product development always involves a core team of people. From developing a business case to defining a clinical need to developing a technical solution to selling a new medical technology, dozens (and sometimes hundreds) of people will work on all the intermediate activities, such as product design, bench testing, clinical testing, manufacturing, obtaining regulatory approvals, patenting, marketing, distributing, and selling. All these activities require specialized expertise, yet most of these tasks themselves are inherently team-based. Some individual tasks can be completed in parallel, thus saving time in a project schedule. For example, literature reviews, clinical observations, product dissection, and patent assessments can be performed concurrently by different members of the team.

The makeup and size of industry teams will depend on the desired outcome, working environment, available resources, and culture of the organization. Having team members with complementary skillsets enables many of the tasks associated with design to be accomplished simultaneously by different members of the team. Despite a wider project team having perhaps hundreds of people, they may be grouped into smaller teams that play a more specific role in executing the overall project goals. Teams in industry vary in size between 2 and 25 with about 5 to 15 being the norm; team members are sometimes distributed around the world.

2.3 Academic Design Teams

In this section we focus on the makeup of academic design teams as well as tools and frameworks to help you establish good habits. These will be critical in navigating the design process as a team.

Design teams are formed in an academic setting for the same reasons as in industry. For example, a team can put in more cumulative time than an individual, and therefore move a project forward more quickly. There are, however, some unique aspects of academic design teams. First, having several students working collaboratively on a project means more cumulative time can be devoted to it and some independent tasks (such as writing, literature searching, and testing) may be performed concurrently by subsets of the team. Second, although the composition of a student team will not be as large or diverse as an industry team, you may find that your program has found ways to mimic an industrial team. For example, some programs have longitudinal design teams that span multiple class levels or include members from other technical or nontechnical majors, or both. Just as in industry, your team will vary depending upon the program size, project needs, and programmatic resources. Third, many project deliverables and elements of an academic design project focus on your learning and growth as a team, following a biomedical engineering design process as well as meeting final project goals.

Beyond your core student team, you will likely have faculty mentors, advisors, and sponsors who may be considered part of the broader team, especially as the project moves along. Recognize that nonstudent members may assist the core team by serving as advisors or providing expertise, but generally do not participate in day-to-day activities of the student team. They should play an advisory role for the team but should not assist the team in creating concepts or design solutions. These professionals usually interact with the core team at intervals agreed upon by both parties. These intervals vary in length considerably (e.g., from once a week to once a semester). Those considered part of the extended team, the frequency of interactions, and the type of interactions with advisors and experts will vary by design program and by design project.

2.3.1 Forming Academic Design Teams

Industry design teams are typically formed by someone in a technical management role based on the available personnel and expected expertise needed for a project. There are many ways to form student design teams, and team formation is a function of the design program at your institution. Whichever method is used, it is important that you accept your project and team assignment and begin to evolve from a group of budding engineers into a biomedical engineering design team.

Educational psychologist Bruce Tuckman studied group dynamics and teams for about 10 years in the mid-1960s. He created one of the most used models of team development that is used by industry and academia alike. His model has some similarity to a developing biological organism. At first, it is a somewhat indistinguishable blob of cells (the original state of a team when they are still a group of individuals). Then there is a series of differentiation steps as the organism matures into a unique life-form able to function on its own in the world. Tuckman articulated five steps, shown as a block diagram in Figure 2.1, using this analogy:

Each stage is briefly described below.

Forming—A group comes together to learn about the opportunities and challenges and agrees upon (or accepts from the outside) a unifying goal. This is usually the stage where group members are most amenable to creating mechanisms to move forward in a unified fashion. Included in those coordinated behaviors are setting common goals, initializing roles within the team, and agreeing on how the group will make or modify decisions.

Storming—Through exposure and shared experiences, the individual team members build trust with, and respect for, each other; the team then creates rituals and internal processes and roles. During this stage, conflicts may arise as team members assess the team and vie for power, roles, and status. As conflicts arise, it is key that members try to resolve them in a way that is positive, professional, and cordial with a focus on problem solving.

Norming—The roles, responsibilities, and rhythms of the team become more established. Individuals begin to put the group goal before their own individual goals. Minor conflicts occur but compromises are easily found. A risk at this phase is groupthink—when the team comes to agreement without needed discussion out of fear of questioning or challenging a team member with a strong personality, or in the interest of moving on and not delaying the project. This risk is increased if a team member with a strong personality is unchallenged.

Performing—The team makes decisions, takes actions, and begins to achieve outcomes that are directly related to the overall team goal. It is during this phase that a highly functional team will embody the desirable characteristics articulated in the next section.

Adjourning/Mourning—The team completes a final step, either achieving the original goal or not, that terminates its reason for being together. For highly functional teams, this can come with the mixed emotions of accomplishment and the potential loss of regular professional and social connections.

A later stage was added called *re-norming*. Because a project is over does not mean that a team, or portion of it, will not work together again. Former teammates might rejoin for a new project, in which case they must go back to at least the norming phase. Even if you have worked with a team member or members in the past, you will need to rearrange roles and responsibilities for the new project. The advantage of working with former team members is that you know each other's work ethic, style of work, as well as strengths and weaknesses.

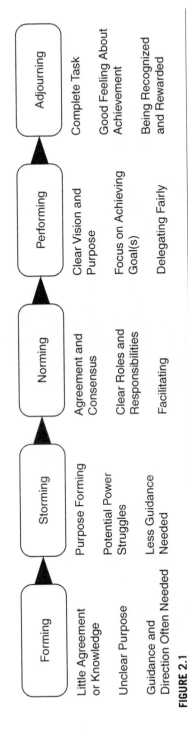

FIGURE 2.1

Tuckman's five-stage model of team development. Although stages are depicted as unidirectional, there are occasions when a stage might be revisited (e.g., late addition of a new team member). (Modified from Nelson, D.L. & Quick, J.C. (2019). ORGB[6] (6th edition). Cengage Publishing.)

Tuckman did not view the process as a linear one, nor did he prescribe specific timing for moving through phases. Teams may experience internal or external events (either helpful or harmful to the team) that move them back to a previous step in development. There may be times that your team needs to re-storm or re-norm. In some ways, it is similar to the product development cycle where some steps may need to be revisited.

2.3.2 Desirable Characteristics of Team Members

Being able to navigate Tuckman's team cycle and work well together requires traits that can be developed and adapted. Below are some guiding ideas that can help your team thrive.

Commitment

Every team needs something that binds them together. In the context of a design project, this should include a commitment to the project. Team members who are committed to the successful outcome of the project will work as hard as possible toward its completion. Sharing a common interest in the project is one of the keys to having team members committed to the project.

Equitable Team Contributions

How a team works together is unique to each team. Equitable work distribution is a reasonable goal to strive for; however, measuring work effort is difficult in practice. In an ideal world, equitable work distribution would mean that each team member devotes the same amount of energy, time, attention, and resources to the project. However, life is not ideal, nor is the medical device design process. There may be times when a team member can put in more hours and be more productive than usual. Similarly, there are times when a team member will not be able to contribute as much as others, or at all, due to host of valid reasons such as multiple job or professional school interviews, multiple exams in the same week, religious holidays, or needing to go home for family reasons. What is crucial is that atypical work patterns (decrease in or absence from work) be communicated to your team in advance (when possible) along with a plan for how work can be made up. Your team should also decide how to best accommodate a temporary reduction in effort, analogous to a sports team compensating for an injured player. Rugby and ice hockey teams at times score goals even when a team member is penalized, and the team is temporarily short-handed. This is due in part to the team being immediately aware that a teammate will be off the field or ice for between 2 and 10 minutes, thereby adjusting its efforts to make up the temporary loss of a player. Communications about temporary disruptions in work demonstrate a level of maturity of your team; they also indicate a commensurate commitment to the project by each team member.

Perseverance

The teams that displays perseverance will be able to recover from setbacks. In a survey of mostly biomedical engineering faculty at a VentureWell Open conference in 2013, perseverance was voted the character trait most needed by budding innovators. The reason for this is that obstacles—most of which will be unforeseen—will develop as any design project progresses. Having or developing perseverance (the will to proceed and not give up when faced with obstacles) will enable your team to repeatedly overcome obstacles and move the project along.

Your design team will encounter setbacks. These may include:

- a similar product becomes commercially available months into your project,
- a critical test reveals that the design is not performing as intended,
- a parts supplier stops production, or a manufacturer has an unexpected delay in delivery of a crucial part your team needs,

- a team member takes a leave of absence or elects to drop the course,
- a sponsor moves changes the project scope late in the project,
- a clinical sponsor is no longer available, and
- your team is working together from disparate locations and away from university resources.

Perseverance is a characteristic that helps keep your team going beyond these hurdles. It provides your team the ability to view the obstacles in the road not as hurdles; rather, the obstacles are the path.

Trust

Trust is the basis for all good relationships. If a team member commits to a task, other team members need to trust that the work will get done. If a team member with the task is not finished by the due date, it should be explained why, and your team should be understanding if the reason given is acceptable. It must be equally recognized that each member will approach a task or assignment differently, and each student should respect the honest efforts of other team members. It is difficult to build trust in a team, and it is a fragile entity. Once trust is violated (i.e., not delivering on time with poor explanation), it is difficult to be restored, negatively affects team morale, and makes it difficult to continue to function effectively as a team.

Mutual Accountability

In a similar vein, team members also need to develop a sense of mutual accountability. This is where all team members feel that they owe it to their fellow team members to meet the commitments each member has made to the team and the project. With mutual accountability, team members will feel like they will let their teammates down if they don't do what they committed to do. More than that, mutual accountability ensures that team members will support each other (e.g., help a team member meet a project task on time) without judgement.

Work Ethic

A common work ethic among team members is another characteristic of a successful design team. While there is considerable variation in work ethic among students, part of the Tuckman's norming process is establishing a common work ethic for a project. Recognize that hard work on a project is not necessarily a function of student academic standing or general work ethic but a commitment to work on the project. Independent of academic standing, a good work ethic for all biomedical engineering design team members includes a high level of effort, cooperating with others, dedication to the project, prioritizing project work, sharing values and goals, and exhibiting integrity (i.e., doing what you say you will do).

2.3.3 Team Culture

Just as a biological organism often needs nurturing during its development, a team needs to establish a culture of accommodation, trust, and respect. Such a culture does not happen by accident. No matter how functional a team may seem, it is these elements that will enable it to adapt, and even thrive, when problems are encountered.

There are some simple things a team can do while forming that can aid in building a culture of trust and respect:

- Have a shared social experience at the start that you continue to have occasionally for the duration of the project. Bowling, watching and discussing a movie or discussing a book, having a potluck dinner, playing a game, hiking, biking, and running are among the many activities that can be shared. Timing some of these activities just after a major project milestone helps team morale.

- Hold informal meetings throughout the project as needed, where each member explicitly states her or his own values, goals and hopes for your team and project. These may evolve over the duration of the project. There are three typical goals for the project: (1) goals based on what each team member wants to get out of the design team experience, (2) goals for each individual team member, and (3) team goals. Acknowledge that some team member goals may not all be aligned with the goals of other team members and might at times come into conflict. An extreme example of such a conflict is one member's individual goal to put in the least work possible, and another team member's personal goal of obtaining the highest grade possible.
- Draft a team commitment document together. An example is presented in Breakout Box 2.2.

2.3.4 Core Principles

Core principles are the guiding light to many successful entities. Many successful companies and institutions were founded on core principles that help them make critical decisions. You can learn a lot about how to develop your own core principles by studying the principles of these companies. For example, Johnson & Johnson's core principal established in the 19th century is to alleviate pain and disease, and its hierarchy of responsibilities to stakeholders is:

1. doctors, nurses, patients and all who use Johnson & Johnson products,
2. employees,
3. communities, and
4. stockholders.

Interestingly, stockholders are listed last. This is unusual in that for-profit businesses must generate profit to be successful; their bottom line must be positive, or they cease to exist. Often, profit is the major driving force behind business decisions. Johnson & Johnson believes that if the people it serves are satisfied, profits will follow. That philosophy has worked for well over a century.

How does that core principal manifest itself in day-to-day activities? The 1982 Tylenol crisis is a prime example, also presented in the ethics chapter, in Breakout Box 13.10. In brief, during the crisis, seven people were poisoned and died in the Chicago area due to tampering of Tylenol pills that were laced with cyanide. Johnson & Johnson immediately removed all Tylenol capsules in the Chicago area and the *entire country*, at a cost of around $100M. Despite the cost, the action stemmed from the application of its core principle that customers are their top priority. Johnson & Johnson's response to this crisis created a new *de facto* standard for tamper resistant pill bottles.

The top ranked hospital in 2020 was the Mayo Clinic according to *US News & World Report*. It started as a clinic over 100 years ago with two interrelated core principles: placing the patients' interests above all others and practicing teamwork medicine. That is why, today, new patients are welcomed by professional greeters who walk them through the registration process. Returning patients are greeted by name. Teamwork medicine involves cross-disciplinary communication, which often involves a primary care physician, two or more (often different) specialists, and nursing staff. The medical team determines the most likely diagnosis and optimal treatment plan, and a primary provider discusses options with the patient.

It can be helpful to develop core principles for your team. Consider all aspects of your design project, including faculty, sponsors, clients, relationships, teammates, and the project itself. Core principles are important in guiding your team through major decisions, as demonstrated by the Johnson & Johnson Tylenol case. For example, assume one core principle of a team is the personal well-being of each member. If one member is injured or requires hospitalization for a few days or both, having well-being

Breakout Box 2.2 Commitment Document

One useful tool that teams are encouraged to use is a Commitment Document where all team members agree to a set of behaviors and work ethic on your team. A sample team commitment document is offered in Figure 2.2; variations abound. This document would be signed by all team members. Requiring signatures indicates a commitment to these expected behaviors and is key to maintaining these behaviors as the project progresses through the year.

Sample Team Commitment Document

Project:		Date:
Team Member	Roles	

Team Behavior Expectations

- Meet deadlines, be reliable (deliver what you promise)
- Show commitment to project
- Work well with and respect team members
- Share ideas, contribute to team meetings and project
- Be willing to do more than fair share of work if needed
- Listen attentively to other's comments during meetings
- Accept and give criticism in a constructive manner
- Provide adequate notice of short term commitment problems
- Attend and participate in all scheduled group meetings
- Respect contributions of fellow team members
- Complete all tasks to the best of ability

Additional Expectations (if any, should be appended)

Team signatures (signifies team members' commitment to team and project):

_____   _____

_____   _____

_____

FIGURE 2.2

Sample team commitment document with commonly agreed upon expectations. The list can be shortened or lengthened (i.e., more normative behavior expectations can be added).

as a core principle would lead to your teammate's work being completed seamlessly by other team members. In addition, that team member will likely be visited (physically if possible, or virtually if not) by most or all other members.

As a project moves along, such a document is useful in situations where living up to those expectations may be difficult. For example, assume a task (such as designing a circuit) is needed to be completed before another task (such as creating a printed circuit board) can start. If the team member(s) responsible for the circuit design is (are) late, human nature may cause other team members to be disappointed, complain about it, and perhaps complain about your team member(s) as well. Demonstration of the behaviors included in the commitment document, specifically expectations involving respect for others, a willingness to do more work when needed, and providing constructive criticism will help maintain a positive attitude among the team. In the example above, such attitudes could also accelerate the printed circuit board development.

Developing strategies to overcome problems when the team is formed is also recommended. In the example above, developing a strategy to manage late task completions would enable the team to complete the circuit design and minimize delays in task completion.

A team that starts out well can still run into problems. To maintain a healthy culture and develop a shared identity, you should continue to invest energy into the relationships within your team. Some team health maintenance ideas are:

* Start each team meeting with a brief (<5 minute) entertaining activity at the start. Team members should rotate taking the lead on this.
* Consider establishing and using a unique gesture such as an unusual greeting or handshake.
* Create a team logo, a team name, hat, T-shirt, or other swag. Coordinate with other teams to avoid duplication.
* Customize the space you work in if you have one. For example, make up (or find) a quote or image to post in your design space that will inspire your team. Supplement or supplant it with other quotes and images as the project progresses.

Building and maintaining trust, respect, and camaraderie will help your team navigate the inevitable problems that will occur in tackling any worthwhile challenge.

2.3.5 Virtual Design Teams

Design teams operating from different geographic locations have worked collaboratively for over 25 years. Collaborative design has become more common over the years due to two primary reasons: improvement in digital communication technology and geographic distance separating design teams.

All of the desirable characteristics outlined above are achievable in a virtual team as well. For example, establishing trust is demonstrated by work product. For example, if you say you are going to perform a task (e.g., a literature search) by a certain date, and deliver a summary document that satisfies your team, delivering the document will begin the process of building trust. Doing what you say you will do on time is an important step in building trust on a virtual team, and is also a step toward developing a good reputation among your team members and toward becoming a professional.

Your team will need to agree on the suite of communication tools (e.g., email, chat rooms, video-conferencing) to use. Brainstorming, for example, would require a graphical interface, whereas a typical team meeting would not.

Many activities (e.g., all types of team meetings, developing good teamwork skills) that you can achieve in person, you can also achieve virtually. The loss of personal interaction should not be an impediment to good teaming.

2.4 Peer and Self-Evaluations

Improvement starts with recognizing a weakness or opportunity to improve, and then being willing to change. Active listening is a critical communication skill. These can be combined with multiple methods of evaluation. In industry, these evaluations may be individual (in the form of performance reviews) or team-based (often as part of a design review). Evaluations are sometimes based on peer reviews as well. The intent is to praise and reinforce good performance, help employees identify areas for improvement, and help management determine salaries and promotions. Sometimes, consistent poor performance reviews in industry could lead to job transfer or demotion, no pay raise, or termination.

Many of these same mechanisms of evaluation are used in an academic setting. A major difference between outcomes of industry and academic performance evaluations is the assignment of grades in the university as opposed to salary increases in companies. Grades are used to evaluate individual performances, yet design milestones are often achieved as a team. Instructors therefore generally use two mechanisms for determining individual contributions: individual efforts (e.g., homework, exams) and peer evaluations (i.e., evaluating the contributions and performance of team members by fellow team members). Team members can be the best evaluators of their teammates because they observe their work and behaviors up close. Each team member is in a position to provide objective summative evaluations on categories that include quality of work, quantity of work, level of effort, attitude, and timeliness.

The key to writing a good evaluation is to be as objective as possible. The list below provides some guidance on how to best communicate feedback most effectively to your team to ensure that you maintain a healthy team balance and stay focused on growth.

- Emphasize positive behaviors, and then indicate problematic ones, if any. Positive feedback reinforces continued good work. Focus on the positive.
- Be as specific as possible; provide examples of the type of performance you are observing. If observing a problem, try to address it in-person beforehand and use the written peer evaluation to elaborate.
- In general, feedback from peer evaluations should not come as surprise. If you have observed a problem, it is best to bring it up in conversation beforehand. The sooner a team member is made aware of the problem, the more time your team member may use to improve.
- Avoid being negative when providing constructive feedback. If you have constructive criticism, say something positive first. This is sometimes referred to as the "I like, I like, I wish" paradigm. The format is two positive statements followed by a critique couched in terms of a wish. For example, "I like your intensity when working, I like your thoroughness, I wish you stayed in a working mode long enough to be more effective."
- It is human nature to become defensive when receiving negative feedback. If constructive feedback is given to you as described above, this defense mechanism should be diminished, and you should acknowledge the critique and appreciate the feedback. Even if criticized in a negative way, try to be receptive and understand the source of the criticism before responding.
- Self-evaluations are perhaps the most difficult to write as you need to highlight your unique contributions without coming across as being superior or being too humble. It is important to provide an honest appraisal of your work and not inflate or exaggerate your contribution to your team and project.
- If you will be providing numerical scores and written comments, these should be consistent with each other.

A comparison of two sets of peer evaluations is presented in Table 2.1. One set is more meaningful (and therefore more useful) than the other.

Table 2.1 Table of peer evaluations and self-evaluation that compares less meaningful to more meaningful team member comments.

Less Meaningful Peer or Self-Evaluation	More Meaningful Peer Evaluation	Ranking
Good job. FMEA contributor. Stress can be a problem.	Has done a great job troubleshooting many areas of failure with our prototype. She has also ensured an animal test was completed before final presentation. She has kept us pushing forward and ensured our team did not lose momentum in the final weeks of class. She does occasionally get short with us close to course deadlines, while some of this is deserved, she could work on remaining calmer.	4
Checks out ½ the time.	Is a lot of times late or absent. Brings good ideas about interface design for less than 1/2 hour. Doesn't contribute any longer. Could improve by being more engaged. Perhaps a team meeting dedicated to the lack of progress, where all could contribute, would help.	1
Number one teammate. Involved in almost everything.	Has been amazing these past couple months. He is the lead design and manufacturing member of our team, and predominantly takes the role of designing the prototypes. He works on the 3-D CAD models and animations. He takes care of printing most of the prototypes and making updates to the design. He also works to secure manufacturing channels for injection molding.	5
Has been mostly working with the research committee and meeting presentations.	Although helpful with logistics, she has lagged behind on design input and helping define clinical need. Did not observe medicine at all. Next grading period, I expect to see more time into the project and technical contributions as we begin the design process.	2
Self-evaluation: I carried my weight.	My primary contributions included leading the prototyping subteam, helping with laboratory testing, and making a committee presentation. I only missed one team meeting. I am working on being more organized and more patient as team member.	3

The differences between the two sets of comments are that more meaningful comments capture the essence of the bulleted characteristics outlined above and are thus more helpful to the teammate. The five-point numerical evaluations (five being best) correspond to and justify the verbal comments.

2.5 Overcoming Common Problems on Design Teams

As you navigate an academic design project, problems sometimes arise that can result in low team performance and morale. There is no way to anticipate every possible conflict, but there are some common problems that arise in team-based academic design projects. This section presents scenarios along with some suggested action steps. Seven common problems are these:

1. A team member drops the course or stops doing all work on assigned tasks early in the project.
2. One team member dominates other members and team meetings.
3. One teammate is attached to one design choice that team agreed not to pursue.
4. Team members avoid dealing with an unpleasant issue.
5. Two members are having frequent conflicts.
6. Team members feeling indefatigable stress.
7. A team member stops putting in her or his fair share of the work well into the project.

Below are some possible ways a team might overcome the problems enumerated above. These potential solutions are offered as guidance and come from decades of experience of biomedical engineering design faculty. While the guidance below provides only one method to address the issue, there are many others, including seeking faculty assistance. These scenarios are presented to help you think more deeply about how you might address issues that will inevitably arise on your team.

1. *A team member drops the course or stops doing all work on assigned tasks early in the project.*

While cajoling and encouraging a team member who is dropping the course or is not doing a fair share of the work can sometimes be effective, someone who has decided to drop the course or to not work on a project is not likely to change. These team members understand the consequences (very poor peer evaluations and a likely lower grade than the rest of the team) and have accepted this. If early enough in the project, reach out to the course faculty and project advisor so they are aware of the problem and can provide additional resources (e.g., add a member) to your team. Practically, a team may need to revise its plans or reduce the scope of the project. These changes should be made under the review of the faculty overseeing your team's work. If the project scope needs to be reduced due to one less team member, faculty generally understand this.

2. *One team member dominates other members and team meetings.*

In standardized personality tests, people are often characterized psychologically into four pairs of personalities. Some team members are extraverted while others are introverted. Some are emotional while others are cerebral. Some are intuitive while others rely on sensation. Some like to plan while others are spontaneous. If team members are extraverted, emotional, intuitive and spontaneous, they tend to be domineering. If others are introverted, cerebral, rely on sensation and like to plan, they tend to keep quiet. These personality traits should be recognized early, discussed openly, and worked with to improve communication as the project progresses. During a meeting, one technique to limit imbalance in speaking time is to time every team member each time they speak. When finished, compare total speaking times of each member and discuss at the end of meeting or soon after the meeting. If one team member is spending a lot more time speaking than other team members, that team member will realize it and likely self-adjust. Likewise, someone who is not speaking up much may adjust to speak more next time. Sometimes this requires a third team member to help an introverted quiet member with a prompt such as: "Pat, what are your thoughts on this?"

3. *One team member is attached to one design choice that the other team members agreed not to pursue.*

"Dinosaur baby" is the informal term that designers refer to a proposed design that all other members of the team have discarded from consideration as potentially viable. At times, a single member of a team may remain committed to a dinosaur baby and thus impede the team achieving consensus on a final design solution. This should be handled as a team in a nonconfrontational manner. The team member should be given the chance to clearly and objectively explain why this "dinosaur baby" is better than the design choices that your team wishes to pursue. Your team needs to listen to and acknowledge your teammate's position and understand the design rationale. Similarly, your team needs to explain why it is in favor of the other choices, and the team member needs to acknowledge and understand your team's position. This approach alone may in fact solve the problem simply by gaining more clarity. If not, an alternative might be for your team and the dissenting member to quickly create a prototype of one idea that might better demonstrate efficacy. Such an approach brings more information into the decision-making process. Only if resolution cannot be found within your team should guidance be sought from a faculty member, project advisor, or mentor.

4. *Avoiding an unpleasant issue.*

The design process always contains peaks and valleys. One valley is being on a team where one member is performing poorly. There can be a tendency for the rest of your team to feel down but continue without discussing its disappointment. If your team wants to improve, however, your team must acknowledge the problem and discuss it openly with the team member. As a start, there may be temporary personal circumstances that are affecting one team member. Just having the team aware of it, without necessarily sharing personal details, will allow some leeway for the poorly performing student. Trying to understand the root cause of a poorly performing team member is the key to deciding on a way to best improve the poor peformance.

5. *Two team members are having frequent conflicts.*

Consider the last time a loved one angrily disagreed with you. It is natural to defend yourself and sometimes extend beyond the issue that caused anger. It is important to resist this tendency. Diversity of opinions as well as the need for many decisions will result in conflicts. Conflicts can also be a source for creativity if harnessed correctly. How conflict is managed is the difference between a creative process and one that exacts energy and time and is unhelpful. Research into teams found that merely viewing someone dissenting emboldens individuals to think for themselves, express their disagreements, and break from the majority. If managed well, conflict can lead to divergent thinking, a search for alternatives, and increased risk taking. When two (or more) team members are in conflict, it can be helpful to return to some of the techniques described in this chapter. In particular, look back at Tuckman's Team Development Diagram. Focusing on the project and the tasks at hand, rather than the relationships on your team, can sometimes also reset a disruptive team dynamic. Lastly, recognize that almost all personal conflicts are due to a deficit of empathy, respect, or trust. Sometimes simply listening will help resolve a conflict.

6. *Feeling indefatigable stress.*

A bit of stress is natural, but too much—especially if chronic—is a problem. In a design project, stress most often derives from unrealistic expectations and deadlines. For example, the stress level is likely to be high if your team is preparing to submit a project assignment by a deadline imposed externally and time is short. Two options exist if the submission is incomplete as the deadline approaches: finish the work needed and submit it late with a penalty, or submit by the deadline as unfinished, also likely with a penalty. Neither outcome is good, and worrying about it will not be beneficial; you cannot always control outcomes. Accept your situation rather than wallow in regret and learn from the experience by processing it as a team to identify what may have caused the missed deadline. Then use your project management skills to mitigate the risk of an incomplete or late submission of the next assignment.

7. *A team member virtually stops doing her or his fair share of the work well into the project.*

In this case, a teammate or team leader should discuss this with the offending team member privately. It is important to focus on the offending behavior, and not the person. The conversation might start with a statement of the undesired behavior, followed by an explanation of how the behavior impacts the rest of the team, and the consequences related to project progress and relationships with other team members. For example, a team member can be told he or she is not pulling his or her weight on the team (behavior). When this occurs, it causes the rest of the team to increase its workload to keep the project on schedule (impact). In addition, the other team members resent the additional workload, potentially slowing progress and negatively affecting their relationship with the team member who is not completing the assigned tasks (consequences).

During this conversation, it is important not to focus on the student (i.e., resist characterizing the team member negatively) but to focus on the undesired behavior in a nonthreatening way. One example of that is asking something like, "Your work has been acceptable in the past, so the team is concerned that something may be going on with you that is causing you to miss deadlines and not contribute your fair share. Is there something affecting your work output?" The goal here is to try to improve the future performance of the member who has not been meeting expectations.

We reiterate that these approaches offer one way to solve a team problem; there are many more, including getting help from faculty and external sources. Six basic concepts for preventing problems are summarized below. Some of these can also be applied to resolve conflicts outside of design team.

1. Address the core issue as early as possible.
2. Engage in open, direct, nonjudgmental dialogue, with a focus on behavior and not on the individual team member.
3. Listen, even when what you are listening to might be critical of you, or unpleasant.
4. Occasional stress is normal; chronic stress is not. One way to reduce it is to have realistic expectations. Another way is to take time to reflect upon the work accomplished to date.
5. If a result is subpar, accept the things you cannot change, learn from them, and control what you can.
6. Keep conversations as objective as possible. When discussing poor performance, the less emotion the better.

2.6 Project Management

Project management is the coordination of people, resources, and actions to ensure that timely and efficient progress is being made toward achieving an end goal. When designing a medical device, this end goal is to create a technology that can impact patient care, the healthcare system, or both. As discussed in Chapter 1, there are thousands of technical and nontechnical actions and decisions that must take place to reach this goal. To provide context, a broad view of project management is depicted in Figure 2.3. It is at the center of all project activities and is used to oversee project activities, coordinate work between teams, maintain a project plan and schedule, and track progress and expenses.

An industry project might involve hundreds of people working in different divisions, perhaps all over the world, over many years. Project management is the adhesive that holds an entire project together in a coordinated way.

This section presents the basic elements of project management, including scope definition, project and tasks, risk management, budgeting, and communication. Multiple chapters in this text build upon these aspects of project management.

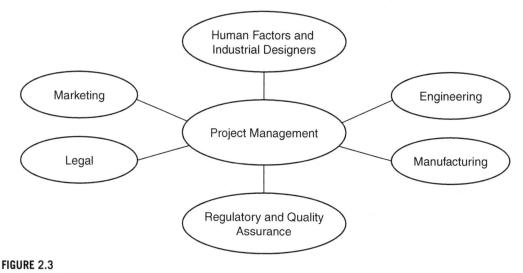

FIGURE 2.3

Typical broad structure of the technical development teams. In addition, finance, sales, and distribution (not shown) are an essential part of commercial medical product development.

2.6.1 Project Objective Statement

The Project Objective Statement (POS) is the highest level description of what a project will accomplish (scope), when it is to be completed (schedule), and what it will take to complete the project (resources). The scope of a project addresses the desired results (deliverables) required for the project to be successful. A classic example of a visionary project scope was President Kennedy's foresight in 1961 to "land a man on the moon and return him safely…" It was visionary then, in that much of the technology (e.g., computing, rocket technology) did not exist at the time. The schedule for the NASA project was addressed with the statement "…by the end of the decade." The resources for the project address what was available then, and what was to be developed, to complete the project. This is typically expressed through project funding, personnel assigned to the project, or both. The resources provided for the NASA project included a budget of over $2B (in 2020 dollars).

Using this information, the following POS can be constructed for the NASA project:

Put a man on the moon and return him safely by December 31, 1969, at a cost of $2 billion.

Typical resources at a university include students, faculty, project sponsors, advisors, university staff, project funds, design space, supplies, suppliers, existing laboratories, equipment, and other resources. A successful project achieves desired goals on time with maximum productivity, minimum cost, and—in academia—excellent grades and preparation for a successful career.

There are many project management "gurus," tools, books, and online resources. Independent of the source(s) you use, all project management constructs are based on three fundamental ideas; scope, schedule, and resources (the triple constraints of project management). From a planning perspective, the scope of the project should be achievable within a certain cost, schedule, and specified resources. These parameters are interlinked. In a well-planned project, the scope cannot be broadened without increasing the costs or resources or both. Similarly, to shorten the schedule, scope can be reduced, resources increased, or a combination of these two changes can be implemented.

In academia, since time is usually set by the academic calendar and resources are limited, an academic design team has greatest control over project scope (i.e., what is going to be delivered at the end of the project [Chapter 4] based on the problems to be solved [Chapter 3]). This will also affect subsequent project activities such as generating and selecting ideas (Chapters 5 and 6), proto-typing and detailed design (Chapters 7 and 8), and testing, toward the end of the project (Chapters 9 and 10). Industry design teams often have additional options to shorten project schedules. These include reducing project scope and adding resources such as personnel, funding, or equipment. Companies may delay the product launch date if scope changes or additional resources are not available.

Specifically, five broad management components of an academic medical technology project include:

1. Scope definition,
2. Project planning,
3. Assessing and managing risk,
4. Budgeting and resources,
5. Tracking and communicating progress.

In industry, budgeting is usually included in project planning.

In the following sections we present the first four aspects of project management and intrateam communication. Communicating in general, which is embedded within the other four management components, is presented separately in Sections 2.8 and 2.9.

2.6.2 Scope Definition

In both industry and academia, activities need to be planned while considering resources and external constraints. For example, critical deadlines are often fixed and then used to determine time frames for performing intermediate tasks. In industry, there may be a deadline set by senior management, such as a new product line to be launched by a specific date. Ideally, an industry team would be allowed to determine the staff needed to meet the launch date after careful consideration of all required project tasks and realistic task completion times. Your academic design team will need to determine the tasks, completion times, and resources required to complete the project. If the required project completion date is deemed unrealistic, then your team would be responsible for recommending changes in project scope and resources needed to shorten the project schedule.

In academia, adding more resources is usually a limited option. Some project deadlines and mile-stones are imposed by the faculty as well as the academic calendar. The scope of the project is key to meeting the course requirements and project deadlines. It is important to limit scope so that technical goals are achievable in an academic setting. For example, a clinical need to reduce the incidence of hospital-acquired infections would be too broad a project scope. It would not be feasible to achieve this goal in an academic year due to the need to design and conduct a clinical test to show that infections were reduced. A more achievable scope for an academic design team would be to focus on a specific source of infection (e.g., inserting a central line, surgery), ascertain the likely causes of infection, and develop a technology or device to address those specific causes. Measurable goals can include reducing the contact time with devices, maintaining sterility where required, or reducing handling time. All of these are indirect ways to reduce infections.

Effective coordinated planning is a complex process. To break it down, planning should begin with creation of a Work Breakdown Structure (WBS), which is a hierarchical decomposition of the scope of work to be carried out by the project team to accomplish all required project tasks. This involves hierarchically listing all tasks and subtasks required to complete the project. A sample portion of a WBS is shown in Figure 2.4.

There are many online resources and software tools to help create a WBS. One effective way at the start is to use Post-It® notes or an electronic equivalent to list deliverables and activities. By doing so, tasks can be readily moved around on a wall or screen, allowing the grouping, hierarchy, and sequence of activities to be easily modified as needed.

Analyzing the WBS more finely, tasks involved in the observation of a medical procedure would include the following: (These are expanded upon in Chapter 3.)

1. Learning about the medical procedure well ahead of observation,
2. Viewing videotapes of similar procedures if available,
3. Creating a document of the team's understanding of the procedure,
4. Formulating questions to ask and expectations of what will be observed (e.g., who will be in the room),
5. Thinking about what information will be learned from a live observation and what is needed,
6. Anticipating what you will observe,
7. Watching the procedure,
8. Documenting what was observed and learned soon after the procedure, and
9. Generating questions about what additional information might be needed.

The actual observation is 1/9th (Task 7) of the work needed to observe effectively. These subtasks would be included in the workplan (see Figure 2.4) for the project under the task of clinical observation, which is part of the heading "learning the medicine," which means acquiring enough medical knowledge to have a deep conversation with a medical specialist about a clinical problem. Each team will have more specific activities, or subtasks, as outlined above. For legacy projects (those designed and prototyped by a previous design team), broad additional design activities include reviewing previous project documentation, gaining new and deeper understanding of the medicine, and testing and evaluating the prototype(s). All of this will be required before typical design activities begin. Not all tasks will be known at the start of a project, and it is expected that the WBS will have a degree of flexibility in adding, deleting, and modifying tasks throughout the design process.

If working to improve an existing commercial product, tasks will include finding out as much as you can about the product's design, development, clinical use(s), and associated problems. If possible, physically obtain an existing product, disassemble it, and determine its inner workings. This is often referred to as reverse engineering, further explained in Section 5.10.

In contrast to academia, industry front end design activities often include starting with a business case. Without a business case (i.e., a way to make a profit or break even), many potential new products will not advance within the stage-gate process. If the return on investment (ROI) is projected to be too low or negative, the company will not recover its initial investment if the project is commercialized, and the project will not be pursued in most cases. In rare circumstances, a company may invest in developing technology at a financial loss for a charitable or global health cause. This is often through a nonprofit division of a business. Examples of nonprofit organizations within for-profit companies include the Johnson & Johnson Foundation and Laerdal Global Health.

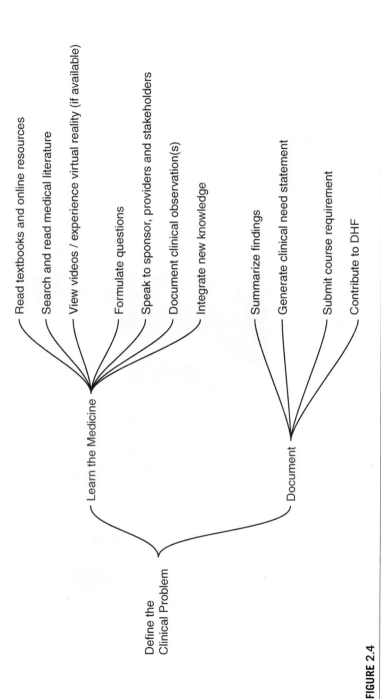

FIGURE 2.4

Portion of a Work Breakdown Structure for defining the clinical problem. Each of the activities can be further divided into more specific tasks.

2.6.3 Project Planning

The design process requires planning to coordinate team members, and task generating, timing and sequencing. There are many tools that can be used to plan tasks, track progress, and document completion times. A Gantt chart is a venerable way to graphically show the tasks from a WBS and include timing and personnel. A sample Gantt chart is shown in Figure 2.5. Many different styles exist. In Figure 2.5, all tasks, task durations (how long it takes to complete the task), task dependencies (e.g., which tasks must be completed before others can start), task start and end dates, and team member responsibilities (who is assigned to each task) are shown in one place. Countless variations of project plans exist, and many helpful resources can be found online and in software tools (e.g., Clickup, Excel, Instagannt, RedBooth).

The Gantt chart in Figure 2.5 shows project tasks for a team working on a neonatal head cooling project to mitigate the devastating effects of Hypoxic Ischemic Encephalopathy (HIE) at birth. This specific chart is focused on the tasks associated with "learning the medicine," which in this situation

Project: Neonatal Brain Cooling to Mitigate Hypoxic Ishemic Encephalopathy (HIE) in the Developing World

ID No.	Task Name	Assigned to	2-Oct	9-Oct	16-Oct	23-Oct	30-Oct	Preceeding steps
	Define the problem	Team	■	■	■			
1	Literature search	Judy, Jason	■	■	■			N/A
2	Meet with sponsor	Jack, Jill, Joe		■	■			1
3	Meet with stakeholders	Team		■	■			1,2
4	Integrate & Summarize Results	Jill, Judy			■	■		1-3
5	Refine user population	Jack				■		1-4
6	Find screens for HIE	Jill, Judy				■		1-5
7	Refine patient population	Jason, Joe				■		1-6
8	Generate first project statement & clinical need	Team				■	■	1-7
9	Draft report	Jack					■	1-8
10	Review report	Team				■		9
11	Finalize report and submit	Jack, Team					■	10

FIGURE 2.5

View of a sample 4-week Gantt chart to identify clinical need as initial work by the J team on a neonatal brain cooling project, comprised of Jack, Jill, Joe, Judy, and Jason.

means acquiring enough medical knowledge to have a deep conversation with a neonatologist about the clinical problem. The imposed deadlines (e.g., course requirements, submission of a report) are useful to establish the schedule for the length of the project. These set dates, which have no flexibility, can then be used to establish the many internal team-determined deadlines for specific tasks. Many of the tasks can have flexible finish dates, as demonstrated by wide time ranges in Figure 2.5. For example, it may take only 10 days complete the task, instead of 2 weeks' worth of meetings with sponsor and stakeholders.

Gantt charts are helpful for sharing your work plan with team members and others for an overview of activities. More importantly, they are most helpful internally to your team if used properly. First, there may be an overall Gantt chart for a project, and more refined Gantt charts for individual phases of a project. For example, a task such as "define the problem" may be one entry in the overall Gantt chart, but—as indicated in Figure 2.5 and expanded on in Chapter 3—is composed of many subtasks that include memorizing relevant anatomy, reading, understanding, and interpreting medical textbook chapters and journal articles, learning from relevant reputable websites, observing medical videos (if available) or actual procedures (if possible), and documenting knowledge learned, each of which takes considerable time. An even more detailed Gantt Chart would show each of these subtasks. Second, a Gantt chart is a living document and will change as the project progresses and changes. For example, you may discover that a task is unnecessary, completed early or late, or has transformed into a different task. Likewise, an entirely new task or tasks may need to be added as unforeseen issues arise or new requirements are discovered. These changes can be made to the Gantt chart for all to see. Fourth, you may find ways to highlight externally imposed tasks or deadlines that cannot be changed. For example, Item 11 in Figure 2.5, "finalize and submit report," is a course or program requirement that must be completed by October 30.

Task numbers and persons assigned are indicated on the left, and the duration of each task is indicated by the horizontal bars in Figure 2.5. The activity of defining the problem is further divided into specific tasks such as meeting with the sponsor, meeting with others, searching for relevant literature, and compiling and integrating information. The focus of this Gantt chart is to learn the medicine as indicated by the activities listed.

Generating a Gantt chart is also an exercise that can help teams better understand and graphically visualize how specific project tasks are dependent upon other tasks. Some tasks can be completed in parallel (as can be seen in Figure 2.5), whereas some tasks are dependent on others having started or been completed. For example, if a task involves viewing a medical procedure, it is crucial to learn some of the medicine surrounding that procedure to the extent possible *before* observing. By having that knowledge and some expectation of what will occur, observations become much more insightful. This makes the learning some medicine task a predecessor to the "viewing a medical procedure" task, even if learning the medicine is ongoing, as is should be throughout the project.

After establishing dependencies for each task in a schedule, a next step is to determine the expected completion time of any predecessor tasks. These are critical as predecessor delays can add up and delay the entire project. If two tasks have the same required finish date and the first task will take less time than the second task, the first task can be started later than the second task and still be completed on time. This allowable delay in the start time of the first task is called *float* or *slack time*. It indicates the amount of time allowed between the early start (ES) and late start (LS) time for a task (ES – LS) while still finishing the task on time. Tasks with no float have no flexibility in their start times; they must be started on time to be able to finish on time, or the entire project will be delayed. Float may also be reflected in finish times. If it does not need to be completed until 4 weeks from now, then it has an allowable late finish (LF) time of 4 weeks from now. For example, if a task that requires 2 weeks to complete is started now, it has an early finish (EF) time that is 2 weeks from now. This allowable delay in the finish time (EF – LF) represents the float for this task. Completion of this task within the allowable float time will not delay the project.

Gantt charts are helpful in identifying the *critical path*, which is the longest sequence of dependent tasks with no float. In general, the overall project duration will be determined by this critical path. Any delay along this critical path with cause a delay in the entire project completion date. You should therefore pay careful attention to any problems or delays that occur for tasks on the critical path for your project to keep the project on schedule. To do this requires you to first identify the critical path of your project. One critical path aspect for all academic projects is meeting course requirements when they are due.

2.6.4 Risk Assessment and Management

Risk is a concept you are all familiar with. There are risks associated with driving a car or crossing the street and there are risks of complications during most medical procedures. In many cases, risks can be quantified. When a teenager is first licensed to drive, for example, parents may lament that their car insurance premium becomes more expensive because of a newly licensed driver being added to the policy. Insurance companies base their rates on the relatively higher accident rate (i.e., higher risk) for 16- to 18-year-old drivers compared to the general population.

Technical risks, which are those associated with devices not working as expected, are always present in an engineering design process because the steps of the design process are project dependent, and the solution is not known until the design is completed. As such, there is no (or little) history upon which to base risk assessment. Technical risks can be thought of as the effect of uncertainty on achieving objectives. Throughout the design process, however, you should periodically assess the technical risks of your project.

One example of a technical risk associated with a medical problem is the occasional disconnection and clogging of feeding tubes that cut off or limit nutrients to patients. The current medical solution to this clinical problem is to either replace the disconnected tubes or clear the clog in the existing tube, if possible. Either procedure must be performed by a gastroenterologist. To reduce the incidence of feeding tube disconnection and the frequency of clogged feeding tubes, one possible technical solution might involve an expanding tube and spring. Determining the risk of mechanical failure of the spring alone is challenging, even when material properties and geometry are known. Spring failure, expanding tube failure, and user error are only some of the many technical failure modes, many of which may have little to do with the material properties. In addition, as demonstrated in Figure 2.6, assessing risk is often a judgment call (e.g., will the helmet protect the mouse from the mousetrap?). A formal method for identifying and managing technical risks (e.g., Failure Modes and Effects Analysis, or FMEA) is presented in detail in Sections 8.9, 9.6, and 10.6.

For an academic design project, the most common forms of risk are project risks and technical risks; without those being mitigated, project completion with a functional prototype is unlikely.

As hinted at in the previous section, some tasks have more uncertainty than others because you do not know how long the task will take to complete and whether you can accomplish the task on time. As design is inherently about the new and uncertain, there are a fair number of types of project risk. These

FIGURE 2.6

Mitigating risk. One way to mitigate the risk of a mouse trap is for a mouse to wear a protective helmet. If you were the mouse, would you go for the cheese?

include scheduling risk (threats to on-time completion of the project), technical risk (successful project completion and safe device performance), and commercial risks in industry, such as legal risk (product liability and freedom to operate), regulatory risk (e.g., clearance to market), and marketing risk (market acceptance of a new product). Throughout this text we encourage you to periodically identify, evaluate, and mitigate (take actions to reduce) risks to your project.

One helpful strategy during design selection to reduce risk is to try efficient experiments along the way that will reveal possible design flaws. For example, you are encouraged in Chapters 6 and 7 to build simple prototypes as early as possible to assess the *technical feasibility* of a design idea under consideration. The reason for this is to rule out problematic solutions under consideration early in the process. By doing so, you eliminate the risk of spending time and resources on a design that a simple experiment has demonstrated is fundamentally flawed. If the test result is successful, continue with that design idea with increased confidence. However, if the results are unfavorable, the concept should be eliminated from further consideration.

Consider that in a typical nondesign course, you are individually responsible for homework, individual project work, and exams. In a design program, however, the entire team is responsible for accomplishing all required tasks. In this situation, miscommunication between team members can lead to missed deadlines. This presents a scheduling risk to the project. Scheduling tools like a Gantt chart (e.g., Figure 2.5) are one way to mitigate this risk. There are other kinds of scheduling risks as well. For example, there may be tasks dependent on the availability of resources (e.g., parts delivery, limited use of specialized resources, access to a testing lab), or other work being completed first (e.g., prototype completion being required before testing). Common threats to on-time completion of the project (scheduling risks) and possible mitigation plans for reducing risk are presented in Table 2.2.

When threats to project schedules have been identified, mitigation plans often consist of either preventive actions or contingency plans. A *preventive action* is one that is taken prior to a problem occurring to reduce the probability that it occurs, thus reducing the risk. A *contingency plan* consists of

Table 2.2 Common threats to schedules and associated mitigation plans	
Threat	**Possible Mitigation Plans**
Unable to view a medical procedure or problem in real time.	Search for videos or animation of medical procedures and confirm with clinical advisor if realistic.
	Same for virtual reality simulations, if available.
Industry of clinical sponsor not available for consultation; project delayed.	At the start of the project, establish alternate contacts for sponsor.
	If necessary, move ahead by consulting with faculty advisor.
Late delivery of components, delaying prototyping or testing.	Follow up with vendor prior to delivery date.
	Order from alternate vendor, expecting to cancel or return an order.
3D printing capabilities are backlogged or temporarily not working.	Find alternative 3D printing capabilities.
	Create prototype using alternative methods.
Prototype fails testing; redesign required.	Pursue alternate concepts not selected previously.
	Anticipate test failure and include extra design, prototyping and testing time in schedule.
	Determine potential design changes ahead of time.
	If unable to prototype, propose to develop computer simulation or software code, and a quality animation of intended use of technology.

actions to be taken after a problem occurs. For example, as shown in Table 2.2, following up with a vendor prior to the expected delivery date for ordered parts is a preventive action that can reduce the probability of the parts being delivered late, or prevent a late delivery. Being prepared to pursue an alternate design after the team's first choice fails is an example of a contingency plan; one that is implemented if the problem actually occurs.

In an academic design project, you will have requirements beyond the project that may be imposed upon you, such as other courses, homework, assignments, projects or presentations. You may also experience delays in delivery of parts or have limited access to specialized resources. Although you cannot anticipate all threats to your project, it is helpful to periodically consider threats to future phases of the design process.

There may be situations early in an academic design project where certain risks can be determined definitively. For example, one problem in orthopaedic surgery and neurosurgery is misdirected screws, or screws that penetrate too far into bone. Possible solutions include a force-sensing surgical drill or a smart screw redesign. A surgical drill is an operating tool in contact with the patient for far less than 5 minutes. A new screw is an implant that is intended to be permanently implanted. From a regulation standpoint, a drill is a Class I device, where an implanted screw is Class II (greater risk than class I) because it is continuous contact with body tissue. Regulatory clearance requirements for implants are stricter (resulting in more time and higher cost) than those of a directional drill. This knowledge may or may not factor into the solution that a team ultimately chooses to pursue. Although this knowledge may not affect the design choice, it is good to be aware of it early. Regulatory risks are presented throughout the text, and Chapter 12 is devoted to regulatory requirements for medical devices.

2.6.5 Budgeting and Resources

A critical aspect of project management for a design team is budgeting and access to important resources that are needed to complete the project. Preparing and monitoring a budget is a critical skill for both project planning and communication. In industry, it is common for an initial budget estimate to be prepared before a project starts. Such an estimate is used to determine financial viability; whether a new device can be designed, manufactured and sold, and generate profit for the company over time. As a project proceeds, the project budget is monitored to ensure that expenditures do not exceed set amounts; budgets can also be used as a communication tool to show progress toward completion of project milestones relative to expenditures. The budget is often consulted when making critical decisions and adjustments might be required to keep the project moving forward. For example, an industry project might be accelerated (because a competitor is believed to be developing a similar product) by increasing the budget (i.e., increasing resources).

Budgets include different types of expenses depending on the stage of the design process and the environment in which the budgeted funds are to be used. The four-quadrant diagram shown in Figure 2.7 helps clarify some of these differences between academic industry design activities. The horizontal axis of "Project" to "Product" illustrates the differences in budget expenditures as a project evolves from a medical problem to a final product. The vertical axis of "Industry" to "Academia" is meant to highlight the differences between the types of expenses included in budgets used in academic design projects and those used in industry design projects.

A number of key characteristics about Figure 2.7 are presented below:

- In academic projects and in industry products, documentation is a key aspect. It is ubiquitous and therefore not specifically identified as separate activity.
- The industry budget items are in addition to those in the academic quadrants.

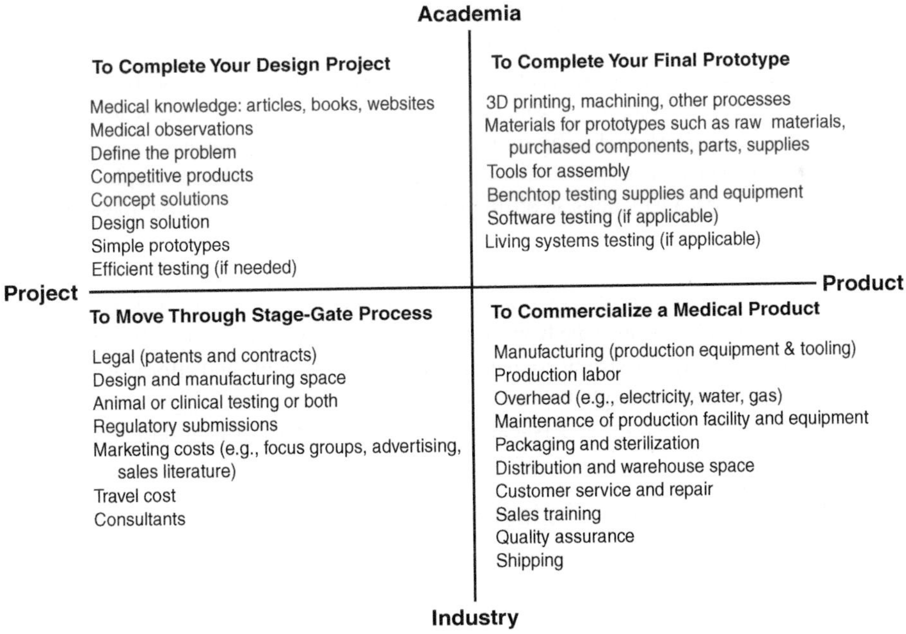

Academia

To Complete Your Design Project

Medical knowledge: articles, books, websites
Medical observations
Define the problem
Competitive products
Concept solutions
Design solution
Simple prototypes
Efficient testing (if needed)

To Complete Your Final Prototype

3D printing, machining, other processes
Materials for prototypes such as raw materials,
      purchased components, parts, supplies
Tools for assembly
Benchtop testing supplies and equipment
Software testing (if applicable)
Living systems testing (if applicable)

Project ———————————————————————— **Product**

To Move Through Stage-Gate Process

Legal (patents and contracts)
Design and manufacturing space
Animal or clinical testing or both
Regulatory submissions
Marketing costs (e.g., focus groups, advertising,
      sales literature)
Travel cost
Consultants

To Commercialize a Medical Product

Manufacturing (production equipment & tooling)
Production labor
Overhead (e.g., electricity, water, gas)
Maintenance of production facility and equipment
Packaging and sterilization
Distribution and warehouse space
Customer service and repair
Sales training
Quality assurance
Shipping

Industry

FIGURE 2.7

Four-quadrant view of typical academic design project issues and industry product issues.

- The final deliverable of an academic design project is typically a functional prototype, whereas in an industry design project it is a fully functional, manufactured medical product ready to be commercialized.
- The items listed are illustrative examples; the exact budget line items will vary depending upon the project and final product.
- For academic projects, typical budget costs are those associated with obtaining competitive products, prototyping, tools and supplies, professional services, testing, and travel (if justified).
- The budget for a manufactured product will be affected by the components listed in a Bill of Materials (BOM) and will include indirect costs (e.g., overhead, maintenance, rent).
- Industry design project budgets include costs that go beyond the purview of the design engineer; examples include clinical testing, marketing materials, sales, training, and shipping expenses. It is common in industry for there to be an overall project budget, of which the design team is allocated a percentage sufficient for required design and development costs.
- During the course of your academic design project, you may be asked to consider, or perhaps even include in your budget, expenses that appear in the industry quadrants (e.g., manufacturing costs, presented in Section 8.4). This is excellent practice in understanding how budgets are established and used for both planning and communication.

In an academic design project, many intellectual and physical resources may be provided to you at no direct cost to your team. These usually include the time you and your teammates spend on the project, access to faculty and staff, subject experts, project sponsors, advisors, and laboratories. It will also include supplies, materials, software, design space, and university facilities, such as a library, 3D printing capabilities, machine shops, electronic facilities and computing facilities, for which the team does not pay. In addition, an external sponsor may donate time or other resources at no, or reduced, cost. All of these resources, however, do incur real costs. A skilled machinist's time is expensive; often more than $100/hour outside of an academic setting. Inside the university, these costs are usually reduced or available at no charge as a university resource. It is therefore important to be aware of these costs and to be judicious in how you use your university's resources.

In addition to academic resources, you may have also been provided with a separate budget for your project for parts and services that are not otherwise available. How those funds are used is often at your team's discretion, within limits and required approvals. This mirrors industry, where project budgets always have some limit. In both settings, it is important to make judicious use of project funds, which is key to keeping costs within your budget. For example, if your team requires an off-the-shelf part for a prototype or experiment, you will need to decide whether to build it yourselves or purchase it. Cost and time are often the key factors in this decision.

Academic design teams tend to save their budget funds until the prototyping stage late into the project. The reason for this is that teams don't know how much they will need to make their final prototype and test it. There are times, however, when purchasing helpful items early in the project is a good investment and a wise decision. One example is to buy an existing competitive product, if available and affordable, early in the project. Another helpful early purchase might be a needed technical standard. Preparing and monitoring a budget are critical skills for a project engineer; understanding a project budget is a critical skill for all engineers.

2.6.6 Effective Team Meeting Strategies

An important part of project management is clear and timely communication, both within your design team and with others who support your project. This section outlines team meetings as a critical internal method for communication and coordination of team activities. Section 2.8 provides additional guidance on communicating with others outside of your team.

Team meetings are an essential part of product design and are primarily used to:

- have discussions and conduct activities that result in team decisions or progress,
- problem solve (e.g., brainstorm) at various phases of the design process,
- divide work among team members, and
- share critical information that requires discussion.

Although meetings may come in many forms (e.g., Breakout Boxes 2.3 and 2.4) and there usually is someone leading a meeting, all team members are responsible for the general flow and outcomes of a well-run meeting. In that spirit, below are general recommendations for formal team meetings.

Logistics

- Create a standing meeting time (e.g., every Monday from 8 to 9 am) that all members can attend. A regularly scheduled meeting creates an expectation that team members block out that time for team business. Although early morning slots may not be desirable, coordinating schedules among several team members may lead to that time slot being the only one during the day everyone can attend.
- Meetings should generally remain under one hour and should have an agenda; long, open-ended meetings can lose focus.
- Start on time unless there is a compelling reason to delay. Waiting for late team members wastes the time of all team members who arrived on time.
- End on time or early. No one will be upset if the meeting ends early; team members will appreciate that the organizer is respecting their time.
- After the meeting, send meeting minutes that summarize who attended, what was discussed, decisions that were made, and action points assigned to various team members. Updating progress on these action points should be part of the next scheduled meeting.

Breakout Box 2.3 Standup Meetings and Informal Meetings

Typical 30- or 60-minute meetings can be very effective for longer discussion topics. Sometimes, however, it can be more effective to also have short informal meetings with your team. If short enough (<10 minutes), they may even fit in between classes. These short meetings could be for updates or to have a short discussion on one topic, such as agreeing on logistics of observing a medical procedure. The usual result is that an issue is resolved and can be removed from discussion during a regular meeting.

Meetings intended for brainstorming, problem identification, and problem definition activities are often more effective as informal meetings. When defining the problem, the team can collectively list all its issues or questions relating to a clinical problem. For example, pressure sores are a major clinical problem. If considering this as the subject of a possible design project, a meeting to identify the specific issues is more effective when run as a collaborative group. Patient populations at risk, types of sores, causes, existing prevention methods, and healthcare costs are some of the issues that can be identified in an informal or a formal meeting. Brainstorming meetings are used to generate conceptual ideas. Each idea can be quickly indicated on a whiteboard, Post-It® note, chart, or computer screen. As the ideas accumulate, themes usually emerge, and concepts often can be grouped into categories. The outcomes of these meetings must be documented.

Spontaneous, informal meetings often take place in a location mostly used for pleasurable purposes (e.g., restaurant, living room). These usually take place after hours. Most of the formalities do not apply; however, there should be a good reason for the meeting. This can be to discuss a simple issue, such as selecting a team logo that is currently undecided. Although a tangible outcome is desired, it is not required. The intangible benefits support team culture and morale and enhance relationships between team members.

Breakout Box 2.4 Videoconferences

Virtual teams with members located in different locations have been used in industry for decades. Videoconferencing has been used for personal and business communications to accommodate differences in personal schedules, time zones, and geographic locations. When using this technology for virtual team meeting, plan as you would for a face-to-face meeting, with goals and agendas.

   If a presentation is part of the meeting, it should be the only thing that attendees can see during the presentation. Questions during a presentation may be asked orally, typed into a chat feature, or asked after the presentation has ended. Those not speaking should mute their microphones. If there is more than one speaker, a transition method should be agreed upon so that a change in speakers is as seamless as possible.

   Most importantly, virtual meeting attendees should be prepared for technical difficulties (e.g., video frozen or not appearing when needed, sound muted when unintended, or Internet slow or down). Although infrequent, technical difficulties can and do occur.

Design team meetings are most effective when they are planned. Several preparation steps are presented below.

Preparation for meetings

- Distinguish between issues that can be managed through written communication (e.g., announcing a deadline) from those that would require a meeting to discuss (i.e., whether to move an internal deadline or not). Announcements can be made at meetings, although these are often shared electronically beforehand so as not to require much time for discussion during the meeting.
- Every meeting should have at least one specific purpose (e.g., to brainstorm or make a decision); however, having more than four can cause a team to go beyond the allotted time or cut short discussion on other issues that need longer discussion.
- Create an agenda that states the goals of the meeting and distribute it well before the meeting (24 hours). Any pre-meeting actions (e.g., reading, reviewing documents, completing project tasks, or preparing to present), especially specific ones assigned to individual team members, should be included as reminders. An electronic copy of the agenda can also serve as a good place to record meeting minutes, updates, decisions, and action items during the meeting. Using a standard template for an agenda, such as the example presented in Figure 2.8, is a good practice.
- After the meeting it is critical to capture what was discussed and decided, including action items and who is assigned to each. In a design project, it is useful to create a post-meeting action plan from the meeting minutes, which includes a summary of what was decided and action items (along with who is responsible for completing each action item) for the next meeting. A sample post-meeting action plan is presented in Figure 2.9. This can be included as part of your Design History File.

Meeting Flow

- Limit interruptions during the meeting whether face-to-face or virtual and have only one conversation at a time. One effective method to limit disruptions in a meeting is to agree to have each team members' cell phone turned off during the meeting.
- Try to recognize when someone is being quiet and not contributing and find ways to encourage that teammate's participation. Asking a question such as, "What do you think?" is usually effective and nonjudgmental.
- If you find that someone is very passionate about a particular topic, consider assigning the team member an action item that involves that topic.

Team Meeting Agenda		
Topic(s):		
Objective(s):		
Attendees:		
Location, Date and Time Started:		
Time Completed:		
Summary of Previous Meeting:		
Agenda Item	**Discussion Lead**	**Time Allotted**
Outcome	Action Items for Next Meeting	Team Member Assigned

FIGURE 2.8

One sample meeting agenda template; many different styles are available. Key elements include meeting details (who, where, when, why), agenda times, agenda lead, and time allotted. The Action Items are the individual team members tasks as a result of the meeting. The outcome, task assigned, and team member responsible can also be part of the post-meeting documentation as can be seen in Figure 2.9.

- It is sometimes appropriate to have some portion of a meeting used for parallel work, at which time each member is working on a different task. Working together helps with accountability and encourages communication about interrelated issues.
- Each meeting should generate individual and team tasks that will move the design process forward. Using the template in Figure 2.8 or one similar (e.g., when evaluating project opportunities), team members may be delegated tasks to learn more about specific project topics. In a subsequent meeting, each team member would then report on the findings as they relate to agreed-upon criteria, and the team can begin prioritizing project choices.
- There are times when it is appropriate for a task to span multiple meetings (e.g., prototype creation and refinement).

The decisions and tasks generated during meetings should be documented and included in your Design History File (DHF), introduced in Section 2.9. You should also include sketches, schedules,

Post-Meeting Action Plan	
Topic(s):	
Date:	
Attendees:	
Objectives:	
Agenda Item #1:	
Discussion:	
Decision:	
Agenda Item #2:	
Discussion:	
Decision:	
Agenda Item #3:	
Discussion:	
Decision:	

Action Items		
Tasks	**Person(s) Responsible**	**Due Date**

Additional Comments:

FIGURE 2.9

One template for a post-meeting documentation; many others exist. Assigning tasks as part of the action plan encourages members to complete them by the next meeting, especially each member designated to be discussion leader on the topic assigned.

critical updates, or results from research or testing. Documentation of meetings is a critical part of the record that shows how you navigated the design process. A common method of capturing the events at a meeting is to take minutes, essentially providing enough detail that a teammate who may have missed a meeting could read the minutes to understand what occurred during the meeting. Meeting minutes serve as a record of what was discussed and decided during a meeting and can help clarify misunderstandings regarding which team members committed to completing which action items.

A summary of the meeting should be drafted and used to help prepare the next agenda, as well as provide documentation for the DHF.

2.7 Team Leadership

Every engineering design project requires project management. In some cases, there is a designated leader (often called the Project Manager in industry), whereas in other cases team members will have moments when they step into project management roles. Irrespective of your official role on your academic design project, each team member will likely be involved in some aspect of project management. These activities include, but are not limited to, planning, monitoring, and adjusting the project schedule. Other leadership activities include helping teammates overcome hurdles, motivating others, and coaching. Although being a leader is often a rewarding experience, recognize that it is an added responsibility to your technical responsibilities as a team member. The long-term benefits of taking on a leadership role on a design team is presented from the perspective of an academic design team alumnus in Breakout Box 2.5.

Breakout Box 2.5 One Alumnus' Perceptions of Having Been a Design Team Leader

I had my first experience in Design team in spring term as a freshman; I subsequently led a Design team as a senior and had an equally powerful but very different experience. As a freshman, I had an accelerated learning because I could model my own behavior from upperclassmen (I grew up on a dairy farm in upstate NY and was fully unprepared for the academic rigors of a top tier university). Our first semester was pass-fail, and after midterms I was struggling to pass. My team leader invested time in my development and helped me make a few key decisions (I dropped calculus, applied to do research for NASA, and started playing intramural hockey). My senior year I applied these learnings to give back to others and worked with a team of 10 students to develop a medical device. After 2 weeks of building a detailed project plan, we learned that a similar product already existed and was for sale. We immediately went back to the drawing board and came up with an idea to detect kidney failure in the MICU. We developed the concept, evolved with clinician discussions, and built a business case and early prototype. A venture capitalist later told me "you don't know anything about business—go up to a tech center and meet some people who do," so that's what I did. We built a team, incorporated as a company, and secured seed funding for lab space.

Design team has been an important part of my training and continues to provide benefits 20 years later. Fundamentally, it provided a structured way of thinking that allowed me to be more successful in graduate training and research (leading to publications in PNAS and Nature Reviews Drug Discovery) and in my professional career (at McKinsey and Novartis). In my Design team experience, I had the opportunity to lead teams in a safe environment and apply technical skills to manage deliverables on short timelines while engaging and motivating my team. What was magical about the design team experience was that it fostered leadership capabilities that mattered, not only in science, but also in my personal life. I had real-time training in taking calculated risks, creative thinking, dealing with setbacks and failure (lots of them), and most of all, how to adapt and learn. These were foundational skills that I've continued to refine and hone through my career, which I've brought with me into 40+ countries across different industries and, most importantly, to different people and cultures. With time, I had learned how powerful it can be for a leader to set a vision and be optimistic; seeds that were planted years ago during my time in design team.

As I reflect back over 15 years, there were several key ingredients to creating this secret sauce and a learning environment: (1) the content and context of learning had to be challenging, as some of the most powerful lessons come from stressful situations; (2) team diversity was important as I navigated across language and cultural barriers and learned that each team member communicated and was motivated differently; and (3) having top-notch mentors and faculty who could provide coaching and guidance along the way.

Seth Townsend, PhD, MBA
Thermo Fisher Scientific

While much has been written on leading a team in different contexts, an excellent example of the benefits of having a team hierarchy in a design setting is "The Bakeoff" experiment published in the New Yorker. In brief, three teams were formed to develop the world's best-tasting cookie within 6 months. The design constraints were less than 130 calories and less than 2 g saturated fat per cookie. One team involved two food industry experts (the Pair); a second team involved a hierarchical team including a product development leader and several food industry experts (the Hierarchical Team); and a third team used the open-source model of 15 of the top food industry bakers and scientists (the Dream Team). The Pair team took the least time and generated an oatmeal chocolate chip hybrid. The Dream Team generated 34 ideas, but had trouble selecting the final choice (a chewy oatmeal cookie with caramel and chocolate glaze). The Hierarchical Team applied lateral thinking by using a concept for tortilla-making and applied it to make a strawberry cobbler cookie with sugar crystal glazing. The survey results from 3000 households showed that 14% preferred the Pair team's cookie, 41% preferred the Dream Team cookie, and 44% preferred the Hierarchical team. Four key observations (regarding design teams) from this experiment are:

1. With only a two-person team, creativity was inherently limited.
2. The Dream Team could not agree on the best way do things, limiting the team's progress.
3. There was a tendency to overestimate the importance of expertise and underestimate the value of leadership and organization.
4. The winning edge, albeit thin, ultimately was the team's ability to apply lateral thinking to two previously unconnected ideas—a tortilla chip and a cookie. Applying lateral thinking is more prevalent when a manager encourages it. Weaker sports teams sometimes win because of good coaching, not better players.

In addition to living up the Commitment Document (see Breakout Box 2.2), which is expected of all team members, effective leaders require excellent interpersonal and communication skills and perseverance. Leadership is often taught through case studies of famous leaders and how they responded when faced with a crisis or setback. Leaders often find strength and advice in studying other leaders.

One example is Steve Jobs, who was successful at overcoming failures, some of which were of his own doing. He was fired from Apple nine years after he founded the company in 1976. Two years later, he started NExT computing, which merged with Apple in 1997, where he led the company through a revolution of life-changing new products. He demonstrated perseverance, which is a trait that designers need to have. Failures will occur during the design process and moving on from them is necessary. John Maxwell's quote at the chapter beginning applies to many aspects of the design process.

Abraham Lincoln is another example. He was a master at managing interpersonal issues. He had the ability to have an accurate appraisal of himself and of others. He also had the ability to put himself in other's shoes. It was these traits that enabled Lincoln to bring his disgruntled opponents (three of whom were candidates opposing Lincoln for the presidency—Bates, Chase, and Seward) together, create a team of rivals, and marshal their talents to the task of preserving the Union. Preserving the Union as the overriding goal for all cabinet members was a key factor in his success. In a design team, if project completion and adherence to the Commitment Document are the paramount goals for all team members, a good leader will help navigate interpersonal issues as they arise.

A list of interpersonal skills important in leadership are as follows:

- problem-solving, collaboratively as well as individually,
- actively listening and understanding what is being said,
- aggregating differing viewpoints,

- maintaining faith in the team,
- motivating team members,
- overcoming technical and team member obstacles,
- communicating well orally, graphically, and in writing,
- tailoring communication to the audience being addressed,
- being positive and socially aware,
- displaying good manners,
- asserting self-control of emotion,
- resolving conflicts, and
- recognizing when to be assertive and when not to be.

Even though you may not have all or many of these traits at the start of a project, the project team experience provides you with an opportunity to develop, apply, and strengthen some of them and sharpen your leadership skills. A first critical step is to be aware of your own limitations. A second critical step is to be willing to change. Even if you are not the designated leader of a team, you can still practice all of these skills.

2.8 Effective External Communication

A design process always requires input and feedback from people outside of the core team. Whereas Section 2.6.5 focused on communication within your team, it is equally important to engage professionally with those who are willing to support your process. These people may include teaching assistants, faculty, advisors, healthcare professionals, industry sponsors, and others inside and outside of academia. While your internal communications may at times be informal, your external communications should be formal. Communication skills will only grow in importance as you advance in your career because you will interact with a more diverse and wider range of technical and nontechnical people. Practicing good communication skills now will not only streamline your design process but also help you as you advance in your career. This section reviews common forms of professional communication and provides some suggestions on how to make them as effective as possible.

Communication is a critical part of teamwork and is especially important when one member of the team interacts with professionals inside and outside of academia.

There are many options for design teams to communicate meaningfully. It is good practice to ask someone outside your team how they would prefer to communicate. It may be a platform (e.g., a pager) that your team is not using. Respect their request to the extent possible. This section reviews common forms of professional communication and provides some suggestions on how to make them as effective as possible.

2.8.1 Effective Use of Email

Email is ubiquitous. Annoyingly, many email messages that we receive are unsolicited, unwanted, and some perhaps even virus-laden. Even so, email is an important tool for communicating within your team as well as with advisors, sponsors, and others external to your team. As such, it is important to become skilled at crafting effective emails. Although internal emails within your team may be informal (e.g., first names, colloquial language) and convey information only relevant to the team (e.g., agendas, minutes, or documents), they are considered part of the documentation of your project. In some circumstances, especially in industry, emails may be used as evidence if there are questions about a design process. Your emails may be intended for busy people whose inbox fills up quickly. As a result, it is important for a designer to become skilled at crafting effective emails.

Critical aspects of an effective formal email include relevant subject line, salutation, introduction, purpose, closing, and sign-off. There are some specific ways to increase the likelihood of having your emails read:

- The subject line is key as it needs to be brief and sufficiently descriptive to increase the probability that the recipient will open the email. A good subject line should be brief and descriptive. If writing to a faculty member, include the name of the project; to an industry representative, a product name would be a good subject line.
- Use an appropriate salutation at the beginning of the email (e.g., Dr., Professor, Ms.) followed by a last name. A terminal degree designation (e.g., MD, PhD) should not be used in this case.
- Provide a single line about who you are, especially if this is a first contact.
- State the specific purpose(s) for the email in a single line as the first or second sentence. You may elaborate on this purpose later in the email or in an attachment.
- Keep the body of the email as brief as possible. If large sections of text are needed, use an attachment or link to a website. Remember to include attachments and reference them in the email.
- Bold any requests or important dates.
- Include a last sentence with a thank you for previous help or anticipation of future help.
- End with a signature.
- Send the email only to those who need to see the email.

Examples of many of these good practices are shown in Breakout Boxes 2.6 (email to a mentor/sponsor) and 2.7 (email to a faculty advisor). Breakout Box 2.8 shows an example of a bad faculty email and how to fix it.

Design team members often get to know one another very well, and internal team emails are thus generally informal (e.g., colloquial language). They might be used to convey information, schedule a meeting, share an agenda or minutes of a previous meeting, assign tasks, or share multimedia (e.g., presentations, videos). Informal language is acceptable.

Breakout Box 2.6 presents an example on one effective email to a project mentor or sponsor. Another email example (Breakout Box 2.7) is one written for a project advisor and follows the format above. It is one example of a proper email and comments have been added to highlight critical components. There are virtually limitless other ways to craft an equally effective email message.

Breakout Box 2.8 presents two email examples from a faculty member to a dean. The first one does not follow the model above. It is poorly written and is an example of what not to write in an email for reasons noted in the Breakout Box. (Even faculty can write poor emails.)

Breakout Box 2.6 Example Email to Mentor/Sponsor

Subject: Pressure Sore Project: Request for Meeting

Dr. Smith,

My name is Al Jones, and I am on an undergraduate design team of five students in our biomedical engineering design program. Our team is interested in the project you submitted to the program about mitigating pressure ulcers.

I write to inquire about an initial meeting to share with you what we have learned about the problem thus far, and to understand more about it from you. Our schedules are flexible and all or most team members could meet at a time and place convenient for you. Please advise as to the best times for this meeting.

Thank you. We look forward to working with you for the academic year.

Best wishes,

Al Jones Class of 20xx,

Breakout Box 2.7　Sample Email to Faculty Advisor

Subject: Fall Prevention Project: Request for Meeting

Professor Goldberg:

　　Thank you for serving as our faculty advisor. I am a member of The Fall Prevention Team in the Biomedical Engineering Design Project class taught by Professor Taylor. The project we are working on this year is to develop technology to reduce the risk of falling in the elderly population, which appears to align well with your biomechanics expertise.

　　We request an introductory meeting (<30 minutes) at your convenience to introduce ourselves, to share with you our progress to date, and to establish a regular meeting time for the year ahead. Thursdays and Fridays are good days for us to meet as all team members have no class or one class those days; other days are possible as well. We look forward to meeting you.

　　We very much appreciate your willingness to be our faculty advisor and look forward to the year ahead.

　　　　　　　　　　　　　　　　　　　　　　　　　　　　　Jane Doe　Class of 20xx,

Breakout Box 2.8　Example of One Poor and One Effective Email

Hey Deen Buff Jonesy

Can we get some additional funding so our student's can get really cool projects done? It would be really great and it would make thisintro design class I'm supposed to teach much easier for me to get through. Oh, it would be nice if you could come to my class and talk to the student so they know that I'm serieous about this and that there is potential for them to get some extra cash for their work on the projects that are part of senior design and really important.

Sorry I kicked your butt in poker last night, we'll have to do that again some time!

The email above is far too informal. Problems with it are many, and include:

1. Starting the email with "Hey" (this should **never** be done);
2. Containing numerous spelling mistakes ("Deen," "thisintro," "serieous");
3. Employing extraneous words ("Oh, it would be nice");
4. Being unfocused and unspecific ("extra cash," "come to my class");
5. Utilizing informal and unprofessional language ("Buff-Jonesy," "really cool," "really great," "kicked your butt"); and
6. Being unclear (intro class or senior design).

　　Can you find other problems with it?

Below is a formal, professional version of an email for the same request.

Subject: New class on freshman design

Dean Jones,

　　Greetings. As you may be aware, I am slotted to develop and teach BME Freshmen Design for the first time this fall term and am excited about the opportunity.

　　I am considering having students experience prototyping in their first semester, for which we would like to have additional resources. Specifically, those resources would be mostly for prototyping costs and some funds would go to awards for high performing students. The estimated cost is between $2K and $3K. I would be grateful if you would consider allocating funds for this and am happy to meet with you to further explain details. I certainly understand if funds are not available, and I would be able to develop the class in a different way with existing departmental resources.

　　Independent of additional resources, I ask that you consider making a cameo appearance to the first class, schedule permitting, to demonstrate support and interest in our freshmen class. Thanks for considering this.

　　I look forward to our usual poker game next week.

2.8.2 Digital Communication Tools

Many software applications (apps) exist that can span multiple platforms (laptops, tablets, cell phones) and enable groups to create alternative channels to email. The advantage is that messages will not become buried in your email inbox. Although some less formal apps exist (e.g., GroupMe, Snapchat), most companies use more professional products (e.g., Slack, Trello, Asana as of 2021), some of which have free versions that you can download. These apps can usually link to other shared documents, organize and manage various conversation threads and channels, enable subgroups to be created, allow a project manager to review team progress, and keep a record of communication.

If you choose to use a digital communication tool, you should consider the following.

- Ask your faculty mentors and clinical advisor if using a digital communication tool to communicate with them is acceptable.
- Keep your app threads organized and professional. These could become a matter of public record should you wish to protect potential intellectual property (e.g., patents).
- Create a private channel within your app to discuss more general items that do not fit into a specific thread.

2.8.3 Communicating with Advisors, Mentors, Sponsors, and Staff

To accomplish the goals of any complex design project requires interacting with experts outside of your team. At the start of an academic project, establishing good lines of communication with course faculty, administrative and technical staff, project advisors and sponsors is critical. Along the way you may also be reaching out to others including end users, hospital administrators, faculty outside of the class, industry experts, consultants, government personnel, insurance representatives and more stakeholders. Recognize that many of these external experts have years of training and experience and are generally busy. Helping your team is in addition to performing their usual duties. You should respect their valuable time and experience and be grateful for it. For example, a clinical or industry sponsor may only be able to meet with you (physically or virtually) before or after a shift that may be very early in the morning or late in the evening, and at a location that is convenient for them. Many busy people can best be reached by contacting their administrative staff. Before asking for a meeting, consider if you can obtain the information you need through written communication instead. When making requests for meetings or information, recognize that it may take several days or longer to hear back.

Meetings with sponsors, mentors, and advisors should follow the same guidelines described previously, such as distributing an agenda prior to the meeting that outlines the goals of the meeting and coming prepared to listen and take notes. You should also do as much homework ahead of time as possible and come with specific questions to ask. Generating open-ended and personalized questions will help guide your discussion. For example, assume your team is working on a project attempting to reduce the incidence of pressure ulcers, and your team is trying to find out more about the incidence of pressure ulcers. You have done some preliminary research. You might ask a clinician a specific question such as, "We understand that the incidence of pressure ulcers in the general hospital population is about x%; what is your personal experience?"

2.9 Design History File and Documentation

As indicated in Chapter 1, documentation is a major part of design, typically occupying 40% or more of most designers' time. Documentation was required for some of the project-related activities discussed earlier in this chapter, such as your POS, team commitment document, meeting agendas, and minutes of meetings.

For the team formation aspect of your project, the type of information to be documented includes team name or number (if one assigned), team member names (and roles, if decided), team rules (if created), current title of project or problem (if known), relevant communications, and requirements of the design program. These early project documents may serve as the introduction to your DHF. One reason to include information regarding the design process in a DHF is to provide documentation that is required to receive regulatory clearance to sell a medical device in the US.

As the name implies, the DHF in industry is a complete description of the project from team formation through production; in academics, the technical endpoint is prototype testing results. Each subsequent chapter indicates what elements from that chapter should be included in a DHF. This is standard practice in medical device design.

Companies and academic design programs will have different requirements for the DHF. Your design team DHF must follow the file structure defined by guidelines provided. Typical components follow the design process model in Figure 1.4, and can include:

- an executive summary of a project (one page, drafted at end of project),
- design input (e.g., literature and patent search output, competitive benchmarking, clinical need statement, project statement, design specifications),
- design process (e.g., solution concept generation, design selection, design and prototype development, risk analysis, design change history),
- design output (e.g., analytical models, design drawings, sketches, CAD files, source code, final models, materials, prototypes, assembly directions, bills of materials, product cost information),
- verification and validation (e.g., verification testing plan and results, validation results),
- project summaries (e.g., presentations, design review summaries, critical project meeting summaries, relevant external communications), and
- in industry, additional manufacturing information is included (e.g., vendor information, production and assembly processes, sterilization and packaging processes, maintenance and repair documents).

There is no prescribed format or structure for a DHF—these factors will be determined by your company or design program. The structure (i.e., folder hierarchy and names) might be prescribed or you might have flexibility in its organization. There are some qualities that all DHFs should possess, such as:

- *Contemporaneous entries.* Add documentation in real time or soon after it is created. It is much easier to document at the moment rather than recollect it later.
- *Easy accessibility for the entire team and all information contained in one location.* This is helpful for your team's entry and viewing, as well as design reviews by those outside your team.
- *Logical organization,* such that it is easy to find previous entries and clear where new entries will go. A table of contents is a useful guide for this purpose. Common DHF folder organizations are mostly chronological and functional (e.g., clinical need, engineering problem, specifications, design, tests, budget).
- *Summary information at the entry point of the DHF* (e.g., executive summary, problem statements, labeled image of the device) such that someone not involved in the project could quickly understand the goals of the project.
- *Dated entries,* such that a timeline of design activities and the evolution of the design could be reconstructed.

Each company and design program have different DHF requirements, and prescribed guidelines should be followed. If not required, you are encouraged to create your own DHF organization based on the information presented here and dependent on project management resources.

Creating a DHF may sound daunting, but each chapter of this book indicates which elements from that chapter should go in a DHF, along with more specific organizational suggestions. From this chapter, for example, a title page (including course information and your names) could be the first document in the DHF. If you have created a team commitment document, for example, that would also be appropriate. If you have made notes or solved exercises, those could be included as well.

Key Points

- Teamwork is an essential part of the design process. Teams go through stages of development. One becomes better at teamwork through experience and with accepting constructive feedback.
- Socializing among team members fosters a culture for collaboration and teamwork.
- Commitment, shared work responsibility, perseverance, trust, work ethic, and mutual accountability are characteristics of effective teammates and effective teams.
- Interpersonal problems may arise between teammates. Dealing with them early, openly, honestly, and as objectively as possible reduces the risk of worsening problems as the project moves forward.
- Vigilant project planning and risk management are necessary to optimize team performance, deal effectively with unforeseeable events, and keep projects on schedule. Gantt charts are a venerable, useful tool in project planning.
- Threats to project schedules will exist. Teams should anticipate them and develop plans to prevent them from occurring.
- Written communication to those outside your team should be formal and succinct to be most effective.
- A Design History File (DHF) is a legally required component of all medical device projects regulatory approval and clearance. Creating and maintaining a DHF for your project throughout the design process is good experience before starting your career

Exercises

1. It is important for the team to start the semester with team building activities to develop trust among members. Try some of the examples presented here (or others) and have fun while doing them. If on campus, simple activities can include sharing a meal, having a pot luck dinner, playing games, tossing a Frisbee, bowling, or playing laser tag. Virtual teams can share personal stories, share a virtual meal, and demonstrate unusual talents (i.e., juggle, perform virtual magic). This can be done before the semester starts if teams are known in advance. These types of activities should continue during the semester on a periodic basis (e.g., after each major project milestone is completed).
2. Visit the website of a well-known medical device company (e.g., Medtronic, Zimmer, Abbott, Siemens) and find its core values. Use what you find to write core values for your team. Post these values in a visible place in your design space if you have one. Plan to reexamine your core values in a few weeks. Revise if core values have changed.
3. Have each team member create a somewhat detailed individual plan for a full week. Start at a specific date and time. A 2-hour interval is a reasonable time unit while awake; there will typically be 8–9 entries in a day. Include academic and nonacademic activities, as well as those unplanned (e.g., spontaneous sporting event). During the week, record what you actually did during each interval. At the end of the week, compare what you planned with what you actually did. Discuss

how activities aligned or were not aligned. What can be learned from this effort of project planning for yourself?

4. Create a Gantt chart for your design project, spanning the length of your project. Enter all known deadlines. Entries at this point will likely be team formation and due dates based on course requirements (which most likely will be part of the critical path). It will change over time. Be sure to update it as needed and at least once a week. Post the chart both electronically and in the DHF.

5. When your clinical problem to solve is known, develop a project objective statement for your project (similar to the one that President Kennedy created in 1961 for the moon landing project). It should address the scope of your project (e.g., functional prototype, proof-of-concept), schedule (when the project will be completed), and resources available to the team (e.g., budget, team members, equipment).

6. Create a list of potential threats to your project schedule (i.e., things that could occur that would prevent your project from being completed on time). Include one effect beyond your control, such as being forced to work as a team with each member in a different geographical location.

7. Below is an initial email to a project advisor. Critique it and revise it as needed.

   Subject Senior Design Project

   Hey Professor Smith,

       We are team members in Professor Jones design class, and you have been assigned as our project advisor. We write to request an initial meeting. Please advise us what dates and times are good for you. If we can accommodate one of them, we can meet you in your office, perhaps during your office hours.

   Thanks.

   Team member SK

   Class of 20xx

8. Create a meeting agenda template (see example in Figure 2.8) to be used throughout your project.

9. If you develop a romantic relationship with a teammate, discuss the issues this raises for peer evaluations. Could the personal relationship affect your objective peer evaluation? Do you still fill out a peer evaluation, or choose not to? Do you disclose the relationship to course faculty? This is a primary example of the concept of Conflict of Interest (COI), which is fundamental in all research and is presented in multiple chapters, including a separate section (Section 13.6.3) devoted to the topic.

10. Consider the following team meeting scenario:

    Meeting to Decide on a Problem Statement & Generate Action Plan

    October 9, 6–8 pm Design Studio Conference Room

6:00–6:15	Good, Guy and Gal show up, chat, have refreshments.
6:15–6:20	Average and Mediocre arrive, discuss the new design studio.
6:20–6:45	Senior Itis enters, and we start (even though Arro Gant still hasn't arrived). We discuss four major issues, make decisions, and begin to generate action steps for the first issue.
7:05	Senior Itis remembers another meeting where attendance is expected, and leaves.
7:07–7:10	Arro Gant arrives, leaves to relieve himself and returns.
7:10–7:20	Arro Gant (who has not read agenda and has not read reports distributed to the team since the last meeting) listens to a summary of discussion that went on during the previous 25 minutes.
7:20–7:30	Arro Gant raises objections to two of the four decisions, reminds us of the project vision, and exits. Average and Mediocre leave.

7:40–7:50   Good Guy and Gal and I clean up refreshments, wipe off the table.

The team leader knew that something needed to be done to make these meetings more productive. What could she do to improve attendance, get team members to meaningful way? What would you do if faced with a similar situation?

References and Resources

Allen, R. H., Acharya, S., Jancuk, C., & Shoukas, A. A. (2013). Sharing best practices in teaching biomedical engineering design. *Annals of Biomedical Engineering, 41*(9), 1869–1879. https://doi.org/10.1007/s10439-013-0781-y.

Berger, W. (2010). *CAD Monkeys, Dinosaur Babies, and T-Shaped People: Inside the world of design thinking and how it can spark creativity and innovation.* Penguin Books.

Collins, J. C., & Porras, J. I. (1994). *Built to last: Successful habits of visionary companies.* Harper Collins.

Collins, J. C. (2001). *Good to great: Why some companies make the leap and others don't* (1st ed.). Harper Collins.

Dingel, M., & Wei, W. (2014). Influences on peer evaluation in a group project: An exploration of leadership, demographics and course performance. *Assessment & Evaluation in Higher Education, 39*(6), 729–742. https://doi.org/10.1080/02602938.2013.867477.

Dreyer, B. (2019). *Dreyer's English: An Utterly Correct Guide to Clarity and Style.* Random House.

Gladwell, M. (2005, August 29). *The bakeoff.* The New Yorker. https://www.newyorker.com/magazine/2005/09/05/shiftenterthe-bakeoff.

Goodwin, D. K. (2005). *Team of rivals: The political genius of Abraham Lincoln.* Simon & Schuster.

Graff, G., Birkenstein, C., & Durst, R. (2010). *They say/I say.* W.W. Norton.

Haase, M., & Mortenson, M. (2016). *The secrets of great teamwork.* Harvard Business Review Press. https://hbr.org/2016/06/the-secrets-of-great-teamwork.

Harvard Business School Publishing. (1997). *Project management manual,* 11–12.

Harvard Businesses Review. (2015). *HBR's 10 must reads: On communication.* Harvard Business Review Press.

Harvard Business Review. (2015a). *HBR's 10 must reads: On leadership.* Harvard Business Review Press.

Harvard Business Review. (2015b). *HBR's 10 must reads: On teams.* Harvard Business Review Press.

Harvard Business Review. (2016). *Running virtual meetings (HBR 20-minute manager series).* Harvard Business Review Press.

Kruse, K. (2015). *15 secrets successful people know about time management.* The Kruse Group.

Lee, H. J. (2009). Peer evaluation in blended team project-based learning; what do students find important? In T. Bastiaens, J. Dron, & C. Xin (Eds.), *Proceedings of E-learn 2009: World conference on E-learning in corporate, government, healthcare, and higher education* (pp. 2838–2842).

Nidamarthi, S., Allen, R. H., & Sriram, R. D. (2001). Observations from supplementing the traditional design process via Internet-based collaboration tools. *International Journal of Computer Integrated Manufacturing, 14*(1), 95–107. https://doi.org/10.1080/09511920150214938.

Northouse, P. G. (2016). *Leadership: Theory and practice.* Sage.

Peters, T., & Waterman, R. H. (1982). In *Search of excellence.* Harper Collins.

Pozen, B (n.d.). What's the secret to running effective meetings? IHI – Institute for Healthcare Improvement. http://www.ihi.org/education/IHIOpenSchool/resources/Pages/Activities/PozenMeetings.aspx.

Project Management Institute. (2017). *A guide to the project management body of knowledge (PMBOK® guide)* (6th ed.). Project Management Institute.

Sacks, J. (2015). *Lessons in leadership.* Maggid Books.

Sande, P. S., Neuman, R. P., & Cavanaugh, R. R. (2000). *The six sigma way.* McGraw-Hill.

Salovey, P., & Mayer, J. D. (1990). Emotional intelligence. *Imagination, Cognition and Personality, 9*(3), 185–211.

Strunk, W., & White, E. B. (1999). *The elements of style.* Pearson.

Tuckman, B. W. (1965). Developmental sequence in small groups. *Psychological Bulletin, 63*(6), 384–399.

PROJECT SCOPE AND SPECIFICATIONS

2

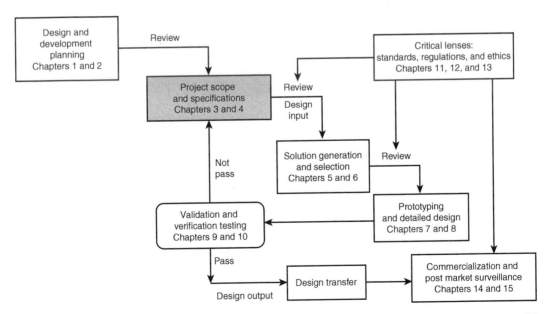

In the next two chapters you will learn techniques for identifying and understanding medical problems and translating them into a project statement with clear technical goals. This is no small task given that the healthcare ecosystem is comprised of many stakeholders with a range of opinions and motivations. Although you will continue to discover the nuances of your problem and refine your technical specifications as you move through the design process, a well-written project statement and list of preliminary specifications is a major design milestone.

Chapter 3 provides tools and techniques for bringing a complex healthcare problem into clearer focus. These techniques include reviewing literature and media, observing behaviors and procedures in medical and industry environments, engaging healthcare providers and other stakeholders through interviews, surveys and focus groups, and reviewing existing solutions. If you are searching for a problem to solve, you will also learn ways to graphically map out and narrow a list of problems that you find most interesting and promising. Your findings will be summarized in a short project statement that includes a definition of the problem and how you intend to meet the needs of stakeholders and add value to the healthcare ecosystem.

Chapter 4 guides you in translating a medical problem or opportunity into an engineering problem through measurable outcomes known as specifications. These specifications ensure that your design will meet your customer needs, clarify the value you intend to create as compared to existing solutions, guide the generation of your design solutions, and provide a way to demonstrate the success of your design solution via testing later in the design process.

By the end of Chapters 3 and 4 you will be able to:

- Identify real-world medical problems and articulate the value to the healthcare ecosystem of solving these problems.
- Identify the needs of a range of stakeholders for a particular problem.
- Find and integrate information from many sources to gain greater insight into a medical problem or opportunity, along with accompanying performance requirements and design constraints.
- Develop a Project Statement that includes the clinical need statement and scope.
- Perform competitive benchmarking to find and evaluate existing solutions.
- Draft a written project statement that summarizes the four points above.
- Create a list of preliminary target specifications.

Defining the Medical Problem

3

People ignore designs that ignore people.
— Frank Chimero, Designer, software developer, and author

The goal of a designer is to listen, observe, understand, sympathize, empathize, synthesize, and glean insights that enable him or her to "make the invisible visible."
— Hillman Curtis, new media designer, author, musician, and filmmaker

No. I don't think the Empire had Wookies in mind when they designed it, Chewie.
— Han Solo to Chewbacca about the Tydirium imperial shuttle, Star Wars Episode VI

Chapter outline

Biomedical Engineering Design. https://doi.org/10.1016/B978-0-12-816444-0.00003-1

3.1 Introduction

Identifying, dissecting, and clarifying a problem in the healthcare ecosystem requires time, effort, creativity, and resourcefulness. The aim of this chapter is to provide you with tools, techniques, and guidance on how to write a brief but comprehensive **project statement** that clearly explains the goals of your design project. This statement will serve as the input to all of your future work.

3.2 Project Statements

A project statement is a high-level yet succinct explanation of the scope of your work and why a problem is worth solving. Most academic programs and companies will ask you to write some version of a project statement. Although a project statement is short in length (typically less than one page), it is typically one of the more challenging writing assignments. In this section, the most common elements of a project statement are briefly discussed. These elements also serve as a framework for this chapter.

3.2.1 The Many Purposes of a Project Statement

A project statement serves many purposes. First, the initial phases of a design process are often disorganized and disorienting. This is a challenge to team dynamics and project management practices. Chapter 2 explains that a having a shared sense of purpose is the key difference between a group and a team. A project statement is the clearest distillation of your team purpose. Second, the iterative process of drafting a project statement is a synthesis of what you have learned. Even if you have been given a project statement, either from a previous design group, project sponsor, or mentor, your own synthesis makes the problem more personal and motivating. In drafting a project statement, you will gain a better sense of the problem, what will count as a solution, and where you need more information. Third, a coherent and concise project statement is often the first major deliverable. As such, it is an early indicator to advisors and mentors of the quality of work they can expect from you. On the other hand, it is often the first time that you will receive substantive feedback on your progress. Finally, a good first draft of a project statement guides the definition of technical requirements that are the focus of the next chapter.

3.2.2 Needs, Problems, and Opportunities

The terms need, problem, and opportunity are used in different ways in the context of the engineering design process. In this text, a **need** is defined as a desire expressed or implied by a user. A **problem** is an impediment to performing a biological, clinical, technical, or logistical function. An **opportunity** is a way to add value by overcoming this impediment. To tie these ideas together, engineers write a *project statement* that frames a *problem* as an *opportunity* to create value by meeting user needs through a technical solution.

3.2.3 Elements of a Project Statement

There is no single formula for a great project statement. There are, however, some common elements that should align and support one another. A project statement generally begins with a **statement of need** as it is experienced by users. Often included are the specific population impacted, incidence rate, and severity, to demonstrate both urgency and significance of the problem. In some cases, the needs of several users are interrelated. For example, if the need is for a less-traumatic way to perform brain surgery, the ultimate beneficiaries are neurosurgical patients; however, neurosurgeons, surgical nurses, and insurance companies are also affected. For this reason, we do not make a distinction between needs and wants; the needs of one user may be the wants of another. The aim of discussing the need is to convince a reader that your project is focused on an important healthcare problem.

When working on a healthcare project, you are likely to hear the terms "medical" or "clinical"; within the phrase "clinical need statement," for example. For the purpose of this text, we will use the term *medical* broadly when referring to any aspect of the healthcare ecosystem. We will reserve the term *clinical* when a project, problem, or device involves patients. A device that sequences the DNA of tissue samples in a lab is a medical device, whereas a surgical tool would be a clinical device.

Current solutions are the means by which users are presently satisfying their need or addressing the problem. When the need is known, a variety of solutions many be available that include procedures, guidelines, technologies, or policies. There is often a *gold standard* solution that is the most widely used or available. On the other hand, if the need has gone unrecognized, there may be no solutions or only ad hoc "work-arounds." The aim of writing about current solutions is to convince a reader that there is an opportunity to add value to the healthcare ecosystem by introducing a new and improved solution.

Technical barriers are the biological, clinical, and engineering impediments that are expected when creating a solution. These barriers are translated into technical specifications that are discussed in the next chapter. Within a project statement, only the highest-level requirements are discussed. The aim is to convince a reader that barriers have been assessed, yet there is a reasonable pathway to overcoming them over the course of your project.

Measurable goals are metrics that, if met, indicate that a need has been met and value has been added to the healthcare ecosystem. Establishing measurable goals is especially important in the medical field because needs often present themselves as symptoms of a deeper problem. Peeling back the layers of the need to get to the real problem is often the best way to capture what will change when the problem is solved. Be mindful, however, that value may be directly or indirectly gained by users. For example, decreasing the cost to manufacture a product is easier to measure but does not directly address a clinical need. A more cost-effective solution that keeps the quality-of-care high may make the solution available to more users. Rather than cost, the important clinical outcome would therefore be an increase in patients served, relative to the current solutions.

A **project objective statement** is one sentence that declares your intended end point (deliverables) along with your timeline and resources (e.g., budget, people). For example, you may make it clear that your endpoint will be a functional prototype that has undergone benchtop testing and the project will be completed in one year and stay within a given budget. The goal is to define what counts as success for your project. As discussed in Section 3.10, your project objective statement is a step toward the real-world measurable goal.

The exact format of your project statement will depend on the nature of your project. The ultimate aim is for a reader to understand the connection between the needs and the value to be gained by the healthcare ecosystem. Breakout Box 3.1 discusses two broad types of value creation that may serve as a helpful framework in drafting your project statement.

Breakout Box 3.1 Blue Ocean, Red Ocean

Chan Kim and Renee Mauborgne coined the terms Red Ocean and Blue Ocean to describe two types of projects that can add value. A Red Ocean project makes incremental improvements to an existing solution; there is intense, blood-red competition. An example is a project to redesign a ureteral stent to make use of a novel biodegradable polymer (presented in Breakout Box 4.1). A Blue Ocean project explores a completely new solution or addresses an unrecognized need; blue indicates a blood-free arena. When it was first commercialized, the lithotripsy device (presented in Section 4.4.4) was a Blue Ocean solution. Table 3.1 describes some of the features that are relevant to engineering design projects.

Table 3.1 Red and Blue Ocean projects

Red Ocean	Blue Ocean
Iterative improvements along particular product attributes	Re-envision what a solution looks like
The need, value, and demand are known	Address a new or unrecognized need
Project inherits known technical constraints and specifications	Project must clarify new technical specifications and navigate new constraints
Risk is generally known and low	Risk is unknown and often high
Competes in an existing market space	Market space is unexplored and uncontested
Goal is to beat the competition	Goal is to make the competition irrelevant

Although no design project fits neatly into these binary categories, most lean toward Red Ocean or Blue Ocean. Different types of new product development projects are presented in Section 15.5.4.

3.2.4 What a Project Statement Does Not Include

You may have noticed that a project statement *does not* specify a design solution. This is to keep the focus on the need, problem, and opportunity to add value, instead of a particular design solution. This is considered good design practice for two practical reasons. First, once you have written a good project statement, it should not change based on how you solve the problem. Second, approaching a project without a solution already in mind allows you to potentially discover a solution that has been previously missed or ignored. It is natural to think of possible design solutions; you should make note of these ideas so that you can revisit them when you are ready to generate solutions.

There are times in both academic and industry design projects when a project statement may hint at a solution. First, a project may be initiated as a redesign of an existing solution. For example, a project to redesign a catheter will assume that the solution will be a catheter. As discussed in Breakout Box 3.1, this is a Red Ocean project that makes an incremental improvement on an existing solution. The project statement for a catheter redesign should mention a catheter in the current solutions, measurable goals, and project objective statement. Second, although many projects begin with a need and progress to a solution, some begin with a technology and then search for problems to solve. More is explained about these types of projects in Breakout Box 3.2. In these cases, project statements have all of the same elements, but the solution may highlight the technology to be used.

Breakout Box 3.2 Technology Push and Market Pull

Much of this chapter is predicated on the idea that a design process begins with an identified healthcare problem or opportunity. This is known as *market pull*—once the problem is identified, the designer begins to develop and recombine technologies that solve the problem. The opposite is also possible and is known as *technology push*. With a particular technology or discovery in hand, one looks out at the world to find applications where it may have an impact. For example, W. L. Gore and associates have built much of the company around a particular material (Gore-Tex®). The technology can be used to solve a wide range of medical problems; for example, problems requiring vascular grafts. As another example, consider that a gecko's feet are sticky because of microstructures that form many small bonds with a surface; they are not sticky in the same way as a glue. Some labs have found ways to manufacture such surfaces with these microstructures and have proposed using this technology to make better band-aids and sutures. A nearly opposite microstructural phenomenon occurs with shark skin which, when mimicked in manufactured materials, can create antimicrobial surfaces (e.g., lining of catheters). Technology push projects generally begin with the technology and then search for problems to solve. The project statement for a technology push project has all of the same elements but the goal is to help a reader understand why the technology is the most promising solution to the problem.

3.2.5 Writing a Project Statement

Writing a succinct and clear project statement is a challenge. As you write it, there are some points to keep in mind. First, if you have been given a project statement, it may be incomplete, unclear, inappropriately scoped for your design experience, or may fail to identify the underlying problem. You should consider it a rough draft. Second, the order and contents of your project statement may deviate from how it is presented in Section 3.2.3. The ultimate goal is to cover all of the elements in a way that flows for the reader. Third, each of the elements described in the previous section should be distilled down to one or two sentences. Remember that your overall goal is to focus a reader on the most critical aspects of your project. Finally, revising your project statement helps you learn more about the problem you are trying to solve, what counts as a solution, and the scope of your project. You should consider it a work in progress. To get you started, Breakout Box 3.3 contains a first draft of a project statement along with points to consider for revisions.

Breakout Box 3.3 First Draft and Faculty Comments on a Project Statement

Imagine that you have created a first draft of a project statement:

Title: Preventing Iatrogenic Injuries Related to Surgery

Of the 200M open surgeries performed in the US annually, 4000 of them require a second surgery to remove items—surgical tools, sponges, towels, pins, screws—inadvertently left in the patient during the first surgery. Being foreign bodies, these items must be removed to prevent infection, fever, reduction in quality of life, and death. Patients are adversely affected during the second surgery because of additional trauma and the potential for more adhesions. At $2K per surgery, the annual additional healthcare cost to the system due to these additional surgical procedures is over $80M, which does not include legal costs due to malpractice. During every surgical procedure, nurses and OR technicians manually count all items at the beginning of and near the end of every surgery. The final count must always be "right" (first count = final count) before the patient is closed. Human error is the only cause of miscounting, which is the only reason why sponges, towels, pins, and screws are left inside the patient after closing. The clinical problem we are trying to solve is the high incidence of items left in patients after surgery, causing harm, and requiring secondary surgical procedures to retrieve them. **To reduce the incidence of preventable second surgeries, healthcare providers need a better way to keep track of counted OR instruments and disposables used in a primary surgery.**

Example feedback from a faculty mentor could be:

Overall, this project statement is a good start. The opportunity is clear and supported by quantitative assessments that demonstrate the significance of the problem. The current gold standard solution (e.g., counting all items before and after a surgery) is provided and no technical solution is implied. The last sentence is bolded because it is a clinical need statement. For a next draft, it would be helpful to consider the following:

- Have there been other attempts at technical solutions? What were they? Why have they either not worked or not been widely adopted?
- Are there any technical barriers you expect to encounter? For example, given that the solution will likely need to be in the OR, you might explicitly state that any solution will need to be sterile, portable, perhaps self-powered, and so on.
- Acronyms such as OR should be spelled out the first time they are used.
- Is the second to the last sentence necessary? Is there any new information being added? This comment can be applied to all of the sentences—what new information do they add?
- The last sentence is a measurable outcome (reduced incidence of unnecessary secondary surgeries). It may not be achievable in an academic design project and therefore should not be part of your project objective statement. It would be helpful to have a separate sentence that states what you hope to achieve. For example, by the end of the Spring semester, we aim to have a functional prototype that will demonstrate the feasibility of flagging errors in surgical tool counts.

3.3 Understanding the Medical Need

A good project statement makes a clear connection between the medical need and the project objective statement. Learning how medical concepts, terminology, and processes are applied in the real world will help you clarify the need, reveal opportunities to create value, and focus your overall end goal. Although you may have taken courses in biology, the academic discipline of biology is different than the practical application of biology to medicine. It is important to note that the medical need is at the forefront of some projects. For example, a project to redesign a surgical tool is easy to connect to a clinical need and measurements that would indicate value have been provided. On the other hand, some design projects are several steps removed from the need. Consider designing a lid for a 96 well plate to rapidly wash Zebrafish embryos. Such a research device is derived from a medical need (e.g., screening new drugs for effectiveness in treating amyotrophic lateral sclerosis), but a measurable impact on the healthcare ecosystem may require years or even decades to determine.

The aim of this section is to prepare you for learning more about the medical or clinical discipline that underlies your design project. Sections 3.4 to 3.7 then provide you with more specific techniques for clarifying the medical need.

3.3.1 Each Medical Discipline is Unique

Designing a medical device requires you to dive deeply into a particular medical area. Each specialty within medicine is unique; cardiology is very different than urology. Furthermore, within the specialties are sub-specialties; within cardiology there are subdisciplines that focus on electrophysiology, surgery, imaging, catheterization, and pediatrics. As you explore, it is helpful to know that every field has critical events, key people, technologies, and specialized vocabulary, procedures, and processes that distinguishes them from other fields. Many of these sub-specialties have evolved to solve particular problems and meet needs that arose throughout the development of the field. As a result, every field is filled with quirks from the past. Some of these may make sense to you, while others may seem strange. Furthermore, you may find that some disciplines use the same devices and tools (e.g., scalpel, O_2 saturation monitor) but in ways that are unique to their discipline. You may also find that some practices have been passed down from generation to generation without question; you may hear phrases such as, "it is just the way it is done," or "we have always done it this way." It is not reasonable or necessary for you to become an expert in an entire field or sub-specialty. Rather, you should become familiar with the particular aspects of the field(s) most closely related to your project.

3.3.2 Adopting an Innovator's Mindset

Approaching a new field can be daunting. Your success in learning can be aided by adopting two related mindsets that nearly all innovators possess. First is a growth mindset, whereby you recognize that learning is a process that does not always proceed in a straight line. Second is to regularly switch your focus from details to the big picture and then back to the details.

Throughout your career, you will encounter many new environments and fields, both professional and personal. The ability to quickly learn a new domain is a critical skill. This is even more challenging when the domain itself is changing, with innovations in technology and processes, additional regulations and laws, and new economic and market forces. Having a growth mindset means approaching new situations with curiosity, as a challenge to be met, and with the belief that, with effort, you can learn. Learning about the nuances of a medical specialty will allow you to develop your growth mindset.

Great innovators are often laser-focused on a specific problem as well as obtaining the knowledge, skills, and resources needed to create a solution. However, they do not become so focused that they ignore the big picture. Although it is tempting to only focus on the information that seems to be relevant to your particular medical need, it is also important at this phase of the design process to take in everything you discover. Understanding the broad context of a need is a first critical step in designing a solution that works in a real context; it is often a seemingly irrelevant detail that leads to an innovative insight. Keeping the big picture in mind can also help you discover latent needs—those that users do not recognize because they have become so accustomed to one way of dealing with a chronic problem.

To provide you some intuition as to how a sharp focus can be combined with a fuzzy focus, consider an analogy at the intersection of neuroscience and art, shown in Figure 3.1. For centuries, viewers have been amazed by the ephemeral nature of the smile of the Leonardo DaVinci's Mona Lisa. It seems that when you look at her mouth, there is no smile; but look away slightly, and the smile appears. Margaret Livingstone, a neuroscience researcher who studies visual perception, was able to demonstrate that this phenomenon is based on two aspects of our visual system. Our center vision, what we are paying attention to, is very good at resolving fine level details. Our peripheral vision, on the other hand, is able to see very broad trends and patterns. What DaVinci did was place the straight mouth in the detail (so that when you look directly at it with center focus you do not see a smile), and the smile in the shadows, which you detect with your peripheral vision. The take-away is that a specific focus will help you clarify the particular problem at hand, whereas the broad focus will help you better understand the context and constraints of a solution.

3.3.3 Documenting Your Research

Collecting information from many sources can be a daunting process. Often, one source leads to another, requiring you to follow several parallel threads. This is especially true when you are conducting research as a team. A theme throughout this text is to document your work. As you gather information for your project statement, you should keep records of everything you find. It is

FIGURE 3.1

Close-up of the Mona Lisa showing that her smile is embedded in the shadows around her mouth (blurred left panel), which only appears in peripheral vision. The actual lines of her mouth (right panel) are straight and are what appears with center vision. The effect is that her smile only appears when you are not looking directly at her mouth.

always easier to capture information in the moment rather than trying to retrace your steps later. At this early stage in the design process, you do not yet know what information will be important later.

3.4 Literature for Learning the Medicine

The availably of medical resources online means that you likely have access to many major medical, clinical, and life-science journals, websites, and training materials. These can be excellent ways to learn about the medical and clinical aspects of the problem you are trying to solve. In this section, we suggest ways to find and sort through the varied resources that will help you clarify the medical need.

3.4.1 Getting Started

A barrier to understanding the medical background of your project is the specialized language used by practitioners in the field. To a novice, it is much like learning a new foreign language. You do not need to become fluent, but you should have a working proficiency that includes basic anatomy and physiology terminology common to all of medicine, as well as specialized terms relevant to your project.

An excellent way to begin learning medical terminology is to find someone in the field who can serve as a guide. Ask a mentor to point you in the direction of good resources. Every medical specialty has journals, conferences, review articles, "classic" introductory books (e.g., Williams' *Obstetrics*, Volpe's *Newborn Neurology*, Abrams' *Interventional Radiology*), and websites that explain the critical ideas, procedures, and vocabulary. You can often find training materials created for medical residents or trainees, nurses, and others entering the medical profession. You can find a quick guide to common medical terms in Breakout Box 3.4

3.4.2 Selecting and Searching Medical Literature

Before you meet with an expert, it is often helpful to explore the available resources on your own. Doing so will help you ask better questions of practitioners in that field and help you make sense of a medical setting. Searching and interpreting medical literature is somewhat of an artform. The internet is a wonderful resource; however, it is not always easy to distinguish quality, evidence-based work from less reliable sources.

There are differences between clinical/medical research and technical/scientific research. While technical research aims for objectivity, some medical research is based upon clinical judgment and is inherently subjective. The quality of the medical data, and the conclusions drawn from these data, are therefore based in large part on the type of study. A double-blind randomized-controlled trial is the gold standard of medical evidence because it eliminates provider bias and patient selection bias. As a rule, prospective studies (keeping track of patient data as patients are treated) are more reliable than retrospective studies (relying on past medical records for data). There are two types of retrospective studies; cohort and case control.

An example of a cohort study is one that looks back at data from a group of similar patients with and without risk factors for a disease (e.g., nonsmoking and smoking males aged 30 to 50 years with lung cancer). If significantly more smokers became lung cancer patients than nonsmokers over a longer

Breakout Box 3.4 Common Medical Terms

Anatomical Terms

Term	Definition
superior/inferior	higher/lower
distal/proximal	further/closer to the midline of the body (medial plane, with patient as reference)
cephalo-/caudal	toward the head/toward the feet
anterior/posterior	front/back
medial/lateral	middle/side
thorax	chest
abdomen	below the rib cage (belly)
flexion	reduced angle of joint
extension	increased angle of joint
supine	laying on one's back

Prefix/Suffix	Meaning	Example	Description
peri-	around	pericardium	tissue around heart
epi-	above	epidermis	top layer of skin
inter-	between	intercostal	between the ribs
retro-	backward	retrograde	reverse flow
ante-	forward	antegrade	forward flow
supra-	above	suprapubic	above the pubis
per-	through	percutaneous	through the skin
trans-	through	transdermal	through the skin
endo-	within, inner	endothelium tissue	within/inside a vessel
sub-	below, under	subdural	below the dura mater (brain)

Medical Terms

Prefix/Suffix	Meaning	Example	Description
a-	without	arrhythmia	without a rhythm
co-	together/both	collateral	both sides
bi-	two	bilateral	two sides
hydro-	water	hydrocephalus	water on the brain
hyper-	excessive	hyperlipidemia	high levels of blood lipids
hypo-	deficient	hypothermia	lower temperature
-genic	created, produced	thrombogenic	produces blood clots
thrombo-	clot	thrombogenic	produces blood clots
-pathy	disease	nephropathy	disease of the kidney
iatro-	treatment, physician	iatrogenic	treatment/physician induced
idio-	unknown	idiopathic	of unknown disease or origin

Breakout Box 3.4 Common Medical Terms—cont'd

Medical Conditions

Prefix/Suffix	Meaning	Example	Description
-itis	inflammation	tonsillitis	inflammation of the tonsils
tachy-	rapid	tachycardia	rapid heart rate
brady-	slow	bradycardia	slow heart rate
-emia	blood	septicemia	blood infection
-uria	urine	hematuria	blood in urine
-algia	pain	myalgia	muscle pain
-osis	abnormal condition	dermatosis	any skin abnormality
-megaly	enlargement	cardiomegaly	enlarged heart
-trophy	shape/size	hypertrophy	enlarged
pneumo-	air	pneumothorax	air in thorax (lung)
-plegia	paralysis	hemiplegia	paralysis of one side of the body

Organs and Structures

Prefix/Suffix	Meaning	Example	Description
arthro-	joint	arthritis	inflammation of the joints
cysto-	bladder	cystoscopy	examination of bladder with cystoscope
gastro-	stomach	gastritis	inflammation of the stomach
cardio-	heart	cardiomegaly	enlarged heart
nephro-	kidney	nephrectomy	surgical removal of the kidney
hepato-	liver	hepatitis	inflammation of the liver
hystero-	uterus	hysterectomy	surgical removal of the uterus
arthro-	joint	arthritis	inflammation of the joints
spleno-	spleen	splenectomy	surgical removal of the spleen
angio-	blood vessel	angiogram	image of blood vessel
rhino-	nose	rhinoplasty	reconstruction of the nose
laparo-	abdominal wall	laparotomy	incision of the abdominal wall
mammo-	breast	mammogram	image of breast tissue
masto-	breast	mastitis	inflammation of the breast
neuro-	nerve	neuropathy	disease of the nerve
uretero-	ureter	ureteroscopy	examination of ureter with ureteroscope
colono-	colon	colonoscopy	examination of colon with colonoscope
trachea-	trachea	tracheotomy	surgical incision of the trachea
derma-	skin	dermatitis	inflammation of the skin
prosta-	prostate	prostatectomy	surgical removal of the prostate
hemato-	blood	hematology	study of the blood
myo-	muscle	myocardium	heart muscle
encephalo-	brain	encephalitis	inflammation of the brain

Continued

Breakout Box 3.4 Common Medical Terms—cont'd

Prefix/Suffix	Meaning	Example	Description
ophthalmo-	eye	ophthalmoscope	instrument to view inside of the eye
oto-	ear	otoscope	instrument to view the inner ear
cutaneous	skin	percutaneous	through the skin
cholecysto-	gall bladder	cholectystectomy	removal of gall bladder
pancreat-	pancreas	pancreatitis	inflammation of the pancreas
osteo-	bone	osteotomy	incision of the bone
procto-	rectum	proctoscope	instrument to examine the rectum
stetho-	chest	stethoscope	instrument to listen to the chest
vaso-	vessel	vasodilator	substance causing vessel dilation
venous	vein	intravenous fluid	fluid delivered through a vein

Procedures

Prefix/Suffix	Meaning	Example	Description
-ectomy	excision, removal	appendectomy	removal of the appendix
-tomy	incision, to cut into	osteotomy	cutting into bone
-plasty	reconstruction	arthroplasty	joint reconstruction, replacement
-scopy	examination, view	ureteroscopy	examination of the ureter with a ureteroscope
-centesis	surgical puncture	amniocentesis	puncture of the amniotic sac

Instrumentation

Prefix/Suffix	Meaning	Example	Description
-gram	record, picture	electrocardiogram	record electrical activity of the heart
-graph	recording instrument	electromyograph	electrical activity of the muscles
-scope	instrument for viewing	cystoscope	instrument for viewing the bladder
-meter	device for measuring	oximeter	device to measure oxygen concentration

---

period of time, one conclusion from such a study is that smoking is a risk factor for lung cancer in men aged 30 to 50 years. A case control study is one where data from a population starts with data from specific patients (e.g., aged 30 to 50 years) with a disease (e.g., lung cancer), and draws correlations with risk factors (e.g., smoking). If significantly more smokers became lung cancer patients, a similar conclusion to the cohort study can be made. The results of a cohort study are considered to be stronger medical evidence than a case control study. The reason is that the exposure (smoking) and outcome (lung cancer) are already known; case control studies can lead more directly to medical advice on risk factors.

A wide range of search engines can help you find medical literature, with Google Scholar, Pubmed, EMBASE, and Web of Science being commonly used. You should meet with a university librarian or mentor to learn what additional search engines are available to you. For example, you may have access to a medical library through a clinical mentor.

When using search engines, you should consider search terms and the reliability of the sources you find. In general, a search return of thousands or millions of hits is not useful. As you begin your search, you may not know exactly what terms will yield a focused list. Your ability to choose good search terms, however, will increase as you learn more. You may find it helpful to use the National Library of Medicine MeSH (Medical Subject Headings) as a way to get started. Likewise, some search engines allow you to limit your search to only the abstract or title of an article. For example, assume that you are looking for articles on shunt replacements in the brain. A search for "shunt replacement" on Google Scholar yields >500K hits while on Pubmed the return is >1K hits. When the search is limited to only titles, however, there are <100 publications using "shunt replacement." You may then notice that a number of these articles are for arterial or cardiac shunts. By eliminating titles with the terms "aorta" or "arterial" or both, the search yields <50 articles. More guidance on search terms can be found in Sections 5.8.1 to 5.8.3.

From a narrowed list of sources, you should assess the reliability of each source and relevance to your project. The better sources of online information are those that have been reviewed by others, often going by the terms "refereed" or "peer-reviewed" (e.g., WebMD, textbooks, most disciplinary journal articles). Articles published by nonprofit organizations (e.g., diabetes.org) are often informative and trustworthy, albeit written by non-experts. Wikipedia articles are typically written by non-experts but often contain references for further exploration. If an article is >10 years old, you should check if the information is still relevant. A strategy is to use a search engine (e.g., Google Scholar) that lists forward citations, allowing you to determine if an older article continues to be cited. Typically, >2 citations per year is good, 5 is very good, and >10 is outstanding. Keep in mind, however, that some older studies are highly cited because they are invalid or incorrect. For example, one paper linking vaccines to autism, which was retracted, has >2500 citations. On the other hand, if a publication is from the past two years, it may not yet have many citations.

3.4.3 Videos and Patient Perspectives

Videos can be especially helpful in understanding the flow and context of a medical procedure. Although YouTube is often a first resource, there are more reliable websites. Typical clinical peer-reviewed video databases include accessmedicine.mhmedical.com, *Journal of Medical Insight*, *Journal of Visual Experiments*, *New England Journal of Medicine's* online videos, and *Harrison's Internal Medicine*. You may also find helpful videos at the following websites:

- BroadcastMed—ORLive (https://www.broadcastmed.com/orlive)
- Medtube (https://medtube.net/)
- CSurgeries (https://www.csurgeries.com)
- eMedTV (http://www.emedtv.com/video.html)
- MySurgeon (https://mysurgeon.net)

Videos can offer advantages over live observation (discussed in Section 3.5). For example, due to the nature of the procedure, orientation and size of the room, or number of people in the room, a video may provide a better view than a live observation. A video also affords you the possibility of multiple viewings so that you can continue to extract new insights.

You may also find the first-person perspectives of patients and caregivers on the websites of support and activist groups. Public-facing physician websites also exist, often targeted toward particular diseases or clinical specialties. Company websites contain medical information and testimonials as well as literature on their products and services. Some of these sites contain online chats, blogs, and webinars that can provide you with even more information.

3.4.4 Documenting and Organizing Information

As in all phases of design, you should document your work. Many bibliography managers exist (e.g., Reference Manager, RefWorks, Write-n-Cite), can be integrated with text editors (e.g., Microsoft Word, Google Docs), and are supported by most university libraries. Likewise, you may find it helpful to maintain a glossary of terms whose definitions you can either look up online or obtain by asking a mentor.

As with all research, it is important for you to organize the resources you find. This is especially true if you are working on a team that has delegated search responsibilities to you. A good first approach is to collect a small number (20 to 30) of resources and only read the abstracts. You can then choose a smaller number of sources that look promising and read the entire article more carefully. From this review, you can develop 3 to 5 categories, which become folders into which you can organize future findings. This is an iterative process that evolves as you sharpen your search terms and build a bibliography of relevant sources.

3.4.5 Summarizing Results and Disease Mapping

A next task is to curate what you have found based upon its relevance to your project and the reliability of the information. Summarizing what you have found is a good way to make the information your own and can also help communicate your findings to others. Such a summary may take many forms, including a presentation or a short review article. There are some additional creative ways to graphically display what you know about the medical problem and need.

Most clinical needs are related to the diagnosis, prevention, and treatment of a disease. As you gather information regarding the progression of a disease state, you may want to create a graphical flow map, as shown in Figure 3.2. When applicable, it is helpful to begin with biological origins and the first discovery of a problem and end with terminal outcomes. Along the way, the various diagnostics, preventative measures, and interventions (e.g., surgical, pharmaceutical, behavioral) may be noted. Rehabilitation, remission, readmittance, or cure may also be included, in addition to the death of the patient. Such an exercise can be shared with a medical expert to ensure that you have a good understanding of the disease pathways and diagnostic and treatment options.

3.5 Interactions With Medical Personnel and Clinical Environments

Interacting with experts in a field, especially in their work environment, can yield insights that are difficult or impossible to gain through online research or videos. For example, an initial problem may be expressed as the need to change patient positioning mid-procedure due to an emergency such as a

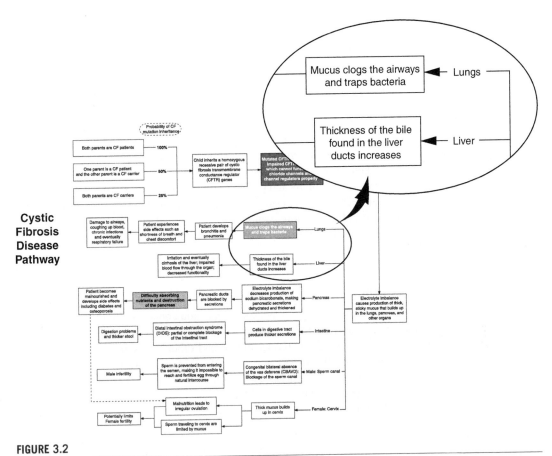

FIGURE 3.2

A student-created systems flow map of the origins and progression of cystic fibrosis.

hemorrhage. A physician may have one view of this emergency, but a surgical nurse may have a very different perspective and set of needs. The orientation of the table could be crucial to understanding the underlying problem, information you might only learn by talking to a surgical nurse. Likewise, observing the clinical environment (e.g., form of the table, number of people in the room) can help you better understand the context of the problems and needs. Similar insights may be gained in an industry-sponsored project by meeting with personnel from a company who are familiar with the needs of the user. This section prepares you for interacting with personnel and environments that you may encounter. Sections 3.6 and 3.7 provide you with some structure techniques for your interactions.

3.5.1 Clinical and Industry Etiquette

An excellent way to begin understanding a discipline is to visit the places where that discipline is practiced. When entering a new environment, remember that you are a guest. Unlike at your university, where learning is a high priority, this is not the case in a clinical or industrial setting, unless you are visiting a teaching hospital. Compared to a classroom, the pace of change is typically much faster and the urgency is far greater; life-threatening emergencies (e.g., hemorrhage, anaphylaxis) can arise in the clinic very quickly and acting within seconds can make a difference.

Medical and industry personnel take pride in the practical and professional nature of their work. When entering any professional environment, you should try to mirror the behaviors of those in the environment. You may be required to take a short course, watch a video, sign visitor forms, or wear a special badge. Professional attire should be worn unless otherwise instructed. Above all, you are an ambassador of your program, school, and the engineering profession. You should act accordingly.

Academic medical centers also double as training sites for medical students, residents, and fellows. In some cases, you may be mistaken for a medical student or resident. In an academic medical center, it is common for trainees to interact with patients and perform some procedures. Unless you are a medical student or resident, or are licensed to assist in a surgical procedure, at no time should you assist with or participate in a surgical procedure. Furthermore, you should not interact directly with patients (even verbally), unless you have gone through the proper training and filled out the proper paperwork discussed in Section 3.5.3.

3.5.2 Preparing for the Experience

When entering a new environment, it is natural to have some fears and concerns. Table 3.2 shows common worries of design students that may mirror your own.

You may enter environments where organs, tissue, blood, and other fluids are in full view. From blood labs to surgery suites, some clinical settings are not for the faint of heart. Eating carbohydrates and drinking a small amount of water about an hour before you enter the clinic can help you maintain your composure. Given the new sounds, sights, and smells, it is not unusual for a newcomer to these settings to feel nauseated, uncomfortable, or faint. If you are feeling uncomfortable, be proactive and

Table 3.2 Engineering design student worries about clinical observations

Questions	Common Answers
What fears do you have about your clinical visit?	• Overwhelmed by environment • Feeling faint or nauseated • Getting lost or being late • Not knowing what to expect • Overlooking important information • Making a good first impression • Observing death of a patient
What concerns or challenges do you see working with clinicians?	• Interrupting workflow • Knowing when to ask questions • Intimidation by authority • Being in the way, knowing where to stand • Understanding medical terminology
What concerns or challenges do you see about making observations?	• Distinguish between observations and opinions • Overlook important details • Miscommunication and misinterpretation • Understand objectives of clinical environment

ask to leave the room. If that is not possible, you should squat down with your head angled between your legs. This will help take images out of your line of sight and move blood back to your head. Another technique if you are feeling faint is to redirect your attention. Focus on the devices or amount of lighting in the room. How many people are in the room? Are they moving around or mostly station-ary? Becoming an active observer can distract you and also has the added benefit that you may notice something you would not have noticed otherwise.

When observing a procedure, it is important to know when it is acceptable to ask questions. The person hosting your visit may discuss this with you ahead of time. Before you ask a question, try to read the room. For example, is everyone tense, moving quickly, or only talking about the procedure? If so, this is a clue that it is not a good time for questions, and you should write down or remember your ques-tions for later. On the other hand, if the atmosphere is more relaxed, you may begin with "is this a good time to ask a question?"

You may enter environments that vary in the degree of sterility and may be asked to "scrub in" before entering. In the clinic, the intention is to reduce the risk of infection to the patient. Draping in light blue (Figure 3.3) indicates a sterile surface and merely touching or grazing it makes it unsterile. If this occurs, the sterile field will need to be reestablished, causing a significant delay in the procedure. In industry, tape on the floor or physical barriers often indicate an area in which some critical process (e.g., chip manufacturing) is being protected. A good way to avoid an accidental touch is keeping one's hands behind the back. Another way to avoid accidentally backing into a protected area is to never stand with your back to a sterile field. Typically, observers are asked to maintain a three- to four-foot "cush-ion" around the sterile field to prevent contamination.

3.5.3 HIPAA and Privacy

As a visitor to a medical environment, you may need to complete some training or take courses to keep yourself and others safe. In addition, all personnel who work in medical environments in the United States must abide by the Health Insurance Portability and Accountability Act (HIPAA) of 1996, which

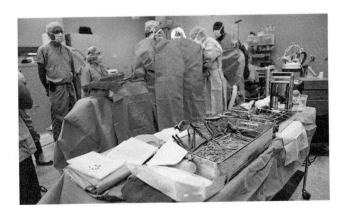

FIGURE 3.3

Example surgical suite draped in blue to indicate a sterile environment, tools, and devices. To gain access to these sterile spaces, you will likely need to scrub in.

protects the privacy and security of individual healthcare data. The general guidelines are relatively simple; do not reveal any specific patient data that you may become aware of during your site visit. This includes not discussing cases while in any public areas within a hospital. No photos or video recordings should be taken without permission. HIPAA is discussed in more detail in Sections 10.5.1.2 and 13.4.3, and you should be aware of the policies of any medical environment you enter.

3.6 Observations and Ethnographic Methods

Observing a problem as it occurs in an actual environment can provide important insights that often cannot be obtained from other sources. As you may have a limited time for live observations, it is important to be prepared. In this section we discuss ethnographic methods that you may find helpful in being a careful observer.

3.6.1 Ethnographic Methods

Observing an unfamiliar environment or culture has historically been the domain of anthropologists, who have created a powerful set of ethnographic research methods. Most anthropologists approach a new setting by first passively observing to identify a preliminary set of actors, events, and environments that seem to be most important. They then try to uncover the causal relationships between these elements, often forming hypotheses that are then verified through more focused observation. Over time, a coherent narrative emerges of how all of the elements fit together. In fact, many anthropologists report what they find through stories that are similar to engineering case studies.

In pure ethnographic research, the researcher only observes. Any interaction is viewed as a disruption and many anthropologists attempt to blend into the environment by adopting an already defined role within the system. Although this may not be possible in a clinical or industrial setting, you should disrupt as little as possible.

3.6.2 Ethnographic Preparation and Advice

When preparing to observe a procedure, you should determine what information you aim to collect. Research the procedure ahead of time, perhaps even by viewing videos, so you have a general idea of what to expect. Communicate what you hope to gain to those performing the procedure and ask for suggestions on safety, how best to take notes, and good perspectives from which to view the procedure. For example, a broader perspective may be best if you want to understand workflow. On the other hand, a closer view may be better if there is a specific detail you want to see, such as how a surgical instrument interacts with the patient's body. Before taking pictures or notes, you must obtain the proper permissions.

If you are allowed to take notes, a good strategy is to write down what you see and questions that you have. The clinical observation report in Breakout Box 3.5 may give you some ideas. You may also be able to get permission to make an audio recording, to be analyzed later. Each member of a team should take their own notes so that you may compare what you saw and heard.

It is best to debrief on a visit as soon as possible afterward, taking even more notes on themes or trends that were observed. The ride back from a visit may be a good time for this post-observation discussion. If possible, you may then follow up with questions for specific personnel in the room.

Breakout Box 3.5 Clinical Observation Report

When observing a procedure, it is helpful to use a framework to document the details of your observation. The following is an example of the type and organization of information to include in a Clinical Observation Report that documents what you learned during your clinical observation.

I. Introduction
   A. Type of procedure observed
   B. Date of observation
   C. Location of procedure (in the hospital or other facility)
   D. Name(s) of physician or surgeon conducting procedure
   E. Support personnel present
   F. Gender and age of patient
   G. Indications for procedure
   H. Goal of procedure
II. Brief description of procedure (what was done)
   A. Anesthesia/sedation, patient positioning
   B. Short summary of procedure (one paragraph)
   C. Time taken to complete procedure
III. Technology used during procedure
   A. Imaging equipment used, if any
   B. Procedure-specific devices, special surgical instrumentation
   C. Other technology
IV. Problems with procedure and technology
   A. Problems articulated by surgeon, nurse, or other medical personnel
   B. Problems observed by student or design engineer
   Examples:
   • Procedure-related problems (access to or visualization of operative site, time consuming procedures, other problems)
   • Technology-related problems (difficult to use devices, device failures, time consuming devices, other problems)
V. Opportunities for new product development
   A. Stated needs from surgeon, nurses, others
   B. Unarticulated (latent) needs identified by student or design engineer
   C. Potential applications of technology to meet needs/solve clinical problems

If you obtain permission to videotape or take photographs, you must be sure you understand what can and cannot be recorded. In general, identifying patient information, such as faces, is not allowed. The same is true for healthcare providers. If you have approval, you should establish the acceptable parameters for videography and photography, such as the use of a tripod, best allowable perspectives, acceptable lighting methods, and how images or movies should be recorded in a secure manner.

It is important to maintain your focus before and after a procedure. In the clinic, observing what occurs in waiting rooms, hallways, storage rooms, and office spaces can yield important insights. Sometimes, overhearing a conversation or learning about the role of someone not directly interacting with patients can be the key to understanding an unmet need.

3.6.3 Seeing the Unseen

Preparation to enter an unfamiliar environment goes beyond logistics. You should also go in with the right mindset. Findings from psychology and neuroscience in the past few decades have altered how we understand perception. It has become clear that we can only take in a tiny fraction of the sensory information that is available. We generally focus our attention on something in particular and then do not notice other elements in the environment.

The guides for your observation may suggest where to look. Over time, however, we all develop mental habits that we rely on in familiar environments. Often, the only things noticed are drastic differences or changes. For this reason, those who work in a professional setting may have become so accustomed to how things normally are, and where they place their own attention, that they can miss obvious problems. This is the origin of the well-known joke—one fish turns to another and asks, "how's the water?" The other fish replies, "what is water?" The point is that if you are immersed in an environment all the time, you begin to block out all but the surprising parts.

Your greatest asset as a new observer is that you can enter without any opinions, biases, or preconceptions. You may see problems and work-arounds that employees no longer see. As an engineer, you may find yourself gravitating toward devices or processes. That is a familiar place to start and can help you begin to understand the flows and rhythms of the environment. You should, however, stretch yourself to go beyond what is familiar to you. Some techniques to shift your focus to notice new aspects of an environment are:

- Close your eyes and listen. An environment is often filled with sounds that are not registered until you are no longer bombarded by visual stimuli.
- Write down everything you see that is a particular color, perhaps blue or green. Is there some reason why an object is that color?
- How many staff members are there in a location? Who are they and what jobs do you suspect they perform? Are they interacting with one another? If so, how?
- What signs are posted? Do they look official or unofficial? Where are they posted and for whom do you think they are intended?
- Do you see evidence of "work-arounds"—unofficial homemade solutions to a problem?
- Look for difficult to use instruments and equipment, tedious tasks, confusing user interfaces, and other potentially problematic technology. You can determine this by paying careful attention to facial expressions, body language, and the use of emotionally charged words (e.g., swearing), as clues to likes and dislikes. These may be clues to unmet needs that are not consciously recognized.

When you feel as though you are saturated using one of these techniques, you should switch to another. Likewise, you may find that recording unstructured thoughts or drawing concept map diagrams in a notebook may work best for you. For others, it may work better to have a formal protocol with the exact information you wish to collect. Note that your approach may also vary depending upon the phase of the design process and the nature of the observation. As practice, you may wish to try some of these techniques on the exercise in Breakout Box 3.6.

3.7 Direct Interactions With People

Surveys, interviews, and focus groups are ways of gathering information about the opinions, feelings, actions, habits, or ideas of a specific population. Unlike observations, which are typically passive, these methods actively engage people. To avoid ambiguity, you should always identify yourself as a student whenever contacting someone external to your university for the first time. You should also check with

Breakout Box 3.6 Seeing Problems

It is important as an engineer to be able to enter a new environment and find potential problems. Of course, you need to validate that these really are problems by gaining more information. View Figure 3.4 for one minute and write down all potential problems that you see.

FIGURE 3.4

Seeing Problems in Photographs.

Now look again at the background, small details, and context. What else do you see? What questions would you have? This technique is known as *painstorming*; essentially, searching for problems embedded within an environment that impact people living and working in that environment. You can take this idea of painstorming into medical or industrial settings, as well as in processing your notes, photographs, and videos after a visit.

a mentor or advisor to determine if any approvals are required before you begin contacting anyone outside of your team or program. This section introduces how to design and conduct informal surveys, interviews, and focus groups as you refine and clarify your project statement. We return to these methods in Chapters 9 and 10 as they are often helpful during the validation and verification of a solution.

3.7.1 Research Versus Design Studies

The design of a survey, interview, or focus group involves a rigorous process that has been well developed by the social sciences. Developing a study, gaining the proper approvals, validating questions, executing the study, analyzing the results, and determining reliability of the conclusions can take months or sometimes years. Some academics make such studies the core of their research. Some companies have entire departments or hire external consultants to design and conduct surveys, interviews, and focus groups.

At this early stage of a design process, it is unlikely that you need to conduct a rigorous study. Rather, you seek information to clarify the significance of the need by verifying what you learned from other sources, identifying gaps or inconsistencies, and adding additional perspective and considerations. If you determine that you require more rigorous methods at this phase of your design process, you should consult Section 10.5.2.

Imagine that you have identified the need to mitigate the risk of infection in home dialysis patients. Interviewing only one patient or caregiver may reveal insights that could become critical to developing a solution that works in the home environment. A formal study of hundreds or thousands of home dialysis patients form the basis for an academic paper but is not necessary for writing a good project statement. Guidance from a clinical mentor, industry advisor, or someone else familiar with the field can help you determine the best way to develop questions, recruit participants, interpret the results, and determine if any training or approvals are needed.

3.7.2 Selecting a Population

When directly contacting people, it is important to ensure that you are targeting the right population. A best practice is to identify the information that you hope to gain, and then identify the population(s) that you think can best provide that information. Two concerns are their ability to answer the questions, and whether their answers are representative of the demographic group. Keep in mind that sometimes it is important to identify *extreme users*. Extreme users are those who push the limit of the design along some dimension. For example, a drill used continuously in a long surgery may overheat and compromise the functionality of the drill. Sometimes, a device does not work or is not adopted because it fails in extreme cases. Hearing from extreme users, and designing for them, can help you think more deeply about how your solution will work in real situations. In developing hypotheses and questions, you may discover that no one population can get you all the information that you need.

In general, the number of participants you contact depends on several factors. The most practical is how much time you have to collect data. From a more theoretical perspective, the ideal sample size is related to the diversity of views held by the population. For a population that holds homogenous views, very few samples are needed. On the other hand, populations that are heterogeneous require a larger sample size. It is sometimes best to recruit participants on a rolling basis, known as a "snowball" method. The recruitment can end when you are not receiving any new or surprising information.

3.7.3 Developing Questions

Generating good questions is an artform and is the basis for surveys, interviews, and focus groups. The general goal of a question is to obtain some particular desired information. For example, you may guess from the literature that it requires about 30 minutes to close a surgical site and that the surgeon does the suturing. In an interview, survey, or focus group, rather than asking "does it take you 30 minutes to close the site?" (which leads the subject), you should instead ask, "who closes the site and how long does it typically take?"

Some additional considerations in developing questions are:

- Learn as much as you can about your target population so that you ask questions that they are best suited to answer
- Ask open-ended questions that do not prompt a simple yes or no answer, but rather elicit a nuanced judgement or explanation
- Be careful not to bias your questions by assuming that there is a "correct" answer. Likewise, be mindful of how you phrase a question; stating that a surgery has a 10% risk of death is perceived very differently than stating that it has a 90% success rate
- Generate questions using a tried-and-true method used by journalists; the Five W's (Who, What, Where, When, and Why)
- Develop a long list of questions and then refine, recombine, and rephrase. Make sure each question targets some desired information
- Arrange your questions in order of the most critical, which you should ask first
- It is often helpful to begin with an easier question that breaks the ice
- Reserve time toward the end of the survey, interview, or focus group for participants to add their own thoughts on how to solve the problem.

Breakout Box 3.7 contains examples of questions that were generated for a survey of patients to better understand noncompliance. Make a similar table to plan questions for your own survey, focus group, or interview.

3.7.4 Surveys

Surveys are an efficient means of gaining information from many participants at the same time.

There are several practical survey design principles that can help your participants complete your survey and provide you with valuable information. First, the directions and flow should be self-evident yet short (e.g., a few sentences). Although most of your surveys will be taken when you are not around to answer questions, this is not the place to explain everything about your project. Second, many participants are busy, and your survey is one more thing for them to do. Ideally, your survey can be completed in 10 minutes or less. It is good practice to let participants know when you recruit them how much time they should allocate for taking your survey. Third, the easiest questions should come first. Often these introductory questions are multiple choice or simple numerical rankings; it is best to keep the choices to five or fewer (e.g., do not use a 20-point scale). Short-answer questions typically appear at the end of the survey and are limited in number. Fourth, the most important short-answer questions should appear first. It is not unusual for some participants to finish your survey only partially. Even in these cases, you can still gain valuable information. Fifth, it is helpful if you group together questions that have a similar theme. Finally, the last page of the survey should thank the participant. As in engineering design, you often need to make tradeoffs between these principles. It can be helpful to have a few others outside of your team take your survey (even better if they are members of the target population) to better evaluate the clarity of your questions and instructions as well as estimate of the length of time needed to take the survey.

Three popular online survey instruments are Survey Monkey (www.surveymonkey.com), Google Forms, and Qualtrics (www.qualtrics.com). Each has advantages and disadvantages. One important functionality they all share is the ability to make a survey anonymous, meaning you cannot match

---

Breakout Box 3.7 Designing Questions to Study Noncompliance

Imagine you are working with a cardiologist who is studying noncompliance among patients prescribed to use a wearable defibrillator while they wait to have one implanted. The goal of your project is to improve the current design so patients become more compliant. To understand the problem, it is critical to know why patients are noncompliant. Table 3.3 is an example of how you could plan the questions that you could ask in a survey, interview, or focus group.

Table 3.3 Planning for a patient survey

Question	Rationale	Form of Expected Response	Interpretation
Do you use the device? If so, how often have you been using it, and for how long each time?	Establish if patient uses the device.	Yes or no, numeric (hours, days, months)	Was device tried or not? Was it used at all? Is patient compliant?
Does the device alter your lifestyle? How does it affect your lifestyle?	Is change in lifestyle a reason for noncompliance? What changes in lifestyle are a reason for noncompliance?	Yes or no Textual	If answers are mostly positive, see next question. What lifestyle change is a factor in noncompliance for most patients?
What would you wish could be changed to increase your compliance?	Find faults in current device that lead to noncompliance.	Textual	A theme among responses indicates what can be done to change the design to improve compliance (helps refine the need and establish specifications).
On a scale of 1–5, how strongly would you recommend the device to someone else?	This could also be posed as a yes/no question but asking to rate the device using a scale can lead to more detailed responses. To be diverse.	Numeric	Responses would add an objective measure to previous question.
What are the three most important things that could be improved for the device?	Specific user input to problems is most insightful.	Textual	Themes among responses are important for problem definition.

---

answers to individuals. It is a good practice to let participants know that you will not be able to identify their responses.

3.7.5 Interviews and Focus Groups

Interviews and focus groups provide the opportunity to ask questions, receive answers, and then follow up with clarifying questions. Interviews are typically conducted one-on-one in a quiet space. A focus group (eight or fewer participants is recommended) is less conversational but allows interactions between participants that can often draw out additional insights. Both interviews and focus groups are more time consuming than a survey (often lasting one hour), but when done correctly can yield important and unique insights. For example, a potentially useful type of exercise for interviewees is to have them list the tasks that they regularly perform and the specific roles they play. Follow up by asking them to estimate the percentage of the time they spend engaged in particular activities. If you feel it would be useful, you could also ask them to rate the relative difficulty of the tasks.

Because interviews and focus groups are time consuming, you should be selective in whom you invite. Unlike surveys, you must establish a time and place to meet, although it may be most efficient in some cases to meet virtually. Ideally, the location would either be a neutral site or a place that is comfortable for the interviewees. While you may be busy, you should try to work around the schedules of your participants. It is good practice to send a reminder email one day before the interview or focus group.

You should consider the flow of the conversation. It is customary to use the first few minutes for informal conversation or an icebreaker. This is meant to put everyone at ease and serves as a buffer in case of delays. There should then be a brief overview of the parameters (e.g., timing, privacy) and what information you hope to gain. As you conduct the interview, use your list of questions to initiate conversation and then allow the interviewees to drive the conversation forward. If the conversation stalls or you are not gaining the information you desire, you should be ready to reframe your question or ask a follow-up question. Be wary of receiving only positive or negative comments. For example, if you detect that you are only receiving positive comments, prompt the interviewee to consider the negatives and provide constructive criticism. At the conclusion of the interview, you should thank participants for their insights. If possible, it is also a nice gesture to offer to share your results when they are ready.

3.7.6 Extracting Meaning From Your Results

The processing of results from surveys, interviews, and focus groups depends upon the nature of the questions. Numerical and multiple-choice questions can be summarized using bar and pie charts, statistical analyses, and other quantitative descriptions. On the other hand, open-ended responses typically lend themselves to more qualitative analysis. There is an entire field dedicated to rigorously analyzing qualitative data; however, most techniques aim to extract general themes or trends. For example, pain is difficult to rate on a typical diagnostic scale of 1 to 10. However, in a survey, interview, or focus group, you may find that, without any direct prompting, many participants discussed the pain associated with a particular step in a procedure. Although not a quantitative assessment of the level of pain, the theme of pain being discussed is a strong indicator that pain is a problem.

A method of summarizing and integrating quantitative and qualitative results is to build an *archetypal participant*—one who would be generally representative of the group. In some cases, you may need to create more than one archetype to account for variability. It is common practice to give your archetype a name (e.g., Hurried Harry), some statistics such as age (e.g., 45 years old, lives in Memphis), occupation (e.g., professional truck driver), family status (e.g., married for 21 years with three children), past history (e.g., family history of heart disease, open heart surgery at 41), and so on. Archetypes should be story-based, such that you could imagine them being a character in a movie or book, with details on how they think, act, and feel in particular situations. The more rich and vivid you can make an archetype the better. When making design decisions you can ask, "what would Harry think of this?" as a way to ensure that you are considering his particular perspective. In some cases, you may need to make an archetype not of a person but of an entity, such as an insurance company, supplier, or hospital system from information you have gained in other ways (e.g., your literature research, policy documents, course materials).

3.8 Current Solutions and Technical Barriers

Solutions already exist for many medical problems and unrecognized or latent needs may be currently addressed through work-arounds. For example, in a surgical suite, many cords may have been bundled together with cable ties to mitigate the risk of tripping. This ad hoc solution is not ideal and represents a latent need for a better solution. A critical component of a project statement is a summary of the technical and nontechnical solutions available to users. Although such a summary in a project statement may be short, there are many practical reasons for including a few sentences on existing solutions. First, it subtly hints that others have found the problem worthy of solving. Second, recognizing the barriers encountered in designing existing solutions can help you anticipate the barriers that you can expect to encounter. Third, explaining the shortcomings of existing solutions reveals the added value you intend to provide in your solution. You may use any of the methods described (e.g., interviews, literature searches, observations) to identify and analyze existing solutions. Checking in with a mentor is also a good start, and as indicated in Breakout Box 3.8, even a nontechnical mentor may be able to provide some important technical insights. In the remainder of this section, we discuss two additional methods for learning more about existing solutions.

If a solution is on the market, you may be able to uncover shortcomings and unmet needs by searching websites that contain customer reviews. It is important to note that these reviews are generally only for direct-to-consumer products. For example, you likely can find reviews for an over-the-counter blood pressure cuff, but you are unlikely to find reviews for a clinical-grade blood pressure cuff. You should remember that reviews are the voices of individuals and should be verified by other means, such as those presented elsewhere in this chapter.

Breakout Box 3.8   What's in it for the Client or Sponsor?

I have served numerous times as a clinical mentor for senior design teams. While I hope that my involvement has benefited the students, the benefits to me professionally and personally have been very significant.

Since my first meeting with a design team over a decade ago, I have found the challenge of developing something new and better to be very exciting. An early realization for me was that mentoring a senior design team provided a creative outlet that can be lacking in a clinical setting. Working with engineering students to solve a given problem is, quite simply, a lot of fun.

Engineering students bring a fresh, frequently "out of the box" perspective to solving clinical problems. As an example, I challenged a recent team to take on a perplexing problem pertaining to pediatric resuscitation. After a few weeks of mulling over the problem, they presented a solution that in retrospect was simple and rather obvious. As clinicians, the approach to problem solving can come with a certain momentum gained from collective experience of always doing something a certain way. The engineering students were able to look at the situation with fresh eyes, and an engineer's problem-solving skillset. It is very exciting to be a part of something like that.

My experiences with the students have very definitely changed the way I approach every single day at work. I find myself always wondering, "why do we do it this way?" and "how can we make this more efficient/safe/simple?"

Leo Vollmer MD
Emergency Department, Geisinger Medical Center

To an engineer, there is nothing like being able to take apart an existing solution to understand how it works. If you intend to improve upon an existing commercially available product or a prototype created by another design team, it is important to understand as much as possible about the design decisions made by other engineers. Doing so provides clues as to what has and has not worked in the past and reveals the technical constraints and barriers you may face. If you can obtain any existing products, you can conduct a product dissection, discussed in more detail in Section 5.10.

3.9 Value to the Healthcare Ecosystem and Measurable Goals

Value is a word that is used often yet rarely well defined. For our purposes, we adopt a very broad but simple definition: Value = Benefit/Cost. The benefits in the context of creating a medical device are multidimensional and may be attributes of a design such as ease of use, accuracy, effectiveness, safety, or portability. Likewise, the costs go beyond financial and include increased risk, cognitive load, uncertainty, waste, and repair. Many design decisions aim to find compromises that maximize value given the constraints. Using our definition, maximizing value means increasing the benefits while reducing the costs. However, benefits and costs are often related. For example, making a device easier to use could reduce error and cognitive load, but require greater maintenance and be more difficult to manufacture. Furthermore, not all stakeholders will necessarily agree on the benefits and costs of a design. For all of these reasons, it is critical that your project statement clearly define how you frame the value you intend to create. Often the simplest way to define added value is to propose a measurable outcome. In this section, we explore a variety of ways to clarify the value you intend to add to the healthcare ecosystem.

3.9.1 Benefits and Costs of Improvements to Medical Devices

Value may be added to the healthcare ecosystem in several ways, as shown in Table 3.4.

Four factors are important to consider regarding value. First, significant value may come from a product that either is highly important and desirable to a small number of people or moderately important and desirable to a much larger number. Your project statement should clarify the significance of the value you intend to create. Second, many of the pathways to increasing value are intertwined; increasing a benefit along one dimension may come with a corresponding cost along another. For example, a project that aims to decrease errors by 5%, but becomes far less usable than the current solution, may in fact decrease overall value. However, the intertwined nature of value can also lead to synergies; it is not uncommon to set out to enhance value in one way, and then discover additional ways to add value through the design process. Your project statement should be focused on a particular dimension of value, while acknowledging other dimensions of a solution that may require consideration. For example, you may make a statement such as, "our project aims to decrease the cost by at least 15%, while keeping the portability, battery life, and safety comparable to the existing solutions." Such a statement makes clear that your intentions are to decrease costs while keeping the benefits of the current solutions. Third, all value is relative to the existing solutions discussed in Section 3.8. In the previous example, the 15% reduction in costs is compared to the current solutions. Finally, value in the healthcare ecosystem is often felt by many stakeholders and their needs and wants are often intertwined in complex ways.

Table 3.4 Incomplete list of different ways to add value by decreasing costs and increasing benefits

Ways to Decrease Costs	Ways to Increase Benefits
Reduce invasiveness of procedure	Increase accuracy and reliability of tests
Shorten procedure time (including pre/post)	Standardize procedures and policies
Decrease personnel required for procedure	Increase ease of sterilization and sanitization
Lower readmission rate	Improve tracking of patients and outcomes
Decrease training required	Increase compliance with best practices
Reduce cognitive load	Enhance effectiveness of procedures
Reduce waste or cost	Target intervention more specifically
Streamline patient record entry	Increase longevity and effectiveness of interventions
Detect mistakes, errors, or false positives	Improve dissemination of medical knowledge
Lessen recovery and rehabilitation time	Automate steps in a procedure
Reduce inpatient and waiting time	Improve usability and ergonomics
Reduce infection rate	Encourage use of preventive measures
Simplify procedure or manufacturing process	Increase availability of and access to a medical solution
Reduce pain and discomfort	Enhance training and safety

All of these factors come together in a **value proposition**. It is a single sentence or phrase that clearly states the value to be added by a new product and is typically focused on the benefits to a customer or user relative to existing solutions (if solutions exist). In some industries, the relationship between the company and customer or user is simple (e.g., a customer buys a product and uses it); the value proposition is therefore straightforward. In the medical device industry, however, a value proposition is inherently multidimensional as many stakeholders may see the value (costs and benefits) of a device differently. Different stakeholders may not agree on the relative value to be gained and you should consider your value proposition from several perspectives. We discuss stakeholders in more detail in the next section.

3.9.2 Considering Stakeholder Perspectives

Stakeholders in the healthcare ecosystem often have diverse and sometimes conflicting views and goals. A broad definition of a stakeholder is anyone who is impacted by, can influence, or cares about a solution to the problem. In a clinical setting, this includes patients, caregivers and family members, physicians, nurses, lab technicians, physical therapists, clinical engineers, maintenance staff, and purchasing departments. It also involves others in the healthcare ecosystem such as policy makers, insurance companies, device companies, hospital systems, and employers of patients. As discussed in Breakout Box 3.9 and in later chapters, the healthcare ecosystem plays a major role in society (e.g., economic, political, cultural) and therefore the public at-large can be considered a stakeholder. Understanding how a particular stakeholder intersects, directly or indirectly, with how you intend to add value is critical to successfully navigate the design process.

It is often the case that stakeholders can be classified into two overlapping categories; the impacted, and the influencers. For example, hospital purchasing agents are important influencers on a healthcare

---

Breakout Box 3.9 The Triple Bottom Line and the Triple Aim of Health Care

The US healthcare system is the costliest in the world, accounting for approximately 17% of the Gross National Product (GNP), and growing every year. As a response, the Institute for Healthcare Improvement (IHI) developed the Triple Aim of Health Care to refocus the US healthcare system on three interrelated goals for any new innovation or change to the system:
- Improving the patient experience of care (including quality and satisfaction)
- Improving the health of populations
- Reducing the per capita cost of health care
    Many healthcare systems are adopting the Triple Aim and searching for ways to implement value-based innovations.
    A related idea is the Triple Bottom Line of sustainability that considers the social, economic, and environmental dimensions of a solution. The Triple Bottom Line goes by other names such as The Three Pillars, The Three-Legged Stool, The Three E's (Environment, Ecosystem, Economy), and The Three P's (Planet, People, Profit).
    The intention of both frameworks is to consider value through multiple lenses that may reveal conflicts between stakeholders at different levels (from the individual to the entire society).

---

purchasing decision, but they are generally not personally impacted by a patient's decision. Insurance companies are also influencers through the financial incentives they provide to encourage or require use of certain design solutions and not others, and they can be impacted by a patient's or physician's decision. Nurses, on the other hand, often have direct contact with medical devices, as explained in Breakout Box 3.10, and are impacted by purchasing decisions. Not all stakeholders have equal influence or are impacted in the same way, which is often the origin of their competing interests. This is especially important when there is a large disparity in the ability or authority to make a healthcare decision, particularly when the person receiving care is a child, has a mental or physical disability, or is elderly.

At this phase of your project, it is worthwhile to consider all of the stakeholders who may be impacted or be influencers. Two tools, shown in Figures 3.5 and 3.6, can help you dissect the various perspectives, and impacts to and influence of stakeholders. Filling in these diagrams will help you identify the stakeholders who are most relevant to your project. It may also help you focus on stakeholders to learn more about or contact. You should update these diagrams throughout your design process, periodically checking to ensure that your design decisions meet the needs of the various stakeholders and how they may perceive an increase (or in some cases decrease) in value.

3.9.3 Graphical Mapping of Perspectives, Flows, and Timing

Graphical representations can be powerful tools for both discovery and communicating potential value. The process of building a graphic can help you understand the current state of the art, the perspectives of various stakeholders, and areas where value could be created. A first draft may contain hypotheses that can be validated through interviews, surveys, or observations. Later drafts can show how your proposed project intends to add value by integrating into existing processes and solutions. Three related graphical methods can be helpful in exploring and communicating perspectives and flows.

The perspective of a particular stakeholder may be summarized in a **journey map**. For example, a journey map created for a patient's first treatment visit would include sitting in the waiting room, filling out paperwork, interacting with the front-line staff at the reception desk, and having vital signs checked by a medical assistant or other healthcare providers. A journey map is often laid out in time and is annotated with levels of discomfort, attention, motivation, cognitive load, anticipation, or other

---

Breakout Box 3.10 A Nurse's Perspective on Medical Technology

Individuals who enter the nursing profession often express two main reasons for pursuing a nursing career: caring for patients and making a difference in patient's lives. Managing medical technology is not a reason, yet 99% of nurses interact with some form of technology during their working hours. Nurses are on the front line of patient care and have the ability to identify and solve complex medical problems. Additionally, they are now having to recognize potential errors that may result as technology intersects with their efforts to care for patients.

Nurses possess the innate ability to weave their educational knowledge with their work experiences to effectively detect and treat problems. Consider that the average ICU patient is connected to approximately 10 different medical devices during their stay, including but not limited to patient monitors, ventilators, IV pumps, balloon pumps, etc. If this were not challenging enough, each nurse will also need to recognize the differences in equipment, vendors, and user interfaces. Additionally, it is a common misunderstanding that nurses are trained on all pieces of medical equipment before use. In reality, all too often, this does not happen or occurs only in an abbreviated form. As a result, nurses are spending more time managing/programming these devices and determining how these devices work and/or interact rather than providing direct patient care. This not only adds additional stress to an already stressful environment but may result in medical errors which can endanger patients' lives.

Nurses desire to spend more time caring for patients, not fighting technology. If they are forced to "fight" technology when caring for a patient, it may result in feature underutilization, misuse, errors, or outright abandonment. The key to medical technology success is understanding that medicine is a dynamic environment that continues to evolve with the latest research and available technology. This understanding ensures that the feature/function that was tirelessly developed will be used by a nurse. There is no greater reward, as a technology developer, than to see how something the developer designed made someone's job a little easier and solved a problem!

Simplicity, seamlessness, and intuitiveness are the fundamental pillars in design. Remember to solve the right problem and listen to what the users are saying about the product! It is important to develop what is needed, NOT what is perceived as innovative. Finally, the developer must ensure that whatever technology they develop is an extension of the nurse's hand.

Traci R. Bartolomei BSN, RN
Clinical Development Manager—Visualization and Analytics at GE Healthcare

---

first-person attributes. The goal is to capture the experience in one diagram. This same technique could be used for any stakeholder (e.g., nurses, surgeons) for any particular situation. A more advanced method would create multiple journey maps for the same situation, one for each of the key stakeholders.

Timelining is a way to show how events and decisions in a dynamic situation are related in time. Unlike a journey map it enables you to show how various stakeholders interact with one another. Key attributes to include are the decisions, lengths of time, and stakeholders involved. Such a graphic can often reveal that the source of a problem is inherited from an earlier process or action. For example, a design team was timelining a 4-hour surgical procedure. The insight gained was that the actual procedure only required 15 minutes—most of the surgery consisted of preparation, opening, and closing. It was also found that the surgery can last up to 8 hours in obese patients. Such an insight helped the team focus on faster ways to bind together deep tissue layers. A more advanced graphical method can integrate a timeline with various journey maps.

Resource and personnel mapping is a way to track an individual, resource, or material over time. Mapping out the various activities of a nurse, receptionist, lab technician, caregiver, or surgeon during a typical day can reveal bottlenecks, sources of error, and frustration in their day. Likewise, tracking how a particular material, tool, or drug moves through a clinical environment (from when it enters to when it leaves) can uncover inefficiencies.

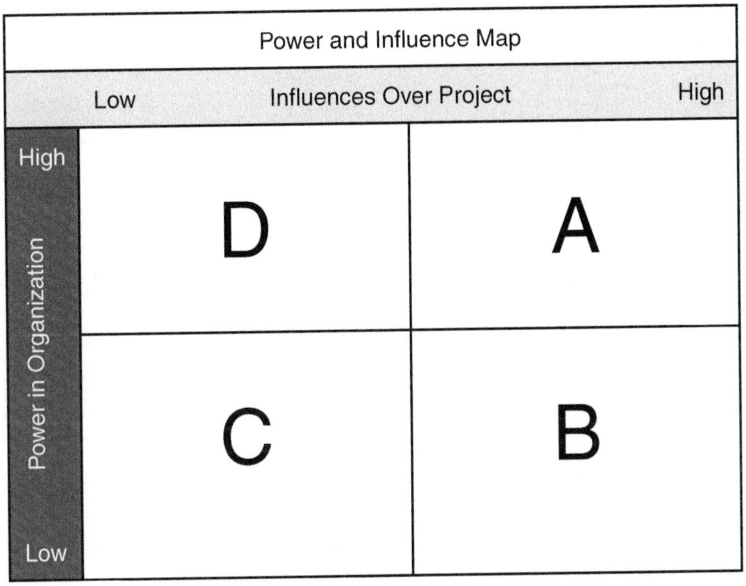

FIGURE 3.5

Power and Influence Maps. Maps list stakeholders along two axes; influence over the project and the power within an organization. For example, the FDA may have a great deal of power but little influence over specific design decisions, and therefore would be listed in quadrant D. On the other hand, a faculty mentor may have much influence over your project direction, but not over adoption of your design concept by the medical community. She would therefore be placed in quadrant B.

The purpose of these types of graphical maps is to summarize your research. They may also be used to show what will change as a result of your project. Adding a graphic to your project statement can be an effective way of communicating the value to be created.

3.10 Project Objective Statements

A project objective statement is typically a single line at the end of your project statement that declares what you hope to realistically accomplish with your time, effort, and resources. As explained in Section 2.6.1, it defines what counts as a successful conclusion to your design project. In both academia and industry, medical device designers often stop short of demonstrating the ultimate clinical goal that can be measured in the real world. In industry, a design team transfers final documentation for the project to manufacturing personnel, a step known as design transfer, after which other personnel will be responsible for demonstrating that the ultimate clinical goal has been met. It is often the case in academic design projects that the project concludes with a functional prototype that has only been partially verified to work as required. Although some parameters of your project objective statement may have been set for you (e.g., timeline, resources), many academic programs grant you some latitude in defining your own project scope (final deliverable).

Source: The Innovator's Toolkit, Stakeholder Management (Technique 8)

					Current/Desired Support					
#	**Key Stakeholder**	**Role in Organization**	**Impact of Project on Stakeholder (H, M, L)**	**Power/ Influence Category**	**Strongly Opposed**	**Opposed**	**Neutral**	**Supportive**	**Strongly Supportive**	**Reasons for Resistance or Support**
1										
2										
3										
4										
5										
6										
7										

Stakeholder Diagnostic

FIGURE 3.6

The Stakeholder Diagnostic. Considering the need in your project statement, begin by listing the key stakeholders and their roles in the healthcare ecosystem. Then move on to estimating the level of impact of the need and power/influence over the adoption of a new solution. Last, indicate the level of support for a new solution along with notes on any resistance that may be encountered. The aim of filling out the table is to better understand the perspectives of all stakeholders who may impact your design decisions.

Scoping a design project is an important skill. Engineers are often asked, "can you do this?", which is a question of project scope. The ability to confidently answer this type of question depends on your experience with the design process and an understanding of the problem to be solved. The tools in this chapter are intended to help you clearly understand the needs and problem to be solved. Your experience with the design process through your current project will help you answer this question and best scope a project considering the applicable time and resource constraints.

3.11 Drafting and Refining Your Project Statement

Synthesizing all that you have learned into a project statement is often the most daunting (and important) writing task in a design process. In this section, we suggest ways to draft and refine your project statement. To get you started, Breakout Box 3.11 contains a template project statement. It is in the form of a MadLib™ because the best project statements tell a story. It also helps you identify missing information. You should feel free to modify the template to better match the constraints of your project and program. Furthermore, the draft is purposely written in colloquial language so that it is easier to fill in. Later drafts should refine the language so that it is more precise and professional.

Breakout Box 3.11 Generating an Initial Project Statement

It is not easy to write the first draft of a project statement. Most, however, have a similar format which can be summed up as a MadLib™.

When we heard from _____ we learned that _____
                    (company/person)   (observations).

This got us thinking about _____ that impacts _____ . As we
                           (medical/clinical need)   (users)

explored more, we found _____
                        (statistics here on significance and magnitude of problem)

Current solutions include _____ , however, they do not adequately
                          (list of solutions)

address _____ , because _____ .
        (medical/clinical need)   (shortcomings)

We don't know how to address this challenge yet, but we expect the technical barriers to be

_____ . However, in meeting the need we would hope
   (technical or other barriers)

to increase/decrease _____ which may also
                     (one or two measurable results)

_____ . We think we are well positioned to address this challenge by
   (additional benefits)

_____
            (one line project objective statement).

3.11.1 Selecting and Scoping Your Project

In academia and industry, design projects may be initiated through many different pathways. As explained in Breakout Box 3.1, some projects are blue sky and therefore afford you flexibility in identifying a problem to solve, clarifying the value to be gained, and scoping the project. Other projects are Red Ocean and may be assigned or selected from a list of potential projects. In academic design projects, you often are granted some latitude in choosing and scoping the project you pursue. In industry, many projects are approved by upper management based on market demand and the expected return on investment, and engineers are then assigned to these projects.

Selecting and scoping a project can be aided by drafting a project statement using the template in Breakout Box 3.11. Having each of your potential projects, even if they are variations on a theme, in the same written format make them easier to compare. Furthermore, the process of writing a project statement often reveals which are most suitable to your skill set, timeline, resources, and interests.

An additional way to narrow your project choices is to create your own canvas; a one-page summary of the various aspects of a potential project. Figure 3.7 shows examples of creating a Project ID canvas.

The topics to consider in the boxes can be brainstormed, perhaps using sticky notes. Several attributes (costs, risks, and benefits to you) can be listed as factors in project selection. For example, you may consider your interests, access to a clinical environment where you can observe the problem, quality of mentorship, and so on. These may vary depending upon which project you select. Considering less than four factors is likely too few, whereas more than ten is too many. Once you have created your canvas, you can try filling out the canvas for each of your candidate projects. The purpose of the canvas is to allow you to compare the advantages and disadvantages of several projects relative to one another.

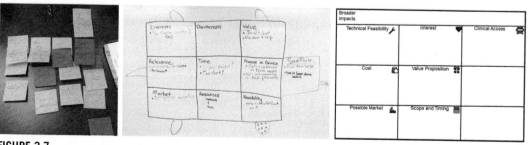

FIGURE 3.7

Brainstorming factors to consider in project selection using sticky notes (left) along with two examples of project ID canvases (middle and right).

There are other ways to modify your canvas as well. For example, in the "interest" box, you may include the level of interest of faculty mentor(s), clinical mentors(s), or an industry contact. Likewise, you may find that you simply do not have enough information about a particular topic. Placing a question mark in a box indicates that more information is needed and can lead to more targeted research efforts.

3.11.2 Refining Your Project Statement

You will continually revise your project statement as you learn more. A helpful process is to proactively identify where more information is needed and then use the techniques presented in this chapter to fill in the gaps. As you refine your project statement, you should consider the following questions:

- Is the medical/clinical problem related to a therapy, diagnostic device, surgical instrument, assistive technology, training simulator, or other solution?
- What will change before and after you solve this problem? Is there a way to measure the impact of these changes?
- Are there specific benefits or costs (see Table 3.4) that may be realized through a solution?
- Which stakeholders are impacted? Which are impacted the most? How? Are there stakeholders who might resist a new solution?
- Who makes decisions about the course of treatment or action? Which stakeholders have influence over these decisions?
- Are there statistics that describe the magnitude or rate of incidence of the problem?
- Are there statistics that describe the societal or healthcare impact of the problem?
- What is the potential market for a solution to this problem?
- What are the current costs to the individual, insurance company, hospital, and healthcare system?
- Are there demographic factors (e.g., patient population characteristics) that increase or decrease the incidence or magnitude of the problem or need?
- How is the problem currently being solved/addressed? Are there alternatives? What are the success rates of existing solutions?
- What are the side effects or complications of current treatments? Who treats the side effects or bears the cost of complications?
- Why is a new solution needed?

Breakout Box 3.12 Project Statement Received From an Industry Sponsor

Imagine receiving the following project statement from an industry sponsor:

Title: Improved Fit Between Modular Hip Components

When using modular hip implants, a separate modular head is paired with a separate modular stem. The head component contains a female Morse taper, and the neck forms a male Morse taper. During surgery, the modular head is impacted onto the modular neck to create a tight, press fit. Sometimes, the tolerances between the two taper surfaces are such that the head can rotate very slightly with respect to the neck with each cycle of loading during walking. This small-scale micromotion can lead to fretting wear and the ingress of bodily fluids present around the taper junction. The better the fit between these components, the less small scale micromotion (fretting) will occur between them, and the less fluid will enter the space between the Morse tapers. This will lead to less wear and corrosion occurring at the modular taper junction. This project involves the development of a method to connect the two modular components that will minimize the size of the gap occurring at the taper surface junction and create a tighter press fit that will allow minimal fretting and fluid ingress.

As this statement was likely written by an industry project manager through a technical lens, you should use the guidance from this chapter to create your own version. You should consider the following points to guide your research and revisions:

- The statement focuses primarily on the technical problems of hip implants. More should be included on the clinical need.
- There is no indication of how users are impacted or the significance of that impact.
- Numbers or statistics on the size of the problem (e.g., How many hip implants of this type are performed each year? How often do they experience excessive wear or corrode?) would help communicate the importance of solving this problem.
- A clear measurable outcome would help make clear the ultimate goal of the project. For example, stating that the redesign would extend the average life of the implant by approximately 3 years would provide a measurable clinical outcome.
- It would be helpful to report if there have been previous attempts at solving this problem. Are there any known or anticipated technical barriers?
- The last line is a version of a project objective statement but is too general. It would be helpful to state that a prototype will be created and tested, by when, and with what resources.

Not all of these questions will be applicable to your project, and you may need to add other questions of your own. We conclude this section with a final example of a project statement, and accompanying commentary, in Breakout Box 3.12.

3.12 Design Reviews

Design projects have many natural milestones, many of them mirroring the chapters of this text, which you will be asked to report on during your progress. A common method is the design review. A **design review** is a live, or sometimes virtual, report on your technical progress and the rationale for design decisions, typically to a panel of experts (e.g., faculty, industry personnel, government employees, healthcare providers, potential customers). In industry, design reviews usually occur at the conclusion of a particular design phase, and as such can be the deciding factor in whether or not a project continues onto the next phase. In an academic design project, a design review may serve as a pivot point for a project, when you refine or redefine the goals of your project. It is therefore not uncommon to have a first design review upon completion of a project statement.

Design reviews serve two broad purposes. First, as they often occur at critical junctures in a design process, they are excellent opportunities to obtain unbiased feedback from others regarding design decisions and the direction of your project. This feedback allows you to see your project through the eyes of others, which can bring about meta-reflection on past decisions and possible future trajectories. Second, design reviews allow experts from outside of your team to contribute to the design process. They can help you make a critical decision, reveal previously unknown information, or connect you to an important resource.

Design reviews may take many forms, depending upon when they take place during the design process. In early phases, a design review is often a presentation supplemented by handouts. Once prototypes have been created, a design review is an opportunity to let your prototype drive questions and feedback. Later in the design process, a design review may focus on the demonstration of a fully functional device.

You should carefully prepare for a design review so that you can gain the most from your limited time with the panel of experts. Preparation usually includes generating materials (slides, handouts, prototypes), creating an agenda, generating a prioritized list of questions, and practicing. You should balance communicating progress and reserving time for discussion. For this reason, it is best to get to your main points as quickly as possible. For example, if you reviewed 40 journal articles, you may only present a summary of the literature landscape, perhaps highlighting one or two articles of greatest significance.

There are two points important to keep in mind during a design review. First, you should resist becoming defensive and seriously consider any suggestions made. The role of an expert panel is to provide constructive criticism. This may be uncomfortable or feel like a setback. One strategy that can help you listen rather than immediately defend your decisions is to schedule a post-review debrief. At this meeting, you can decompress, share what suggestions you heard, and develop a summative set of recommendations for future actions. Second, a design review should result in a record that can be included in your Design History File. One possible format to record discussion items is shown in Breakout Box 3.13. These kinds of standardized forms can help you record and process what you learned from the design review.

Breakout Box 3.13 Example of Design Review Record

Design review #:
Project name:
Date:
Attendees: names, titles
Team member names: identify presenter
   High-level takeaways from the meeting
- Bulleted list of main ideas based on issues raised
   Table of specific issues raised, comments, and actions to be taken

Issue	Comment	Action

   Follow-up action items
- Bulleted list

Key Points

- Drafting a project statement helps you gather relevant information from many sources, synthesize what you have learned, and communicate how your design project intends to add value to the healthcare ecosystem.
- Project statements are typically one or two paragraphs and communicate the need, current solutions, known technical barriers, measurable outcomes, and a project objective statement.
- The significance of a medical need is supported by information (often statistics) from sources that include literature, websites, marketing materials, books, observations, interviews, and surveys.
- Understanding current solutions allows you to highlight the relative value of your project and identify anticipated technical barriers.
- Value can be achieved in many different ways by considering costs and benefits to multiple stakeholders. Often, value to the healthcare ecosystem can most clearly be communicated by stating a measurable medical or clinical outcome that would ultimately demonstrate the positive impact of your design solution.
- Your project objective statement is a declaration of what you hope to achieve through your project. It includes the scope, timing, and resources available to complete your project.
- You should document all of your findings, even those that do not appear in your project shiftenterstatement.
- Design reviews are opportunities to receive feedback that helps move the design process forward.
- A good first draft of a project statement is important to move forward with the design process but will continue to be refined as you learn more.

Exercises to Help Advance Your Design Project

1. Several sources of information were given in this chapter. List all of the possible sources for information that you expect to use throughout your design project. When you are stuck, you can refer to this list to see if some of these sources could help provide additional clarity. You should add this list to your Design History File.
2. Do you suspect your project is blue sky or Red Ocean (Breakout Box 3.1)? Is your project technology push or market pull (Breakout Box 3.2)?
3. Review the ways of generating value in Table 3.4. For your design project, classify each way to gain value as either (1) possible, (2) maybe possible, or (3) unlikely.
4. Choose a critical stakeholder (perhaps using Figure 3.5 or 3.6) and build an archetype of that stakeholder (person or entity). What sources of information (e.g., survey, interview) might you use to validate your archetype?
5. Formulate a list of questions to ask your mentors or sponsors about the problems they encounter with current design solutions.
6. Fill in the MadLib™ in Breakout Box 3.11 and show it to your mentor or sponsor for feedback.
7. Create a Problem ID canvas as suggested in Breakout Box 3.11. Show it to your mentor/sponsor and modify it based upon the feedback you receive.

General Exercises

8. Generate your own list (or add to the list given in Table 3.4) of ways to achieve value in the healthcare ecosystem.

9. A theme of this chapter is to develop a holistic understanding of a medical problem. Generate a list of medical/clinical problems that you have encountered through various information sources. You will likely notice trends when sorting these problems into themes (grouped in some way that makes sense to you, perhaps by organ system, or point of intervention).

10. Read two student-created project statements from a previous year. What information is missing? How would you go about refining these statements?

11. Medical papers sometimes come to varying (sometimes even opposing) conclusions based on similar data. One possible reason is data collection bias. Present two more reasons why conclusions in the medical literature may not always align.

12. Clinical recommendations are sometimes resisted or not followed. For example, Landon et. al. (2009) demonstrated in a randomized study that treatment of mild gestation diabetes (MGD) improved clinical outcomes in pregnancy (e.g., fewer cesarean sections, hypertensive disorders, fetal overgrowth, reducing the risk on neonatal injury). Despite this, the obstetric community currently has not uniformly adopted this (i.e., some hospitals treat for MGD and some do not). List reasons why a proven solution may not be adopted.

13. A common clinical need, which often finds its way into design projects, is to reduce surgery time sufficient enough to reduce patient anesthesia time and potentially freeing up OR time for additional procedures. This is a noble and measurable outcome. From the hospital's perspective, however, the value of this outcome is limited. Why is that? What is a better outcome statement that would be more valuable to providers and to the hospital?

14. In 2017, unintentional drug poisoning deaths involving opioid analgesics had grown by a factor of six since 2009. Half of these deaths were due to drugs obtained by a patient's relative or friend from leftover opioids prescribed for a patient in pain. Knowing only this, craft a need statement for this problem.

15. A classic, and still very relevant, video of a team engaging in a design process is the IDEO shopping cart redesign documentary. Watch this video online (try searching for IDEO Shopping Cart and look for the 22-minute ABC News version). Watch the video and notice that the designers spent at least 20% of their time gaining a deep understanding of shopping and shopping carts. What aspects from the design process in the video might you use in your own design process?

References and Resources

Arthur, W. B. (2009). *The nature of technology: What it is and how it evolves*. New York: Free Press.

Byers, T. H., & Dorf, R. C. (2008). *Technology ventures: From idea to enterprise*. Boston: McGrawHill Higher Education.

Creswell, J. W. (2012). *Qualitative inquiry and research design: Choosing among five approaches* (3rd ed.). SAGE Publications.

De Bono, E. (1995). *Serious creativity*. HarperCollins.

Dym, C. L., Agogino, A. M., Eris, O., Frey, D. D., & Leifer, L. J. (2005). Engineering design thinking, teaching, and learning. *Journal of Engineering Education, 94*(1), 103–120.

Kelley, T., & Littman, J. (2006). *The ten faces of innovation: IDEO's strategies for defeating the devil's advocate and driving creativity throughout your organization.* Penguin Random House.

Menzel, H. C., Aaltio, I., & Ulijn, J. M. (2007). On the way to creativity: Engineers as intrapreneurs in organizations. *Technovation, 27*(12), 732–743.

Muller-Borer, B. J., & George, S. M. (2018). *Designing an interprofessional educational undergraduate clinical experience.* 2018 ASEE annual conference & Exposition Proceedings.

Osterwalder, A., & Pigneur, Y. (2010). *Business model generation: A handbook for visionaries, game changers, and challengers.* John Wiley & Sons.

Otto, K., & Wood, K. (2001). *Product design: Techniques in Reverse engineering and new product development.* Prentice Hall.

Privitera, M. B. (2015). *Contextual inquiry for medical device design.* Academic Press.

Saldana, J. (2009). Introduction to codes and coding. In J. Saldana (Ed.), *The coding manual for quantitative researchers* (1st ed.) (pp. 1–31). SAGE Publications.

Smith, N. J. (Ed.). (2002). *Engineering project management* (2nd ed.) Blackwell Science.

Stanford d.school Bootleg. Retrieved from http://dschool.stanford.edu/wp-content/uploads/2011/03/shiftenterBootcampBootleg2010v2SLIM.pdf.

Ullman, D. G. (1992). *Mechanical design process.* McGraw-Hill Education (ISE Editions).

Defining the Engineering Problem

4

Engineering is design with constraints.
— Samuel Florman, Moral Blueprints

Good engineering is characterized by gradual, stepwise refinement of products that yields increased performance under given constraints and with given resources.
— Niklaus Wirth, Swiss Computer Scientist

Chapter outline

4.1 Introduction

In this chapter, you will learn how to convert the project statement you created and customer needs identified in Chapter 3 into target product specifications. These specifications become your initial design goals (what you hope to achieve) that are impacted by design constraints (things that limit your design). It is important for you to identify desired goals and applicable constraints as early in the project as possible to avoid costly design changes later in the project. You should identify, study, and compare competitive solutions to ensure that your design solution is equal to or better than existing products, procedures, or other ways of solving the same or similar problem. To be successful, a new product design must provide value to the customer (and other stakeholders) and offer benefits beyond what is currently available to solve the customer's problem. Section 4.6 provides an example of how to convert the customer needs and anticipated constraints included in your project statement into quantifiable target product specifications.

4.2 Specifications and Requirements

According to ISO 9000:2015 *Quality management systems—Fundamentals and Vocabulary*, a **specification** is a "document stating requirements." A **requirement** is a "need or expectation that is stated, generally implied or obligatory." The FDA *Quality Systems Regulations* define a specification as "any requirement with which a product, process, service, or other activity must conform." These definitions illustrate the relationship between needs, requirements, and specifications. A need can become a requirement and a specification is a concise statement of that requirement. Written requirements should be complete, measurable, feasible, unambiguous, nonconflicting, and traceable, as described in Table 4.1.

For example, a requirement stated as "device must be portable" is incomplete and ambiguous. A less ambiguous and more complete requirement could be written as "device must weigh no more than 20 lbs and fit into a volume of 10 in³." For more qualitative requirements such as "device must be comfortable for patient," a better quantitative statement would be "patients will assign scores of 3 or higher on a 5-point Likert scale."

International quality management standards, ISO 9001:2015 *Quality Management Systems— Requirements* and ISO 13485:2016 *Medical Devices—Quality Management Systems—Requirements for Regulatory Purposes*, include a section on design controls (discussed further in Section 11.4) which describes the steps that companies must follow in their design processes. These standards define **design input** as "the physical and performance characteristics that form the basis of a device design" (including product specifications), and **design output** as "the results of a design effort at each design phase and at the end of the total design effort." Specifications should be based on needs and constraints so that, if a product performed per specifications, it would meet these needs. Using ISO 9001 and 13485 terminology, we would state this by saying that "design input = design output," or that design output must meet design input requirements.

A specification includes a **metric** that defines a design characteristic (e.g., weight, size, hardness) or performance requirement (e.g., speed, ease of use, level of comfort, battery life) and a **value** that quantifies the metric. Specifications are a way to quantify performance and physical characteristics. They must be measurable and include a numerical value and appropriate units. Establishing quantitative metrics makes it clear to your design team exactly what is expected of the final design. For

Table 4.1 Characteristics of Good Product Requirements.

Characteristic	Description
Complete	Stated in one place with no missing or extraneous information
Measurable	Able to be tested or measured to confirm that requirement has been met
Feasible	Does not violate laws of physics (e.g., perpetual motion machine)
Unambiguous	Concisely stated with no vague wording; expresses objective facts, subject to one interpretation
Nonconflicting	Does not conflict with other requirements
Traceable	Can be linked to a need including the source of the need, which includes the customer, standards, and regulatory requirements

example, if a customer (surgeon) expresses a need for a lighter endoscope—without a quantifiable metric—members of your team (along with customers) might think that "lighter" means an endoscope that weighs less than 10 lbs, while others might think it means less than 5 lbs. An agreed upon metric and associated value with units reduces confusion and communicates to your team (and customer) exactly how the product must perform. During design verification activities (tests presented in Chapters 9 and 10 conducted to ensure that performance specifications have been met), metrics and their associated values allow your team to determine if the specifications have been met. Metric values can be numerical (5 lbs), binary (pass/fail), or based on a subjective scale (ease of use based on a 5-point scale where 5 means easy to use and 1 means difficult to use). To provide a range of acceptable values, **marginal** and **ideal** metric values should be established for characteristics of a design. Ideal metric values define the design characteristics that are preferred by your team (best case). Marginal metric values are minimally acceptable to your team and define minimally acceptable design characteristics.

In an academic design project, desired clinical outcomes are almost never achievable within the time frame of an academic design experience. For example, if a new surgical tool is created, there might not be time to obtain the approvals required to conduct a clinical trial. However, the list of requirements should include those associated with the desired clinical outcomes, even if they might not be achievable during the time frame of the design experience. It is not uncommon for academic design projects to be completed over a period of time that requires more than one project team. For this reason, it is important to accurately represent the end goals at the beginning of the project, including the desired clinical outcomes. However, for practical reasons, the team should focus its work on those needs and related requirements that can be met within the time frame of the design experience. For example, requirements for specific flow rates, load capacity, power requirements, and material characteristics are testable within the time constraints of a design experience. However, reduction in infection rate, which requires clinical testing to determine, is an outcome that may not be directly achievable within these time constraints. A reduction in the number of touches and exposure time that contribute to infections can be used as indirect measures of a reduction in infection rate and may be achievable in this time frame.

4.3 Converting Needs and Constraints to Specifications

When needs and constraints have been determined, they must be converted into target specifications that define how the product must perform and what **characteristics** (defined as "distinguishing features" by ISO 9000:2015) it must display. Early in a project, target specifications are established with the understanding that they may change as the design evolves over the course of the project. Negative test results, changing market needs, and changes in the priority of customer needs may result in new or revised specifications and/or the elimination of some preliminary specifications. In this text, we use four distinct terms to refer to product specifications as they evolve through a design process:

- **Target specifications** are the initial preliminary product specifications that represent your original design goals. This chapter focuses on the establishment of target specifications.
- **Transitional specifications** are revised target specifications that result from and reflect changes to the design as it evolves through the design process. As the design changes, some original specifications may be revised or eliminated, and some may be added. Transitional specifications (used when generating solution concepts in Chapter 5) often change several times prior to design verification and validation activities. The most recent version of transitional specifications describes

a design that you expect to meet all needs and performance requirements (to achieve the project outcome defined in the project statement) based on appropriate analytical modeling and preliminary testing. Prototypes built per these specifications are tested (discussed in Chapters 9 and 10) to determine if performance requirements and customer needs have been met.

- **Final specifications** describe a design that has been tested to confirm that performance requirements have been met. They incorporate changes that test results indicated were needed to meet performance requirements. These are the specifications that many academic design teams establish at the end of the project. Due to the typical time and resource constraints of academic design projects, clinical studies of prototype devices are rarely possible. For this reason, final specifications may change depending on the results of animal or clinical studies.

- **Frozen specifications** represent the true final design after all verification and validation testing has been completed and the design is ready to be transferred to manufacturing (as shown in Figure 1.3). However, if problems are reported in the field, the design specifications may need to be "thawed" and revised prior to market introduction due to problems discovered during manufacturing or after market introduction. For example, consider a situation in which thousands of surgeons begin using a new handheld surgical instrument. If many surgeons begin reporting that the thumb-controlled power button on the instrument is uncomfortable to use, then the design specifications for the thumb switch would be "thawed" and changed to require a more comfortable power button. Once corrected, the design specifications would be frozen again.

Some needs may be more important to the success of a new product than others. Limited resources may require you to choose the most important needs to be addressed. For these reasons, needs should be prioritized. It is helpful to separate needs into two groups. The first group includes needs that must be met to create a **minimally viable solution**. If a design does not meet these needs, then the new product will not solve the problem, provide value to the customer, or be commercially successful. These needs are often referred to as "musts" or "must haves." The second group includes needs that do not have to be met for a solution to provide value and become a successful new product. Meeting these needs could provide added value through additional features and functions but are not absolutely necessary for the product to be successful. These needs are often referred to as "nice to haves." In an industry design project, marketing product managers typically advise design teams as to which needs are the most important for commercial success. In this text, we do not differentiate between the two groups of needs by name; rather, we consider higher priority needs determined by your design team to be the more important needs to be met. Early in the design process, you might not know which needs are more important than others. Continued research and discussions with stakeholders can help determine the priority of each need.

4.3.1 Competitive Benchmarking

Competitive benchmarking is a method that allows you to establish defining metrics of a design to perform as well as—or better than—competitive solutions. It involves determining the performance characteristics of competitive solutions and comparing them to ideal and marginal values for the new design.

To identify appropriate competitive solutions, you should ask the question "How is the problem being solved now?" A competitive solution can be an existing product, procedure, "work around," or any method used to solve the problem. These are often discovered during needs identification activities such as customer interviews, observations of customers using products, and literature searches (presented in Section 3.8). Identification of existing solutions prevents you from "reinventing the wheel" or creating a design that infringes a patented design (discussed in Section 15.3). If an identified competitive solution is in the form of a commercially available product, then you ideally would obtain samples of the product (if available and affordable) for inspection and testing (and to reverse engineer as discussed in Section 5.10) to quantify the product's performance. If your project is sponsored by a physician or a medical device company, you may be able to obtain samples of competitive products from them at no cost. If this is not feasible, then you can conduct online searches to find performance data that may be available on manufacturers' websites or contained in sales literature. Industry design project teams are often able to obtain product samples through physicians or hospitals by swapping products manufactured by their companies for competitive products, or by reimbursing hospitals for samples.

4.3.2 Incorporating Design Constraints and Requirements Into Specifications

Samuel Florman states "engineering is design with constraints." In his book *Moral Blueprints*, he explains the tradeoffs resulting from design constraints:

> ...all designs contain some element of danger. In seeking to minimize danger, one usually increases cost, and therein lies a dilemma. There is no trick to making an automobile that is as strong as a tank and safe to ride in. The challenge is to make an automobile that ordinary people can afford and is as safe as the community thinks it should be. There are also considerations of style, economy of use, and effect on the environment.

His message is that any engineer can design almost anything if there are no constraints on resources, time, cost, or performance. A **constraint** is defined as "a condition, agency, or force" that "imposes a limit or restriction or that prevents something from occurring." It limits design choices and can impact customer needs, which are used to determine requirements and specifications. There are many technical, economic, legal, safety, regulatory, social/cultural, and environmental constraints that can affect a final design. These should be identified as early in the project as possible, ideally when identifying customer needs, to mitigate costly and time-consuming redesign efforts later in the project. For example, just before your project is completed, it is too late for you to learn that your client or sponsor cannot accept a new device if it contains a specific material. This represents a limit or restriction imposed on the design that could limit your design choices, thus meeting the definition of a design constraint. This new constraint would be translated into a new need and eventual requirement for the design to be free of the undesired material. According to the FDA:

> The term requirement is meant in the broadest sense, to encompass any internally or externally imposed requirements such as safety, customer-related, and regulatory requirements. All of these requirements must be considered as design inputs.

This suggests that if "any internally or externally imposed requirement" restricts or limits your design choices, then it would fit the definition of a constraint. Design constraints include limitations imposed on you by external factors as well as self-imposed limitations to improve a design. They often include time and budget constraints, functional and nonfunctional requirements, compliance with standards and regulations, style, and usability. Requirements reflect constraints and constraints often inform requirements. For these reasons, in this text, we consider some constraints to also be requirements. The following constraints limit design choices and will influence design requirements and specifications.

Time constraints are the biggest and most common constraints for all projects. All projects have deadlines. In industry design projects, these are often self-imposed and based on market conditions and windows of opportunity. In many academic design projects, time constraints inherent in an academic design experience typically make it impossible to conduct a clinical study, and teams are often unable to determine if specifications related to clinical performance have been met. Some of these specifications can be verified using alternatives such as simulations, phantoms, and cadavers. Chapter 9 and part of Chapter 10 are devoted to laboratory, animal, and human testing, respectively. Need No. 9 in Table 4.3, "acceptable service life," is an example of a clinical need, whose associated clinical specification can only be measured through a clinical trial and may not be able to be assessed as part of an academic design project. Time constraints and their impact on projects are graphically displayed in Gantt charts (presented in Figures 2.5 and 2.6).

Technical design constraints impact performance requirements to ensure proper function and usability (ease of use). The required service life of a product (how long it must function properly) is limited by the service environment (conditions in which the product must function). For example, most orthopedic implants are expected to function for at least 15 years. After implantation, the body presents a hostile service environment that tries to destroy the implant (discussed further in Chapter 8). In this example, the service environment presents design constraints, and the expectation to function for a minimum of 15 years is a requirement.

In industry, technical requirements include manufacturability (must be manufacturable in large quantities at a reasonable cost) and serviceability (can be easily serviced and maintained). The sterile environment inside the body limits design choices regarding materials, geometry, and other factors. This design constraint requires most medical devices to be sterilizable (able to be sterilized without compromising product function). Device packaging must protect a sterile product and maintain sterility while in shipment and storage. These topics are discussed further in Chapter 8. It is important to be aware of all technical constraints and requirements as they may impact your design.

Financial design constraints include budgetary and profitability constraints which are different for academic and industry design projects. In an academic design project, you will have a project budget that will constrain some of your design choices. You will need to determine the cost to produce your prototype and ensure that it will not exceed your project budget. For example, if the project budget is $1000 but required parts cost $1100, then alternate parts or a new design using less costly parts may be necessary. In industry, budgetary constraints may be less restrictive if the project is expected to be profitable and provide a significant return on investment.

Profitability is determined by manufacturing costs (cost to produce a single unit at production quantities) and selling price (how much the customer will be willing to pay for the product). Selling price is affected (constrained) by profitability requirements, price competition (how much competitive products sell for), and insurance reimbursement (how much insurance companies and government agencies will reimburse hospitals for medical devices). Your design choices may be limited by constraints imposed by the market. For example, market conditions and competition with existing products may create cost constraints that limit design choices to those that will allow for a competitive selling price and still be profitable for the company. If it is determined that a selling price of $150 is needed for these reasons, the production cost might need to be $100 per unit. This cost constraint will require designers to choose materials, components, and manufacturing processes that will result in a production cost of no more than $100 per unit. This cost constraint may similarly be expressed as the requirement "cost to produce must be ≤ $100/unit." Many of the constraints presented here are revisited in Section 6.3, where they are considered when narrowing the list of potential design solutions and selecting the best solution.

Legal and ethical design constraints result in requirements aimed at reducing risk, potential liability (making it safe to use), and patent infringement. They may also affect patentability. A design that infringes an existing patent is generally not acceptable unless a company is willing to license the technology from the patent owner to avoid a lawsuit. Some companies require new products to be patentable, creating an internal design constraint which may limit the choices of acceptable designs. From a legal and ethical perspective, it is important to anticipate how your design could possibly be used (or misused) in a way that could cause harm. Ebola protective gear is one example; how it is put on and taken off varies from one healthcare provider to another and can result in Ebola infection if not used properly.

Regulatory design constraints come from regulatory agencies such as the U.S. Food and Drug Administration (FDA), Environmental Protection Agency (EPA), and the Occupational Safety and Health Agency (OSHA). The FDA and other international regulatory agencies presented in Chapter 12 have rules that apply to the design of medical devices that can impact the choice of acceptable designs being considered by a medical device company. An example of how these rules (constraints) can impact the design of a medical device is presented in Breakout Box 4.1 and are discussed in more detail in Chapter 12.

Breakout Box 4.1 Effect of Regulatory Constraints on the Design of a Ureteral Stent

Ureteral stents are used to maintain urinary drainage between the kidney and bladder, to allow the ureter to heal after surgery, bypass a ureteral stone, or open a narrowed ureter (see Figures 4.1 and 4.2). They are inserted through the urethra, bladder, ureter, and into the kidney. This is done using a flexible fiber optic cystoscope with a topical anesthetic applied to the urethra while the patient is under twilight sedation (not unconscious, but sedated). Once the ureter is healed and normal drainage is established, the stent is removed using a similar cystoscopic procedure in which the distal portion of the stent is used to pull the stent out of the kidney, then the ureter, bladder, and urethra.

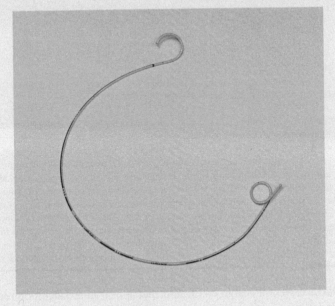

FIGURE 4.1

Image of ureteral stent with proximal and distal curls.

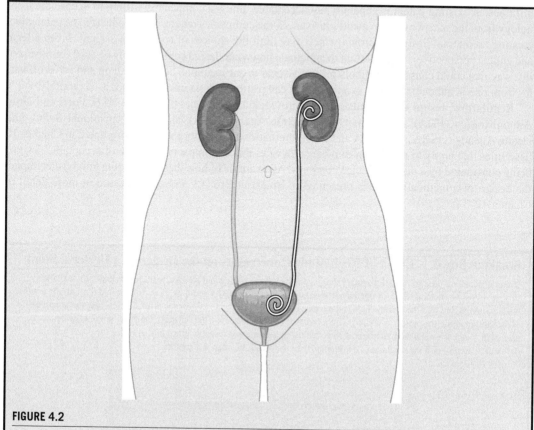

FIGURE 4.2

Location of stent while in the patient's body. In this image, the proximal curl is located in the left kidney (upper half of figure) and the distal curl is located in the bladder (lower half of figure).

A ureteral stent that makes use of a novel polymer able to dissolve in the body would represent an improvement in ureteral stent design by eliminating the need for a secondary, hospital-based, cystoscopic stent removal procedure and associated anesthesia. It would reduce the cost of using a stent, eliminate potential trauma from cystoscopy, increase patient safety, and decrease potential liability. However, depending on how novel the dissolvable stent material was and whether it was used in similar applications, a regulatory agency might require significant safety and efficacy testing in the form of animal and clinical studies. These studies would significantly increase the cost and time requirements to obtain regulatory clearance to market the new device. This might force the company to use another, less innovative material previously used safely in medical devices to be able to introduce the product earlier. In this case, regulatory requirements would impact the design due to tradeoffs between the market impact of a novel new material and time to market. It is important for you to be aware of the different classes of devices and the impact of these requirements on the likely regulatory pathway to market that your device might follow. Device classes are discussed in greater detail in Sections 12.4 and 12.5.

Industry standards represent design constraints that appear in product requirements. Compliance with these standards often impacts design choices and costs. Requirements to comply with standards should be confirmed early in the project. Redesign, retesting, and delayed market introduction of the new product may be required when an industry project team discovers late in the project that they must comply with a standard. More information on industry standards is presented in Chapter 11.

Social and cultural constraints reflect cultural norms and may affect the types of medical devices used in a particular country or region of the world. For example, heart valve replacements can be made of synthetic materials or xenografts, such as bovine (cow) or porcine (pig) tissues. However, due to cultural constraints resulting from religion-based restrictions, bovine heart valves would not be used in countries such as India, where cows are considered sacred. Porcine heart valves may not be used in many patients in the Middle East, where pigs are considered religiously unclean. This is especially true for global health projects in which you might design solutions for customers with different cultural and social norms.

Environmental design constraints may originate from international standards, customers, or actions of regulatory agencies such as the EPA. These address pollution, hazardous materials, recyclability, landfill disposal, and sustainability, and support the Triple Bottom Line presented in Breakout Box 3.8. For example, in the 1990s, medical devices containing polyvinylchloride (PVC) tubing could be disposed of in landfills in the United States. However, this was not allowed in the European Union. This design constraint required the redesign of devices to eliminate the use of PVC materials. It is important to be aware of any environmental design constraints that could impact your design and to consider how your device will be disposed of at the end of its serviceable life.

4.4 Creating Value for the Customer and Other Stakeholders

Customer needs and design constraints dictate metrics and metric values. The goal of new product development is to create value for the customer. New products must be equal to or better than existing products in some way. If they are not, then customers will have little reason to purchase the new product. There are several ways to define "better." In the medical device industry, there are multiple ways in which new products can provide benefits over competitive devices. Value can mean lower cost, time savings, novel method of treatment or diagnosis, improved performance and more effective treatment, and improved quality of care and patient outcomes. If a new product fails to provide at least one of these benefits, your design team or company should seriously consider if it is worth the time, effort, and money that will be spent to develop a prototype and introduce the new product. Academic design projects may include goals involving improved performance, time savings, and decreased cost, all of which can be translated into a specification and measured through testing. Demonstrating more effective treatment and improved quality of care and patient outcomes is typically beyond the scope of a design course due to required longer-term clinical evaluations.

In Section 3.9, ways to add value to the healthcare ecosystem are introduced. Sections 4.4.1 to 4.4.5 expand on these and present specific examples of how medical devices can create value for the customer.

4.4.1 Improved Performance and More Effective Treatment

A device that results in improved outcomes or a more effective treatment can increase the quality of care. This result can be achieved through a new medical device or technology that performs better than current products. The new device may be easier to use making it safer than competitive devices. A previous example, presented in Breakout Box 4.1, describes a novel biodegradable ureteral stent that would dissolve over a period of a few weeks, thus eliminating the need for a retrieval procedure while reducing costs and patient risk. This stent could provide improved performance and safety as well as a more effective treatment. It has the potential to become the standard of care in ureteral stents if it provides a clear benefit over competitive ureteral stents.

4.4.2 Time Savings

The adage "time equals money" applies to medical device design. New devices that can reduce the amount of time it takes to complete diagnostic or therapeutic procedures can lower cost and increase patient safety. For example, if a new surgical instrument allows an orthopedic surgeon to perform a total knee replacement in 60 minutes instead of 90 minutes, this will result in a time savings of 30 minutes, which can save $1980 per procedure at current average rates for operating room (OR) time of $66/min. Shortening the procedure time also reduces anesthesia time, in turn reducing risk to patients. Depending on how much time is saved, shorter procedure times may allow hospitals and surgeons to schedule additional procedures into the surgical schedule, increasing productivity and revenue for the hospital. Although the benefits presented here are clear, realizing these benefits in the healthcare system is a far more complicated process.

4.4.3 Cost Savings

Third party payers—such as Medicare and private insurance companies in the US, and national health systems in other countries—exert significant pressure on hospitals and healthcare providers to reduce the cost of healthcare, one of the components of the Triple Aim of Healthcare discussed in Breakout Box 3.8. This includes lowering the cost of pharmaceuticals and medical devices. Companies developing new devices and technologies that can lower the cost of healthcare and save hospitals money will find a significant market for these products.

Commodities are highly price-sensitive products and are not differentiated from their competitors by product features. A tongue depressor is an example of a commodity in the medical device industry. All tongue depressors perform the same function; to hold down the tongue so that it does not prevent a physician from examining the back of a patient's throat. Cost is typically what makes a hospital purchase Company A's tongue depressor instead of Company B's product. So, if a company can sell its tongue depressors for less than a competitor's, and the hospital purchases thousands of these each year, the hospital can realize significant cost savings by purchasing from the lower cost producer.

A cost reduction project that allows price-sensitive products to be sold for less than the competition often helps companies maintain or gain market share from competitors. For example, hospitals purchase millions of urinary drainage bags each year. Due to cost containment pressures, they want lower cost products that will provide significant savings. Low-cost urinary drainage bags from different manufacturers perform similarly and offer similar features; thus, hospitals began purchasing drainage bags based on cost and not features, making drainage bags a commodity. In response, drainage bag manufacturers initiated design projects to lower the cost of their urinary drainage bags. Engineers considered ways to reduce manufacturing costs including reducing the number of components, finding lower cost suppliers, and identifying alternative lower cost materials. Companies were able to reduce their selling price by reducing the manufacturing cost by just a few cents per drainage bag, allowing hospitals to realize significant cost savings. This allowed companies to compete in this price-sensitive market.

4.4.4 Novel Method of Treatment or Diagnosis

Truly novel, new-to-the-world medical technologies can improve the quality of care, improve patient safety, lower costs, and increase revenues for hospitals and physicians. Sometimes, however, they can increase costs. Only about 10% of all new product introductions involve truly new-to-the-world products (those that do not currently exist). In the 1980s, German aerospace company Dornier introduced extracorporeal shockwave lithotripsy (ESWL). It provided a noninvasive method for eliminating kidney stones. Prior to its introduction, kidney stone removal required an open surgical procedure that involved trauma to muscles, other soft tissue, and the kidney, as well as weeks of postoperative recovery, lost work time, and risk of infection and other potential complications. The ESWL procedure used acoustic energy created through cavitation produced by a spark plug in a water bath, as shown in Figure 4.3. The patient was sedated but general anesthesia was not required. Acoustic waves were reflected using two elliptical reflectors that allowed the energy to be focused on a point in space. By adjusting the focal point to coincide with the kidney stone in the patient's kidney, a highly localized level of acoustic energy could be delivered to the kidney stone, causing it to disintegrate into fine sand particles that could then be passed through the kidney, ureter, bladder, and out the urethra.

There are times, however, when a new-to-the-world technology can produce unwanted outcomes. For example, implantable defibrillators have been known to activate accidentally. When this occurs, the patient feels an internal shock that has been compared to being "kicked in the chest by a horse." The role of the designer is to prevent unwanted outcomes and maximize the improvements and resulting benefits of a new device or technology.

4.4.5 Improved Quality of Care

The ESWL procedure allowed the removal of kidney stones without the need for any incision or insult to the body, making it safer and less traumatic. It improved the quality of patient care. The ability to use twilight sedation instead of general anesthesia improved patient safety, thereby reducing liability. As an outpatient procedure, hospital stays were avoided, reducing costs and allowing patients to return to work much sooner than with the more invasive traditional procedure. This new technology offered a new reimbursable procedure that was not previously available and resulted in additional incremental revenue to the hospital and surgeons. ESWL is an excellent example of a technology that improves several clinical metrics simultaneously. You can also try to find solutions that will result in more than one improvement while also lowering costs, as discussed above and in Table 3.1.

4.5 Tradeoffs Between Metrics

The process of designing a medical device requires choices that will optimize value based on ideal metrics. However, tradeoffs between higher and lower priority needs and metrics are acceptable. It may be acceptable to exceed a cost goal if it can be shown that a customer is receiving something of value to justify the increased cost. Similar tradeoffs can be made between other metrics involving performance characteristics such as lower infection rates, ease of use, increased safety, lower power consumption, and others.

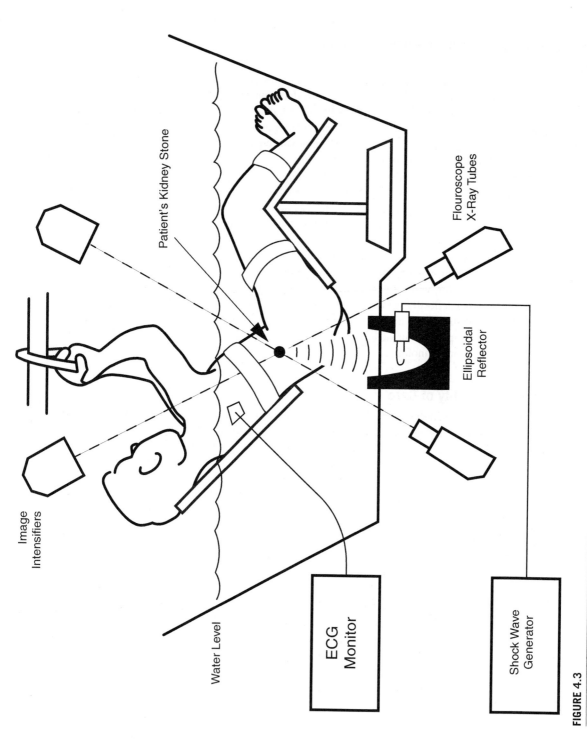

FIGURE 4.3

Extracorporeal shockwave lithotripsy (ESWL) system for treating kidney and ureteral stones. The water bath allowed cavitation to occur and provided the acoustic impedance needed to transmit the resulting acoustic wave to the patient's body. This new-to-the-world technology provided a minimally invasive alternative to open surgery.

For example, urinary drainage bags are connected to Foley catheters to collect urine as it collects in a patient's bladder. Urine flows from the bladder, through the catheter inserted in the patient's urethra and held in place by a saline filled balloon, and out into the drainage bag (Figure 4.4). During insertion of the catheter, the balloon is deflated to prevent trauma and discomfort. Once in the bladder, the balloon is inflated with saline injected through a port and lumen along the length of the catheter. The inflated balloon serves as an anchor to prevent the catheter from migrating out of the urethra. Previous studies have shown that for every day past three days that a patient uses a drainage bag, the probability of urinary tract infection (UTI) occurrence can increase by approximately 10% per day. Most UTIs are treatable but can be fatal. Insurance companies typically do not reimburse hospitals for treatment of these infections or for additional days in the hospital required to treat the infection.

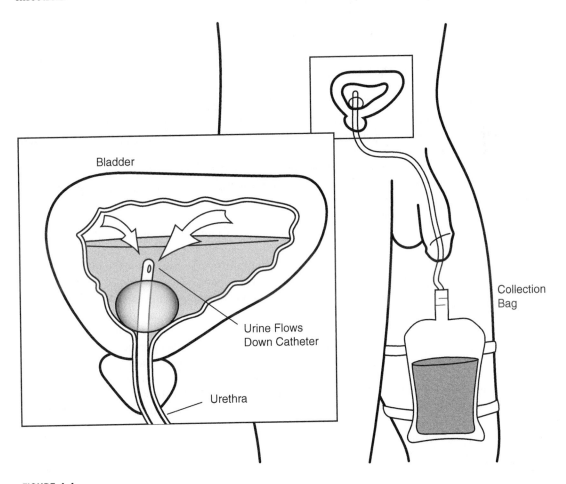

FIGURE 4.4

Foley catheter and urinary drainage bag used to drain urine from bladder. Foley catheters are often used in surgical procedures requiring general anesthesia.

To address this problem, an antimicrobial urinary drainage bag system was developed to reduce UTIs. The bag contained a packet of powder that would slowly release a gas that would kill microbes and reduce bag-related infection rates. It was connected to a Foley catheter with a silver oxide antimicrobial coating that was shown to kill bacteria within the urethra. The cost of this bag/catheter system was higher than traditional drainage bags and catheters, but the manufacturer was able to show that by spending five dollars more for its antimicrobial bag, hospitals could reduce infection rates and save money on infections that they no longer needed to treat. This was an example of an acceptable tradeoff between metrics involving increased cost and lower infections rates.

4.6 Documentation of Specifications

There are many ways of documenting specifications such as needs/metrics tables and competitive benchmarking tables, which present important information in one place. These documents and the information they contain are all part of the Design History File (DHF) described in Chapter 2. The following example demonstrates how they are used to convert needs into specifications.

Urinary incontinence is a condition where a person cannot control the flow of urine out of their bladder. It affects millions of people with disabilities, spinal cord injuries, spina bifida, and other conditions. Many elderly people are incontinent, as are younger women after multiple childbirths. The problem is currently managed nonoperatively using diapers, Foley catheters, and intermittent catheters, shown in Figure 4.5.

After learning of existing solutions and the benefits and drawbacks of each, the design team decided to develop a new type of catheter that would provide benefits over intermittent and Foley catheters. Table 4.2 contains a list of needs for the novel catheter design obtained from interviews with incontinent patients and their physicians.

The next step is to translate the needs (qualitative goals) in Table 4.2 (identified while creating your project statement) into specifications by creating a list of metrics and associated values with units. A list of marginal and ideal values chosen to address each of the metrics for the most important needs is shown in Table 4.3. Ideal values represent the desired level of performance. Marginal values represent the minimum acceptable level of performance. The far-right column indicates the sources of the needs and/or metrics, which can include standards, sponsors, clients, patients, caregivers, physicians, surgeons, nurses, physical therapists, regulatory agencies, company personnel, or benchmarking data. Note that these metrics are numbered. This can be helpful later in the design process when you refer to your specifications during testing. It can also be helpful to create a metric hierarchy. For example, a high-level metric might be labeled as 1, whereas lower-level sub-metrics related to the higher-level metric might be labeled as 1.1, 1.2, and so on, as shown in Table 4.3 for the need "Ease of use." With your design team, categorize those specifications that you think can be met during your design project (i.e., one semester, two semesters, or other time period).

As discussed in Section 4.3, some needs may be more important than others to the success of a new product. Tradeoffs between metrics are discussed in Section 4.5. To indicate the more important needs that must be met and those needs whose tradeoffs may be acceptable, asterisks (*) appear after each need in Table 4.3. A single asterisk (*) indicates a higher priority and two asterisks (**) indicate that tradeoffs might be acceptable.

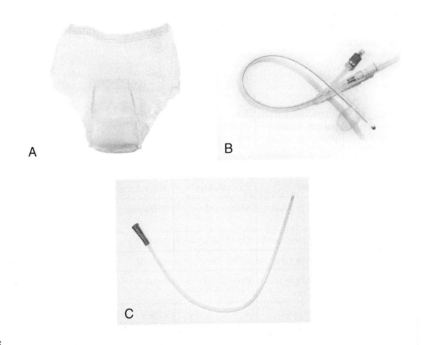

FIGURE 4.5

Some existing solutions used to manage incontinence including an adult diaper (*A*), a Foley catheter (*B*), and an intermittent catheter (*C*).

Table 4.2 List of Needs for Design of Novel Incontinence Catheter.	
Category	**Need**
Safety	- Must not damage tissues - Must not migrate out of bladder - Low infection rate
Performance	- Adequate flow rate - Comfortable to patient during indwelling period - Reliable - Function for 30 days
Ease of use	- Allow patient to operate without assistance - Easy to insert in physician's office or patient's home - Easy to remove in physician's office or patient's home
Cost	- Affordable - Reimbursable by insurance company

Table 4.3 Needs and Corresponding Metrics Including Ideal and Marginal Values and Source of Needs and/or Metrics.

Metric No./ Customer Need/Priority	Metric	Ideal Value	Marginal Value	Source of Need and/or Metric
1/Ease of use*				
1.1/Ease of insertion	Physician, nurse, patient feedback	5/5 (5 pt. scale)	3/5 (5 pt. scale)	Nurses, patients
1.2/Ease of removal	Physician, nurse, patient feedback	5/5 (5 pt. scale)	3/5 (5 pt. scale)	Nurses, patients
1.3/Ease of operation	Patient feedback	5/5 (5 pt. scale)	3/5 (5 pt. scale)	Nurses, patients
2/Adequate flow rate**	Flow rate	200 mL/min	100 mL/min	ASTM standard
3/Reasonable cost**	Cost to produce	$50/month	$100/month	Company, benchmarking data
4/Comfort*	Patient feedback	5/5 (5 pt. scale)	3/5 (5 pt. scale)	Physicians, patients
5/Must not migrate*	Migration rate	<5%	<8%	Physicians, patients
6/Will not damage tissue*	Cystoscopic evaluation	Pass	Pass	Physicians, patients
7/Low infection rate*	Infection rate	<3%	<5%	Physicians, FDA
8/Reliable*	Patient/physician feedback	Pass	Pass	Physicians, patients
9/Acceptable service life**	Service life	30 days	27 days	Physicians

A single asterisk (*) indicates a higher priority and two asterisks (**) indicate that tradeoffs may be acceptable.
ASTM, American Society for Testing and Materials; FDA, The U.S. Food and Drug Administration.

Table 4.4 includes columns that define and allow quantitative comparison of the values for the new design to those of the competitive solutions (intermittent catheters and Foley catheters). The need is shown in the first column and the metrics (what will be measured) are shown in the second column. Ideal and marginal values are shown in the third and fourth columns, respectively. If used properly, all customer needs should be addressed by at least one metric; likewise, all metrics should correspond to at least one need. If a need has no metric associated with it, the need will not be addressed by the specifications. If a metric does not correspond to a need, then it may not be necessary, as it does not support a need. Actual values for existing competitive solutions are shown in the two right hand columns. These values can be determined from testing of competitive devices, surveys of users of these devices, or information available on a manufacturer's website. Some information will not be available, indicated by "?", and other information is not applicable or relevant, indicated by "N/A." Competitive benchmarking tables allow designers to ensure that their target specifications describe a product that will perform as well as or better than existing competitive products or procedures. Competitive benchmarking and product dissection are also discussed in Sections 5.9 and 5.10.

Table 4.4 Competitive Benchmarking Table for Novel Incontinence Catheter.

Metric No./ Need	Metric	New Product		Competitive Solutions	
		Ideal Value	Marginal Value	Foley Catheter	Intermittent Catheter
1/Ease of use	Patient feedback	5/5	3/5	3/5 (must be skilled)	4/5 (patient performs procedure)
2/Adequate flow rate	Flow rate	200 mL/min	100 mL/min	100 mL/min (per ASTM standard)	100 mL/min
3/Reasonable cost	Cost to produce	$50/month	$100/month	$10 per use $100/month	$6/day = $180/month (diapers = $50/month)
4/Comfort	Patient feedback	5/5	3/5	1/5	2/5
5/Must not migrate	Migration rate	<5%	<8%	<5%	N/A
6/No tissue damage	Cystoscopic evaluation	Pass	Pass	?	?
7/Low infection rate	Infection rate	<3%	<5%	10% per indwelling day (past 3 days)	10%
8/Reliable	Patient/physician feedback	Pass	Pass	?	?
9/Ease of insertion	Physician feedback	5/5	3/5	3/5	5/5
10/Ease of removal	Physician feedback	5/5	3/5	4/5	5/5
11/Functions for at least 30 days	Acceptable service life	30 days	27 days	<7 days	Must function during voiding period

N/A, not applicable; ?, information not available.

The specifications you create using the tools and methods presented in this Chapter are preliminary target specifications. They are created before design solutions have been generated, narrowed, and prototyped. As you proceed through the design process, specifications will be added, deleted, or revised. For these reasons, maintaining versions of tables such as Tables 4.3 and 4.4 for your project will provide an added benefit of documenting how your understanding (and meeting) of key specifications evolves over the course of the design process.

Key Points

- Needs are often qualitative and must be converted into quantitative specifications that can be tested.
- Metrics describe what will be measured to verify that a need has been met.
- Ideal values represent what you prefer based on customer feedback and competitive benchmarking. Marginal values represent what can be accepted as a minimum.

- Some needs are more important to the basic function of a product than others and should be prioritized accordingly. There may be tradeoffs between needs.
- Competitive benchmarking is a way to learn about current solutions to the problem and ensure that a new design will be as good as or better than existing products.
- Technical, financial, legal, regulatory, social, cultural, and environmental constraints limit the number of acceptable designs.
- There are several ways to create value for the customer including cost and time savings, improved performance and safety, improved quality of life, and increased effectiveness of a new technology.
- Needs/metrics tables are one tool that can be used to create a list of specifications based on customer needs.

Exercises to Help Advance Your Design Project

1. Using Table 4.3 as a template, create a needs/metrics table for the device you are designing.
2. Using Table 4.4 as a template, create a competitive benchmarking table for the device you are designing.
3. Section 4.4 presents different ways a new product can provide value. Does the device you are designing provide value in any of the ways presented? What value does it provide to a customer or user?
4. Name at least six different types of design constraints. Which of these apply to your design?
5. Are some of your specifications more important to meet in your final design than others? What tradeoffs will you be willing to make regarding the specifications?
6. Find another design team and exchange your needs/metrics tables with each other. Each team should critique each other's specifications and provide constructive comments.
7. The first three chapters stressed the importance of documenting the various dimensions of the medical problem and current solutions. At this point in the design process, you likely have many documents (e.g., electronic notes, articles, sketches) and perhaps samples of some existing products. Discuss with your team how your organizational system is working. All team members should be able to readily find disparate information that has been collected. By organizing this as a team, it should be relatively clear whether it should be placed in your DHF for others to see.
8. If you are using a software platform (e.g., Slack, Google Docs), is it working well? It is easy to set up an organizational system when you have no real data to place into it. It is more difficult when you have real information (which you should now have) coming from many sources. It is not as much the system or structure you choose, but rather the discussion you have as a team that will allow you to use whichever system you agree will work for everyone. Discuss with your team whether your software platform is working well.
9. As discussed in Chapters 2 and 3, your team can be thought of as an organism that is developing alongside your evolving design problem and solution. The two co-evolve together and will impact one another. Events that occur in the course of the design process, both celebratory and disappointing, will change the dynamics of your team. Likewise, team or individual events can

also impact your project progress. Keeping your team healthy is critical to navigating the difficult decisions presented in Chapters 5–8. This is a reminder to invest in your team.

- Have you established healthy rituals?
- Are there any personal issues arising that should be addressed before they grow?
- If you could change one thing about your team, what would it be? Remember that no team (or individual) is perfect and, just like your project, will require time and attention.

General Exercises

10. What are the differences between ideal and marginal values?

11. Search "Apollo 13 air filter" and view the video from the movie *Apollo 13*. It depicts how engineers working on the Apollo 13 mission had to deal with some challenging design constraints when dealing with a malfunctioning air filter that threatened their mission. What constraints did they have to deal with?

References and Resources

Booz, Allen, & Hamilton. (1982). *New products management for the 1980s*. Booz, Allen, and Hamilton.

FDA Rules and Regulations. (1996). *Federal Register*, *61*(195), 52608.

Florman, S. (1983). *Moral blueprints: On regulating the ethics of engineers*. Engineering Professionalism and Ethics. John Wiley and Sons.

Hoffman, M. (2012). *Design input: Designing the right device*. AAMI Leading Practices. Association for the Advancement of Medical Instrumentation.

International Organization for Standardization. (2015). *Quality management systems—fundamentals and vocabulary*. (ISO standard no. 9000:2015). https://www.iso.org/standard/45481.html.

International Organization for Standardization. (2015). *Quality management systems—requirements*. (ISO standard no. 9001:2015). https://www.iso.org/obp/ui/#iso:std:iso:9001:ed-5:v1:en.

International Organization for Standardization. (2016). *Medical devices*. (ISO standard no. 13485:2016). https://www.iso.org/iso-13485-medical-devices.html.

Kano, N., Nobuhiku, S., Fumio, T., & Shinichi, T. (1984). Attractive quality and must-be quality. *Journal of the Japanese Society for Quality Control (in Japanese)*, *2*(14), 39–48.

Nicolle, L. E. (2014). Catheter associated urinary tract infections. *Antimicrobial Resistance and Infection Control*, *3*(23). https://doi.org/10.1186/2047-2994-3-23.

Spacey, J. (2016). *Nine types of design constraints*. Simplicable. https://simplicable.com/new/design-constraints.

Shippert, R. D. (2005). A study of time-dependent operating room fees and how to save $100,000 by using time-saving products. *American Journal of Cosmetic Surgery*, *22*(1), 25–34. https://doi.org/10/1177/074880680502200104.

Ulrich, K., & Eppinger, S. (2012). *Product design and development* (5th ed.). McGraw-Hill Irwin.

SOLUTION GENERATION AND SELECTION

SECTION 3

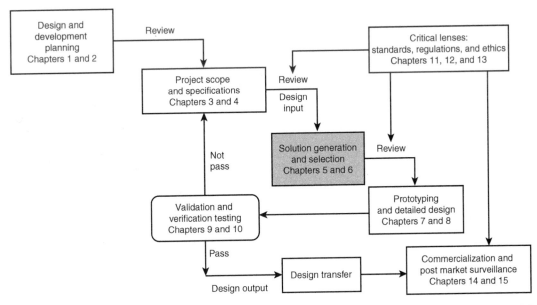

In the next two chapters you will generate solution concepts and ultimately narrow your list to one design solution that you will pursue further. Engineering design is inherently a creative endeavor that shows up most clearly in the ideation phase of the design process. Ideation is composed of two core activities; the generation of ideas and the selection of the most promising ones. A robust ideation process will alternate between these two modes, keeping ideas that have survived the selection process and using them as the input to another round of idea generation. Not all idea generation or selection techniques work all the time, so you are encouraged to try multiple techniques.

Chapter 5 introduces individual and group methods of generating ideas. These include places to look for inspiration as well as brainstorming methods, such as Concept Mapping, C-sketch, and recombining your existing ideas. The goal is to help you systematically generate and develop solution concepts (abstracted or vague hints at a solution) and then translate them into design solutions (detailed drawings and/or physical embodiments).

Chapter 6 introduces several qualitative and quantitative techniques for selecting the most promising solution concepts and designs. The limitations of resources and time allow you to only move one (or two) design concepts to the next phase of the design process. You will learn how to progressively apply more restrictive technical and nontechnical filters to determine which solution has the best chance of being realized in the form of a prototype. Direction is provided on engaging others in concept selection, documenting your process, and justifying your decisions to others.

By the end of Chapters 5 and 6 you will be able to:

- Apply multiple methods of individual and group idea generation techniques to formulate and iterate upon solution concepts.
- Translate general solution concepts to more specific design solutions.
- Apply qualitative and quantitative selection methods to reduce the number of concepts and designs.
- Select the most promising design solution to be prototyped.
- Add specifications or modify existing ones that apply to your chosen design solution.
- Document the generation and selection of concepts and designs, the decisions made, and the rationale for your choices.

Generating Solution Concepts and Preliminary Designs

5

Good design in fact is like good poetry…a beautiful design always contains some unexpected combination that shocks us with its appropriateness.
— Brian Arthur, Author of The Nature of Technology: What It Is and How It Evolves

Chapter outline

Biomedical Engineering Design. https://doi.org/10.1016/B978-0-12-816444-0.00005-5

5.1 Introduction

The early phase of a design process focuses on understanding needs and opportunities that are transformed into a project statement and technical specifications. You were encouraged not to let bias influence this discovery process by considering solutions. The next step in the design process is to consider many possible solutions and then select a single design solution to pursue further. This chapter is focused on generating ideas; the next chapter provides methods to help you narrow to the most promising. Although the sections are presented in a way that you can follow sequentially, you should mix and match techniques as you iterate upon your ideas. A wide range of techniques are presented because some will work well for particular problems or at particular phases of the design process. Likewise, some teams may find that a particular technique aligns well with their internal dynamics. Find what works best for you. If you get stuck, try another technique. Many of these same techniques may be helpful as you encounter sub-problems later in the design process.

5.2 The Transition From Problem to Possible Design Solutions

In this section we provide you with terms and analogies to guide your transition from problem to solution. First, we discuss the difference between a solution concept and a design solution. Second, we introduce ideation as the process of generating many possible designs and then narrowing to one design that you will pursue further.

5.2.1 Solution Concepts and Design Solutions

Technical problems generally consist of a core problem, surrounded by layers of context, constraints and sub-problems that are revealed during the design process. Solutions are often structured in the same way; an idea begins as a solution concept and is then transformed through iteration into a more specific design solution.

A **solution concept** (sometimes called a conceptual design) is a framework for how a problem could be solved. For example, if the problem is a clogged artery, solutions might include a very small flexible

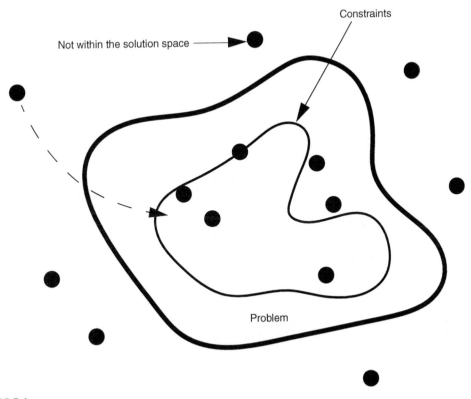

FIGURE 5.1

A graphical way of thinking of the relationship between problems (thick outer line), constraints (thin inner line), and solution concepts (dots). The process of transforming a solution concept into a specific design solution moves a dot (indicated by the arrow) to be within the constraints by making design decisions.

spinning rotor, ultrasonic methods, or the release of a chemical that dissolves the clog. These are solution concepts because there are no specifics on materials, mechanisms, power sources, dimensions, or usage.

One way to visualize the abstract relationship between the problem and solution concepts is shown in Figure 5.1. Chapter 3 considers the nature of the problem (thicker outer line) and Chapter 4 defines the constraints of a technical solution (thinner inner line). Possible solution concepts are represented in the figure as dots. In this abstract diagram, the shapes of the two lines are intentionally convoluted because you will continue to refine your understanding of the problem and constraints as you move deeper into the design process. As indicated in Figure 5.1, some solution concepts (dots) may not initially fall within the constraints, or in some cases, seem relevant to the problem. For example, the concept of using a spinning rotor to unblock an artery may not seem promising as an abstract idea, yet it could help you better define what is or is not a viable solution. During ideation, generating many solution concepts (i.e., adding dots to the diagram) can help you refine your understanding of both the problem and constraints.

A **design solution** is a specific implementation of a solution concept. For example, there are several versions of spinning rotors, as well as speeds, housings, and materials. Moving from a solution concept to a design solution requires you to answer hundreds of questions that transform your general solution idea into a design. What material should be used? How should this part be attached to other parts? What energy source is best? How might energy be generated and delivered? What safety features could prevent or minimize adverse outcomes? Systematically answering these types of questions moves a concept to a design. Varying the answers to these questions is also how one solution concept can lead to several related design solutions. Furthermore, each of these decisions may move the concept closer to being within the constraints (movement indicated by the arrow in Figure 5.1), thus being a viable solution to the problem. The general idea of a spinning rotor may not have seemed attractive at first, but may become a leading design solution upon seeing a more specific design.

It is important that the generation of solution concepts and design solutions remain separate steps for several reasons. First, generating many possible solution concepts is a demonstration that you engaged in a thorough exploration of the problem and solution spaces. In fact, this kind of exploration often reveals that the exact problem and solution criteria were not quite right, leading to a refinement of the project statement. Second, it is not uncommon for a concept that initially seemed to fall outside of the solution space to be iterated upon until it has the potential to solve the problem. This occurs in Figure 5.1 when a solution concept (represented by a dot) moves from outside to inside the solution space as a design solution emerges. Sometimes it is these designs that are the most novel and viable. Third, following a systematic process—especially during the most ill-defined step in the design process—more often than not leads to a viable and robust solution.

5.2.2 Ideation

Populating the solution space with solution concepts and transitioning to design solutions is generally achieved through a process known as ideation. **Ideation** is a multi-step process that ultimately aims to arrive at a singular "best" idea. A helpful graphical analogy for ideation is the sideways Christmas Tree shown in Figure 5.2. One begins at the stem of the tree with the problem to be solved. There is a **divergence** phase to generate possible solutions followed by a **convergence** phase to select ideas that seem most feasible. From these selected ideas, another round of converge/diverge can be used for further refinement of ideas. Through this iterative process, a "star" idea emerges.

An important step is missing from the Christmas Tree analogy. After each diverge session, it is helpful to have a **confluence** session. This is when you should pause to consider recombining, sorting, and grouping of ideas. You should remind yourself of the problem and perhaps even refine your definition of the problem. A confluence stage helps prepare you for another round of ideation.

5.2.3 Keeping Notes

Ideation sessions are more free-flowing than structured meetings. It is tempting to try to remember the most memorable ideas after the session, yet it is often the less memorable ideas that, on iteration, can grow into your most promising solutions. For this reason, you should record all of your ideas—this also shows evidence of the rigor of your process. On the other hand, recording every idea in detail can impede the free flow of thought and discussion that is a hallmark of a good ideation session. To strike a balance, consider working at a white board or with sticky notes. The added benefit of movable notes

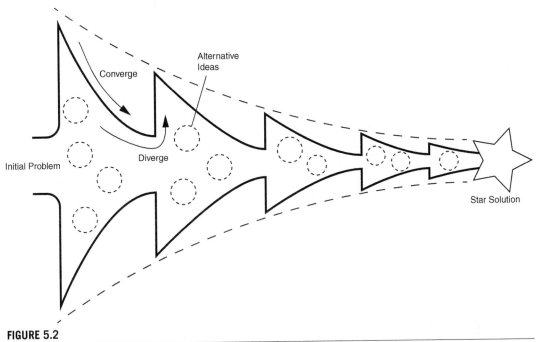

FIGURE 5.2

The sideways Christmas Tree model of the converge-diverge cycle of ideation. A star solution is eventually developed through continued converging and diverging.

is that you can later categorize ideas by moving the notes around. Documentation of this process can then be as simple as taking a photo to include in your Design History File (DHF). Another technique is to use one large piece of paper (e.g., butcher paper or a large sticky pad). Rolling up this paper when you are finished allows you to continue ideating at another time. Figure 5.3 shows examples documentation of an ideation session.

5.3 Divergence and Generating Solution Concepts

Generating solution concepts is the first step toward solving a problem. The goal is to add as many dots (solution concepts) to Figure 5.1 as possible. The generation of solution concepts is a specific instance of what many would call creativity. Popular culture often portrays creativity as a single moment (e.g., "eureka," "light bulb moment") and creative work as the product of a lone "creative genius." This section aims to dispel these assumptions and replace them with the idea that creativity is a muscle to be exercised—practice makes you better. Furthermore, you should think of idea generation as a process, much like the design process, that is systematic, iterative, and aided by techniques and tools. In this section we explore idea generation techniques that can be used by an individual or in a group. The following section discusses techniques specifically design for groups. Many more creativity tools are included in the resources listed at the end of the chapter.

FIGURE 5.3

Taking handwritten notes helps document ideation sessions without impeding the flow of ideas.

5.3.1 Concept Mapping

During the problem identification phase, you may have generated ideas for solutions. The advice at that time was to record these solutions, and to then move them to the back of your mind to prevent influence over your discovery process. Concept mapping is a way to bring back these initial solution ideas and build upon them.

A concept map is a graphical representation of how ideas are connected. In the context of solution generation, it is a way to begin mapping out the solution space by grouping and connecting your ideas. Figure 5.4 shows a solution concept map for the problem of gaining better access to a surgical site. Note that this map is incomplete (much more could be added), some ideas are expanded upon more than others, and there is a mix of solutions and what is known about the problem. It is in many ways a mess, but that is the point of a concept map; building one is a way of collecting your thoughts in a single place in the nonlinear manner in which ideas typically spring to mind.

In Figure 5.4, even outlandish ideas have been included. For example, one possible solution concept is to perform the surgery in space under microgravity conditions. Even though this idea would seem to initially be well outside of the problem (i.e., on the far edges of Figure 5.1), it could spark a new idea or be recombined with other ideas. A later confluence step, discussed in Section 5.5, involves discussing your ideas and better organizing your concept map.

To engage in a concept mapping session, read over your project statement and consult your specifications. This is to remind you of the scope and constraints of your problem and is helpful in focusing your session (see Breakout Box 5.1). Be sure you have plenty of space, with either a whiteboard, writable screen, or a blank wall and stickies. It is particularly helpful to engage in several rounds of concept mapping, separated in time by as little as 10 minutes or as much as a few days. You should periodically take pictures to document the evolution of your concept map. If you would prefer to work digitally, mindmapping software programs include InsightMaker, XMind, and Google's Coggle.

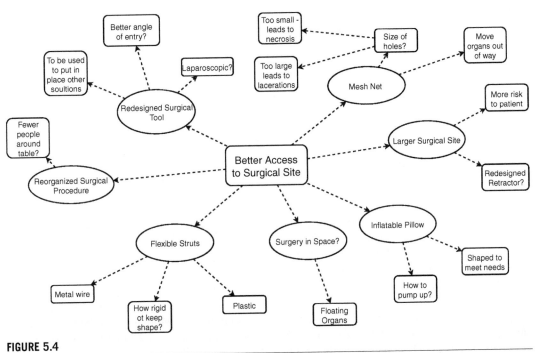

FIGURE 5.4

An example concept map for gaining better access to a surgical site created using InsightMaker.com.

Breakout Box 5.1 The Power of Constraints

It may come as a surprise that constraints can in fact help you develop solutions. As a simple demonstration, time yourself on the following two tasks, taking only 30 seconds for each:
1. Name as many white things as you can.
2. Name as many white things as you can in a refrigerator.
   This exercise comes from Keith Sawyer's book *Group Genius*. In a wide-ranging study, Sawyer found that for most people the lists were approximately the same length, and in several cases the second list was longer. He attributed this to two factors. First, we prime our brain when we create the first list to only consider white things. Second, by constraining ourselves to only the refrigerator, we are able to create a clearer mental picture in our mind. Of course, your results may differ for this single exercise, but your specifications may help you more quickly generate viable solutions.

5.3.2 Delphi Methods

Delphi is a technique that was originally created to help make predictions about the future of complicated systems. A group of knowledgeable people are asked to independently answer some questions on a survey or during an interview. In fact, the participants generally do not know who else is answering or even how many others are part of the surveyed group. This independence helps participants stay focused on ideas rather than who is proposing the idea. After collecting, anonymizing, and organizing the responses, the results are recirculated back to the participants for further comment or additions.

Breakout Box 5.2 The Power of Constraints (Part II)

One way to keep going during a divergent session is to impose constraints that are not explicitly stated in your specifications. For example, you may temporarily impose the following constraints while asking, "What if our solution must...":
- be built for under $5?
- work with no electrical power?
- be created from parts found only at a hardware store or big retailer?
- be operable by an untrained nonmedical person?
- work in the developing world?
- work in a veterinary hospital setting?
- be constructed in the next 30 minutes?
- be operable by a battlefield medic?

Often some form of voting takes place. This cycle continues until only a few (or possibly one) ideas survives. There are many variations of the Delphi method that can be used in solution generation.

During your exploration of the problem in Chapters 3 and 4, you may have interacted with experts in the field who could now help you in generating solution concepts. Using the Delphi method, the questions asked are carefully designed so as not to bias the results. Open ended prompts are often best for idea generation. For example, you could ask, "What is a solution that would work in a low-resource setting?" or "What is the wildest possible solution you can think of?" More ideas for building prompts can be found in Breakout Box 5.2. You should be critical of the results of a Delphi ideation session, as your participants may be too close to the problem. This is especially true for subspecialist physicians (e.g., pediatric surgeons, interventional radiologists) who have years of additional training in one subspecialty. Treat them as a starting point for your own ideation process.

5.4 Brainstorming

When the divergent phase of ideation occurs in a group, it is often called **brainstorming**. In this section we first discuss how to set the stage for effective brainstorming sessions and then provide you with some techniques to try with your team.

5.4.1 Preparation

While tempting to jump right into brainstorming, four important steps should be completed prior to a productive brainstorming session. First, you should consider who should be in the room. This may be your team, or some subset of your team. Note that brainstorming typically works best with three to six people. Fewer than three people often inhibits the free flow of ideas (as discussed in the Bakeoff experiment in Section 2.7), while more than six can make it difficult for everyone to have their ideas heard. If you have more than six people on your team, consider splitting into teams during brainstorming. It may be helpful in some cases to invite guests who are not familiar with your project to join you. Be careful, however, as advisors, mentors, or supervisors can make team members self-conscious and hesitant to contribute ideas. Second, you should reserve a space that is conducive to idea generation. An ideal space would be large enough to move around yet comfortable and private enough that you are not

---

Breakout Box 5.3 The 6-3-5 Method

The 6-3-5 method, also known as brainwriting, is a brainstorming method whereby an individual builds off of the ideas of their teammates. The technique begins with six people each writing down three ideas in about 5 minutes, hence the name 6-3-5. At the end of the 5 minutes, everyone rotates their paper to the person next to them. A new round of 5 minutes begins with each individual either building on the idea in front of them or starting with a new idea. This cycle continues four more times for a total of 30 minutes. If you do not have six people, you can still use the technique by modifying the total number of passes. Typically, a 6-3-5 session is silent, but it is customary to have a debrief session afterward to review the ideas generated. The goal is to generate as many possible solution concepts as possible.

Breakout Box 5.4 Six Hats

Brainstorming is all too often a free-for-all and the group drifts in many different directions. Although this can be productive at times, there are frameworks that can help bring some structure to your brainstorming sessions. One of these methods is for everyone to take on a particular role so that the conversation is balanced. A classic technique that falls into this category is Edward De Bono's Six Hats. Each person takes on one of the following roles:

White: Information known or needed. Just the facts.
Yellow: Explore the positives and probe for value and benefit.
Black: Devil's advocate. I like, I like, I wish. Prevent groupthink.
Red: Feelings, hunches, share fears, dislikes, loves.
Green: Possibilities, alternatives, new concepts, perceptions.
Blue: Manager, timer, ensure on task.

If you do not have six people, combine the roles of the Blue and White hats and, if needed, the Yellow and Green hats. Special note should be taken of the Black hat role, which aims to keep conversation going and ideas flowing. The role is not about criticizing or trying to eliminate ideas from consideration. A common phrase of the Black hat should be "I wish…" It can help to have actual hats, slips of paper, or some other colored object to remind everyone which roles are being played. Your group should switch roles periodically (perhaps every 5 minutes) to keep the discussion going.

You may find yourself being more comfortable with some roles and less comfortable with others. Likewise, you may find that you are particularly good at some roles but not others. You should take note of this and continue in that role for the project duration if the team agrees. Remember this when you talk about your strengths and weaknesses at interviews.

---

distracted by others. You should consider ways to capture your work and ensure participation from everyone (e.g., note paper, post-it notes, white boards, enough writing utensils, camera). Third, it is helpful to find a time when your team is most alert and creative. It is difficult to stay focused if you are tired or under stress. In addition, it is recommended that you also schedule time either immediately after brainstorming or soon afterward to process your ideas in a confluence session. Fourth, have some sharable snacks available. Last, it is helpful to assign a team member to facilitate the session and perhaps choose a brainstorming method. Any of the methods could be modified to better suit your purposes. Additional methods are presented in Breakout Boxes 5.3 and 5.4.

The effectiveness of a brainstorming session can often be enhanced by some individual preparation time before the session. First, some participants may feel more comfortable having time to develop ideas on their own before sharing them in the haphazard dynamic of a brainstorming session. Individual thinking beforehand usually generates a greater diversity of starting points and provides everyone with something to contribute early in the session. Second, individuals should perform research ahead of time so that that they have something to share. For example, a session on the various types of materials that may be used could be primed by assigning a specific material for each individual to research.

5.4.2 Setting the Tone

It can be daunting to start populating a blank solution space with ideas. Three simple steps can be taken at the beginning of a session to set the tone and overcome the barriers to idea generation. First, the group should review the project statement and specifications to ensure that everyone has the same overall goal in mind. Second, it is important to clearly state the narrower goals of the session (e.g., develop a first draft of a solution concept map) and the timing (e.g., 30 minutes). Third, starting the sessions with an icebreaker, such as a silly fact or telling a short story, can set a fun and relaxed tone. These three steps only take a few minutes but will make your brainstorming session much more productive.

5.4.3 Facilitating a Session

Brainstorming done correctly engages the mind, emotions, body, environment, and social surroundings. As a facilitator, you should consider the following general guidelines for keeping a session moving forward:

- Explain that participants must reserve judgement of ideas until the session is over.
- Stand up and be sure everyone can see one another. It has been shown that being on your feet raises attention levels. Being able to see everyone enhances the social connections necessary for a productive session.
- Notice if someone is not contributing and ask their opinion. Note that there are a variety of roles played during brainstorming. Some may feel comfortable while others may not. For example, some may throw out ideas while others may stay in the background and look for synergies or point out commonalities. Both roles are important.
- Try to recognize when the group has fallen into *groupthink*—when you have converged too quickly and get stuck discussing the details of one idea. The goal of a good brainstorming session is to generate many ideas. Later sessions can dive more deeply into the ideas that show promise.
- Watch out for *dinosaur babies*—ideas to which one team member becomes attached. This is most common when someone falls in love with their own idea and uses a brainstorming session to try to convince the team that it is "the best" idea. Remind everyone that a debrief afterward is the time to discuss ideas that show the most promise.
- Finish strong. As you near the end of your time, it is common to feel that you are running out of ideas. That is okay; you should persist anyway. Often the most novel modifications or even entirely new ideas come after you have exhausted the obvious possibilities.
- Take a short break (~5 minutes) every 20 to 30 minutes to avoid burnout. For example, if you have a 45-minute session, break it into two 20-minute sessions with a 5-minute break between the sessions. Effective use of a break may include a walk around the building, getting a snack, telling a joke, or doing something fun together.
- Remind everyone of the post-brainstorming session during which you will process your ideas and deconstruct how the session went.

5.4.4 Rules of Brainstorming

There are several general rules for brainstorming that can be made explicit during the session. The term "brainstorming" was invented by the advertisement executive Alex Osborn in 1948. Osborn outlined four rules for brainstorming:

1. Go for quantity.
2. Withhold criticism (defer judgment).
3. Welcome wild ideas.
4. Combine and improve ideas (recombination).

In general, these are good guidelines, although researchers have shown over the years that they do not lead to the highest quantity or quality of ideas.

The product design firm IDEO has their own rules that expand on Osborn's and have been adopted by many organizations:

- Defer judgement
- Encourage wild ideas
- Build on the ideas of others
- Stay focused on the topic
- One conversation at a time
- Be visual
- Go for quantity

Almost any framework has been shown to help, as it aligns a team around the same set of ground rules. Two additional methods are presented in Breakout Boxes 5.5 and 5.6.

Breakout Box 5.5 Yellow Cards

A playful way to remind all team members of your brainstorming rules is to create a Yellow Card, similar to that shown in Figure 5.5. Based on the soccer analogy, a team member would get a yellow card when failing to defer judgement by saying one of the phrases. This is only meant for fun and should not be used to shut down the free flow of a good brainstorming session.

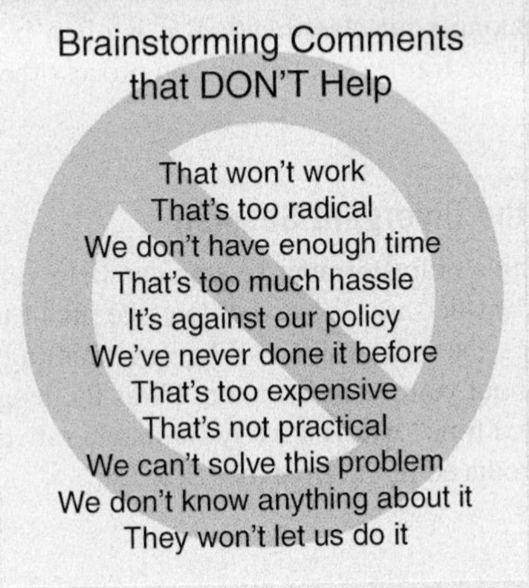

Brainstorming Comments
that DON'T Help

That won't work
That's too radical
We don't have enough time
That's too much hassle
It's against our policy
We've never done it before
That's too expensive
That's not practical
We can't solve this problem
We don't know anything about it
They won't let us do it

FIGURE 5.5

Brainstorming Yellow Card.

Breakout Box 5.6 Yes, and...

A very simple technique that originated in the improvisation world is the attitude of "yes, and." When someone offers an idea, say "yes and..." and add to the idea. This can become outlandish and that is okay—remember you are trying to generate a rich set of ideas. For example, if someone were to say "Let's paint green dots all over the ceiling," you might respond with "yes, and then rainbows on the walls." This technique is meant to not only keep the flow of ideas going but also to build positive affirmation for wild ideas. The opposite of "yes and..." is "yes but," which generally shuts down an idea. Watch the language you use during brainstorming to avoid shutting down the flow of ideas.

5.5 Post-Processing and Confluence of Ideas

The free flow of a good divergent session often results in some disorganization. Confluence is a step that brings back some organization. A good confluent session has two related purposes. First, the word confluence is used to indicate a merger of flows, often between two rivers. In this merger no water is lost, only mixed. In the context of ideation, confluence is about the merging of ideas. It is not about narrowing, but rather stepping back to see patterns that can flow into your next round of divergent thinking. Second is to deconstruct the divergent session itself so that you can make future brainstorming sessions more effective. You should set aside time after each divergent session to engage in confluence and capture your results in your DHF.

5.5.1 Classifying and Grouping Ideas

The results of a divergent session can generally be sorted quickly into categories. The goal of sorting is to help you communicate your ideas to others in a logical way and to help you iterate upon those ideas in a systematic manner. Most often some organizing attribute is used, such as solution type, the user need served, or the value to be gained. You may also rely on gut instinct about the feasibility of an idea. A simple classification is often best, such as "Promising," "Iterate Upon," and "Consider Later." Note that these categories do not imply that any ideas are eliminated; you can always move an idea from one category to another. Before making your classification, it is often helpful to discuss each idea first. Breakout Box 5.7 contains a simple framework to help you process the ideas generated during a divergent session.

5.5.2 Deconstruction of the Divergent Session

A few minutes of deconstructing a brainstorming session can help your team become better at brainstorming. Some questions to consider are: What did you like about the session? What do you wish would happen differently during the next session? What was missing? How long did the group stay focused? What did you learn about your team dynamics? Was the length of time appropriate? Should you try a different technique next time? Are you ready for another divergent session? A simple framework for post-processing is introduced in the Breakout Box 5.8.

Breakout Box 5.7 I Like, I Wish

Post-processing the ideas generated during a brainstorming session is an important part of sorting ideas, comparing and contrasting ideas, and getting ready for another potential round of brainstorming. A simple way to engage in post-processing is with the framework of "I Like," and "I Wish," first introduced in Section 2.4 in the context of peer evaluations. For each idea you have generated, have everyone say one or two things they like about it and one thing that they wish for. During this process you may also notice general themes of what you like or wish for in a solution. After discussing what you like and wish for each idea, sort the ideas into categories. Remember to record your discussion as documentation of your process.

Breakout Box 5.8 Plus/Delta

A plus/delta, shown in Table 5.1, can be used at any time during a design process. The Plus is a list of all of the actions or behaviors to continue in the future. Making this list can recognize and reward group members for good work. The Delta is a list of actions for the group to work on and may be targeted at individual members or at the entire group. Such a simple framework could be included at the bottom of the meeting minutes.

Table 5.1 Plus/Delta Method of Finding Areas for Continuous Improvement

Plus (+)	Delta (Δ)
Everyone was engaged	We want on a tangent about the Simpsons
Sally did a great job of keeping us moving from one idea to the next	It probably went on too long. Maybe break into two sessions next time
This was at a good time during the day	This technique worked, but is not the best for a next divergent round
We generated 15 good concepts in 30 minutes	Our confluence after brainstorm took too long

---

5.6 Concept Iteration

A robust divergent process is iterative, using the results of previous sessions to spark ideas in a next round of divergent thinking. In this section we introduce two methods that are helpful in expanding upon existing concepts and discovering novel concepts that may have been missed during previous divergent sessions.

5.6.1 Vertical Thinking

Vertical thinking, also known as piggybacking, is a way to expand upon a concept. For example, several concepts were generated for gaining better access to a surgical site in Figure 5.4. However, each concept could form the center of its own map and be further explored. The analogy to Figure 5.1 is to move a concept that is not within the solution space into a form that could solve the problem. It is often the case that a concept that did not initially seem promising later becomes the most interesting and novel solution. Osborn's checklist for exploring concepts may be helpful in thinking vertically:

Adapt How can this idea be used as is? To what other uses it can be adapted?

Modify Change the meaning, material, color, shape, odor?

Magnify Add a new ingredient? Make it longer, thicker, stronger, higher?

Minify Split up? Take something out? Make it lower or shorter?

Substitute Who else, where else, or what else? Another ingredient, material, or approach?

Rearrange Interchange parts? Other patterns or layouts? Transpose the cause and effect? Change positives to negatives? Reverse roles? Turn it backward or upside down?

Combine Combine parts, units, ideas? Blend? Compromise? Combine from different categories?

5.6.2 Lateral Thinking

Lateral thinking is a set of techniques meant to help you move your thinking "out of the box" to find new concepts that may have been missed during previous divergent sessions. One technique particularly helpful in design is known as Other People's Perspective. The goal is to consider a problem from the perspective of others and imagine what solutions they would want. One example would be for toy designers to crawl around on the floor to view the world from the perspective of a child. The Delphi method is another systematic way to gather feedback from others. You may return to these methods or try the method in Breakout Box 5.9.

5.7 Generating Design Solutions

In this section we begin to explore how to transform abstract concepts into design solutions. It is natural during concept generation to have had concrete ideas in your mind that you then abstracted. For example, you may have envisioned the exact mesh net for gaining better access to a surgical site. This solution concept would be like placing a fishing net around the organ to move it out of the way. During concept generation, however, you simply recorded "mesh net" (see Figure 5.4). We suggested focusing on solution concepts because while your initial concrete idea may not be exactly what you need, the concept may still be a great start. Furthermore, an abstract concept can be used to generate many concrete alternatives (design solutions). The transition from concept to design is driven forward by answering questions about materials, dimensions, mechanisms, power sources, usage, and other design attributes. This section introduces techniques that are particularly effective at moving your ideas from design concepts to design solutions. Sections 5.8–5.10 then aim to help you refine your designs by considering solutions that were developed by others.

5.7.1 Functional Mapping and Blackboxing

A common technique for generating solutions is to use a function map, sometimes also called a function tree. To create the map, list the highest-level need first and then list supporting functions or needs.

Breakout Box 5.9 Getting Out of the Building

Immersion in the technical details of a project can serve you well, but may also cause you to lose sight of the user/customer perspective. A helpful solution generation technique is to mimic as closely as possible what it would be like to live and work with the problem. For example, if you are designing a device for someone without sight, consider walking outside blindfolded with a partner to guide you, as shown in Figure 5.6. Pick a location and a task to achieve once you reach that location. For example, you may choose a student center as the location and ordering coffee as your task. Do not take off the blindfold until you have completed the task. Another example is to borrow a wheelchair and use it in the real world to experience the problems that people who use wheelchairs encounter every day. After completing these types of exercises, be sure to record what challenged you and how you adapted. These adaptations can serve as the basis for new solution concepts.

FIGURE 5.6

Experiencing what it is like to navigate campus while blindfolded.

Such a rudimentary functional map for a pacemaker is shown in Figure 5.7. This as an extension of concept mapping in which the concepts are hierarchically ordered to show dependencies.

Creating such a map for each of your design concepts also shows you any feature overlaps. These are likely necessary components of any solution to the problem you defined in your project statement and specifications. On the other hand, there may be some features that are unique to a particular solution concept. These maps can then be used to spark discussion, compare and contrast various concepts, and serve as a visual aid in communicating your ideas to others. The features or functions listed at the ends of the tree will require design decisions and are generally associated with specifications. For example, in Figure 5.7 "long battery life" would require selecting a battery that could achieve this function within specifications.

A related functional mapping technique known as the black box technique is shown in Figure 5.8. Some inputs and outputs are listed, which may be materials, actions from users, power or many other quantities that are needed to achieve a function that is inside the box. How that transformation takes

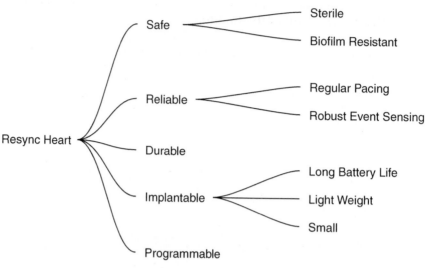

FIGURE 5.7

Function tree for resynchronizing the heart including the clinical needs of being safe, reliable, durable, implantable, and programmable. The tree is only partially completed in that the end of each branch should have an associated specification.

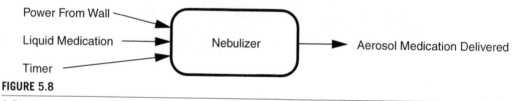

FIGURE 5.8

A Black Box for a nebulizer.

place may not be known, but it can serve as a starting point for identifying functions of a solution concept. For example, a nebulizer requires a source of power that could be electrical, mechanical, nuclear, geothermal, chemical, or some combination. Most nebulizers use an electric motor (a combination of electrical and mechanical power).

The black box approach can also be hierarchical (boxes within boxes). In one glance, such a diagram communicates which functions must be implemented, as well as the various flows of materials, power, and reactions.

5.7.2 Collaborative Sketch (C-Sketch)

The process of creating a simple sketch can transform a concept into a design solution. Collaborative Sketch, sometimes called C-Sketch, is a brainstorming technique that builds off the 6-3-5 method described in Breakout Box 5.3. To run a C-Sketch session, each member of your group should have

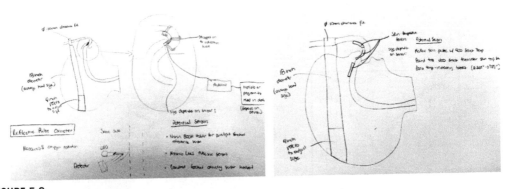

FIGURE 5.9

Examples of drawings generated during a Collaborative Sketch session.

something to write with and a single sheet of blank paper. Each person privately spends approximately two minutes (this should be timed) sketching a possible solution, but not adding any clarifying words or dimensions. Without talking, the papers are then rotated among the group members. Each member has an additional minute to add their own ideas or details that clarify the solution, again not adding any words. If there are three or more team members, it is helpful to continue switching papers until everyone has seen each sketch. Afterward, a single drawing is placed in the center of the group for discussion, during which labels and clarifying words may be added. Every drawing should be discussed, but it should not take more than 2 to 3 minutes to discuss each drawing. In a 10- to 15-minute session, a group of three or four people can quickly generate several design solutions. Figure 5.9 shows examples of C-Sketch drawings. Note that the diagrams were created step-wise and that the text was added during the team discussion.

5.7.3 Theory of Inventive Problem Solving (TRIZ)

Genrich Alshuller was a Russian patent officer throughout the 1940–1970s. After reviewing thousands of patents, he found that there seemed to be a number of patterns to invention. The most basic pattern is that inventors often face a problem that seems specific to their discipline but has in fact been encountered and solved by another discipline. For example, a civil engineer may need to inspect miles of underground sewers to determine where repairs are needed. To solve the problem, however, the engineer could generalize the problem to be about navigating and assessing any kind of pipe-like structure. The solution is then solved more generally before being translated back to a specific discipline. Medical device invention can follow a similar pathway. For example, the problem of seeing inside a sewer system is analogous to imaging the GI tract, as in Figure 5.10. A general solution is a camera that can be advanced and navigated through sewer pipes or tubular structures in the body, respectively. Other historical examples come from the invention of computer punch cards, which were based on loom weaving patterns.

The flow of this ideation method, known as the Theory of Inventive Problem Solving (or TRIZ as translated from Russian), is summarized in Figure 5.11. The essential idea is to find analogies to your problem and then look for solutions in other fields without concern for the scale, scope, or materials

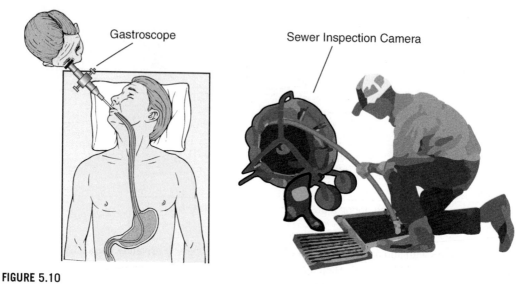

FIGURE 5.10

Similarity in solution concept between an endoscope and sewer imaging system.

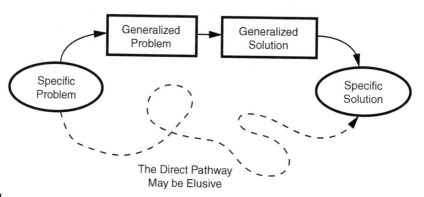

FIGURE 5.11

A general theory of problem solving showing two possible pathways to solution generation. The direct pathway moves from the specific problem to a specific solution. The theory of inventive problem solving explores a second pathway. The problem is generalized and solved in this more general domain before being translated back to the specific problem.

used. Two related ideas are BioTRIZ and Biomimicry, which look toward solutions that nature has evolved. For example, bandages often use an adhesive that is based on methacrylate, a vinal resin. An innovative solution could be based on the nanoscale bonds that are formed between a Gecko's feet and wall surfaces (as in Breakout Box 3.2), enabling this small lizard to stick to surfaces.

A second key aspect of Alshuller's theory is that invention usually occurs when one or more contradictions are overcome. For example, a solution concept may aim to simultaneously reduce loss of energy while also increasing the intensity of a force. Alshuller categorized several of these types of general design contradictions and how various inventions overcame them. In the example of loss of energy and intensity of a force, TRIZ would recommend investigating phase transitions and strong oxidants as ways to overcome the contradiction. A similar contradiction can be found in the functional map for a pacemaker in Figure 5.7; small size and powering the pacemaker for years are at odds with one another and a good design finds a compromise. A number of companies use TRIZ as a way to identify and overcome contradictions in design constraints. Visit the TRIZ website listed in the references for a complete list of the contradictions and possible solutions.

5.7.4 Switching the Laws of Nature

One definition of design is the process by which the laws of nature are exploited to achieve a purpose. For example, mechanics is the application of Newton's Laws while electronics is the application of Maxwell's Equations and quantum effects. What makes engineering a creative endeavor is how these practical applications can be mixed and matched. For example, consider a sub-function of a design that must move fluid from one place to another. Some ways this function could be achieved are:

Acoustic Cavitation
Acoustic Vibrations
Archimedes' Principle
Bernoulli's Theorem
Boiling
Brush Constructions
Capillary Condensation
Capillary Evaporation
Capillary Pressure
Coanda Effect
Condensation
Coulomb's Law
Deformation
Electrocapillary Effect
Electroosmosis
Electrophoresis
Electrostatic Induction
Ellipse
Evaporation
Ferromagnetism
Forced Oscillations
Funnel Effect

Gravity
Inertia
Ionic Exchange
Jet Flow
Lorentz Force
Magnetostriction
Mechanocaloric Effect
Osmosis
Pascal's Law
Resonance
Shock Wave
Spiral
Superconductivity
Superfluidity
Surface Tension
Thermal Expansion
Thermocapillary Effect
Thermomechanical Effect
Ultrasonic Capillary Effect
Ultrasonic Vibrations
Use of foam

Many of these methods can be crossed off the list for any given solution. For example, only some would work for transporting blood, perhaps due to concerns over heating, turbulence, or harmful chemicals, all of which can denature proteins or destroy cells. Once you have abstracted your functions, consider how each could be achieved through electrical, mechanical, optical, fluid, chemical, thermal,

---

Breakout Box 5.10 Getting Out of the Building (Part 2)

Getting out of the building is a great way to collect some real-world data, as discussed in Chapter 3. In the context of moving from solution concept to design solution, getting out of the building can take on a new meaning. For example, you may recognize that you need to fasten two parts to one another, but there is nothing in the design concept that defines exactly how this should be done. A variety of related design solutions could be generated by considering different fasteners. To gather ideas, a trip to the local hardware store (see Figure 5.12) can be illuminating. Wander up and down the aisles with a notepad and write down the various fastening methods. Many teams have found that this exercise, in addition to revealing many more fasteners than they could list in an academic setting, also sparked new ideas. Another helpful place to look for ideas is the toy section of a store. One previous design team was looking for the correct size connector that could provide rotational motion. They found that the shoulder joint of a Barbie doll could do exactly what they needed. Another team hacked a remote control car to design a glove that could control a motor using hand gestures.

FIGURE 5.12

Student exploring options for fasteners and connectors at a hardware store to enhance the ideation process.

---

or other means. Pay particular attention to the physical laws that are most difficult to exploit (e.g., superconductivity), as these are often where innovative approaches and ideas can be found.

One senior design team aimed to temporarily illuminate an area with blue light but had the constraint to avoid the need to plug a device into a wall outlet. As part of their brainstorming, they made a trip to a hardware store, a technique described in Breakout Box 5.10. During their field trip they discovered several ideas for generating blue light. One of those ideas was the simple chemical reaction used in emergency camping glow sticks. This is an example of translating a solution from one field to another. Another example involved the need to centrifuge cells in the developing world. This required a power source, but the team recognized that relying on electrical power (including the use of a battery) would not be feasible. Their solution involved storing mechanical energy in an elastic material in tension.

5.7.5 Biological Dissection

Most solutions to medical problems require you to consider the interface between technology and living systems. As the designer of a medical device, it can sometimes be very helpful to dissect biological tissues and organs. For example, when designing a knee implant, you may have generated several concepts but are unsure how to properly and safely distribute mechanical loads. A dissection of the knee from a cadaver or large mammal may help you not only answer questions, but it can also spark additional considerations for further refinement of your design solutions. Any dissection should be performed in a proper laboratory environment you are approved to use. More detailed guidance is presented in Chapters 9 and 10.

Obtaining cells, tissues, organs, or entire organisms should be guided by your needs and local resources. For cells and some tissues, you should consult research laboratories at your university or a nearby research lab. For animal tissues and organs, a visit to a local butcher may be helpful. If you need human tissue, organs, or organ systems, you could request access to the autopsy room of a medical school or a gross anatomy lab located on or near your campus.

Working with biological materials requires you to follow protocols that aim to protect your safety and the privacy and dignity of the donor. You may need to find special rooms for dissection, determine the proper way to dispose of biological material after your dissection, and complete training programs. Generally, the closer to humans on the evolutionary tree the more regulations there are, whether you need cells or an entire organism. Consulting Section 10.5.1 and a lab technician can be helpful in determining what protocols, approvals, and trainings are required.

To prepare for a dissection:

- Consult anatomy charts and videos to have a mind's eye view of what you will see.
- Find an appropriate space to perform the dissection. Be sure you are familiar with all lab rules and have approval to perform the dissection in this space.
- Assemble all equipment you may require including cutting instruments clamps, gauze, cotton swabs, pins, a microscope (if needed), and any measurement instruments you may require.
- Dissections should be done in a bin with sides to prevent fluid leakage to surrounding surfaces.
- Ensure that you have a way to dispose of any hazardous or biological waste.
- Draft a detailed dissection protocol. A template may be available online. The protocol should include purpose of dissection and benefit to be gained.
- Wear necessary protective gear that may include gloves, smock, and eye protection.
- At least two people should be present. Even if you anticipate one member performing the dissection, having a second person present can be helpful if something goes wrong, serves as another set of eyes, and is a good safety measure.
- Document your dissection with photos and videos and use these images to create a detailed write-up of what you learned.

5.7.6 Design Solutions and Transitional Specifications

Target specifications for your project, defined in Chapter 4, are a first step toward translating a problem into measurable goals. As design concepts are transformed into design solutions, new specifications are added that are unique to that particular design. We refer to these solution-dependent specifications as **transitional specifications** because they will be further refined, some may be eliminated, and new specifications will emerge as you move deeper into the design process. However, these transitional specifications can help you narrow your solution options (discussed in Chapter 6), document your design decisions, and communicate your ideas to others.

Consider the "mesh net" idea for gaining better access to a surgical site, shown in Figure 5.4. To translate this concept into a design, you would need to specify the minimum (to avoid necrosis) and maximum (to avoid lacerations) size of the mesh holes. These would be some of the transitional specifications for the mesh net design concept. A different set of specifications would apply should you pursue a design that uses a balloon. A second example is shown in Figure 5.7; the word "pacemaker" does not appear, as it would imply a solution. Rather, the function tree lists various functions and attributes for resynchronizing a heart. Once you begin to explore a more specific design, you would add transitional specifications regarding battery life, dimensions, materials, and other technical attributes.

5.8 Prior Art

Generating your own solutions is a pathway to understanding the relationship between your problem and possible solutions. It also exercises your ideation skills. It is now time to explore how others have solved the same or a similar problem. Doing so not only prevents you from "reinventing the wheel" but may also spark new ideas. For example, you could combine the ideas of others in a new way or notice gaps in the solution space (as in Figure 5.1) that have not yet been explored. Searching for previous solutions also prevents you from infringing upon the creative work of others.

Publicly disclosed solutions are referred to as **prior art**. Prior art may be found in product catalogs, company websites, advertisements, journal articles, conference presentations, and patents; an idea appearing in one of these sources does not mean it appears in all of them. This section briefly discusses searching for prior art in academic and clinical literature and then focuses on patents. Sections 5.9 and 5.10 discuss how to find and dissect existing products.

5.8.1 Searching Academic and Medical Literature

Chapter 3 recommends searching medical literature to better understand your problem. You should go back to these same sources to search for solutions. In addition, you may find information on specific designs in the *Journal of Medical Devices*, *PLoS One*, *Medical Devices: Evidence and Research*, *Expert Review of Medical Devices*, and *IEEE Transactions on Biomedical Engineering*. If you are working in a particular area, perhaps cardiac stents, you should identify 3 to 4 journals and 3 to 4 conferences where information on the development of novel cardiac stents may have been disclosed. A great place to start is the PubMed database (www.ncbi.nlm.nih.gov/pubmed) that specializes in biomedical and clinical literature. Other sources of information on medical devices include industry-based websites such as mddionline.com and medgadget.com. You should also ask your mentor where they publish their own work.

5.8.2 An Introduction to Intellectual Property and Patents

Intellectual Property (IP) is legal protection for creative work and includes copyrights, trademarks, trade secrets, and patents. Governments grant IP to encourage disclosure of new ideas that could benefit society at large. In exchange for disclosing their ideas, inventors and authors prevent others from using their creation without permission.

Patents are the most common way for engineers to document and disclose their ideas to others. They provide a period of time for the inventors to exclusively "practice their inventions." Drawings and the text of a patent conveys how an idea is "novel, useful, and nonobvious to one skilled in the art." "Novel" means that the invention represents something new; either a completely new idea or an incremental improvement. "Useful" implies that the invention has a helpful purpose. The requirement to be nonobvious means that an engineer or scientist with expertise in the subject matter of the patent would not consider the invention to be an obvious extension of an existing product, process, or invention. More is discussed on patents and other forms of IP in Chapters 6 and 15. The goal of the remainder of this section is to help you search for and understand patents as that may relate to your own design solutions.

It is helpful to understand three practical aspects of patents before beginning your search. First, a patent is a negative right in that it prevents others from implementing the same idea without permission. This legal protection extends for some period of time (generally in the range of 15 to 20 years for devices) depending on the type of invention and country awarding the patent. In the United States, this period is generally 20 years from the date of filing with the patent office. After this time period, the patent is considered expired but it is still part of the public domain. Furthermore, once a patent application has been published, it becomes part of the public domain whether it is granted or not. This means that a patent search will reveal not only active patents but also patent filings and expired patents. Second, there are many products for which there is either no patent or the patent has expired. Third, most patented ideas do not become products. Many start-up companies go out of business before commercializing the invention disclosed in the patent. Universities may file for a patent but not find a commercial partner willing to license the technology. To prevent their competitors from making a product that could threaten their market share, companies sometimes file a patent application with no intent of commercializing the invention.

5.8.3 Searching for Patents

Several resources and tips can help you effectively search for patents. Google Patents (patents.google.com) is a popular search engine and includes patents in many countries. Freepantentsonline.com has powerful text search capabilities (in spite of the ads) and is good when trying to find the first patent relevant to a proposed technology. It is often the case that a patent is granted in one country but not in another; it is therefore a good idea to search broadly for international patents.

Like all searches, the results are only as good as the search terms. You should try variations of the following search methods for your solution concepts. Search by:

- Solution or function (e.g., fill in "device that _____" with "cools blood" or, more generally, "cools fluid")
- Mechanism (e.g., a device that cools fluid by "evaporative cooling")
- Company (e.g., "Medtronic" shows you patents assigned to Medtronic). This can be very useful if you know the major companies manufacturing devices already on the market.
- References cited (many patents list other patents you can use to refine your search)
- Patent number (you may also find patent numbers on company websites or printed on the device or packaging)
- Inventor(s) (an expert in an area often has several related patents)

To highlight one of these methods, consider a search by assignee. For example, if working on a wheelchair project, Invacare Corporation is the global leader in the home and long-term care markets. A search on the US Patent and Trademark Office website (www.uspto.gov) with the assignee being Invacare yields over 300 wheelchair patents. Further refinements can be made by specifying more features or components. For example, limiting the search to wheelchairs that include tilt and suspension features yields 11 patents. Another approach is to search based on product type or category.

When you find a patent, focus on the first page; it generally contains an abstract and summary drawing. These are helpful in determining if the patent is relevant to your search. If you do think it to be relevant, save the file and continue on with your search. You can review the details of the patent later.

Like much of the design process, searching for patents is iterative. The first few search categories get you started. Once you have found some relevant patents, you should then look through the references for cited patent numbers and perform a more directed search. You may also generate additional search terms by reading a patent abstract. As you add new elements to your design, you may need to perform a new patent search. More on patent searches is discussed in in Section 6.3.5.

5.8.4 Documenting Your Patent Search

As in all of your research, you should always document your findings and insights. In the context of patents, this is often accomplished by downloading and including the relevant patents in your DHF. Table 5.2 is an example of a format for summarizing your findings. The last column is critical in that it is explicit about which ideas within the patent are most relevant to your current project. This table is a living document that should be updated throughout your project.

It can be helpful to see how a design has evolved over time. An example of a format for displaying a patent timeline for the pulse oximeter is shown in Figure 5.13. Such a diagram can also show how your design solutions build upon existing solutions.

5.8.5 The Components of a Patent

Most patents include the following sections, shown in Figure 5.14:

- title with inventors, assignees, dates, patent number
- references cited (domestic and foreign patents, other publications)
- abstract briefly describing the problem and solution
- background of the invention
- drawings labeled with parts and components that are referenced in the text
- detailed description
- claims

As an engineer, it is important to note is that there is a difference between the inventors (those who made inventive contributions) and the assignees (those who own the rights). In industry, it is almost always the case that an engineer is listed as an inventor, with the company as the assignee. This is also true at some universities. More is explained on the differences between inventors and assignees in Chapter 15.

Table 5.2 The First Three Entries in a Student-Created Summary of Device Patents to Secure an Intubation Tube. Create a similar table to document the patents most relevant to your project (with more entries than shown here).

Name of Patent	Patent #	Picture	Year	Notes
Tracheal Tube Positioning Device	US 5,474,063		1994	Partial mask, w/ base covering upper portion of face Holds tube out in looped shape Patient's mouth area unhindered
Resilient Nasal Intubation Tube Supporter	US 20060289001A1		2005	Designed for the sinusoidal shape of tube, anesthesia breathing circuits Supportive, flexible body Protective release strip
Multi-purpose Head-mounted Adjustable Medical Tube Holder	US 5,558,090		1996	Holder for endotracheal and nasotracheal tubes Used with different sizes/types of tubes Discourages accidental removal

5.8.6 Understanding Patent Claims

Claims are the legally enforceable part of a patent. In other words, when a patent is challenged, it is the claims that are used to make legal judgements. To help understand the claims in the patents you find, it is helpful to understand how claims are written. Breakout Box 5.11 is an educational example of how claims could be written for a scalpel. More on claims is discussed in Sections 6.3.5 and 15.3.2.

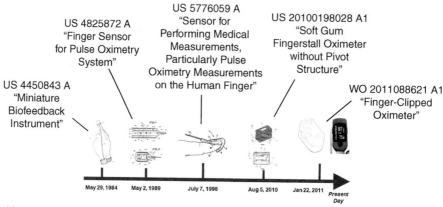

FIGURE 5.13

A student-created summary timeline of recent patents for pulse oximetry devices. There are many more patents and the diagram is only meant to show the evolution of some critical design concepts.

5.8.7 Protecting Your Intellectual Property

At this point in your design process, you do not yet know if you have generated patentable ideas; much more iteration is required to have the specificity needed to write patent claims. Furthermore, it is not uncommon for an engineer's expectations of patentability to be misguided. An idea that seems truly new and nonobvious may have already appeared in other prior art, such as a conference presentation. On the other hand, very strange devices do sometimes receive patents. One amusing example can be found in Breakout Box 5.12. More will be discussed on patentability, ownership, liability, and the application process in Sections 6.3.5 and 15.3.

As you do not yet know if any of your solutions are novel or nonobvious, you should not publicly share your ideas. Such a disclosure would disqualify the idea from being granted a patent in the future. You should not reveal it to anyone outside of your team or program without first consulting with someone at your institution. Furthermore, you may be required you to sign documents for many industry-sponsored projects that legally prevent you from sharing the company's proprietary information with others (nondisclosure documents are discussed in Sections 2.8.3 and 15.3.2).

5.9 Benchmarking

Understanding how others have approached similar problems in the past can help you quickly determine where to make your own contribution. Sections 3.8 and 4.6 discuss benchmarking which is a systematic examination of the existing solutions, typically that are on the market. The aim of benchmarking is to gain deeper insights into how the value of your design solutions compares to that of existing solutions. You may have already completed some preliminary benchmarking of existing products when scoping your problem and creating target specifications. In this section we discuss research-based techniques to find information on existing products that are similar to your solution concepts and designs. Section 5.10 discusses additional methods that can be used if you can obtain products and take them apart.

Breakout Box 5.11 Writing Patent Claims for a Scalpel

The goal of a set of claims is to describe in words the novel elements of a design solution. Imagine that a patent attorney is writing claims for a scalpel. A starting point may be:

1. A scalpel, comprising:
   a handle; and
   a blade having a point thereon

To refine this the claim:

1. A scalpel, comprising:
   a handle; and
   a blade connected to said handle

A further refinement would be:

1. A scalpel, comprising:
   a handle having a first end and a second end; and
   a blade connected to first end of said handle

But we have still lost the original idea of the blade having a point, so:

1. A scalpel, comprising:
   a handle having a first end and a second end; and a blade connected to first end of said handle, said blade having a point thereon opposite said handle

This first claim is known as an independent claim because it stands alone and does not refer to or depend upon another claim. A dependent claim does not stand alone but refers back to and may further clarify other claims. For the scalpel example, dependent claims could include:

2. The scalpel of claim 1, further comprising a grip on said handle.
3. The scalpel of claim 2 wherein the handle is made of aluminum.

   Note that this is for a simple surgical tool. Imagine how much more intricate the claims would be for a complex medical device.

Breakout Box 5.12 **Patently Absurd**

You would be amazed to see some of the ideas that have received patents. A nice collection can be found on the website totallyabsurd.com. In fact, the patent shown in Figure 5.14 is one absurd example; there are other medically related examples on the website as well. As an exercise, you can practice reading patents for these devices before you jump into more complex medical devices. The US patent office has an entire group that reviews nothing but perpetual motion machine patents. Absurd for a different reason is the number of times someone has received a patent for some very common device through the use of clever legal writing. For example, in 2001 John Keogh received an Australian patent (AU2001100012A4) for a "circular transportation facilitation device" (a.k.a. a wheel). Keogh filed the patent application to prove a point and the patent office eventually revoked the patent.

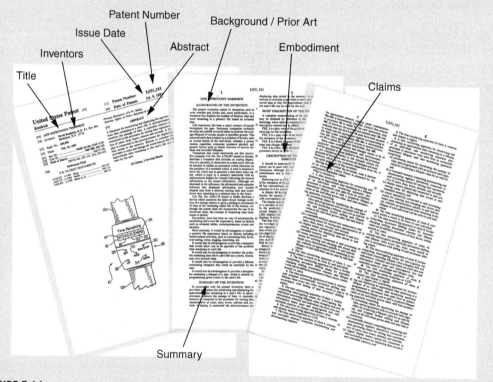

FIGURE 5.14

Selections from US Patent 5031161 for a Life Expectancy Timepiece. The left-most page is typical of the first page of a patent. The middle page is typical of the second page of a patent. The right-most page is the beginning of the numbered claims.

5.9.1 Searching for Similar Products

There are a variety of ways to find information on products that are currently on the market or are likely to be introduced soon. In addition to the advice in Section 4.3.1, you should consider whether a solution is direct-to-customer (a customer can purchase the solution) or business-to-business (the customer is another business). A direct-to-customer product generally can be found at a store or online and you may be able to buy the product. Bandages, breast pumps, digital thermometers, and shoe inserts are all examples of over-the-counter products you could purchase. Note that some medical products require a prescription from a physician, yet may still be found online.

A business-to-business solution may be more difficult to find. These solutions are not sold to individuals but are purchased, often in bulk, by another company. This company is often a hospital or healthcare facility that may need to be licensed to purchase the product. Magnetic resonance imaging (MRI) devices, surgical tools, and pre-filled syringes are all examples. You may still find these products online but not be able to purchase them directly. If you have access to a clinic, the purchasing agent at the hospital may be a great resource for obtaining literature and perhaps samples of these types of devices.

Online sources of information include:

- company websites—find companies that work on the same or similar problems and explore their websites
- marketing databases (e.g., J Walter Thompson, Forrester Marketing)
- Medical Device + Diagnostic Industry (MDDI) (www.mddionline.com)
- US Food and Drug Administration (FDA) (www.fda.gov/MedicalDevices)

It is often the case that patents, journal articles, and these websites can be used synergistically in an iterative fashion. For example, you may find a patent that is issued to a company and then search the company website to determine if the idea was commercialized. If so, the company was likely required to submit information to regulatory agencies that can be found on public websites. From this information, you can gather even more information on specifications, market size, conference or journal publications, clinical trials, and so on.

While performing these searches, you may find that some of your solution concepts have already been disclosed or commercialized in one form or another. It is always better to discover these solutions early rather than late in the design process.

5.9.2 Comparing Solutions

Understanding how your designs compare to the existing solutions is a first important step toward uncovering opportunities to add value. Uniqueness is not necessarily the goal; developing a similar solution superior in one particular dimension or function may be enough to add value. For example, changing a material could enhance one or more specifications. Consult Table 3.4 for the various ways in which value may be increased. In the remainder of this section, we introduce two visual tools for comparing your possible solutions to existing solutions. These tools can also help you communicate your analysis to others.

5.9.2.1 Perception Maps

A common method for visually comparing solutions is to create a *perception map*. An example is shown in Figure 5.15.

To create a perception map, start with a list of the available products. It can help to place them on real or electronic sticky notes. Then develop two axes that you feel represent the distinguishing attributes of the solutions. Often these include cost and some other attribute such as ease-of-use, flexibility,

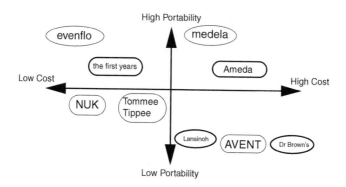

FIGURE 5.15

A student-created perception map for nine commercially available breast pumps organized by cost and porta-bility. The variation in fonts represent the branding of the company that manufactures these breast pumps.

portability, safety, fidelity, or robustness. Create these axes (perhaps on a white board) and then discuss where each product should be placed.

The value of building a perception map is the discussion that takes place during the creation of the map as well as when you show it to others. You can annotate your diagram to reflect your thoughts and uncertainties. For example, you could put a question mark next to a solution if you are unsure of its position, prompting further research. Likewise, your market research may reveal that a company is attempting to move a product in a particular direction (e.g., more cost effective, more portable), which you could indicate with an arrow on the perception map. You may discover that the same company markets several related products with the intention of spanning across the perception map and therefore reaching a wider range of customers.

In some product domains, areas of the perception map may be unoccupied. These are known as **white spaces**. Sometimes this space is undesirable. In other cases, however, it reveals a new market opportunity. For example, in Figure 5.15 you may question the value of the three devices that have low portability and high cost. After more research you may discover that there are other features of these devices (e.g., double outlet, more comfort, durability) that provide value. However, the white space above these devices (i.e., greater portability) is not occupied by an existing product. This white space may indicate a market opportunity for a new product with similar functions and specifications and increased portability.

5.9.2.2 SWOT Analysis

A second visualization that is a SWOT analysis, shown in Figure 5.16. SWOT stands for Strengths, Weaknesses, Opportunities, and Threats. Internal strengths and weaknesses are under your control, while external opportunities and threats are inherited from outside competition and constraints. You can use such a chart to compare your design to existing solutions at almost any phase of your design pro-cess. You can engage in a SWOT analysis of your designs as well as existing solutions. Creating a SWOT chart for each idea should prompt conversation that leads to continued refinement of your solu-tion ideas. It also is a way to clearly communicate the value you aim to realize and the barriers your solution may encounter.

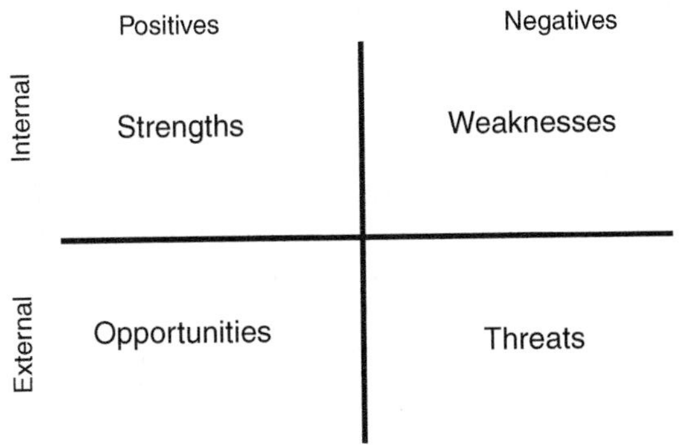

FIGURE 5.16

A blank SWOT chart that explores the internal and external positive and negative attributes of a concept or solution.

Once created, you should annotate a SWOT diagram with your perspective. For example, a solution requiring welding would be an example of a weakness if no one on your team can currently weld. On the other hand, learning to weld may also be listed as an opportunity to learn a new skill. It is not uncommon to list an item in more than one quadrant. One example of a threat would be a company announcing plans to implement a similar solution.

5.10 Dissection and Reverse Engineering

Reading about a product online can provide a great deal of information; however, for an engineer, there is nothing like being able to interact with a device. It is even better if you can take the device apart. This process is sometimes called **reverse engineering** or, more narrowly, **mechanical dissection.** In this section we explore mechanical dissection as a means to learning more about existing products. Similar processes with slight modifications can be used to determine the structure-function relationships in electrical, fluid, thermal, imaging, software, and other systems.

5.10.1 Mechanical Dissection

Mechanical dissection is the process of disassembling a device to understand how structure and function are related. The goal is to understand why particular design decisions were made. Why was a leaf spring used and not a torsional spring? Why was that particular battery chosen? Why is there a large open space between the circuitry and the casing? Why was this material chosen over some other material? Why was this part placed next to this other part? In asking these questions you are attempting to retrace the thought process of the designers and gain insights into constraints, specifications, and decisions that may impact your own project. This section will focus on mechanical dissection. Similar processes with slight modifications can be used to determine the structure-function relationships in electrical, fluid, thermal, imaging, software, and other systems.

Table 5.3 Observations and Hypotheses. These are the first steps in design team's dissection of a digital ear thermometer.

Observation Steps	Hypotheses Development
A. List the needs of the customers who use this device and hypothesize the engineering specifications.	**Customer needs:** To test temperature: needs to be accurate, comfortable, easy to use, clean, cheap, and easy to store
	Engineering specifications: (e.g., performance, safety, economics, ergonomics, reliability): Calibration procedure, warm up with resistor after being turned on for better accuracy, batteries (wireless), anthropometry (comfortable to hold, fit in hand, fits in ear, good button placement). Safety (doesn't go too far into ear), durable (last for a long time)
B. List the sequence of operations to make it work.	Calibration step (factory), remove from case, turn on using power button, attachable lens filter on ear sensor tip, fit tip with lens filter in ear, press start button, wait until the thermometer beeps, remove from ear, read temperature from display, dispose of lens filter
C. Determine the overall function of the device.	Read the user's temperature via the ear canal, display the digital readout for 5 seconds, provide simple animations on the digital readout to guide the user through operation of the device
D. Determine the required inputs and outputs to the system.	**Inputs:** User pushing the power button to activate the thermometer, mechanical input from lens filter being on, user input of pressing measurement button, temperature within user's ear, user pressing the lever to eject the lens
	Outputs: Display of the user's temperature for 5 seconds, simple animations on the digital readout to guide the user through operation of the device
E. Based upon your observations, hypothesize the internal components that make the device work.	Heat sensor that outputs voltage/current converted from analog to digital via circuit. A digital readout for information display. Circuitry powered by batteries. Resistors to warm tip prior to the measurement. Hinge to release lens filter. Infrared sensor

Dissection is often used in industry—sometimes called a "tear down"—as a form of competitive analysis. The results of a tear down assist in competitive benchmarking, as described in Chapter 4. It is not uncommon for a company to dissect competitor's devices to gain design insights, including how a competitor may have achieved a particular function in a different way. Tables 5.3 through 5.6 guide you in a step-by-step process for tearing down a competitor's device. Text in bold indicates the dissection steps while unbolded text is an example from one design team's dissection of an ear thermometer. You may need to modify the steps and questions to better suit the product being dissected.

5.10.2 Obtaining Products

If a direct-to-customer solution concept has been commercialized, you may be able to obtain the product. If you have an industry-sponsored project, the sponsor may be able to provide you with their

Table 5.4 Part List and Dissection Steps. This is the second step in a design team's dissection of a digital thermometer. Only 4 of the 19 disassembly steps and 3 of the 22-part summaries are shown.

Disassemble the device. Keep track of the steps you take so that you can successfully reassemble the device. Take photos during the dissection and document the steps below:
1. Removed battery cover, took out batteries, removed large screw from within the battery chamber
2. Pried off housing (fit with snap fits) using multiple screw drivers: saw that the thermometer is programmed with one large PC board

| |
| 3. Lifted PC board with connected ear tip unit from back housing |

| |
| 4. Removed buttons from the front housing unit (slid out easily) |

15 Additional Steps Were Listed

Continued

Table 5.4 Part List and Dissection Steps. This is the second step in a design team's dissection of a digital thermometer. Only four of the nineteen disassembly steps and three of the twenty-two-part summaries are shown.—cont'd

Dissection Summary

Part #	Part Name	Connected to Parts #	Function	Materials Used in Part	Photo
1	Plastic Front Cover	2, 3, 4, 5, 10, 11, 12	Protective cover of internal components, comfortable surface for user to hold, keeps buttons, screen, and PC board in place	Acrylonitrile butadiene styrene (ABS)	
2	Power Button	1, 14	Pushed by user to turn on the device by engaging with the PC board	ABS	
3	Measurement Button	1,14	Pushed by user to activate measurement by engaging with the PC board	Plastic (C5)	

19 Other Parts Were Listed

Table 5.5 Details for Printed Circuit Board and Display Component. This is the third step in a design team's dissection of a digital thermometer. Only one part is shown, but this same table would be completed for each part.

What decisions were made in the design of this component/module?	Components are laid out to conserve space without crossing connections, and the buttons are positioned to interact with the button depressor.
What are the critical features and dimensions? (It may help to annotate the picture.)	The PC board has to be the right size to fit into the housing, the screen has to be positioned to be seen through the display cover, and the depressed button has to be located at the right spot.
With what other components does this component interact and how?	There are many circuit parts that take care of all of the logic. It is necessary to sense the temperature, convert it to a voltage, communicate that to the display, and convert that voltage to a number that is displayed on the screen.
What measures can we use to evaluate performance?	Circuit board size should be minimized. Simplified circuitry could ease manufacturing and bring down cost. Increased durability of the button depressor would be optimal. An attractive display would make this easier to market. Inexpensive PCB and LCD display would be ideal.

product and perhaps those of their competitors. If you are engaged in a clinically-based project, contact your clinical mentor, supply chain manager, or a purchasing agent for sample products. Another option is to contact the company directly and explain that you are engaged in an academic design project. Some companies will be very generous if they know that your request is for educational purposes. Remember to identify yourselves as students.

5.11 Documenting Concepts and Designs

Generating ideas is one of the least straightforward unstructured parts of a design process. As discussed in Section 5.2.3, that is why it is so important to document all of your research, solution concepts, design rationale, and design solutions in your DHF. This chapter introduced you to several

Table 5.6 A Final Step in Mechanical Dissection. This involves comparing your earlier predictions to your original hypotheses from Table 5.3.

Now that you have dissected and studied your device, compare your earlier predictions to what you have learned through the dissection process.

The group was surprised by how few components are inside of the device. All of the necessary circuitry is located on one PCB. Additionally, most of the space inside of the device is composed of air. However, even though making the device smaller would reduce material and shipping costs, it would also cause the device to be less comfortable to hold. As a result, the group understands the company's decision to make so much of the internal compartment of the device empty space. This also agrees with our hypothesis on the ergonomic design of the device.

Our team made predictions on how the logic in the circuitry works, but we gleaned no new information on this topic through our dissection. The team was not surprised by the presence of a PCB, but we are still not entirely sure how the PCB functions. We searched the names of the chips on the PCB but were unsuccessful in finding their function. However, by looking at how the parts (e.g., the power button, the measurement button, the button depressor, and the ear sensor tip) interact with the PCB, we could gain a general understanding of how the board works. Furthermore, since we could not completely remove the heat sensor from the ear tip without destroying its function, we were not able to fully understand how the sensor warms up or takes measurements. However, we still believe that heat changing the resistance in a wire may be the main principle of how it works.

We did successfully find that there is a hinge and lever mechanism present to eject the lens filters after use, and that the PCB was powered by batteries.

graphical representation methods to help you ideate. These same graphical methods are also effective means of communicating your ideas to others. The range of your ideas can be demonstrated through concept maps (as in Figure 5.4), functional maps (as in Figure 5.7), and sketches (as in Figure 5.9). Your research can be summarized graphically as well. Although Table 5.2 and Figure 5.13 are specific to patent summaries, a similar format could be used to show the evolution of your ideas or research on existing products. Furthermore, the comparison between existing products and your own ideas (as in Figures 5.15 and 5.16) and any insights gained from the mechanical dissection of products (as in Tables 5.3–5.5) can help you clearly differentiate the goals of your project and value you hope to add. By mixing and matching graphical formats, you can simultaneously show the rigor of your process while also focusing attention on designs that you feel are the most promising.

Key Points

- Ideation is an iterative process involving both converging and diverging phases. Many methods of ideation exist. Different methods may work better for particular teams, projects, and different phases of the design process.
- A solution concept is an idea for how to solve a problem. It can serve as a framework for generating many design solutions.
- Design solutions are derived from a solution concept by answering technical questions that specify more clearly how the solution works. These answers are captured in specifications.
- Patents are used to describe and protect novel devices, technologies, and processes. Searching for patents related to your design solutions can help spark ideas and ensure that you do not "reinvent the wheel."

- Benchmarking is a method for comparing the features, attributes, and functional characteristics of existing solutions to your design solutions.
- Mechanical dissection, also called reverse engineering, of current solutions can reveal the decisions made by previous designers.
- The generation of solution concepts and the design solutions should be documented along with your decision-making processes.

Exercises to Help Advance Your Design Project

1. List out the constraints you feel have been inherited from your problem as defined in Chapters 3 and 4. In a separate list, record the constraints inherited from the nature of your design experience (e.g., course structure, mentor requirements).
2. Create a concept map for your problem, using either a whiteboard or stickies. Shortly afterward perform a Plus/Delta or "I like, I wish" analysis of your session.
3. As your team develops its own brainstorming rules based upon ideas from this chapter, document them and post them in your design space (if you have one).
4. Starting with one of your solution concepts, create a functional map. Use this abstract functional map as the input to a brainstorming session that discusses general methods for achieving each function.
5. After developing several possible solution concepts, take a trip to a local store (e.g., hardware, toy store, department store) with a smartphone or notepad. Record what you find as well as your thoughts regarding components that could be helpful in your design. Reserve some time after you are back from your trip to process any insights you may have gained.
6. Perform a patent search on a medical product (similar to what you intend to develop) that yields, at most, 20 patents from a range of domestic and international patent offices. Which 3 to 5 do you feel are most relevant to your project? Why?
7. Create a first draft of a perception map for devices that are similar to those that you are developing. How might you move these ideas into a desirable white space, if there is a white space? Record insights you had while creating this map.
8. If you have obtained an existing solution and can take it apart, use Tables 5.3–5.5 to perform a dissection.

References and Resources

There are many great resources that have influenced the writing of this chapter. Some were already mentioned in the body of the text. Others helpful resources are:

Berger, W. (2010). *CAD monkeys, dinosaur babies, and T-shaped people: Inside the world of design thinking and how it can spark creativity and innovation*. Penguin Books.

Bhamla, M. S., Benson, B., Chai, C., Katsikis, G., Johri, A., & Prakash, M. (2017). The do-it-yourself centrifuge. *Nature Biomedical Engineering, 1*(1), 0026.

De Bono, E. (2010). *Lateral thinking: A textbook of creativity*. Penguin UK.

De Bono, E. (2015). *Serious creativity: How to be creative under pressure and turn ideas into action.* Random House.

Design Heuristics. *77 Cards. (n.d.).* Designheuristics.com. Accessed October 14, 2021, from http://www.designheuristics.com.

Gray, D., Brown, S., & Macanufo, J. (2010). *Gamestorming: A playbook for innovators, rulebreakers, and change-makers.* O'Reilly Media.

Kelley, T., & Kelley, D. (2013). *Creative confidence: Unleashing the creative potential within us all.* Currency.

TRIZ Matrix / 40 principles / TRIZ contradictions table (n.d.). Triz40.com. Accessed October 14, 2021, http://triz40.com.

Michalko, M. (2010). *Thinkertoys: A handbook of creative-thinking techniques.* Ten Speed Press.

Sawyer, K. (2017). *Group genius: The creative power of collaboration.* Basic Books.

Scott Fogler, H., & LeBlanc, S. E. (2011). *Strategies for creative problem solving* (2nd ed.). Prentice Hall.

Seelig, T. (2012). *inGenius: A crash course on creativity.* Hay House, Inc.

Seelig, T. (2015). *Insight out: Get ideas out of your head and into the world.* HarperOne.

Silverstein, D., Samuel, P., & DeCarlo, N. (2013). *The innovator's toolkit: 50+ techniques for predictable and sustainable organic growth.* John Wiley & Sons.

Von Oech, R., & Willett, G. (1990). *A whack on the side of the head: How you can be more creative.* Warner Books.

Selecting a Solution Concept

Excellence is never an accident. It is always the result of high intention, sincere effort, and intelligent execution; it represents the wise choice of many alternatives.
— Anonymous

Choice, not chance, determines your destiny.
— Jean Nidetch, founder, Weight Watchers International

Chapter outline

6.1 Introduction

At the end of Chapter 5, we were left with several broad solution concepts and a few ideas for more detailed designs, **design solutions**. Going from a divergent idea generation phase to a convergent selection phase in the design process is a difficult task as suggested by the quotes above as you are eliminating ideas that you worked hard to produce. Discarded ideas, however, are not squandered as they helped you better understand the problem and what an acceptable solution might look like. In addition, those solution concepts can serve as a backup in case your **chosen design solution** turns out later to be inadequate. Furthermore, as you continue to learn more about your solution through prototyping and detailed design, you may recombine ideas from several solution concepts to generate a new **design solution.**

It can be tempting to rush through this phase and simply select a solution concept that you intuitively feel is best. As noted long ago by Aristotle about destiny, however, decisions should be made with "high intention, sincere effort, and intelligent execution." The main reason for putting effort into selection is that, especially in biomedical engineering design, the vast majority of development cost (both time and resources) is incurred *after* a **design solution** is selected. Furthermore, making good decisions, knowing why you made them, and documenting them are beneficial for several reasons. These include bringing a degree of clarity and confidence with moving forward, communicating your design selection and design rationale to others, and reducing project risks. Concerning reducing risks, the narrowing to one solution can be viewed as a selection of the **design solution** that presents the least risk (or highest probability of success) with moving forward. In addition, doing so eases the regulatory pathway, assists with regularity audits, helps ensure compliance with standards, and may reduce liability. In this chapter, you will be guided through qualitative and quantitative techniques for selecting one **design solution** among the solution concepts that you will move forward to prototyping (Chapter 7), detailed design (Chapter 8), and testing (Chapters 9 and 10).

6.2 Initial Screening and Evaluation

In this chapter, we first explore some initial screening techniques that can be used to readily eliminate solution concepts that are infeasible technically or otherwise problematic. The remaining solutions are then evaluated using progressively more restrictive qualitative and quantitative screening methods until only one is left.

A common challenge throughout the design process is to eliminate solution concepts you worked hard to produce. As in any decision-making process, the challenge is most often to eliminate options based upon limited information. Eliminating an option (or deciding not to pursue that option in the short term), is a significant step. In this section, we explore some formal ways to eliminate design concepts and solutions when the list of possibilities is long. Later sections cover more detailed methods that are helpful when the list is shorter.

You should keep track of those solution concepts you eliminate for three reasons. First, as a designer, it is a good habit to establish a record of all of your ideas, especially those you did not pursue, as well as your rationale for eliminating or keeping ideas. It is not uncommon that during a design review or public presentation that someone asks "why you did not consider a particular solution"; it is helpful to be familiar enough with your design decisions that a team member can confidently answer the question. Second, as design is a nonlinear process, some ideas that you did not consider may come back either in full form or as a hybrid or recombination. It is not unusual during later phases of the design process to mix and match concept solutions. Third, if the chosen concept fails testing, and the team needs to choose an alternate design, documentation of other acceptable concepts (although not first choice) provides the team with a list of previously vetted, potential alternate designs to choose from. This saves the team significant time and minimizes the impact of this testing failure on the project schedule.

6.2.1 Revisiting Previous Convergent Methods

Many of the techniques considered in previous chapters can be used to narrow the list of acceptable designs during the concept selection phase. Creating a Problem ID canvas (Section 3.11) enables the team to compare important dimensions of a problem with which to qualitatively compare some solutions. You may make a similar kind of design canvas that contains the dimensions introduced in the remainder of this section. You were also introduced to a SWOT analysis (see Figure 5.15) which may help to narrow your solutions.

6.2.2 Core Competencies

In some cases, a concept or design may be interesting but is beyond the *current* core competencies of the team. In industry, companies do not usually pursue projects outside their core competencies. For example, Merck, a leading global drug development company, is unlikely to start developing medical devices in the foreseeable future. The design team may rule out a concept because team members do not *currently* have the necessary background or access to the critical information or processes needed to develop a solution.

Remember, however, that in an academic design project you can learn a new technique such as making a printed circuit board, creating a microfluidic device, developing a scaffold for tissue regeneration, or using finite element analysis to explore the effect of variations of a concept. Team members can also get help from faculty, staff, and fellow students. In other words, be judicious in eliminating ideas because the team does not think it is capable of executing one at this stage. A team may choose to pursue a **design solution** that requires members to develop new skills and expand core competencies.

6.2.3 Better Understanding of the Problem

A better understanding of the problem itself often helps a team explore and discuss the merits and drawbacks of a design. This is a good time to revisit your project statement since you have more perspective than earlier on the project. As an example, consider a project that seeks to simplify the performance of cricothyrotomy, which is an emergency surgical procedure to create an airway through the cricothyroid membrane (CTM). It is used for a patient who is unable to breathe due to airway obstruction, trauma, or other reasons, and a traditional breathing tube cannot be used. A detailed medical procedure of the treatment is shown in Figure 6.1.

The clinical need to simplify the procedure arises from a failure rate of 33% among emergency medical personnel (EMTs, paramedics, and combat medics). The primary three reasons for the high failure rate are (1) failure to locate the CTM, (2) the inability to reach the trachea or sometimes going past it into the esophagus, and (3) the infrequent need for such a procedure. After about 5 minutes, a failed procedure results in death due to asphyxia.

Extend the neck whenever possible for better access to the trachea. Immobilize the larynx with your nondominant hand and palpate the cricothyroid membrane with your index finger.

Make a 3 to 5cm vertical midline incision through the skin and subcutaneous tissues. Palpate the membrane through the skin to confirm the anatomy.

Make a <1cm horizontal incision through the cricothyroid membrane. Note that the skin incision is vertical, but the membrane incision is horizontal.

Insert the tracheal hook in the opening of the membrane and rotate it cephalad, while grasping the inferior border of the thyroid cartilage. Ask an assistant to provide upward traction on the hook.

Place the tips of the Trousseau dilator into the opening in the membrane and spread in the longitudinal (vertical) plane.

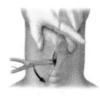

Rotate the handle 90° until the handle is vertical or parallel to the neck

Insert the tube between the blades of the dilator until the flanges rest against the skin of the neck. Keep your thumb on the obturator during tube insertion.

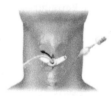

Carefully remove the Trousseau dilator and the obturator.

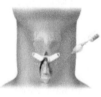

Replace the inner cannula of the tracheostomy tube and inflate the balloon.

Ventilate and confirm tube postion by auscultation and end tidal CO_2.

Secure the tube in place.

FIGURE 6.1

Surgical cricothyrotomy traditional technique.

(Adapted from Custalow, C.B., 2005. Color atlas of emergency department procedures. Philadelphia: Saunders.)

Among possible concepts to reduce the failure rate are:

- a more realistic biofidelic simulator (especially for palpation) for simulation training,
- use of ultrasound to find the CTM,
- an airway depth gage,
- "smart" surgical tools (e.g., a scalpel that can measure pressure),
- alternative surgical tool (i.e., not a traditional scalpel), and
- technology to supplant a cricothyrotomy with a modified tracheostomy.

Creating a tracheostomy, the last concept listed, involves creation of a hole in the trachea as an emergency airway and is normally ruled out quickly for emergency use because the current procedure requires anesthesia. However, further understanding of the tracheostomy procedure may reveal that it can be safely performed with local anesthesia by emergency medical personnel and can be performed using technology yet to be developed. A concept that would not need general anesthesia for tracheostomy should not be ruled out without finding out more about the procedure.

One way to evaluate concepts without detail is to perform a basic technical risk analysis. In this example, the technical risk can be assessed qualitatively without a formal analysis. Table 6.1 is one example of how such a qualitative technical risk analysis can be performed.

As can be seen, each solution concept poses different technical risk levels, which may be used later to help select or eliminate concepts. Great care should be used while exercising that judgment. For example, choosing the lowest risk concept—an airway depth gauge—does not yet answer critical questions. These include: How it would be used? How medically helpful would it be? and Where does fit in the procedure? Answers to these questions and others would be needed to make a "wise choice" among the competing concepts.

In Chapter 5, the idea of sketching was presented as a brainstorming technique. This idea can be extended here to add in the context of the problem as it appears in a real situation. For example, one student-drawn ultrasound solution for an improved cricothyrotomy procedure is shown in use in

Table 6.1 Initial, qualitative top-level technical risk analysis for competing solutions. These evaluations are preliminary and change as the selection process advances.

Concept	Technical risk
Simulator	Medium. Biofidelic simulation difficult to achieve. Human testing with EMTs using the simulator likely needed to demonstrate efficacy.
Ultrasound	High. Components include measurement system, paddle design, and indicator system. Introduces a new step in procedure.
Airway depth gage	Low. Likely fine measuring (+/-<0.1mm) needed. Need to confirm it would be useful clinically.
Smart surgical tool	Medium. Need to choose technology to use to make real-time measurement and integrate into procedural tool. Would need to test for obtaining recommended force values.
Alternative surgical tool	Probably low. Need to determine cutting mechanism. Testing will likely require animal organ or cadaver testing.
Change to tracheostomy	If a possible solution, need to determine if there is an engineering problem to solve. Find out more about the procedure.

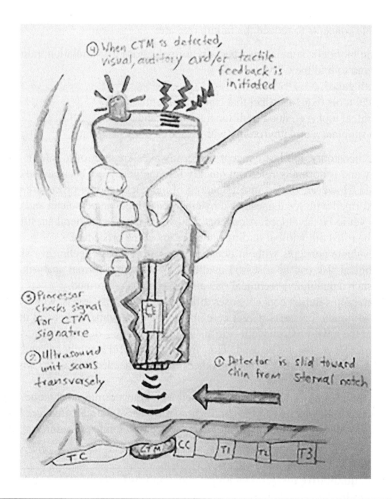

FIGURE 6.2

Hand-drawn sketch of perceived ultrasound solution in use. Verbal instructions within the image describe anticipated sequential use.

Figure 6.2. Showing how it would be used to find the CTM is good practice because it allows the team to visualize the technology in context think: What is the effect of any new technology on current medical practice? Section 8.2 presents human-technology interface design in depth.

The hand-drawn sketch of a conceptual design is not time intensive—the design team envisioned the solution, and one member drew Figure 6.2 in less than a day. The insight gained from such a sketch is useful in comparing solutions and in prompting many questions about the idea. Example questions can be divided into two categories: user-centric and technology-centric, and eight are listed below for this concept:

User-Centric

- How is the head being held so that the neck is extended?
- How much volume does the ultrasound unit occupy and how heavy is it?

- What if there is a high chest wound and the sternal notch is damaged?
- How much pressure by the user is needed to effectuate the desired signal?

Technology Centric

- How is the ultrasound powered?
- Is ultrasound gel needed? If so, when and where is it applied?
- Can the device operate in extreme weather or dusty environments or both?
- How much additional training is needed in using ultrasound for this purpose? (This can be considered user-centric as well.)

Sketching to compare solutions is also a way to iterate on and perhaps combine solution concepts. These types of questions can lead to criteria that can be quantified and compared later. In addition, your questions may provide criteria modifications that can be quantified and compared as you continue to narrow to a single **design solution**.

6.2.4 Technical Feasibility

It is sometimes possible to readily rule out some concepts because they violate a basic physical principle or are inappropriate given the context of the problem. For example, if a project involves transporting blood, some ideas for moving a fluid presented in Section 5.7.4 that included super thermal conductivity or osmosis, can be ruled out due to the medium being transported (biological properties change when having blood pass through a semi-permeable membrane or heating it).

Feasibility can be determined based upon logic and what you may know about biology or the medical situation. Another type of feasibility tool is known as a "back of the napkin" calculation, which is a hand-drawn diagram or approximate calculation readily created on whatever paper is available. Many sketches and calculations have been made during meetings at restaurants to discuss new product ideas, and napkins were the only paper items available during a dinner meeting, hence the term "back of the napkin" calculation. Consider a project to develop a device to help a nurse lift a patient who is unable to stand on her or his own. The clinical need for such a project is to reduce the incidence and severity of nursing back injuries. Nurses suffer from back injuries, much more often than other healthcare workers. This is due in part to nurses being the primary healthcare providers assisting patients to stand. The current practice is for a nurse—sometimes two—to manually lift a patient, who may weigh over 1800 N (>400 lb), usually from the patient's upper body. Using biomechanics, such lifting can induce loads in a nurse's erector spinae—the primary load-carrying group of muscles in the back—8 to 10 times the nurse's bodyweight. Over time, repeated loadings can result in fatigue-related spinal injuries, some of which may be debilitating.

One concept to assist with lifting is to develop a pair of devices to produce a torque at the patient's knees. That torque would be converted to linear motion through a linkage system. The nurse would apply one device to each of the patient's knees and turn them on. A torque would be produced at each knee in a direction to lift the body. An idealized free body diagram (from statics) of the context is shown in Figure 6.3.

One way to determine if this concept is technically feasible is to estimate how much torque would be needed by visualizing the sitting position. Estimation is discussed in detail in Breakout Box 6.1. The greatest torque about the knee is created by the patient's body weight, acting through the moment arm,

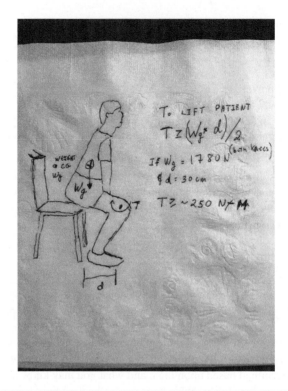

FIGURE 6.3

A hand-drawn idealized free body diagram of a person about to stand. W_g is the weight at the body's center of mass, T is the torque and the moment arm, d is the distance from the knee to the upper body's center of mass. It is assumed initially that the feet are not supporting any load and that the bodyweight is minimally supported by the chair and the upward reaction force from the chair supporting W_g is negligible.

d, which is the distance from one knee to the center of gravity of the patient. This moment arm varies with the height of the patient, but a reasonable estimate for the distance d is 30 cm (~12 in). Assuming a patient weighing 1780 N (400 lb), the total maximum torque produced is ~534 N-m (394 ft-lb). The actual torque needed at each knee is ½ that value, 267 N-m (197 ft-lb).

6.2.6 Development Time

In industry, the time from project initiation to product launch can vary from months to years, depending on the scope and type of project, required technology, and available resources. Beyond design, this includes the development of manufacturing processes, the time needed for regulatory approval, animal and clinical trials (in most cases), patent applications and approvals, marketing, distribution, and sales. The duration for each of these activities is highly variable and delays are inevitable.

On the other hand, the time to complete an academic design project is usually fixed by the academic design program. Depending upon the timeline, you may want to consider this period available to make

Breakout Box 6.1 Estimating Values

The lift example above estimated a torque of 267 N-m. How much is 267 N-m? One ability of a skilled engineer often learned through experience is to have a more intuitive feel for what this kind of torque is like. Notice that the word "like" was used, which implies that one can make an analogy to some other situation. You likely already have some intuition about lengths and can visually perceive the difference between 2 mm and 2 cm. The same idea goes for certain weights. However, in the United States, you are likely to be most comfortable with weight in pounds. For example, if a person weighs 150 lb, approximately how many newtons does s(he) weigh? Knowing that a kilogram is about 2.2 lb, it turns out that 1 lb = 4.45 N. By memorizing an approximate 4.5 conversion factor, 150 lb are equivalent to about 675 N or 68 kg-force, the most common unit used throughout Europe.

Without experience, how you can develop intuition and engineering judgment by making an unfamiliar number (e.g., 267 N-m) feel more familiar? For example, if you need to change a flat tire, you must first loosen the lug nuts (bolts) that keep the car tire rim attached. Assume a bolt was tightened with 267 N-m (197 ft-lb) of torque. Let's assume that the wrench used to unloosen this bolt is 0.3 m (~0.98 ft) long. Then 267 N-m would mean someone must exert 890 N (~200 lb) at the end of the wrench to start loosening. For most people, this is not possible by arm strength alone and is why beginning to loosen this overly tightened bolt could require standing or jumping on the end of the tire iron, or even calling for help. It is also the reason why lug nuts should be attached with less than 70 N-m (~51 ft-lb); this is in the range most drivers can apply with arm strength.

You can also develop a personal frame of reference. For example, what does a gram feel like? Any denomination US bill weighs just under a gram. But a gram is also approximately 1000 strands of hair. By comparing the unfamiliar gram with something familiar like a dollar bill we get a sense of its physical feel.

Some quantities are not as easy to perceive directly. Energy, power, and pressure can be challenging. What does a Pascal feel like? In keeping with the theme of using analogies to make the unfamiliar familiar, see Table 6.2, which provides a range of real-life examples of quantified values of many engineering attributes. Over time, you will begin to develop your own equivalences to help you intuit other measures.

The logic and calculations for this situation provide only a ballpark estimate and may be high because part of the load is supported by the feet in contact with the floor. Nonetheless, the result can be quantified and can allow some concepts to be considered to solve the problem. For example, a solution to the lift problem will likely require a motor that can generate a sizable torque, and we now have an estimate of how large that torque must be. The team can now look at motor suppliers to determine the availability and cost of a motor capable of delivering the required torque and rotational speed desired (rpm). This confirms the feasibility of using a motor to solve the problem.

To determine the approximate motor size, another "back of the napkin" calculation can be used to approximate the relation between output torque and motor size. Horsepower (hp, ~0.75 kW) is sometimes specified to rate motors. The relationship between torque and horsepower is hp =(speed*torque)/7121 where units are rpm for speed and newton-meters for torque. Assuming a slow rotational speed (5 rpm), and a torque of 400 N-m (295 ft-lb), the size of the motor would be ~0.28 kW (~0.37 hp). As an analog, this is about 1/8 the rating of a small push lawnmower. Several motor sizes can accommodate the required torque. The next steps include exploring physical size, weight, and noise level of individual motors.

Similar types of calculations can be made for chemical, optical, thermal, and electrical phenomena. Another example of a "back of the napkin" calculation is provided in Breakout Box 7.3 concerning the battery life of a pacemaker.

progress on a solution as a selection criterion. Some projects that have one or two components (e.g., cricothyrotomy tool (Section 6.2.3) or necrosis-preventing nasal insert (Table 7.4) naturally lend themselves to a clearer path to multiple design iterations. Other projects contain many subsystems (e.g., a blood glucose monitor, patient positioning, and heating system) and generally take longer to design and build a working prototype. More prototyping and testing would normally be required for a design with one or two components rather than with a design with an assembly of subcomponents including

Table 6.2 Mechanical and biomedical analogues (approximate values)

Table of	Mechanical &	Biomedical	Analogues			
Order of magnitude	Distance	Force (kilogram-force)	Pressure	Energy	Power	Volume
10^{-3}	500 stem cell diameters, ~1 mm	dollar bill or jumbo paper clip, ~1 g, #2 pencil ~5 g	pressure differential between outside and inside, 5–8 Pa (~0.04 mm Hg)	10 mJ peak left heart diastolic KE	30–60 mw typical for operating an indicator LED	<1 mm³ miniature LED volume
1	Typical stride length for 2 m person (~6'8") = 1 m	average weight newborn ~3.5 kg, liter of milk ~1 kg	pressure between hand and soda can while lifting, ~25 mm Hg (~3.3 kPa)	1 gram calorie = ~4.2 J; energy needed to light 1 W LED for 1 sec, 1 J	10 W LED = 60 W incandescent bulb	5.6 L average blood supply in and an adult body
100	football field length, 105–110 m	human hair ultimate load 100 g (~1 N) ; weight of typical football player ~100kg	normal systolic blood pressure 120 mm Hg (~16 kPa); laboring patient contraction strength 50–120 mm Hg	100 mi/h (45 m/s) fastball ~145 J KE or soccer ball ~455 J KE	100 W solar panel ~225 W hours (0.8 MJ) enough for three 10 W LEDs for 7 hours	size of many excursion backpacks 100 l
10^3	distance from Washington DC to NYC 1.25 × 10³ km	ambulance weight ~6400 kg	Mean sitting pressure on buttocks 29000 Pa (~218 mm Hg)	1 g of fat = ~3.5 kilojoules of stored energy	space heater 1.5 kW; refrigerator uses 300 hours/year	firetruck capacity ~2 kL water
10^6	25 times around the earth ~10⁶ km	3000 sqm hospital structural weight per floor >1.1M kg	ABS tensile strength 40 Mpa	heat to boil 10L H20 ~4.7 MJ; Ambulance battery ~ 5MJ;	1 MW solar energy can power >150 homes	number of cells in a 10 mm cancerous tumor ~1M cells

Adapted from Halliday, D., & Resnick, R. (1981). *Fundamentals of physics*. John Wiley & Sons. and Otto, K. N., & Krisitin, K. L. (2001). *Product design*. Prentice-Hall.

different modalities (e.g., software interfaces, electro-mechanical components). Testing is presented in detail in Chapters 9 and 10.

In rare circumstances, when the risk of an adverse outcome is negligible, a technology may be tested medically under the supervision of a healthcare provider. Examples of these include projects involving point-of-care devices that only use biological material from the body (e.g., saliva, urine, blood), computer software and smartphone apps for medical purposes, walking aids, rehabilitation equipment used by healthcare providers (e.g., stethoscope, blood pressure monitor, imaging technology). In these cases, especially if the nature of the project requires some medical or animal testing, work with your clinical advisor to develop a plan. If the plan involves living systems testing, you must consider the additional time required for obtaining necessary approvals for testing in living systems presented in Section 10.4 for animals and Section 10.5 for humans.

6.2.7 User Needs

You were introduced to many tools and techniques in Chapter 3 to understand the needs of your target users. However, your understanding of these needs often change as the design evolves. Simply asking the question, "What would a user think of this solution?" can reveal issues that may either eliminate a solution or help you iterate on your idea. A related question is, "How would this solution be used in medical practice?" Even if you cannot yet show your solution ideas to a user or someone familiar with users, you can still imagine how they would respond. The sketch in Figure 6.2 is a representation of how the team members imagined an EMT would use an ultrasound solution to find the cricothyrotomy membrane. You are encouraged to keep the perspective of your users in mind throughout the design process as their view grows in importance, concluding informal validation testing. If you can discuss sketches like Figure 6.2 with a potential user, do so as it is one of the best forms of subjective feedback on a design idea.

6.2.8 Additional External Feedback

If you are able, it is often helpful to engage key mentors, sponsors, and advisors—herein referred collectively to as experts—in reviewing your solution concepts or designs. This may take a variety of forms including interviews, informal meetings, design reviews, or inviting them to a design team meeting. You should have questions prepared ahead of time that aim to gain critical information to help you narrow your choices. You should not simply ask experts which option they would choose. This short-circuits one of the most critical learning objectives of a design process—how to make good decisions. Even if an expert expresses one design choice, you should always consider the information you receive from a single person or stakeholder as one expert opinion. That perspective is valuable because experts look at the design in a broader context. Consider the expert's design choice; however, avoid allowing it to choose for you. Teams should speak with lead users who perform many procedures each year (anesthesiologists), as well as infrequent users (such as EMTs) or trainees, who may perform a small number of procedures each year, or may be performing a procedure for the first time. If only lead users are consulted, then the team may design a product that only expert users can use effectively.

6.3 More Detailed Qualitative Screening

The previous section covered methods that can readily eliminate some options. In this section, we discuss several nontechnical requirements that every successful medical product must meet to be adopted by the healthcare system. These include economic feasibility, regulation, compliance with standards, patentability, and possible commercialization. These constraints were presented in more depth in Section 4.3.2.

6.3.1 Regulatory Pathway

The anticipated regulatory pathway can sometimes be used as a criterion for selecting a concept or design. Regulatory bodies exist to ensure that medical devices are safe and effective. Most countries have a medical regulatory body. For example, the United States has the Food and Drug Administration (FDA), Europe has the European Commission (CE), and Japan has the Pharmaceuticals and Medical Devices Agency (PDMA). Much more information is presented in Chapter 12 and can be found on the website of each regulatory agency.

In brief, the United States, the FDA has three classifications for devices:

Class I—minimal risk (e.g., bandage, tongue depressor, examination gloves)
Class II—higher risk (e.g., powered wheelchairs, infusion pumps, surgical drapes)
Class III—high risk (e.g., implants, cardiac stents, pacemakers)

The European Commission refines the classification further for Class I and II:

Class I—noninvasive low-risk
a) nonmeasuring (exam gloves, nasal cannula)
b) measuring (thermometer, eye droppers)
Class II—medium to high risk
a) implant device is for <30 days (catheters, transfusion tubes)
b) >30 days implant device (tracheostomy tubes, a Hickman port)

The lower the risk classification of your solution, the less time is typically needed to receive regulatory approval and clearance to market, which reduces product development costs. For example, to postoperatively measure blood flow continuously near a surgical area, two possible solutions include (1) implanting a nano-scale flow meter or (2) developing technology to measure blood flow externally using Doppler ultrasonography. The implant would likely be Class III in the US and Class II(a) in Europe. Using Doppler technology or its equivalent would likely be Class II in the US and Class I(b) in Europe. The level of risk also dictates the requirements for testing your solution in an animal or human model.

Reconsidering the cricothyrotomy problem presented in Section 6.3.2, we can update Table 6.1 to include commentary about regulatory risk. This is shown in Table 6.3. As more qualitative screening is considered, additional columns can be added to the table.

6.3.2 Resistance From Other Stakeholders

In Section 3.9.2, you were prompted to broadly consider the various stakeholders in problem identification. You may use Figures 3.5 and 3.6 but now view it in the context of which stakeholders consider a concept or design as a win, and which will likely resist it. It is straightforward to think narrowly about your client, who may be your medical advisor, one stakeholder. One of the challenges of a new medical product is overcoming the inertia in the wider healthcare system. For example, when introducing new technology that changes or disrupts the standard of care—what is normally performed in clinical practice in a particular situation—there will be opposition from providers. The reason for this resistance is that once a healthcare provider is trained in a certain way to perform a procedure, a change in medical practice due to new technology will likely involve retraining. This is

Table 6.3 Qualitative risk analysis that includes technical risks outlined in Table 6.1 and initial regulatory risk assessments.

Concept	Technical risk	Regulatory risk assessment
Simulator	Medium. Biofidelic simulation difficult to achieve. Human testing of using the simulator may be needed.	None. No regulatory approval is needed.
Ultrasound	High. Components include measurement system, paddle design, and indicator system. Introducing new step in procedure.	Class II—based on existing ultrasound technology
Airway depth gage	Low. Likely fine measuring (+/-<0.1mm) needed. Need to confirm it would be useful medically.	Class I—based on approved similar device
Smart surgical tool	Medium. Need to choose technology to use to make real-time measurement and integrate into procedural tool. Would need to test for obtaining recommended force values.	Class II—if used to make incision; scalpels are Class II devices
Alternative surgical tool	?? Probably low. Need to determine cutting mechanism. Testing will likely require animal organ or cadaver testing.	If not used to make incision, likely Class I. If used to incise, risk is Class II device as above.
Change to tracheostomy	??? If a possible solution, need to determine if there is an engineering problem to solve. Find out more.	Class I because tools used today are already approved

why new designs that require deviating from standard practice sometimes do not receive wide market acceptance until well after they are introduced. The cardiac stent is one example of this. Originally invented ~1981 to treat acute myocardial infarction, it was only considered a success after 1200 implants by 1989. Stenting did not become the standard of care to treat narrowed coronary arteries until the late 1990s.

6.3.3 Economic Feasibility

In the context of design, economic considerations act as a constraint in two ways: the cost to develop a medical technology and (in industry) the cost to manufacture, market, deliver, and sell it. Development cost is always a constraint for an engineering design project. In industry, projects are usually given budgets that consider both direct and indirect costs. Direct costs usually include technical salary, materials, prototypes, access to commercial databases or software, medical observations, and in some cases compensation for interviewees and focus groups (Section 3.7). Indirect costs usually include supervisory salaries, administration costs, outsourced contractors on the project team, administration, and other support such as rent, property ownership, phone, internet, lights, and heat (overhead). These are usual distributions of expenses; many variations exist. How to estimate these costs appears in Section 15.2.

In an academic project, many of these costs (e.g., salaries, overhead) do not apply. Rather, teams are typically given a modest budget and expected to make use of many shared resources. Project funds are usually used for purchasing parts, prototyping, testing, outsourcing (if needed), delivery expenses, and travel (if warranted). Some teams may not consider a design concept because the development cost becomes prohibitive. This can occur if you lack (or have limited access to) a critical and costly component (titanium surgical screws), or another process (e.g., laser sintering) that would be needed for building prototypes or testing.

Material choices are a key element of the cost both during development and in a final product. The materials used to create prototypes are rarely the same as those used for a final product; however, many specifications (and therefore verification testing) are related to the choice of material. Depending upon the nature of your remaining solution concepts and designs, you may need to consider the material choices earlier rather than later. More detail on material selection is provided in Section 8.3.

Knowing who the buyer is and who would pay for your design are factors worth considering. Two key questions are:

1. Who decides to purchase your solution?
2. Who pays for your solution?

For medical devices, the answers to these two questions are complicated and often different. The complexity of the healthcare ecosystem in the US is one reason the answers not being straightforward.

If your project is to ultimately become a product sold over the counter in a drug store (e.g., support braces, walkers, canes) and the consumer is the buyer who pays for the product directly. If a new technology is a hospital supply (e.g., OR sponges), which are expendable, or a device or system that is purchased one time, capital equipment, (e.g., an MRI machine), the hospital is the buyer, with no expectation of being directly reimbursed. Convincing a hospital purchase agent to buy a new technology is a complicated process. Before a purchasing agent can act, however, hospital committees first must evaluate the new technology (judge its value proposition) and make a recommendation to purchase it or not to purchase it. This sometimes involves more than one committee making such an evaluation.

If a medical device is to be used in treatment (e.g., a knee implant) or diagnosis (e.g., imaging equipment), the hospital or healthcare provider would be the buyer. However, a private healthcare insurance company or government insurance program (e.g., Medicare) will likely reimburse for each use in the clinic. To be reimbursable, a healthcare provider or facility must submit for each patient a disease code, ICD (*International Classification of Diseases*), and an associated procedure code, CPT (*Current Procedural Terminology*®). For example, for a novel point-of-care Covid-19 test, healthcare providers or institutions would purchase the kits and perform the diagnostic test on patients. For each test, a provider would submit an ICD code (Z03.818 to rule out an infection, or Z20.828 for a confirmed infection) and a CPT code (87635 with a verbal descriptor) to be reimbursed for the test procedure often from an insurance provider. Without both the ICD and CPT codes, insurance companies or government insurance programs will not reimburse a treatment or diagnosis, nor the associated technology. More on insurance and reimbursement is presented in Section 15.7 on healthcare reimbursement.

6.3.4 Compliance With Standards

Although standards are presented throughout the text and are presented in detail in Chapter 11, having more knowledge of them at this stage is beneficial in making design decisions. A useful way to think about a standard is that it is a documented agreement containing technical guidelines to ensure that materials, products, processes, representations, and services are fit for purpose. Standards ensure that medical products perform as required, consistently made with suitable quality, and help ensure products are used properly. Standards are critical for economic development and global trade. They are also needed when exchanging technical information between engineers, between manufacturers and suppliers, and between the government and the private sector. Standards are relevant to biomedical engineering design in that they guide most aspects of medical product development: design, testing, sterilization, manufacture, inspection, labeling, and packaging. Failure to comply with applicable standards can be a major impediment to introducing a **design solution** to the market.

At this stage in the design process, identifying which standards are most relevant to your project is useful in concept selection. For example, if working on a project related to electrocardiographic (ECG) equipment, the FDA database points to AAMI standards EC11 and AAMI 60601-2-25, which provide minimum safety and performance requirements for ECG measurement technology. These requirements help select a design, specifically, the characteristics of a new ECG technology that must comply with specifications contained in the standard. Table 2 in Chapter 11 on standards provides a database of FDA-recommended standards that apply to medical products. Figure 11.5 demonstrates how to find an FDA-recognized standard for any existing medical technology.

6.3.5 Patentability

In some projects, the potential for a patent may be a criterion for narrowing the list of potential solutions; however, this should be only one of several factors that you use to narrow your list of possible solutions. Section 5.8 presents guidance on how to search for patents and you were encouraged to create a database of relevant patents. If patentability is a goal for your project, you want to make a preliminary judgment as to which remaining ideas may be sufficiently novel to warrant a patent application. This may involve a more targeted search that goes beyond medical devices. In this section, we provide guidelines on how to enhance your patent search capabilities to determine if there are additional relevant patents and whether your remaining solution concepts may be patentable. More on patent infringement and the freedom to operate is explained in Section 15.3.

Most importantly, you should not explicitly eliminate a concept or design because you think it is previously published or patented. There are nuances to each disclosure that even though a technology appears similar, it legally is not. It is not an uncommon occurrence that what may seem to be a novel concept, is not when reviewed by a patent examiner or patent lawyer. On the other hand, there are many cases when a concept seems similar to an existing patent but contains one or more new claims that make it patentable.

When searching for ideas for technology similar to your project, you were introduced to patent searching in Section 5.8 and your team perhaps created a summary of relevant patents similar to Table 5.2 There are ways to search beyond this.

As indicated in Section 5.8, the keys to a successful patent search are the actual search terms and using a robust search engine. While Google Patents is readily accessible, the website freepantentsonline.com has powerful text search capabilities (despite the ads) and is good when trying to find the first patent relevant to the team's proposed technology. The USPTO website allows one to search in more than 35 categories—including inventor, assignee, filing date—in the advanced search section.

Searching internationally may be helpful. Two foreign patent offices are the European Patent Office (espacenet is the search engine) and the Japanese Patent Office, where the search engine is readily accessible. Although these databases all have access to the same patents, their search algorithms are different, and therefore you gain a more comprehensive search by using them together.

Another way to search is by patent class and subclass. The US Patent Office arranges similar patents by numbered classes, which number over 400. Most numbered classes have dozens of subclasses for specific areas within the technology; there are over 125,000 subclasses. The combination of text searching and searching by classification forms a complete search. For example, in the timepiece example in Figure 5.13 line (52) reads "**U.S. CL .... 368/280**, **368/108,**" which indicates class 368 (time measuring systems) and subclasses 280 (of a particular material) and 108 (with alarms). Searching on this class and subclasses yields close to 200 other patents with the same classification and subclassifications. Each one of these should be perused. Most can be eliminated because they are not for biomedical applications. Of those that remain, a review is needed.

Common biomedical engineering classes in the US are as follows:

- **Class 128**: Body supports, restraints, respiratory therapy
- **Class 434**: Education (medical simulators)
- **Class 600**: Diagnostic and biological testing, tissue sampling, endoscopy, specula
- **Class 601**: Kinesiotherapy (moving, vibrating, or massaging the body)
- **Class 602**: Splints, bandages, braces
- **Class 604**: Controlled release, needles, adding or removing from the body
- **Class 606**: Surgical instruments
- **Class 607**: Ablation, ultrasound, light, thermal or electrical
- **Class 623**: Prosthesis
- **Classes 700-707**: Data processing
- **Class D12**: Walkers, wheelchairs

In the patent shown in Figure 5.13, line 51 reads in part "**Int. CL G04B 19/24; G04F 8/00.**" This is the international classification, which can also be found on the USPTO website. The designation begins with a letter followed by a two-digit number. In the example, GO4 is Horology. The letter following the designation further refines the category. In this case, G04B are mechanically driven timepieces and G04F are time-interval measuring. The following numbers are further refinements. In the example, 19/24 includes those timepieces with a calendar; 8/00 includes those timepieces with measuring unknown intervals.

There are patent classes that are joint with the European Union that have a Cooperative Patent Classification (CPC) designation. These start with a letter followed by a two-digit number. Most medical applications are under CPC class A61, Human Necessities.

Recall from Section 5.8 that a patent is only issued if an idea is novel, useful, and nonobvious to one skilled in the art. Novelty is represented in a patent in the *claims*, which appear at the end of each patent. Each patent must have at least one independent claim; most have more, some many more. An independent claim does not depend on any other described technology; a dependent claim provides details for the independent claim.

For example, the first independent claim in the Life Expectancy Timepiece patent shown in Figure 5.13 is this:

"A time monitoring apparatus for monitoring and displaying an approximate time remaining in a lifespan of an individual, said monitoring apparatus programmed to decrement time units from an actuarially determined lifespan and to shift a projected lifespan value as the individual grows older, said time monitoring apparatus comprising: …"

The independent claim goes on to list eight dependent claims (e.g., it is wearable), which are characteristics of the first claim. If you were working in a similar domain, your independent claim must be different from the one above. The determination is ultimately determined by a patent examiner. Recognize that patents are legal documents and applications are usually drafted by attorneys or patent agents for patent examiners. One way for engineers and engineering students to assess novelty themselves is to compare drawings in an existing patent to concept drawings. If sufficiently different as judged by others, then the concept may well be novel.

A straightforward way to locate patents for technology that already exists is to search on the USPTO category, Assignee. In most cases, inventors assign their patents to companies that employ them to help develop and commercialize the technology. When working for a company, it is usually mandated that any potential new technology must be assigned to the company. For example, 3M Company is a major manufacturer of stethoscopes. Searching the USPTO website, with "3M" as Assignee and stethoscope in all fields, yields more than 25 stethoscope-related patents. Most of these have different inventors, who were at the time employees or contractors of 3M. Assignees can also be companies that purchased the patents from the inventor(s).

6.4 Quantitative Concept Screening

At some point, a small number of possible concepts remains, and you need to have some other means of further narrowing. This section describes several screening techniques that can be used to quantitatively rank solutions relative to one another. There are over 70 Multi-Criteria Decision Methods (MCDMs) and no one method is appropriate for all concept selection problems—all have strengths and weaknesses of their own. Among the more than 70 methods, there are five broad categories of MCDMs. Listed alphabetically, these are:

- axiomatic design,
- decision support methods,
- hierarchical methods,
- probabilistic design methods, and
- utility theory.

Each of these methods can be helpful at this phase of the design process, but also as you encounter other decisions points in the design process when. You should read about the methods below and then choose one to apply to all of your remaining potential solutions.

In this section, we delve into some of the methods.

6.4.1 Decision Support Methods

The simplest decision tool to master is the matrix decision support method. This method builds directly from technical specifications that you developed in Chapter 4. You likely have already identified a few desired design characteristics and several critical functions and specifications similar to those shown in Table 4.4. These can form the basis for comparing candidate solutions. In addition, you may also want to consider some of the common qualitative criteria, presented alphabetically below.

Accurate	High pressure
Affordable, reasonable cost	Lightweight
Biocompatible	Long shelf life
Broad temperature range	Long battery life
Comfortable for user	Low power
Comfortable to patient	Low pressure
Disposable	Minimal changes to workflow
Disposable as medical waste	Portable
Does not change workflow	Precise
Easily manufactured	Reduced time
Easily prototyped	Reusable
Easy to use interfaces	Short development time
Easily understood	Sterilizable
Easy to use	Stiff
Electrically safe	Technically feasible
Flexible	User friendly

For some devices there may be as few as two or three criteria to consider, others may involve more than 10 that are most applicable to your device. These can also be used to distinguish your remaining solutions from one another. Most importantly, remember to include technical feasibility, as most academic design projects require a working prototype of a device to be constructed.

Technically, feasibility is the most important criterion when developing a commercial product or an academic prototype. Because it is essential—all possible solutions must satisfy this criterion—it is better to not consider it as a separate criterion. A medical product must work safely before it can be of any value commercially. In a student design project, creating a functional prototype is usually the project goal. As such, assessing your team's ability to create a working prototype should be completed early in the selection process. Whether you can or not is related to the constraint of time—a prototype that is possible to construct but not in the allotted time should be reconsidered as an option moving forward with that solution concept. Your team should be careful when making such a decision.

The next step is to begin quantifying the criteria for each remaining solution. Some scores are a mutually exclusive choice (e.g., disposable vs. reusable, introduces workflow changes or not). Others are subjective; however, additional research or bench testing could make your assessment more objective.

Once estimates are made of between 2 and 10 criteria, it may be possible to rank solutions in numerical order as 1, 2, 3 ... with higher numbers being the better score. In ranking solutions using simple numerical ratings, it is a good practice to compare more quantitative specifications to arrive at those rankings. For example, if you are comparing various manual wheelchair solutions by weight, having an estimate of the weights can inform your rankings. Perhaps you have three designs that you estimate would weigh 19.5 kg, 20.0 kg, and 28.1 kg. The first two estimated values are virtually indistinguishable, and each would be given a score of three or four on a five-point scale (where one is the lowest weight), whereas the last solution would likely be given a score of one (highest weight).

A simple total of all scores reveals which solution—when considered holistically—has the highest score and therefore is the best solution to pursue (at least initially). The totals also help determine how close solutions are to one another. It is not uncommon for two or more solutions to rise to the top, as all have closely clustered high scores. These may then move to the next round of narrowing. For example, it may be that wheelchair designs 1 and 2 have similar overall scores but achieved those high scores by scoring well along different dimensions. An example of using the decision matrix is presented in the next section as well as with the example in Section 6.7.

6.4.2 The Weighted Decision Matrix

A more sophisticated decision matrix may use weight factors that are applied to criteria to reflect the relative importance of the needs. It is helpful to express these weights as percentages. For example, assume there are five criteria that reflect the most important customer needs to be addressed by a solution. These appear in tabular form in Table 6.4 along with assigned weighted percentages.

Weights are often agreed upon and reflect the team's perception of the relative importance of each criterion. A more robust approach would include validating the weights by asking the opinions of an advisor, expert, user, or customer. Once established, these weights can be used with the scoring system through multiplication. In the previous wheelchair example, the first and second solutions would yield a score of 3 or 4 $*20\% = 0.6$ or 0.8, whereas the last solution would have a score of $1*20\% = 0.2$. A similar multiplication would occur for the other criteria and the scores totaled, again with the highest score being the top choice.

A weighted decision matrix becomes powerful when there are tied or close scores in an unweighted matrix. Referring to the wheelchair example, solutions 1 and 2 might tie in the unweighted calculation because all criteria were weighted the same. It may be discovered, however, that wheelchair 2 performed better on the second criteria (easy to use), which is weighted higher than the criteria of lightweight.

6.4.3 The Decision Support Matrix

The Decision Support Matrix (DSM) is a more comprehensive version of a weighted matrix. The criteria are listed in the first column, the weights in the second column, and the concepts in subsequent columns. It is helpful to add another column that serves as a baseline to which the other candidate solutions are compared. This baseline solution is called the *datum* (or reference) and could be from an

Table 6.4 Selected qualitative criteria and corresponding weights associated with them

Criteria	Weight (%)
Does not change workflow	35
Easy to use	25
Lightweight	20
Reusable	10
Easily manufactured	10

Table 6.5 Generic form of the decision support matrix, with m concepts and n criteria. A completed decision support matrix is presented in the example presented in Section 6.7.

Concepts/ criteria	Weight	Datum	Concept$_1$	Concept$_2$	Concept$_3$	Concept$_4$	...	Concept$_m$
Criteria 1	Weight$_1$							
...								
Criteria n	Weight$_n$							
Total	-		Σ criteria$_1$					
Weighted Total	1.0		= weight$_1$ $* \Sigma$ concept$_1$	...				

existing product (sometimes considered the gold standard) or an idealized solution where all specifications are met. The datum solution has its relative scores placed in a column as well. Generically, the format would appear similar to that in Table 6.5 form m concepts and n criteria.

As in the weighted matrix approach, each criteria score is multiplied by the corresponding weight and the total score for a given concept computed. The concept of receiving the highest score is typically considered the best concept of the group. The advantage of this approach is that the highest-scoring concept can be compared not only to other concepts but also to the gold standard (the *datum*) or an ideal score (computed as the highest possible score). The table also reveals the criteria in which this high-scoring concept scores well against the gold standard and where it may fall short. Such information can be used to refine the **design solution**. It is important to note, however, that often the scores for the datum are known because it is an existing solution, whereas the scores for your concepts are estimates. As is discussed below and in future chapters, feasibility tests can confirm or refute your assessments.

6.4.4 Scoring Systems

Scoring systems have already been presented including an expanded Likert scale in Chapter 3, as well as a Boolean (yes/no, +/-) approach. There are several other possible scoring systems, a few of which are presented below. These can be helpful when navigating design decisions. Each has benefits and drawbacks. When faced with a decision, you should select a scoring system that helps the team best move forward with the design; this may mean changing your technique and scoring system during the selection process. There are three heuristics you may find helpful in selecting a scoring system. These are:

1. Remember that the purpose of a ranking system is to narrow options and to be able to defend your choices made.
2. When narrowing a large list of choices, having only a few options in the scoring system (e.g., mutually-exclusive or multi-point) is helpful to identify clear losers, clear winners, and those in between.
3. It is helpful to define what is meant by a score for a criterion before performing any rankings (e.g., on a Likert scale, 5 = ideal concept, 0 = likely will not work).

Early in a design process, it is best to include in the scoring system a score that indicates an unknown, such as "?" in the context of selecting criteria. This should prompt the same response from the design team—more information is needed to make an assessment and turn a question mark into a numerical estimate.

6.4.5 The House of Quality

A common issue that arises in ranking concepts is that two or more two criteria could support or oppose one another. These criteria are interdependent and therefore correlated (or anti-correlated). For example, "lightweight" and "portable" would support one another, but "long battery life" and "low battery indicator light" would oppose each other because additional power is needed to monitor the voltage and for a low battery indicator.

The decision support method that supports coupled criteria is known as House of Quality, which is part of a wider design methodology known as Quality Function Deployment (QFD). QFD was introduced in the 1960s and was made famous by Toyota which introduced the method across all company divisions in 1969. Over the next decade, Toyota cut production costs by 60%, reduced design time by months and manufacturing reduced the number of bad parts to three or fewer per million. A significant part of the improvement was the result of far fewer design changes being needed during prototyping, production, and manufacturing.

Although QFD is a much more complex management system, House of Quality is the component of QFD generally relevant to biomedical engineers. The method involves making a judgment about which criteria are coupled (if any) and indicating whether one supports or opposes another. The adaptation to the decision matrix is to add an area above the matrix that reflects the relationships between criteria. When depicted graphically, as in Figure 6.4, this added section is analogous to a gabled roof

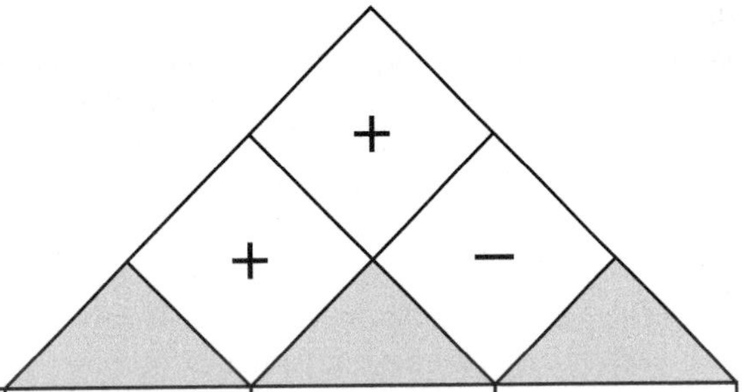

Criteria / Concepts	Concept 1	Concept 2	Concept 3
Criterion 1	1	2	2
Criterion 2	2	2	1
Criterion 3	1	1	2
Weighted Total	4	5	5
Modified Weighted Total	6	5	5

FIGURE 6.4

Decision matrix with roof for three concepts and three criteria that are coupled. The two plus signs are added to the weighted total of concept one, thereby increasing its weighted score from four to six. The weighted scores of concepts two and three remain the same because the plus sign and the minus sign in each case cancel each other out.

and is the origin of the name House of Quality. In Figure 6.4, three concepts are being compared with each of three solutions; the weighted totals of each concept are 4, 5, and 5, respectively.

In this example, the criteria for the concepts are coupled. Criteria 1 (C1) is supported by C2 and C3; C3 opposes C2. The weighted total is recalculated by adding the supporting concept plusses and subtracting opposing concept minuses. Applying this to Figure 6.4, concept 1, has two plus signs, one each from criteria 2 and 3. Concepts 2 and 3 have one minus sign and one plus sign each; there is no change in either weighted total (fourth row in Figure 6.4). The weighted scores of concepts two and three remain the same because the plus sign and minus sign in each case cancels each other out. In this case, the modified weighted totals (last row in the figure) become 6, 5, 5; and Concept 1 would be the top choice. When criteria are coupled and weighted totals are close, this type of analysis (creating a gabled roof) is helpful to accommodate coupled criteria and help make a design choice.

6.5 Prototyping

Prototypes can be very useful in making early design decisions for many reasons. A quote from Tom and Dave Kelly of the design firm IDEO sums up the idea, "If a picture is worth 1000 words, a prototype is worth 1000 meetings." Simple prototypes often reveal issues not previously realized and allow stakeholders to hold a tangible artifact and give feedback that would not otherwise be possible.

Two forms of early prototypes can help narrow **design solutions**. *Proof-of-concept* prototypes (also known as feasibility prototypes) aim to demonstrate that some key specification or function is possible to achieve, but without creating the entire solution. On the other hand, *looks-like* prototypes (mockups) are meant to communicate shape, size, and dimensions, and gain feedback on how a user would interact with the product, but without creating a functional prototype. Depending on the complexity of the design and goals of the prototype, both types can usually be made in less than a day out of common materials such as wood, foam core, cardboard, adhesives, simple electrical components, or by combining items that can be purchased online or from a local store. Although most early prototypes do not take long to create, each one should have a particular purpose; namely, what specific question does the prototype answer. A computer simulation or interface is also possible as a prototype, perhaps interfacing with a physical prototype. If dimensions and connectivity are approximated, a CAD drawing may be generated to create a 3D printed prototype.

6.5.1 Proof-of-Concept and Feasibility Tests

Often there are a few critical specifications that determine if a concept can be chosen. This means that if these specifications cannot be met, the design will not solve the problem. In these cases, it is important to determine early if it is possible to meet the most critical specification to save on time and resources. These kinds of feasibility tests may be a simple calculation, computer simulation, or benchtop experiment that shows that a potential solution shows promise in meeting a specification. A team may be able to create testable prototypes that can address each specification independently. At this stage, experiments should be those that are easily performed and can eliminate a design choice without investing more time in it. For example, noncompliance is a major problem when geriatric adults are prescribed more than a few prescribed drugs. One design proposed to increase compliance requires patients to use an app. After creating a rudimentary user interface, it could be shown to potential users (e.g., patients in a waiting room who permit to be asked) to learn if they would be amenable to using such a technology. If most answers were no and the reasons were understood, this concept needs redesigning or should no longer be pursued. A similar experiment with a simple prototype is presented in Section 6.5.3.

Showing that it is possible to perform a test that can verify a specification is a type of project feasibility. There are also some projects where it is not the solution itself that is in question (e.g., a self-powered contraction monitor, Breakout Box 10.5), but the testing of that solution that is in question. For some simple solutions, the most technically challenging part of the project becomes validation testing, as covered in Sections 9.6 and 10.5.

6.5.2 Looks-Like Prototypes

A looks-like prototype is usually nonfunctional or barely functional. As such, prototypes often have a similar feel, shape, and/or user interface as the functional prototype envisioned. The benefit of creating a prototype is that the tactile sensation alone may reveal features not previously envisioned.

For concept selection, a looks-like prototype is often useful to help decide among different concepts. By showing these prototypes to potential users, qualitative information can be gained on which is better, or which is best if more than two. Anecdotal comments from intended users are also very useful. If users are asked for feedback, quantifying that information is possible with survey methods, outlined in Section 3.7 and presented in Section 10.6.4. This would provide numerical feedback that serves as early support of one design concept, or as a comparison between competing concepts.

Another reason for creating simple prototypes is to test feasibility. As an example, an initial prototype of the concept shown for the patient-lift problem in Figure 6.3 is shown in Figure 6.5. The prototype was made from foam core and duct tape in less than half a day. The purpose of this simple prototype was threefold: (1) to envision the likely form for the frame of a torque-generating system located at the knee, (2) how standing affected the motion of the prototype, and (3) how it would be

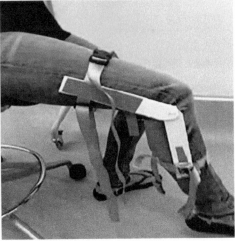

FIGURE 6.5

Initial prototype of device to lift a patient from sitting to standing.

attached to the patient. By doing this, the team pursued the concept further. Had it not worked as expected, that would have eliminated the concept from further consideration.

The type and number of looks-like prototype(s) depend on several factors including the project requirements, value added by a prototype, and the time needed to create one.

6.5.3 Scaling Up and Down

An important consideration for both and looks-like and work-like prototypes is the option to scale the solution. This may be geometric scaling (e.g. scaling down the design for a new type of hospital bed) or functional scaling (e.g. scaling up millimeter-sized tubes to represent a blood vessel but preserving the Reynolds number associated with a particular blood flow). Below are two examples to illustrate the idea; more ideas are presented in Chapter 7.

Consider the problem of the cricothyrotomy presented in Section 6.2.3 and Figure 6.2. Figure 6.6 shows that the initial prototype was used to test feasibility of a design in the form of an alternate surgical tool. The size of the prototype is likely double that of the envisioned handheld device; this size was much faster and easier to create than a handheld model-sized prototype. The second reason for the test was to assess the force needed to insert the prototype through a slit in the through typical of the incision

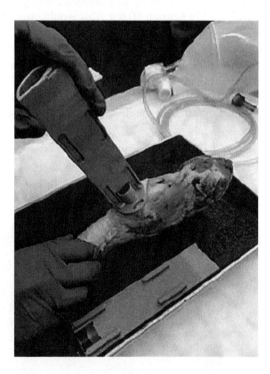

FIGURE 6.6

Initial cricothyrotomy prototype demonstrating it can be used to enter the incision to the bovine trachea without trauma. It also provided the feel of the tissue resistance upon inserting the prototype. If this test is unsuccessful, the concept of this alternate surgical tool would no longer be under consideration.

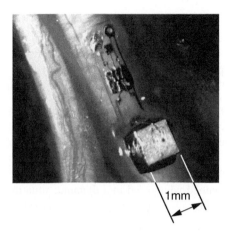

1mm

FIGURE 6.7

Photograph of a miniature brain implant.

used today (steps 3 and 4 in Figure 6.1) in a bovine neck. A third reason for the prototype is to obtain feedback from healthcare personnel in the military. This is a good experiment because it answered fundamental questions about whether this alternate surgical tool was a viable concept.

As a second example, consider a miniature neural implant, similar to the one shown in vivo in Figure 6.7. Medically, these are typically used in the brain-machine interface field and for neural stimulation and must be of the scale of $1 \times 1 \times 3$ mm.

One question about the feasibility of such a device involves the ability to transmit signals and to deliver stimuli, *in vivo*, when subjected to mechanical and thermal stresses. One approach would be to create a simple prototype of the same size ($1 \times 1 \times 3$ mm) and composed of a similar or the same material as the implant (but without the circuitry). The team could then conduct mechanical and thermal tests on this nonfunctional prototype using theoretical calculations or finite element analyses, thereby demonstrating whether or not the solution is feasible.

As a reminder, prototypes at this stage of the design process should only be constructed to help the team choose among alternatives. These types of prototypes should not require much time to create and be limited to using design studio or lab equipment, available materials, and supplies. Far more detail about prototyping and the actual making of a prototype are presented in detail in Chapter 7.

6.6 Communicating Solutions and Soliciting Feedback

As you develop your technical **design solution**, it is important to remember that your design is to be used in context. Before refining sketches, redesigning, and testing, it is important to pause and ask the question: How does this technology interact with the healthcare provider, patient, or both?

Figure 6.2 is a sketch of a possible ultrasound solution to the cricothyrotomy problem as imagined in medical use. Figure 6.6 is a mock-up of the same solution. Both would help communicate your solution to others and may prompt questions such as:

- How does the EMT or paramedic access the device?
- How does the EMT position the patient so technology can be used?
- What are the following steps for the procedure to provide oxygen to the patient?
- How is the device eventually removed after a breathing tube is implanted?

Discussing these and other questions with an EMT or paramedic with this image provides much insight related to technical design. In this particular case, Figure 6.2 was eliminated as a possible **design solution** because of the emergency physician sponsor's opinion that it would be impractical to use ultrasound in a life-saving situation with only minutes to treat the patient.

Figure 6.2 was created to show the essential idea of a solution in a clinical context. However, it did not capture how value can be gained when used in a dynamic situation. A powerful way to show the dynamic use of your solution is through a technique known as *storyboarding*.

6.6.1 Medical Context and Storyboarding

A storyboard is an excellent way to engage others—specifically healthcare providers—in your decision-making process. The technique was first used by Walt Disney to sketch out the basic framing of the plot of animated shorts and movies. It consisted of a series of images that would portray the sequence of events that tell the story discreetly, after which the animators could go back and fill in the gaps.

A similar idea can be used to show how a device is used in the clinic. In this case, the story would be a series of images to show step-by-step, how the device is envisioned to be used. These can be created by hand or via software.

A patient lift device solution was explored throughout this chapter. Figure 6.3 and Breakout Box 6.1 are early "back of the napkin" calculations and Figure 6.5 is a mockup. However, neither communicates how the device would be used and could prompt the following questions:

- Where does the healthcare provider retrieve the lift device from?
- Is it light enough to carry and manipulate?
- How does the device attach to the patient?
- How does the nurse control the device to raise the patient and stop when standing or sitting?

A 10-image computer-generated storyboard was created, shown in Figure 6.8, to generate many of these questions. Text for each image is typically written underneath or within the image to explain that particular step. Creating a detailed storyboard can take time; you will likely only create one as you near your **design solution**. To consider the human factors aspects (Section 8.2) of this **design solution**, this storyboard also could have been generated earlier in the selection process.

6.6.2 Receiving and Incorporating Feedback

Sometimes, sponsors, mentors, or advisors have been a part of the narrowing process, in which case they have already given you important feedback. In other cases, they may only see a single design (or perhaps a few) that remain. Feedback from a sponsor at this stage is critical to ensure that your team understands the actual medical scenario, can address potential misunderstandings, and answer questions generated during the selection process. Obtaining this information during concept selection in the

The nurse retrieves two knee braces

The nurse adjusts the braces to fit the patient's legs

The nurse places the knee braces on the patient and pulls straps snug

The nurse prepares the patient to sit up

The patient assumes ETF position

The nurse operates the braces using a remote

The braces stand the patient up with minimal support from the nurse

The nurse pivots to settle the patient into the wheelchair and lowers them using the brace

The nurse loosens the straps to remove the braces

The patient can now be transferred to the desired location

FIGURE 6.8

Multi-image storyboard. Includes all the major steps the nurse needs to perform to raise the patient from a bed with proposed technology. Although images are not refined, how the procedure is envisioned to be performed is clear. ETF (in storyboard 5) is the abbreviation for exaggerated trunk flexion.

design process saves a team from moving ahead with an incomplete or misunderstood understanding of the medical context.

For a clinical sponsor, it is important to show nontechnical drawings or photos of prototypes, perhaps accompanied by a storyboard of how the device is to be used. Feedback from a clinical sponsor is often most helpful in determining how a solution may fit into the existing medical workflow. It can also be very helpful to explain the rationale behind the selection process as this may also reveal possible misunderstandings of the most important criteria, the weightings of those criteria, or how teams estimated scores.

Feedback from a technical expert, industry or medical sponsor, or faculty mentor often occurs during formal or informal design reviews. It is important to share the technical aspects of your concept as well as the rationale underlying your decisions. Sketches, prototypes, and storyboards can all aid in communicating design rationale, concept selection, design details, and obtaining feedback on areas to focus on for the next design iteration.

All feedback received should be documented and included in the Design History File (DHF). Most especially, teams should note if a change in direction or iteration of a design was prompted by external feedback.

6.7 Diagnosing Bile Duct Cancer

In this section, we present an example of how to use techniques described in this chapter for narrowing down a set of solutions. When a patient's symptoms include jaundice and itchy skin, it raises the suspicion of bile duct cancer. In some of those cases, healthcare providers sometimes access the bile duct percutaneously to perform brush biopsies to collect cells to pathologically determine if cancerous cells exist, and if so, what is the stage of cancer. Brush cytology attempts to mechanically exfoliate cells from the ductal epithelium into the pancreatic juice and is intended to increase cellular yield.

Percutaneous access to the bile duct is typically performed by an interventional radiologist. When at the suspicious location, the interventional radiologist moves and twists the brush bristles against the tumor or lesion. This contact usually enables the bristles to collect a sufficient number of cells so that pathologists can make a diagnosis.

It is not uncommon in some biopsy procedures for the number of undamaged collected cells to be insufficient to make a pathological diagnosis. In those cases, another sample is sometimes taken, or a bigger tissue sample is taken for another biopsy. This is costlier, more traumatic, and increases the risk of spreading cancer cells. The medical need is clear: providers need a more effective means of obtaining cells via the brush biopsy to increase the cell yield count and reduce the need for a repeat or more invasive procedure.

A set of specifications for this medical problem and the rationale for them, along with the relative importance weights, are indicated in Table 6.6.

With these criteria in mind, the brush itself seemed to be the best design option. Research into the cell collection probe demonstrated that increased pressure between the bristles and the interior wall and a greater contact surface area would increase cell collection. As a result, some design concepts focused on supplanting the existing cytology brush bristles with a different shape. Several student-drawn concepts were generated; six are shown in Figure 6.9. One unlabeled image is the expanding wire concept; the other unlabeled sketch is the tampon concept.

Table 6.6 Table of specifications and reasons for them to increase the probability of correct diagnosis for cells collected with single brush biopsy. The importance of each specification is indicated by the percent weight associated with each one.

Specification	Rationale	Weight
Collect 50–100 cells	Minimum number of cells required for pathologic evaluation	50%
1.5 mm < diameter < 5.0 mm	Required for access through a variety of orifice sizes	30%
90% sample success	Required to reduce need for repeat brush biopsy or more invasive and risky cell collection methods	15%
Workflow unchanged	Required to reduce resistance for medical acceptance	5%

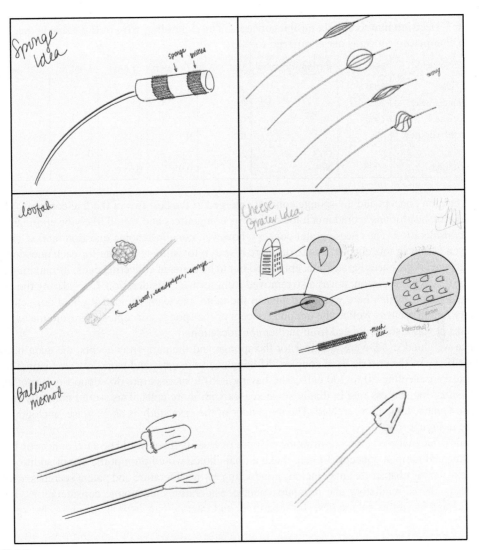

FIGURE 6.9

Six concept sketches for biopsy brush modifications to collect more cells than existing brushes.

With the criteria set, weights established, and the solution concepts generated, a decision support matrix was generated, as shown in Table 6.7. The scale used for comparison included three levels: 1 means better, 0 means same, and -1 means worse than the datum, which is the expanding wire in the example. It could be another alternative or a commercial product. The datum choice is arbitrary and can be a current product or one of the design choices or both. By generating the decision matrix two or three times with different datum columns, you will reduce the risk for potential bias when making comparisons.

Table 6.7 Decision matrix for six design concepts. The expanding wire in this table is the datum; competing concepts are compared to that.

Criteria/concept	Weight	Expanding wire	Balloon	Tampon	Sponge	Loofah	Cheese grater
50–100 cells	50%	-	-1	1	1	1	?
1.5–5 mm diameter	30%	-	0	1	1	1	-1
90% successful	15%	-	0	1	1	0	0
Unchanged workflow	5%	-	0	0	0	0	0
Total	-	N/A	-1	3	3	0	?
Weighted total	100%	N/A	-0.5	0.95	0.95	0.3	?

The tampon concept and the sponge concept emerged as the best two of the five concepts. The balloon and loofah solutions scored much lower than its competitors and would likely be eliminated from further consideration. The cheese grater concept, however, contained some question marks. The guidance given earlier in this chapter was to do more research to estimate a value for each question mark. However, even if the cheese grater concept turned out to be better at collecting cells, its maximum score would be 0.2. For this reason, it was also removed from further consideration. Considering the expanding wire solution, which here was the datum in the table, the team decided it could not achieve the geometry requirement as well as the tampon concept or the sponge concept. As a result, the expanding wire concept was also eliminated from further consideration.

Since weighted totals were identical for the sponge and the tampon concepts, the team needed to consider if any of the criteria were coupled. Although cell diameter and workflow are distinct, uncoupled criteria, collecting 50 to 100 cells, and having a 90% success rate do support one another. The reason is that the success rate in diagnosis increases with more cells. This would normally require a House of Quality table to be created. The remainder of the case study is an exercise, analogous to the example in Figure 6.4.

It is possible that even after constructing a House of Quality there would be a tie, or near tie between the sponge and tampon concept. To help make a final choice would prompt the use of additional suggestions from this chapter that may include conducting further literature and patent searches, exploring relevant standards, regulatory and reimbursement considerations, economic considerations, building looks-like and feasibility prototypes, storyboarding, and interviewing experts, mentors, and sponsors.

6.8 Design History File

As indicated in Section 6.6, documenting the narrowing of your solution list to a single **design solution** helps you justify your choices to others as well as help you gain confidence in moving forward to prototyping and detailed design. It is essential to document aspects of work generated from this chapter as part of the DHF. Some of this was presented in Section 6.6, where communication of design ideas was a part of soliciting feedback from storyboarding. More is needed. Specifically, the following should be recorded in the Design Output section of the DHF:

- text to describe selection method(s)
- all sketches and relevant images

- selection method(s)
- selected concept and rationale for the decision
- prototype images (if any)
- storyboard images (if any)
- text to summarize results

It should be presented in such a way that a non-team member classmate could read it and understand what the team did and how and why it specifically chose the **design solution**.

Key Points

- Concept selection is an iterative process that is necessary to choose one concept, a **design solution**, from multiple design concepts in a formal way.
- There are several qualitative ways to eliminate certain designs including stakeholder resistance, economic considerations, regulatory hurdles, compliance with standards, and patentability.
- Quantitative ways to eliminate some design concepts involve the application of fundamental principles using an appropriate "back of the napkin" type of analysis or similar technique. Making reasonable assumptions and estimations of key values are a necessary part of this method.
- Sketching even low-quality concepts in medical use is advantageous for four reasons: (1) understanding how technology interacts with the provider, patient, or both of them, (2) raising technical and practical questions that could not be answered out of context, (3) conveying to healthcare professionals how you perceive how the proposed technology would be used medically, and (4) as preparation for a storyboard.
- There are dozens of quantitative decision support methods that can be applied to concept selection. Most involve concept comparisons of criteria, which are weighted in terms of importance.
- The Decision Support Matrix is one method readily adaptable to biomedical engineering design. If concepts are coupled (one is in support of another or one works counter to another), the House of Quality method can be used, which involves constructing a gabled roof on the Decision Support Matrix.
- Simple prototypes—usually not at all or barely functional at this point—may be created at this stage to describe technology and to share with potential users. When coupled with a storyboard of the procedure, an influential visual is created and relays much information to the team and to others, and more insight is gained into the design concept.
- Feedback from healthcare providers at the concept selection stage (at times, with simple prototypes) most often leads to a better understanding of the medical problem and further design modifications.
- Key elements to document in the DHF include the design selection process including design rationale, methods used, sketches, prototype images (if any), and one or more storyboards.

Exercises to Advance Your Design Project

1. Create a sketch of one concept from your project and show it in a medical context.
2. For the medical procedure associated with your project create a storyboard starting where your **design solution** enters the flow of the procedure. Incorporate the steps needed to perform that task. If you learn something from it, state what it is. Does it affect your design?

3. Construct a table using one of the Decision Support Matrix approaches presented in this chapter, for your project. If starting with a large number of concepts, use a screening and evaluation approach (Section 6.2) or a qualitative approach, or both, to reduce the number to a few concepts, and then use a scoring approach (Section 6.4) to select the best concept.

4. After presenting the set of concept solutions to your project sponsor(s), document how feedback provided further narrowing or modifying specific solutions.

General Exercises

1. Create a House of Quality table for the Case Study in Section 6.7. Determine which solution is chosen to pursue and explain why.

2. Using Table 6.2 as a guide, list comparable analogs for the following quantities:
   - average newborn head (~8 N, 25% body weight)
   - college-aged normal urine output (~20 mL/sec)
   - college-aged femur fracture limit in bending (~40 Mpa)
   - bicycle external power (~110 J/sec)

3. After asking n nurses whether they would use a proposed new way to clear surgical drains, there is no consensus; 55% said yes and 45% said no. What would you do at this point?

4. In Section 6.2.4, a calculation was performed to estimate the size of a motor (~0.4 hp) required to help lift an obese patient from a bed to a chair. There are many different brands and types of motors at or near that hp rating. Of the other criteria (e.g., physical shape, method of attachment, weight, noise), which factor should be considered most important for user acceptance. Why?

References and Resources

Bertrand, J. (1878). Sur l'homogénéité dans les formules de physique (On homogeneity of physical systems). *Comptes Rendus, 86*(15), 916–920.

Fiod-Neto, M., & Back, N. (1994). Assessment of product conception: a critical review. In *Proceedings of the 1994 Lancaster international workshop on engineering design* (pp. 35–45).

Halliday, D., & Resnick, R. (1981). *Fundamentals of physics*. John Wiley & Sons.

Hauser, J., & Clausing, D. (1988). *The house of quality*. Harvard Business Review.

King, A. M., & Sivaloganathan, S. (1999). Development of a methodology for concept selection in flexible design strategies. *Journal of Engineering Design, 10*(4), 329–349.

MEDtube. https://medtube.net/.

Ming, Z., Wang, G., Yan, Y., et al. (2018). Ontology-based representation of design decision hierarchies. *Journal of Computing and Information Sciences in Engineering, 18*(1), 011001.

Okudan, G. E., & Tauhid, S. (2008). Concept selection methods—A literature review from 1980 to 2008. *International Journal of Design Engineering, 1*(3), 243–277.

Otto, K. N., & Krisitin, K. L. (2001). *Product design*. Prentice-Hall.

Pugh, S. (1991). *Total design: Integrated methods for successful product engineering*. Addison-Wesley.

Pugh, S., Clausing, D., & Andrade, R. (1996). *Creating innovative products using total design*. Addison Wesley Longman.

Rayleigh. (1892). On the question of the stability of the flow of liquids. *Philosophical Magazine, 34*, 59–70.

Roam, D. (2013). *The back of the napkin*. Penguin Group.

Roman, F., Rolander, N., Morales, F., Fernandez, M., Bras, B., Allen, J., et al. (2004). Selection without reflection is a risky business. In *10th AIAA/ISSMO multidisciplinary analysis and optimization conference* (pp. 2004–4429).

Suh, N. P. (2001). *Axiomatic design: Advances and applications*. Oxford University Press.

PROTOTYPING AND DETAILED DESIGN

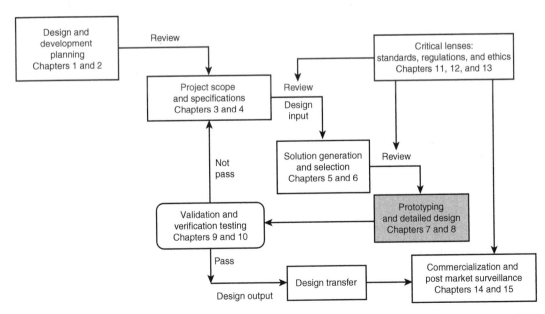

In the next two chapters you will learn how to move closer to your final design by building and iterating upon a solution that will meet your technical specifications and satisfy customer needs. Seeing an idea transform into a functional medical device is one of the most exciting milestones in the design process. Realizing your solution involves an iterative process which can be frustrating as new problems reveal themselves. However, this is an opportunity for you to exercise your creative problem solving skills and technical know-how. When overcoming these hurdles, it is often the case that the problem is further clarified and new opportunities to add value are revealed.

Chapter 7 introduces prototyping as a way to move your design solution toward a functional artifact and to share your design with others for their feedback, as well as document your design iterations. Several techniques are discussed so that you can choose the appropriate set of tools and processes to suit your design solution and phase of the design process. These techniques include crafting, rapid prototyping, traditional machining, molding, microfluidics, electronics, and computer simulations.

Chapter 8 introduces several considerations that will enhance the overall quality, value, and commercial viability of your design. These include cost, usability and human factors, safety and failure prevention, manufacturing, packaging, sterilization, material selection, and maintenance. Although an industry design process would iterate upon all of these factors, you may only have the time and resources to iterate on a few that are most appropriate for your project. You should keep in mind, however, that all aspects of a design are interrelated—sometimes in unanticipated ways—so you should therefore avoid optimizing your design along only one aspect at the expense of other aspects.

By the end of Chapters 7 and 8 you will be able to:

- Select and use prototyping techniques that are appropriate to your design solution.
- Evaluate your design through multiple lenses such as cost, safety, manufacturability, and human factors, for the purpose of enhancing overall quality, value, and commercial viability.
- Document and justify changes as your design solution evolves through multiple prototypes.
- Create a functional prototype that is ready for verification and validation testing.

Prototyping

7

Have no fear of perfection—you'll never reach it.
– Salvador Dali, Spanish surrealist artist

If you're not prepared to be wrong, you'll never come up with anything original.
– Sir Ken Robinson, British author and education advisor

Chapter outline

Biomedical Engineering Design. https://doi.org/10.1016/B978-0-12-816444-0.00007-9

7.1 Introduction

Prototyping is a collection of techniques for transforming solution concepts into tangible artifacts. Creating a physical (and sometimes software) deliverable is a powerful way to communicate your ideas to others, determine the feasibility of meeting critical specifications, and gain valuable feedback from mentors, advisors, users, and customers. Prototyping techniques can also create a unique testing apparatus or other ancillary devices during verification testing. The techniques vary greatly and include crafting (e.g., glue, popsicle sticks, duct tape), machining (e.g., lathe, band saw, drill press), rapid prototyping (e.g., fusion deposition modeling), molding and casting, microprocessor programming, and circuit board layout, smartphone app design, and microfluidics. As an engineer, you should not be afraid to "get your hands dirty" to build a design concept prototype.

Prototyping is also a mindset. This mindset includes being curious and creative, testing your assumptions, and viewing failure as a pathway to deep learning. Prototyping is similar in many ways to the iterative nature of the scientific method—a hypothesis is made, experiments are designed and conducted to test that hypothesis, and then results are analyzed to determine what to do next. Prototyping follows this same paradigm, except the hypothesis is replaced with a design concept or solution. Just as there is a thrill in discovering something new, there is a great sense of accomplishment in seeing one of your ideas finally take shape in physical form.

This chapter aims to introduce you to the prototyping techniques most commonly used in the design and testing of a medical device. As most academic projects end with a one-of-a-kind "final" prototype, we focus primarily on prototyping rather than manufacturing techniques for scaling production. We intend to present you with several options so that you can decide which methods best suit your device and phase of design; you are encouraged to seek additional resources and training that may help advance your design.

7.1.1 Looks-Like and Works-Like Prototypes

Prototyping can be helpful at almost all design phases and should be used to obtain feedback and make design decisions. Throughout this chapter we distinguish between two types of prototypes. A **looks-like** prototype is meant to demonstrate how a device could look (form, shape, color) and eventually be used (function). For example, a cardboard cutout with buttons and knobs painted on the surface could be made to show how a device might look. Another example is shown in Figure 6.5. A looks-like prototype, sometimes also called a "mock-up," can be very helpful in communicating the expected value proposition to various stakeholders. It is not uncommon to create several looks-like prototypes that together communicate your design concept.

A **works-like** prototype shows how one or more functions may be embodied in a device. Works-like prototypes help demonstrate the feasibility of a technical solution and demonstrate how technical specifications can be met. For example, a works-like prototype may be used with a benchtop demonstration to show that the pump you have chosen provides the right combination of power consumption, pressure, and fluid flow to meet a critical specification.

7.1.2 Minimally Viable Products and Prototypes

A **Minimally Viable Product** (MVP) contains the minimum set of features needed for customers to fully understand the value proposition relative to other solutions and to want to purchase the product. It is often associated with smartphone app development, where it is considered a best practice to put an MVP into customers' hands as quickly as possible. Later releases of the product can fine-tune existing functions and add new features based upon user feedback. This iterative process allows for shorter prototyping and feedback cycles between product revisions. It also engages real customers as a complete product emerges. The idea of an MVP has expanded beyond apps and is now a familiar term in many industries, including the medical device industry. Most medical devices must pass various regulatory, legal, financial, and ethical barriers before being commercialized. These barriers often prevent a medical device MVP from being sold or evaluated by users until the device is proven safe and effective through testing.

Most academic design projects aim to demonstrate the technical feasibility and the potential to create value for the healthcare ecosystem, not to commercialize a finished medical product. In this regard, a **Minimally Viable Prototype** (MVPr) can be helpful. For the purpose of this text, an MVPr refers to the works-like prototype that includes the minimum set of functions of the device, even if all specifications have not yet been met. Creating an MVPr is a significant milestone in a design process because it is the first proof that all functions can be performed together in a single device. For example, after iterating on several separate feasibility prototypes, you may have assembled them into your first integrated device. This MVPr, however, may have used an available pump that performs a needed function but does not meet all specifications for power, pressure, and flow. In a later iteration, you would find a pump that would meet your specifications and integrate it into your device. The prototyping techniques presented in this chapter will help you create an MVPr and then iterate until you meet your specifications.

Engineering designers use many terms and acronyms as they navigate the design process. Works-like, looks-like, MVP, and MVPr are all excellent examples of design terminology. Throughout this text we introduce commonly used terms so that you can be conversant with other engineering designers.

However, some terms are not consistently used, and there are variations between designers, companies, textbooks, and industries. Breakout Box 7.1 discusses the prototyping terms most commonly encountered in a design project.

7.1.3 Principles of Prototyping

As you construct your prototypes, keep these principles in mind:

- Early prototypes are often looks-like and incrementally become more sophisticated works-like prototypes.
- Prototypes can reveal problems with the design that were not anticipated. Early detection of these flaws should be welcomed; it will help you identify how to improve your design and prevent significant changes later in the project. This mindset is summed up in the common phrases "fail fast" and "fail forward."

Breakout Box 7.1 Prototyping Terms Used During the Design Process

There are many terms that you might hear during prototyping, each with slightly different meanings. All designers do not agree upon these terms.

Early Prototypes

These prototypes aim to narrow the possible solutions concepts.
- *Feasibility Prototype* aims to demonstrate that a solution can, in principle, solve the technical problem.
- *Proof-of-Concept Prototype* is usually synonymous with a feasibility prototype.

Design Iteration Prototypes

These prototypes aim to help designers iterate upon a chosen solution.
- *Looks-like Prototype* is a nonfunctional approximation of the final solution, often to clarify the imagined user experience.
- *Mock-up* is similar to the looks-like prototype but is often less refined.
- *Works-like Prototype* is a functional prototype that meets one or more of the critical specifications.
- *Minimally Viable Prototype* is a functional prototype that captures all of the most important specifications in the same physical embodiment but may not yet meet all specifications.
- *Final Prototype* is the last prototype a team will build, often because a hard deadline (e.g., end of an academic design experience) has been reached. In an industry design project, it is typically the most refined prototype (closest to the final design) prior to production.

Production Prototypes

These prototypes aim to communicate to nonengineering stakeholders, both internal and external, how close a design is to being completed and ready for market introduction.
- *Minimally Viable Product* (MVP) is a complete solution (technical and nontechnical) that will help users determine if they would like to adopt a new product. It includes the most essential features needed to sell the product.
- *Alpha, Beta, etc.* Products are progressively refined, MVPs tested by select customers. Their feedback is used to improve the product before a broader market introduction.
- *Frozen Design* refers to the design ready for design transfer (defined in more detail in Chapters 9 and 10), which engineers feel is ready for production and market introduction. Prototypes that represent a frozen design are used in final product testing.

The proliferation of different terms used by engineers in different companies can be confusing. When you hear a term being used in a different way than you use it, ask to have it defined to better understand what is expected of you and your team.

- It is generally helpful to iterate upon one aspect of a prototype at a time. Doing so allows you to determine if a single change has positively or negatively impacted the design. It also decreases the time between prototype iterations and enables you to document the evolution of your design idea more clearly. Chapter 8 introduces common considerations for iterating on the design of a medical device.
- Avoid iterating already strong characteristics of a design. A useful method is to list all weaknesses of the current prototype and rank them in terms of which are most critical to address.
- Material selection is an essential aspect of prototype creation. In most cases, prototypes are not constructed from the same materials or with the same techniques used for production units. For example, to save time and money, you may make a rapid prototype housing out of plastic while knowing it will be made out of stainless steel in the final device.
- Devices that must be very small (e.g., stents, microcircuitry, micro-implants) or very large (e.g., large clinical or laboratory equipment) can be scaled in both looks-like and works-like prototypes. Miniaturization (or scaling up) of the design can occur later in the design process. Section 6.5.3 and Breakout Box 7.2 discuss scaling of prototypes.
- Making prototypes can sometimes come with safety hazards. You should carefully follow the rules associated with the tools and spaces you use. At a minimum, you should wear eye protection, long pants, and closed-toed shoes.

7.1.4 Prototyping Tools

The tools used to create prototypes typically bend, cut, join, extrude, deform, smooth, or remove materials. Choose the proper tools to suit the type and scale of your prototype and materials. A selection of generic hand tools (Figure 7.1) and power tools (Figure 7.2) are useful in nearly all types and phases of prototyping. There are more specialized tools (e.g., tube bender, soldering iron), rapid prototyping devices, complex computer-controlled machines, and other tools that can aid you in creating and

---

Breakout Box 7.2 Scale Models

A common method for creating both works-like and looks-like prototypes is to scale the prototype up or down. For example, designing a very small stent may be made much easier by working with prototypes that are five times as large. Many design iterations could then take place using this scaled model. To be a true scale model, all relevant design aspects should be scaled as well. This may include material geometry, flow rates, mechanical loading, range of motion, and so on. The formal term for equivalence is *similitude*; if achieved, the prototype is predictive of how the actual design will perform and respond at actual size. It is usually impossible for a team to scale every aspect of a design, so you will need to justify how and why you scaled a prototype.

There are many scaling methods used in diverse fields, from software to structural engineering. The most appropriate for medical device design is the Buckingham π method, which shows that physical responses have a similitude if output parameters can be expressed as dimensionless variables. Some dimensionless parameters from different engineering disciplines include strain and Poisson's ratio in solid mechanics, Reynold's number in fluid mechanics, aperture in optics, and Nusselt number in heat transfer.

One way to make output variables dimensionless is to normalize them with base parameters. For example, bending stress, σ in a beam is expressed as Mc/I, where M is the moment, c is the distance from the neutral axis, and I is the area moment of inertia. If we create a dimensionless parameter (π_1, defined as σ/E where E is the modulus of elasticity), we can apply the Buckingham π theorem and conduct analyses on the scaled system to predict the stress response of the desired prototype, independent of the material used.

FIGURE 7.1

Common handheld tools that are useful in prototyping. Labels are shown to the right of each tool. (a) Hammer, (b) Safety glasses, (c) Clip, (d) Tape measure, (e) Exacto knife, (f) Adjustable wrench, (g) Adjustable pliers, (h) Clamp, (i and j) Combination wrenches, (k) Needle-nose pliers, (l) Wire cutter, (m) Small flathead screwdriver, (n) Large flathead screwdriver, (o) Tiny flathead screwdriver, (p) Scissors, (q) Allen wrench set, (r) Large Philips head screwdriver, (s) Hacksaw, (t) File, (u) Small Philips head screwdriver, (v) Sandpaper.

FIGURE 7.2

Common power tools that are useful in prototyping. Labels are shown to the right of each tool. (a) Circular saw, (b) Sander, (c) Jigsaw, (d) Cordless drill, (e) Battery for drill, (f) Plug-in drill, (g) Safety glasses, (h) Dremel®, (i) Drill bits, (j) Dremel® attachments.

iterating your prototypes. Remember that using any tool requires some level of training and appropriate safety precautions.

7.2 Crafting

Crafting was a mainstay of prototyping long before the current maker space movement. While it may be tempting to become caught up in cutting-edge technologies and fancy new spaces, crafting supplies are often more cost-effective and readily available, and the techniques are easier to learn. If you have a craft store nearby, it is worth a visit as the staff are likely to have a wealth of knowledge of processes, materials, and tools not commonly found in engineering and science buildings. A craft shop may have sewing machines, devices for cutting glass (for making stained glass), a kiln, a darkroom, a kitchen, and other potentially helpful resources. A nearby art studio will have additional materials, tools, and artists who are constantly building and iterating on their creations; a sculptor knows a lot about materials and processes that engineers do not often use and may not be aware of. Likewise, if you plan to work with wood, the carpenters in a set building shop in a theater department may be very helpful. If you need to work with fabric, you should visit a costume or upholstery store, a local tailor, or a seamstress shop. Each of these spaces may also have scrap materials that you can obtain for free or at a significantly reduced price.

7.2.1 Classic Prototype Materials

With minimal time, effort, and cost, you can create your own simple craft supply by visiting a local department, home improvement, or hardware store. A brief list of helpful prototyping materials might include duct tape, masking tape, coat hangers, popsicle sticks, a few types of glue, pipe cleaners, paper clips, cardboard, clay, Play-Doh, string, construction paper, a glue gun (with glue sticks), rulers, pencil, kids' paint sets, scissors, staplers, 3 × 5 index cards, 8" × 11" paper, rubber bands, LEGOs®, plastic bags, paper cups, balsa wood, Exacto knives, Styrofoam, sandpaper, and foam board. You may already have many of these items in a design studio, laboratory, machine shop, or department office. Displaying prototyping materials in plain view can often spark ideas.

7.2.2 Foam Core, Balsa Wood, and Styrofoam

Architecture, industrial design, and other fields often use foam core (sometimes called foamboard), balsa wood, and Styrofoam to create small-scale mock-ups of rooms, buildings, and entire landscapes. These materials are lightweight, relatively inexpensive, easy to find at craft stores, easy to cut and join, and can be painted to mock up a user interface. The lift prototype shown in Figure 6.5 was first mocked-up using foam core.

   Foam core often comes in sheets that can be cut with an Exacto knife. To cut foam board, three kinds of cuts are possible: (1) cut through the top layer, (2) cut through the top paper layer and foam, and (3) cut through all material, including the bottom paper layer. Combinations of these three cuts can produce a variety of angles and joints using hot glue from a glue gun. For example, you can make corners by removing a small thickness of core material in a line but keeping the backing paper layer intact. The foam core can then easily be folded through any desired angle. To create a particular angle, you could add hot glue along the joint. To keep the joint flexible, you could add packing tape to the backing paper layer to give it more stability. A helpful resource is FoamWerks, a company that sells its products online at sites such as Amazon.com.

Balsa wood can be cut with a small handsaw or Exacto knife and attached using hot-melt glue or carpenter's glue. Mini-clamps can be helpful to hold joints in place while the glue sets. The wood usually comes in various sizes, including sheets, square stock, and cylindrical dowels.

Styrofoam can be cut in the same way as balsa wood and joined using glue, or in some cases, pins. It can be shaped by starting with a block, cylinder, or sphere and then removing material as needed using a file or course grit sandpaper. Fine grit sandpaper can be used to smooth the surface for further finishing, such as painting.

7.3 Materials, Attachments, and Parts

In constructing prototypes, you should choose materials, attachments, and parts appropriate for your prototype's aims. There are several considerations in selecting a material. First, you should consider the phase of design and intent of your prototype. In crafting (often used for looks-like prototypes), the choice of materials is less important. On the other hand, a later-stage prototype requires careful selection of materials when material properties become more important for prototype testing. Second, there are often important specifications (e.g., strength, wettability) that only certain materials can meet. Third, the costs of materials can vary widely, from a few cents per kilogram for homemade Play-Doh to $2000/kg for polyether ether ketone (PEEK). Generally, you would choose the most cost-effective prototyping material. Fourth, some prototyping techniques only work with particular materials. For example, only a relatively small selection of materials can be used in most rapid prototyping machines. As you move toward a final prototype, you may wish to consult the list of common medical device materials in Table 8.1.

Medical devices often have multiple parts that must be connected through on operation, often called **assembly**. Picking a joining method is based on (1) the materials that require attachment, (2) the desired strength of the bond or connection, (3) the need for parts to move relative to one another and in which dimensions, and (4) if the attachment is meant to be reversible. Table 7.1 lists several common attachment methods.

At some point, you will need to purchase specific parts, stock material, tools, or lab equipment. It can be helpful to visit stores that have a wide range of supplies and knowledgeable sales associates. When visiting, use your cell phone or camera to document what you have found. Good places to visit include:

- Hardware stores (e.g., Lowes, Home Depot, Ace Hardware) and craft stores (e.g., Michael's)
- Fishing, bike, and auto parts stores often have unique parts that you will not find elsewhere
- Toy and kitchen sections of department stores (e.g., Walmart, Target)

Online resources often have a more comprehensive selection of materials, parts, and equipment. You are encouraged to spend some time becoming familiar with the types of products these companies provide. Online retailers of general supplies and equipment include:

- McMaster-Carr (https://www.mcmaster.com)
- Cole-Parmer (https://www.coleparmer.com)

- Newark Electronics (https://www.newark.com)
- Avantor (https://www.vwr.com)
- MSC Industrial Supplies (https://www.mscdirect.com)
- Fisher Scientific (https://www.fishersci.com)
- Grainger (https://www.grainger.com)

Table 7.1 Various joining methods used to create attachments between parts and materials	
Joining Method	**Description**
Glue, epoxy, tape	There are many types of glues and epoxies, usually rated based upon strength of the bond and the range of materials to be bonded (e.g., wood, plastic, metals). Some can be applied to make a deeper-than-surface bond.
Press fittings	A hole is created and reamed (to a tolerance <0.025 mm). A pin pressed into the hole is held in place due to friction and compressive forces.
Bearings	Used to reduce friction between parts (often with a lubricant) that move in one direction while constraining motion in other directions. Can be used to constrain both linear and rotational motion. Often used with shafts and axles.
Tube connectors and fittings	Barbed, compression, and other types of connectors and fittings are often used to join parts of a fluid or pneumatic system.
Soldering and welding	Soldering is often used to connect conductive metal components. The connection is not meant to be mechanically robust, but to allow for good electrical conductivity. Various forms of welding are used to mechanically attach metals, producing weld joints that are stronger than the metals themselves.
Thermal bonds	Many polymers and other materials can be melted and fused together as is done with many rapid prototyping techniques. Heat can be used to melt polymeric materials to seal plastic bags and attach adhesive coated Tyvekc® lids to medical device trays. Often, a rigid plastic tray is thermally sealed to permanently enclose a device.
Screws, nails, rivets, and seams	These fasteners are particularly helpful when dissimilar materials must be joined. First, a hole is created through which the screw, nail, or rivet is placed. Generally, screws are used when a service person needs to undo the connection to gain access to internal components. Nails are used when the connection is meant to be semipermanent. Rivets are most often used when the designer does not want a user to be able to take apart the device. Seams are usually sewn to attach pieces of fabric.
Hinges, ball joints	Used to connect two parts that must be able to move relative to one another. Hinges allow for movement in one direction whereas ball joints allow for motion in multiple dimensions.
Pins, clips, and rings	Used to prevent other attachments from moving out of alignment, often by filling in space around a groove or through a hole. These attachments often provide safety and some are meant to make assembly and disassembly easier. Examples include retaining rings, hairpins, Cotter pins, snap rings, and e-clips.
Magnets, snaps, Velcro®, buttons, zippers, suction cups, and snap lids	Used when a user needs to be able to repeatedly open and close a connection as part of the function of the device.

Several additional online sites sell more specialized materials or chemicals, including:

- MilliporeSigma (https://www.sigmaaldrich.com)
- Acros Organics (https://www.acros.com)
- Oakwood Chemicals (https://www.oakwoodchemical.com)
- US Composites (www.uscomposites.com)
- ePlastics (https://www.eplastics.com)

Later sections of this chapter list online retailers that specialize in materials and tools for particular prototyping methods.

7.4 Three-Dimensional Drawings and Files

The most common first prototype is a sketch, as discussed in Chapter 5. However, you should create a more detailed professional technical drawing at some point in the design process. In most cases, a physical device has a complex three-dimensional (3D) structure. However, paper and computer displays are two-dimensional (2D). ASME Y 14.5 (1994/2009) *Geometric Dimensioning and Tolerancing* is a standard that defines how dimensions and tolerances should be used to help you communicate the details of your design. It includes several important basic requirements:

- Each necessary dimension of a device shall be shown in consistent units. No more dimensions than those essential for complete definition shall be given.
- Dimensions shall be selected and arranged to suit the function and mating relationship of a part and shall not be subject to more than one interpretation.
- Each dimension shall have a tolerance. A tighter tolerance typically results in increased cost and longer manufacturing times.
- Dimensions should be arranged to prevent tolerance stack-ups and provide the required information for optimum readability.
- The drawing should define a part without specifying manufacturing methods.

A variety of **Computer-Aided Design** (CAD) Packages (e.g., Solidworks, AutoCad, Creo) can help you create unique 3D objects and generate 2D renderings that satisfy these requirements. In general, you begin working with basic shapes, and then through operations such as revolve, extrude, join, and cut, create the exact shape you need. A relatively simple example is shown in Figure 7.3.

The details of complex geometries can become very difficult to recreate using basic objects and the operations available in CAD programs. There are additional techniques and resources that can be helpful in representing complex geometries. For example, scanning technologies can be used to digitally represent an object you wish to duplicate (e.g., surgical tool). Some are benchtop devices that can perform a 3D scan, mapping out the various surfaces using a laser-based method. Others are devices such as CT or MRI scanners that can generate 3D patient-specific geometries. ITK-SNAP (www.itksnap.org) and ImageJ (https://imagej.nih.gov/ij/) are free programs that can digitally extract particular features or organs from 3D medical images (known as segmentation).

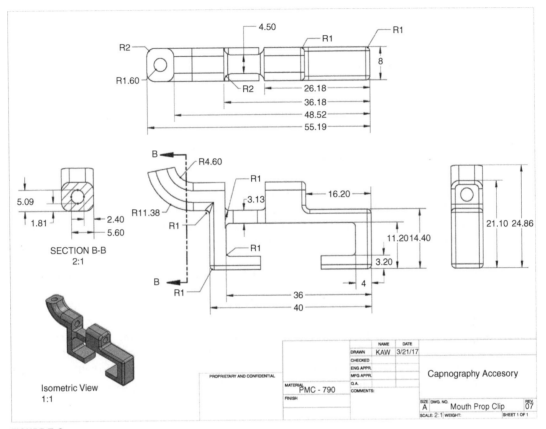

FIGURE 7.3

Student-generated example of a production drawing created in SolidWorks. The physical creation of this part is shown in Figure 7.5. All units are in millimeters.

When designing a medical device, it is often helpful to have access to a 3D shape of an organ or part of the body. There are databases where you can download models of animal and human parts in STL (Standard Tessellation Language) format (more on this format is in the next section and Section 11.2.1 and Figure 11.3). To download anatomical geometries, some helpful online resources are:

- Yeggi (https://www.yeggi.com/q/human+organ/)
- embodi3D (https://www.embodi3d.com/files/category/10-organs/)
- Yobi3D (www.yobi3d.com)
- NIH 3D Print Exchange (https://3dprint.nih.gov/)

Furthermore, many of the sites contain libraries of medical devices; on the Yeggi site, for example, a search for "medical syringe" yields several downloadable results.

7.5 Rapid Prototyping

The design of engineered products has fundamentally changed over the last few decades with the rise of rapid prototyping techniques. Often generically called "3D printing," rapid prototyping is analogous to 2D printing in that a user sends a computer file to a device that renders a drawing as an object. Almost all rapid prototyping methods are computer-controlled to some degree, meaning that there is little (or no) direct user intervention during a print. Unlike a 2D representation, a 3D object can often more quickly and intuitively communicate a design concept. In the past decade, there has been an explosion in improved techniques, fidelity, and materials to be printed, as well as a decrease in cost. The prevalence of rapid prototyping techniques has also allowed objects to be more easily shared across physical barriers.

Rapid prototyping is often referred to as an **additive** process because most of these methods create an object by adding one layer at a time. The nature of additive processes means that complex internal features can be created in a single step. The same object created using **subtractive** processes (discussed in Section 7.6) may require the assembly of multiple parts. Rapid, however, is a relative term. Printing from a drawing may require many hours or even days, depending upon the complexity and size of the object. Therefore, crafting, both additive and subtractive, can sometimes yield a similar prototype in much less time.

The discussion of rapid prototyping techniques in this section is only meant to help you better understand their potential. Each has unique strengths and weaknesses and varies in resolution, print speed, maximum physical print dimensions, material choice, postprint processing types, and other factors. Furthermore, the cost of the devices that use the same technology can vary over an order of magnitude based upon the build size, build speed, and ability to print different materials. To better understand your options, you should consult a technician or one of the many online user communities and tutorials.

7.5.1 Preparing an STL File for Printing

Before discussing the techniques, it is important to know that each follows a general flow from an idea to a printed object. 3D geometry is created using one of the methods described in Section 7.4. The geometry is then converted to STL format. This format (with the file extension ".stl") has become a standard and arose from Stereolithography (the first rapid prototyping technique to be developed and commercialized). The format involves a series of slices through the object, each with a particular thickness (e.g., 0.01 mm, 0.7 mm). The geometry is approximated by a series of triangles within each slice. The STL file is a list of the vertices and outward normal vector of each triangle. Representing complex, especially curved, surfaces requires more triangles at the cost of a larger file size. The STL file is then processed before the print. In this step, the print parameters of the object are optimized. Prototypes are usually weakest and least accurate in the z (vertical) direction. Furthermore, it can be helpful to place the shortest dimension along the z-direction to reduce the number of layers to be printed, and therefore reduce the time to print. Additional support structures are often added to print delicate features such as overhangs, internal cavities, and thin sections. Many rapid prototyping devices come with a proprietary algorithm that aids in these optimizations.

7.5.2 Additive Manufacturing Techniques

Many new rapid prototyping techniques were commercialized over the past several decades, leading to hundreds of acronyms, terms, and definitions. ISO/ASTM52900:15 *Additive Manufacturing—General Principles—Terminology* brought some order to the field. Table 7.2 summarizes the most common naming conventions, advantages, and materials for the seven types of additive manufacturing techniques. More detailed information can be found in the references and resources listed at the end of this chapter.

7.5.3 Laser and Vinyl Cutting

Some rapid prototyping techniques are subtractive, starting with stock and removing material to achieve a desired part or shape. What makes them "rapid" is that the process is automated. Two common techniques are laser cutting and vinyl cutting, both of which work with sheets of material and a 2D vector graphic file. In vinyl cutting, a knife moves side to side while the vinyl is moved beneath the knife. Then the process of weeding begins where excess material is removed. This process can be used to make stamps, signs, and masks from vinyl, fabric, paper, and thin magnetic sheets. A laser cutter works in much the same way, with the knife replaced by a powerful laser. The material can be cardboard, cork, felt, leather, rubber, glass, stainless steel, acrylic, ceramic, or fiberglass, depending upon the thickness of the stock and power of the laser. However, the most common material is a thin sheet of wood because it is generally inexpensive and can be cut quickly. Due to the precision of a laser, a laser cutter can also be used to engrave, where the laser cuts only partway through the material.

7.5.4 Prototype Finishing

After a print is completed, many rapid prototyped models require finishing. Some models may have support structures that were used during the print. These generally need to be removed, either using tools or with chemical treatments. In some cases, a curing or surface treatment (sanding, sealing, painting) can be applied to improve appearance, durability, or water resistance.

7.5.5 The Future of Rapid Prototyping

Rapid prototyping is still a developing area with many new manufacturing techniques, materials, final treatments, and applications made available every month. For example, some new techniques mix additive and subtractive methods. Rapid prototyping methods are also being used to create final products. Some medical products, such as dental implants, hearing aids, stents, and parts of prosthetics are already being created through rapid prototyping.

One of the most promising rapid prototyping technologies is Biological Printing. Although not yet in the mainstream, many research labs are pioneering ways to print 3D living tissues. Generally, the process involves printing a porous scaffold of a biodegradable material using a rapid prototyping technique. Cells are deposited onto the scaffold and allowed to adhere to the internal structures. After some time, these cells lay down their own organic scaffold as the printed material slowly degrades. In some more advanced biological prints, it is possible to leave space for blood vessels, cavities, and other

Table 7.2 Types of additive manufacturing techniques

	Vat Photopolymerization	Powder Bed Fusion	Binder Jetting	Material Jetting	Sheet Lamination	Material Extrusion	Direct Energy Deposition
Alternative names	SLA™—Stereolithography Apparatus; DLP™—Digital Light Processing; 3SP™—Scan, Spin, and Selectively Photocure; CLIP™—Continuous Liquid Interface Production	SLS™—Selective Laser Sintering; DMLS™—Direct Metal Laser Sintering; SLM™—Selective Laser Melting; EBM™—Electron Beam Melting; SHS™—Selective Heat Sintering; MJF™—Multi-jet Fusion	3DP™—3D Printing; ExOne; Voxeljet	Polyjet™; SCP™—Smooth Curvatures Printing; MJM—Multi-jet Modeling; Projet™	LOM—Laminated Object Manufacturing; SDL—Selective Deposition Lamination; UAM—Ultrasonic Additive Manufacturing	FFF—Fused Filament Fabrication; FDM™—Fused Deposition Modeling	LMD—Laser Metal Deposition; LENS™—Laser Engineered Net Shaping
Description	A vat of liquid photo-polymer resin is cured through selective exposure to light (via a laser or projector) which then initiates polymerization and converts the exposed areas to a solid part.	Powdered materials are selectively consolidated by melting using a heat source such as a laser or electron beam. The powder surrounding the part acts as support material for overhanging features.	Liquid bonding agents are selectively applied onto thin layers of powdered material to build up parts layer by layer. Binders can be organic and inorganic materials. Metal or ceramic powdered parts are typically fired in a furnace after they are printed.	Droplets of material are deposited layer by layer to make parts, typically by jetting a photocurable resin and curing it with UV light or jetting thermally molten materials that then solidify in ambient temperatures.	Sheets of material are stacked and laminated using adhesives or chemicals (paper/plastics), ultrasonic welding, or brazing (metals). Unneeded regions are cut out and removed after the object is built.	Material is extruded through a nozzle or orifice into tracks or beads which are then combined into multi-layer models. Common varieties include heated thermoplastic extrusion (similar to a hot glue gun) and syringe dispensing.	Powder on wire is fed into a melt pool which has been generated on the surface of the part where it adheres to the underlying part or layers by using an energy source such as a laser or electron beam. This is essentially a form of automated build-up welding.

Table 7.2 Types of additive manufacturing techniques—cont'd

	Vat Photopolymerization	Powder Bed Fusion	Binder Jetting	Material Jetting	Sheet Lamination	Material Extrusion	Direct Energy Deposition
Pros	• High level of accuracy and complexity • Smooth surface finish • Accommodates large build areas	• Can be used for complex parts • Powder acts as support material • Wide range of materials	• Full color printing • High productivity • Wide range of materials	• High level of accuracy • Full color parts • Enables multiple materials in a single part	• High volumetric build rates • Relatively low cost (nonmetals) • Allows for combination of metal foils, including embedding components	• Inexpensive • Allows for multiple colors • Can be used in an office environment • Parts have good structural properties	• Not limited by direction or axis • Effective for repairs and adding features • Multiple materials in a single part • Highest deposition rates
Cons	• Relatively expensive • Often requires support structures and post curing	• Relatively slow speed • Lack of structural properties	• Not always suitable for structural parts • Post processing significantly adds to build time	• Support material often required • Limited materials	• Finish varies and requires post processing	• Nozzle radius limits final quality • Accuracy and speed are relatively low	• No supporting structures • Low build resolution • Relatively expensive
Materials	UV-curable photopolymer resins	Plastics, metals, ceramic powders, and sand	Powdered plastic, metal, ceramics, glass, and sand	Photopolymers, polymers, and waxes	Paper, plastic sheets, and metal foils/tapes	Thermoplastic filaments and pellets (FFF), liquids and slurries (syringe types)	Metal wire and powder, ceramics

Adapted from ISO/ASTM52900-15 and hybridmanutech.com/resources.

biological structures. Biological printing is a developing area that will have an impact on medical device design in the future.

7.6 Machining

Machining is a collection of techniques that subtract and deform materials. A wide variety of materials can be machined, including metal, rigid plastic, wood, and glass. In general, stock material is in the form of sheets, blocks, or cylinders (solid or hollow). From this stock, the machining technique creates a final object. In this way, machining is similar to sculpting (i.e., starting with raw material and then shaping through a subtractive process). A complex part is often created by mixing and matching several machined parts assembled using one of the attachment methods listed in Table 7.1. Machining can achieve great precision, with tolerances often <0.025 mm (1/1000 inch), and is therefore generally used for making final (or nearly final) prototypes or testing apparatus needed for verification tests.

The goal of this section is not for you to become an expert in machining but rather to understand basic machining operations and effectively communicate with a skilled machinist. You can build and maintain a good working relationship with machinists by keeping the following points in mind:

- Use the same measurement system (Metric or English/Standard) for your entire design.
- Recognize that certain parts require a *fixture* to be made before the part can be made. A fixture is an ancillary holder to ensure that the part does not move while being made.
- Invite suggestions on better ways to design and create a part. Machinists often have suggestions on improving some aspect of your design to make production simpler. Design for manufacturability is discussed in Section 8.4.
- Respect the manufacturing methods chosen. There are often multiple ways to create the same part. Skilled machinists often mix and match techniques and tools as they carefully consider the order of operations. Furthermore, the operations and techniques are generic. For example, the operation of reaming is mentioned, but there are many forms of reaming used for different purposes (e.g., rose, shell, tapered).
- Be mindful that a seemingly simple part could require hundreds of coordinated operations and take hours to complete. Complex parts can take days to create.

7.6.1 Machining Operations

When looking at a design, a machinist determines the operations that need to be performed and the machines that will be used to perform these operations. There is not a one-for-one mapping from operation to machine. For example, a hole can be created by drilling or boring (operations), both of which can be accomplished using a drill press or a mill (machines). Once an operation and machine have been selected, the correct tool or attachment (often called a bit) must be attached to the machine to perform the operation. In the example of drilling, a drill bit is selected based upon the required size of the hole. Before the operation is performed, the material would be aligned and secured. Machinists often

Table 7.3 Common machining operations

Operation	Notes
Cutting	Cutting stock into two or more parts. The nature of the cut depends upon the technique and tool being used.
Sanding	Removing surface material to eliminate burrs and other surface imperfections, smooth edges, and create smooth surfaces.
Drilling	Creating a round hole in a material with a rotating *drill bit*. Often performed in several steps, starting with a smaller hole and making it progressively larger.
Reaming	Smoothing and slightly enlarging the inside of an initially drilled hole. (Tolerances of <0.025 mm are needed for press fits).
Threading	Creating threads in a hole into which a screw will be placed. Required when screws are used in metal parts. Hole size, pitch, and type of thread must be known (it is recommended to pick a standard screw size). Creating the threaded hole is known as *tapping*.
Deformation	Usually bending metal stock. For flat stock a press may be used, sometimes in combination with heating. For tubing, a tube bender is helpful to control the radius of the bend and prevent the inner diameter from collapsing or kinking.
Facing and milling	Cutting of stock to create a flat and finished surface. This is often the first step in machining; it establishes the z = 0 reference point.
Turning	Removing surface material with a sharp tool while rotating (turning) stock (workpiece) about an axis. As the tool is moved closer to the axis, more material is removed, reducing the diameter of the workpiece.
Boring	Enlarging an existing hole by subtracting material from the inner diameter. Typically not used for holes smaller than a few centimeters in diameter but can achieve tight tolerances, similar to the combination of drilling and reaming.

establish an x, y, z coordinate system. For example, in a lathe, the diameter of the spinning workpiece is made smaller by removing material as it spins around a z-axis. Table 7.3 lists common machining operations.

The machines shown in Figure 7.4 can perform several of the operations listed in Table 7.3. As all of these machines subtract material through rapidly rotating stock or tools with sharp edges, there is a significant safety risk to you and others. You should receive proper training before using any devices and carefully follow all safety procedures.

7.6.2 Mill

A mill is a device that holds flat stock in place in the z-direction, while a rotating spindle holding a tool subtracts material in the x- and y-direction. Rotation rates, measured in revolutions per minute (RPM), typically range from 50 to 3000 RPM. The tool performing the subtraction can have a variety of shapes

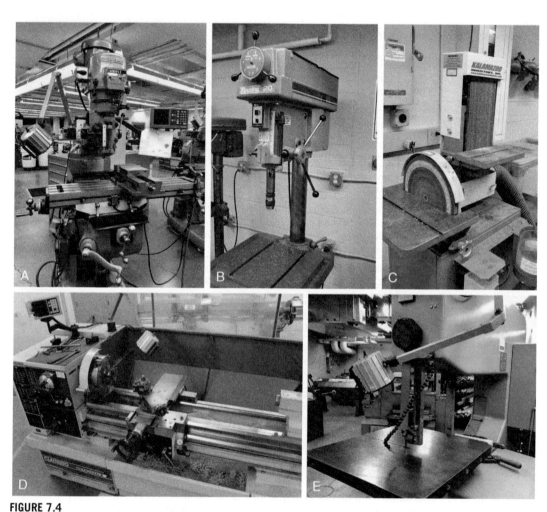

FIGURE 7.4

Common machines found in a machine shop. (A) Mill, (B) drill press, (C) sander, (D) Lathe, (E) Band saw.

and allow for drilling, facing, cutting, boring, and tapping of screw holes. In many mills, the rotating spindle can be tilted, usually by 45 or 90 degrees. Mills can be used on most metals and metal alloys (e.g., stainless steel, aluminum alloy), hard plastics (e.g., fiberglass, high-density polyethylene), and wood. However, some orthopedic alloys (such as CoCrMo and Ti6Al4A) are challenging to mill due to their high surface hardness.

7.6.3 Drill Press

A drill press is a simplified version of a mill and is used to drill holes. It is generally less expensive and requires less training to operate. Its disadvantage is that it is generally not as accurate as a mill because the x-y location is fixed by a clamp and not directly attached to the device itself. However, it is possible to tap holes with the appropriate fixture (this requires experience).

7.6.4 Band Saw

A band saw uses a metal band with a serrated edge that can cut almost any hard material ranging from steel and wood to rubber and fiberglass. It is generally used to cut stock to a rough length or contour and then finished in some other way (e.g., sanding). Many shops have both vertical and horizontal band saws.

7.6.5 Lathe

A lathe is a device that holds round stock and spins it rapidly, often at hundreds or thousands of RPM, around the long axis of the stock (referred to as the z-axis). The stock is held at z = 0 and then a tool (generally a shaped bit of metal) cuts away at the spinning stock, moving progressively down the z-axis while moving perpendicular to the z-axis. The machinist precisely controls the location of the tool by moving a series of wheels. While it is possible to work with square stock, this usually requires a three or four jaw *chuck* for holding and then precise alignment around an axis of rotation. A lathe can be used to perform facing, turning, boring, and threading and produce angled chamfers, with the same materials used when milling. An exception is wood, which generally requires a particular type of lathe that you may find in a craft shop. There are significant safety considerations when using a lathe due to the high rotation rate.

7.6.6 Additional Machining Methods

You may find a wide range of other tools and devices in a machine shop. For example, a sander (Figure 7.4C) can be used to smooth and deburr edges so that a prototype is safe to handle. A sheet metal bender helps create custom boxes and housings for a device.

7.6.7 Power Hand Tools

As shown in Figure 7.2, many powered hand tools can perform the same tasks with less power and precision. For example, a hand drill performs the same operations as a drill press. A Dremel® is a relatively inexpensive tool that, with the right attachments, can perform rudimentary cutting, sanding, boring, and turning. A jigsaw can cut contours in a sheet of stock. Powered hand tools can be beneficial in creating near-final prototypes and often require less training and expertise to use.

7.6.8 Human Powered Tools

Not to be overlooked, hammers, wrenches, hand saws, sandpaper, files, and screwdrivers, shown in Figure 7.1, are often necessary to join parts of a device to one another. Any well-stocked machine shop has hand tools that you can use.

7.6.9 Computer Numerical Control

Most machining operations can be semi-automated through Computer Numerical Control (CNC). Unlike in rapid prototyping, where algorithms directly convert a CAD file into machine instructions, CNC programming is more of an art form. A first step is usually to work with either solids or wireframe surfaces described in CAD. There are hundreds, and in some cases thousands, of tool pathways, rotations, and tool bit changes that must take place to create a part. The machinist must next think about the order of these operations as well as tool lengths, orientations, offsets, rotation speeds, stock feed rates, and other parameters. All of these instructions are contained in a "g-code." Before the automated build can begin, all the tools and stock must be placed and secured in the CNC machine. You may hear the term multi-axis CNC (four or five axes), which means that the table holding the part and the cutting tool can be moved and rotated in space. Multi-axis work allows for more complex parts to be created, but also complicates the creation of the g-code. As in traditional machining, it is sometimes necessary to create a holder for the stock. CNC machines can go beyond traditional machining, including laser cutting, sewing, water jet cutting, and other techniques. The main advantage is that once the g-code is created, many copies can be made of the same part with much less oversight and constant attention from a machinist.

7.7 Molding and Casting

Molding and casting are similar methods of producing a part. Both involve using a cavity (usually called a *mold* or *die*) that represents a negative of the final part to be created. Filling the mold with a soft or liquid material and allowing it to harden results in a positive part with the desired shape. Just as some shapes are easier to create using additive or subtractive manufacturing methods, molding and casting are often the best options when a part is needed with complex surface features or contours that would not be easily created using other prototyping methods. Molding is typically used for plastic materials that are injected, transferred under pressure, or poured into a mold. An example of a molded part is shown in Figure 7.5. Casting is most commonly used for liquid metals or alloys. Most of the materials used in medical devices (shown in Table 8.1) can be molded or cast. You may also find molded epoxy, plaster, or clay helpful in creating parts for prototypes. To reflect industry practice, we will use the term "molding" to describe both molding and casting. This section will step through the process of molding.

The first step is to create a mold that serves as a negative of the desired part. Molds are typically made from a relatively rigid material such as plaster, plastic, rubber, ceramic, or metal. The choice of mold material depends on the desired material properties of the molded part, the temperature required to liquefy and cool the molding material, and the required life of the mold. Molds made of metal will rapidly dissipate heat, allowing molded parts to cool quickly and be removed from the mold. Molds made of more rigid materials, such as steel, will wear less than softer materials such as rubber or aluminum and extend the life of the mold, allowing for more parts to be made before the mold will need to be replaced. However, molds made from harder materials tend to be more expensive. Molds are often made using one of the techniques discussed in this chapter. For example, machining may be used to create an aluminum mold, or a rapid prototyping technique may be used to make a plastic mold. Depending upon the part's complexity, sometimes molds contain multiple parts that are tightly clamped or locked together. The spaces between these parts become the negative cavity used for molding.

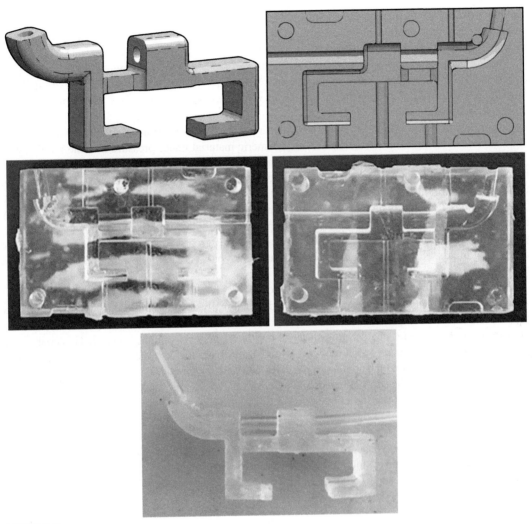

FIGURE 7.5

Creating the complex part in Fig. 7.3 using injection molding. The top left is the desired part to be molded. The top right is a rendering of one half of the mold. The bottom left shows the top and bottom parts of the mold printed in resin. The bottom right shows the molded prototype made with PMC-790 Liquid Rubber.

The second step is molding the part, which involves pouring or injecting an initially soft material into a mold. After some time, either through cooling or a chemical reaction, the material solidifies, taking on the complimentary shape of the mold. Molds must be designed, and specific steps in the molding process must be followed to ensure that the material completely fills the cavity (no short fills), there are no air bubbles trapped in the solidified molded part, and that the part can be easily separated from the mold (no negative draft angles). Runners, sprues, and vents (all types of channels) must be strategically placed to allow air to escape as the molding materials enter and fill the mold cavity. Shrinkage of molding materials (which can occur during cooling and after the part has been "shot") must be considered when designing molds. Molds are designed by die makers who are typically part of a Manufacturing Engineering department in a company.

The last step is to remove the part from the mold and apply any posttreatments. Some molds have release mechanisms built into them to aid in removal. Posttreatments include removing extra material (e.g., flash located in or around runners, sprues, or vents) or chemical treatments that improve surface properties such as wettability or roughness.

Compression, injection, and transfer molding are three common forms of molding. Compression and injection molding involve forcing heated polymeric material under pressure into the mold cavity. Transfer molding is a combination of compression and injection molding. All three methods allow for high-volume production. Multicavity molds produce multiple parts per cycle, allowing production of thousands of parts in less time than other manufacturing processes, reducing the cost of each part. These methods tend to create less waste and can produce complex parts. Injection molds are very expensive and not practical (or cost-effective) for producing single prototype parts. Injection and compression molding are not typically an option in academic design projects due to the high cost and time of creating molds. However, kits are available that allow you to create molds and pour materials into the molds to make a part.

Investment casting (also known as the lost wax process) is commonly used to manufacture hip and knee implants. This process begins by creating a dimensionally accurate mold into which melted wax is poured to create a wax positive. These wax parts are then attached to a wax tree, and the entire assembly is dipped several times in a ceramic slurry to build up layers of ceramic material. This coated tree is then placed into a furnace to melt the wax, which drains from the ceramic coating, creating a series of ceramic shells, each of which contains a positive cavity. Molten metals or alloys are then poured into the ceramic tree to form the parts. When cooled, the ceramic shells are cracked and removed and a positive casting of the desired part remains attached to the tree. The ceramic material is used due to its ability to withstand the heat of the furnace and molten metal. These parts are then cut off the tree and further processed to remove extra material and create a smooth surface. Investment casting is expensive and requires equipment not typically found in academic machine shops or maker spaces, and thus is not feasible for creating a single prototype for an academic design project.

Two helpful sites concerning molding and casting are:

- Smooth-On (https://www.smooth-on.com/; training, stamps, molds, and materials)
- CustomPartNet (www.custompartnet.com/estimate/injection-molding; online cost estimator)

7.8 Microfluidics

Microfluidics is a broad category of technologies involving flow through relatively small channels ranging in size from nano- to micrometers. Mixing and chemical reactions often behave differently when flow diameters become small enough that capillary forces and turbulent flow begin to dominate. Microfluid devices can mimic reactions and fluid flows that occur in real cells and have therefore become a technique for performing assays in the field. The remainder of this section discusses two common materials used to create prototypes of microfluidic systems, polydimethylsiloxane (PDMS), and paper.

The most common prototype material for microfluidic devices is PDMS. It is relatively inexpensive, nontoxic, transparent, and easy to cast into any shape with a wide range of desired feature sizes. PDMS is made by mixing a liquid silicone elastomer base with a liquid curing agent or hardener, which is then poured into a mold. After mixing, bubbles are removed. The mix is baked in an oven, cooled, and then the solid PDMS is peeled off of the mold. Physical properties can be controlled by varying the ratios of the base and curing agent. It is also possible to apply a wide range of treatments to change the hydrophobicity of the surface.

To make very small-scale features, soft lithography can be used to create a mold with the negative of the desired channels. To make such a mold, a silicon wafer is first covered with a thin layer of photoresist material. A patterned mask is then placed over the wafer and photoresist that selectively lets light through some areas but not others. UV light is then applied, which selectively dissolves the photoresist. The resulting master mold can then be used as a negative into which the liquid PDMS is poured. After curing, the PDMS can be peeled off of the master, which can then be used again. To create channels for fluid, one patterned PDMS block is inverted and placed against another flat or patterned block of PDMS.

Some of the same ideas for creating selective barriers and channels can also be created using paper. The key is that capillary action can draw fluid through the porous paper. The advantage of paper is that it is inexpensive, can be cut into any shape, and the wettability can be selectively altered by including hydrophobic barriers within the hydrophilic paper. Because paper can be infused with chemically reactive materials, it is also possible to create paper-based "lab on a chip" chemical labs. If the reaction generates a visible chemical change (similar to a home pregnancy test), a paper-based device can be used as an affordable, portable, and nonpowered diagnostic. Paper-based solutions already exist for detecting glucose, *Salmonella*, and *Escherichia coli*, and several labs are exploring methods to perform more complex disease and environmental screening tests.

7.9 Electronics and Electrical Prototyping

Many medical devices contain electrical circuitry that acts as the "brain" of the device. It is through circuitry and microprocessors combined with algorithms that other components of a device are integrated. Circuitry also is used to control the flow of power and information through the device by making real-time decisions that connect sensors (e.g., flow measurement) to actuators (e.g., pumps, valves). In this section we review some standard tools (shown in Figure 7.6) and techniques for electrical prototyping.

7.9.1 Breadboarding

Electrical prototyping often begins on a breadboard—essentially a board with conductive holes—allowing various electrical components to be easily interconnected. These electrical components may be passive analog parts (e.g., resistors, capacitors, inductors), simple digital logic chips (transistors, diodes, LEDs, AND gates, OR gates), active integrated circuits (e.g., comparators, op amps, filters, sensors) or actuators (e.g., motors, LCD displays). In an electrical prototype (as shown in Figure 7.7), each function is often achieved separately and then integrated together in future iterations. The advantage of breadboarding is that you can troubleshoot connections and the integration between parts. Breadboarding is generally most helpful when used with a power source, function generator, oscilloscope, and logic probe.

FIGURE 7.6

Common equipment and tools that are useful in electrical prototyping. Not shown are electrical components (e.g., resistors, capacitors, digital chips, wire). (a) Digital multimeter, (b) Soldering iron, (c) Digital power supply, (d) Handheld multimeter, (e) Solder, (f) Breadboard, (g) Oscilloscope, (h) Wire strippers, (i) Electrical tape, (j) Small screwdriver, (k) Safety glasses, (l) BNC to alligator cable, (m) BNC to banana cable.

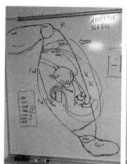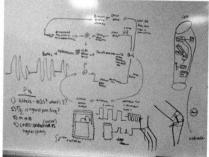

FIGURE 7.7

Breadboards and cost-effective microprocessors allow for electrical prototyping. It is helpful to plan out wiring (left panel) and logic (middle panel) before ordering parts and assembling on a breadboard (right panel).

The exact parts you purchase will depend on your particular project. A variety of online sites sell analog and digital electro-mechanical parts, including:

- Digi-Key (https://www.digikey.com)
- Mouser Electronics (https://www.mouser.com)
- Jameco (https://www.jameco.com)
- Analog Devices (https://www.analog.com)
- MPJA (https://www.mpja.com)
- Precision Microdrives (https://www.precisionmicrodrives.com)
- Pololu (https://www.pololu.com)

You may also find a wide range of physiological sensors at:

- Biopac Systems, Inc. (https://www.biopac.com)
- Vernier (https://www.vernier.com)

Some more specialized but helpful sources for parts can be found at:

- K&J Magnetics (https://www.kjmagnetics.com)
- Battery Kings (https://www.batterykings.com)
- Electric Motor Warehouse (https:///www.electricmotorwarehouse.com)

7.9.2 Power and Power Budgets

When prototyping medical devices that require power, it is good practice to begin with a power supply to provide the necessary current and voltage. When integrating electrical components, remember to use the same ground; this is a common source of problems. Later prototypes would typically use a power supply that is connected to wall power. This generally requires a transformer that converts the wall power to the desired voltage and current for the devices. It is good practice in later prototypes to create components that use the same power source. To improve the electrical safety of a clinical device, it is helpful to isolate this power source (e.g., isolation transformer) so that any spikes in power will not be transmitted to the patient or medical personnel.

While some medical devices are plugged in, others must be battery-powered. This is sometimes a feature included in clinical devices to prevent power from the wall from reaching a person. It is also a means by which a device can become portable or implantable. It is always good practice to reduce power usage, which becomes particularly important when the device is battery powered. Battery selection usually involves two considerations. The first is the maximum amount of power (e.g., product of the supply voltage and current) the device requires at any one time. Batteries are usually rated based upon voltage and a maximum possible current draw. The second is how long a battery can last if run during the normal operation of the device. It is important to note that medical devices often run in one of three modes: (1) background or passive mode during which very little power is used, (2) normal operation during which a power-intensive function is executed, and (3) extreme use during which a very power-intensive function is executed for a brief period. Different types of devices have a different balance (or percentage of time) spent in any one of these three modes. For example, an implantable cardioverter-defibrillator (ICD) operates mainly in the background, powering sensors and minimal detection software. If a cardiac arrhythmia is detected, it may go into pace mode, requiring a significant and persistent amount of power. If this does not work, the device may go into shock mode, requiring short bursts that use a great deal of power. Breakout Box 7.3 shows an example of a simple battery power calculation.

7.9.3 Microprocessors

Some medical devices require more complex flows of information and power distribution than can reasonably be represented on a custom circuit. Microprocessors are mini-computers contained in a single integrated circuit that can be flexibly programmed. Much of working with a

---

Breakout Box 7.3 Back of the Envelope Power Budget Calculation

As shown in Figure 6.3, an approximate calculation can often provide a great deal of insight and help make critical design choices. When designing a battery-powered device, you will eventually need to select a battery. To do so, you should measure the power usage of your prototype during operation while plugged into a power supply. However, the battery you use for an early prototype might not be the battery that would be used in a later prototype. To gain insight into what you might need, it is important to understand the two specifications printed on most batteries. The first is the maximum draw current. All batteries have a limit on the rate at which charge can flow out to another device or component (e.g., current). For example, an implantable cardioverter-defibrillator (ICD) will need no more than 20 mA to produce a pulse. The second is a measure of the amount of charge contained within the battery, usually reported in Ampere∗hours (or Ah). An Ampere is charge/time, so multiplying by a unit of time will determine the charge that can be drawn out of the battery. As charge is drawn from the battery, the voltage will begin to drop, affecting power levels and potentially the function of digital components. For this reason, the rating on a battery is usually 50% of the total charge; a 2Ah battery will actually contain 4Ah of charge. To convert to Faradays (F):

$$4Ah \ast \ (60 \ \text{sec/min}) \ \ast \ (60 \ \text{min/hour}) \ = \ 14,400 \ F$$

Only about half of this charge (~7200 F) will be truly useable due to the voltage drop. Knowing this value will help estimate the battery's service life. Almost half of the size of an implantable pacemaker is due to the battery. As the charge density of batteries has risen, the batteries (and therefore pacemaker) have become smaller.

Consider that in a typical pacing event, a stimulus pulse of 10 mA is delivered at 5 V for 0.5 msec at 70 beats per minute for one minute. We can first estimate the charge required for each pulse. This is simply the duration of the pulse times the current, or 10 mA∗0.5 msec = 5 F. But we know that we require 70 beats over the course of one minute. So, the charge required for this pacing event will be 350 F. If we consider that this pacing event draws current from a battery rated at 2Ah (~7200 F) we can compute how much one minute of pacing will drain this battery.

Real pacemakers are much more complex, having multiple modes (e.g., sensing, reprogramming, charging) that all require different amounts of power. For example, in background mode, the pacemaker might always be running, and draining small but persistent amounts of charge from the battery. The inclusion of a defibrillation capability, the ability to recharge batteries through the skin, multiple pacing wires, and other capabilities make the true calculation much more difficult. For an average patient, it is estimated that over the device's lifetime, about half of the battery drain is due to background mode, and the other half is due to pacing mode. Current ICDs function for a decade or more without recharging.

---

microprocessor involves programming in a language native to that specific processor. Unlike circuitry on a breadboard, complex logical flows can be represented in the code embedded within the microprocessor. As a result, logical flows can be changed through programming rather than rewiring.

For the purposes of microprocessor prototyping, it is important to clarify how the logical flows control the actions of the components outside of the microprocessor. A helpful method is to draw a block diagram with the microprocessor at the center and represent communication with other devices as inputs and outputs. The next step would be to independently test (usually through the writing of functions) the communication with each component. For example, imagine a device that senses pH every 30 seconds and then, depending upon set thresholds, opens a valve and injects a precise quantity of a solution. Such a system would require a microprocessor to coordinate the actions of the pH sensor, valve, and motor that drives the injection device. Each of these devices could be integrated with the microprocessor separately. After working separately, more logic would be added to coordinate the actions of the sensor, valve, and motor.

As you build your prototype, you should keep two points in mind. First, microprocessors have a limited number of inputs and outputs. It can be helpful to revisit your drawing to ensure that you have enough input and output pins available on your microprocessor. Such a count may also help you select a microprocessor. Second, microprocessors generally run on a small amount of power and therefore cannot be used to power external devices. It is often possible to power simple sensors, LEDs, or even small display screens. However, devices that require greater power, such as a pump or motor, require external power sources. Third, different parts often have different power requirements, necessitating some power conversions, broadly referred to as power management.

The Maker's movement has catalyzed the availability of low-cost, powerful, and easy-to-use microprocessors such as Arduino (www.arduino.cc), Raspberry Pi Foundation (www.raspberrypi.org), and Beagle Board (beagleboard.org). Each has supportive online communities, add-on features (e.g., "Shields" for Arduino), and often come in kits with standard parts and self-guided projects to help you quickly learn. Many come in a variety of forms to best suit your needs. For example, Arduino comes in the Uno (standard), Duo (more powerful with more features), and LilyPad (can be used for wearables and even can be washed in a washing machine).

There is also a wide range of maker kits and pre-made boards for microprocessors at:

- Maker Shed (https://www.makershed.com)
- Hobby Partz (https://www.hobbypartz.com)
- HobbyKing (https://www.hobbyking.com)
- Adafruit (https://www.adafruit.com)
- SparkFun Electronics (https://www.sparkfun.com)

Although your final prototype may contain an off-the-shelf microprocessor, these would generally not be used in a commercialized medical device. At some point, a more powerful or custom microprocessor would be used.

7.9.4 Data Acquisition

Microprocessors can take in data from the outside but are often limited to a small number of signals and a relatively slow sampling rate. You may need a Data Acquisition (DAQ) board and the associated software to add many channels or increase the sampling rate. Such is the case when actions must take place quickly (on the order of milliseconds or less), as occurs in imaging, laser, and some electro-mechanical control systems. For medical devices with these requirements, specialized hardware and software may need to be created. For prototyping, it is helpful to explore hardware and software developed by National Instruments Inc. The graphical programming environment LabView (also a National Instruments product) is the gold standard language for data acquisition. There is a great deal of online support and training (both formal and informal) for LabView programming.

7.9.5 Circuit Board Layout

Production-quality printed circuit boards (PCB) can be created after a design has been tested on a breadboard. A PCB is referred to as "printed" because the electrical connections (called traces) are sandwiched

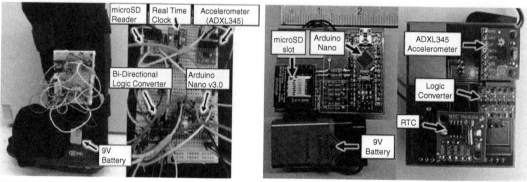

FIGURE 7.8

Student-created prototyping of circuitry. The left side is a preliminary prototype using a breadboard and Arduino Nano. The right side is a later generation prototype with multiple printed circuit boards.

in between nonconductive plates (the board). The various electrical components are then placed into holes in the board and then soldered to make the connections. The advantages over a breadboard are that wires are eliminated, and the connections are permanent. Using a printed circuit board can help simplify complex wiring and result in a more professional-looking prototype. To create relatively simple boards, you can use ExpressPCB (www.expresspcb.com), a free downloadable program that allows you to lay out a board (similar in many ways to CAD). You can then send out the file for printing, or your campus may have an in-house printing facility. Figure 7.8 shows a version of a prototype on a breadboard (left side) and the same design but transferred to multiple PCBs (right side). The smaller PCBs can be more easily contained in a custom-created box that hides the circuitry and only shows the user interface.

7.10 Programming, Connectivity, and Simulations

Advances in hardware and software have enabled programming, connectivity, and simulations to become influential contributors to medical devices and medical device design. The algorithms embedded within infusion pumps and pacemakers, telemetry equipment within a hospital, enhanced connectivity enabled by smartphones, digital patient records, surgical simulators, and many other devices have sparked a rethinking of where, how, and to whom health care is administered. Programming, connectivity, and simulations also played an important role in aiding the engineering design process, particularly the prototyping phase of design. In this section we explore two ways you can create software prototypes in the form of apps and computer simulations.

7.10.1 App Prototyping

Smartphone apps have moved past the novelty stage and promise to impact the healthcare ecosystem in exciting and unpredictable ways. Regulatory bodies have cleared only a handful of app-based applications. However, this area is expected to grow as apps play a more prominent role in medical diagnosis and treatment. Although private medical data must be shared carefully, new security techniques may allow a more accessible and timely flow of medical data between patients, physicians, and providers. Apps are becoming the front-end interface or dashboard for some medical devices, allowing patients and caregivers to access vital signs, dosages, and other critical information. Apps are also being combined with clinical databases to make diagnoses and suggest treatments. For example, Tissue Analytics is a company that has created an app that, given a user-generated image of a wound, can make a diagnosis and suggest a treatment.

The widespread adoption of smartphones in the developing world is driving innovative healthcare apps for low-resource settings. The entrepreneurial community has also recognized the potential impact of health-related apps. Business incubators often support app development, and some incubators, such as RockHealth, provide advice, resources, and policy updates regarding health care apps.

Whether the core of your idea is an app, an app and a device, or simply a new feature for an existing device, you should still create and iterate several prototypes. To create a looks-like prototype, a simple solution is to sketch the interface on paper. You could create multiple screens that are numbered to demonstrate the flow of your app. A similar (and more professional looking) solution would be to move these drawings to a graphics program such as Microsoft PowerPoint or Adobe Illustrator. To move to the next level, you may wish to consider online resources such as proto.io (https://proto.io/) or sketch. io (https://sketch.io/). Using these services (which are often free), you can create the front end of your app (what is sometimes called a wireframe) that is not functional (meaning it cannot connect to the internet or other databases) but can show how a user will navigate the interface. Some online services (such as proto.io) allow for a download to a smartphone to experience what it is like to navigate the prototype.

Going deeper into building an app requires coding. There are essentially two available platforms. Programming for Apple iOS means coding in Xcode (part of the Mac OSx operating system, but also a free download). In contrast, programming for Android can be done in Java, C++, Python, and several other object-oriented languages.

7.10.2 Modeling Devices and Systems

There are instances where computational code or a software package can simulate the desired function or attribute to help determine feasibility or provide guidance during the design process. Often a Computational Fluid Dynamics (CFD) model or simulation of mechanical force and stress distribution within a device can predict device performance and determine technical feasibility. In both cases, Finite Element Analysis (FEA) is often the tool of choice, using a variety of software packages (e.g., COMSOL, CREO, NASTRAN). For example, an undergraduate team explored various geometries for trapping bubbles that often result when fluids are mixed in a cardiac catheter line. Before building physical prototypes, the team created a finite element model to explore the impact of various geometries. This computational study helped the team narrow down possible design options. While the model required time and effort to set up and run, it allowed the team to avoid wasting time and materials on prototypes that would not work and iterate much faster once they began to build physical prototypes. Other possible computational models may be used to simulate electrical, thermal, optical, or chemical systems.

7.11 Soliciting Feedback and Testing

Prototypes serve many purposes in moving a design process forward. This section discusses how to gain insights from prototypes that can help you make future design decisions. The first is to show your prototype to others as a preliminary form of validation. The second is to test your prototype against specifications, a form of verification. Although verification and validation are formally discussed in Chapters 9 and 10, you should use feedback from many sources to inform your next steps once you have a prototype.

7.11.1 Interviews and Surveys

Interviews and surveys can be used to obtain information on prototypes, as in Chapter 3. The goal of soliciting feedback on a prototype from a user or customer is to validate that your design will ultimately lead to a solution that addresses the underlying problem you are trying to solve. With early prototypes, you can have your mentor or advisor represent the voice of the user or customer. As the design advances, you should add more subjects and diversify the groups to ensure that multiple stakeholders' feedback is obtained and considered. It can be helpful to include some extreme users (those who test the full range of functionality). They may find flaws that a regular user would not. It may also be helpful to ask a user to intentionally try to break or misuse a spare prototype. This can help identify potential design flaws that could result in safety issues.

7.11.2 Demonstrations

Prototypes can be one of the most effective ways of communicating both your process and the functionality of your device. In this regard, a prototype can be very helpful during design expositions, design panel reviews, and presentations. In a design review you should summarize the changes since the last prototype, why the changes were made, and what you have learned from the current prototype. It can help to pass around the prototype, and in some cases, allow panel members to try out the functions. Showing a prototype often flows naturally into a discussion of where to focus your design efforts next. There are cases when you may not be able to demonstrate your prototype fully (for example, if it takes too long or if you are not confident that the demonstration will work in the moment). In these cases, it can be helpful to create a video (perhaps edited for time) showing how your device works. If you feel that your device might be patentable, then public disclosures (such as presentations during design expositions) of your design must be avoided. Chapter 15 discusses ways to protect your intellectual property to avoid losing your patent rights.

7.11.3 Testing Prototypes

As you build and iterate prototypes, keep your specifications in mind. Testing your prototypes against your specifications is an excellent way to keep your technical progress moving forward. Early on, you may test particular specifications to determine feasibility. These tests are generally informal and aim to help you make a design decision or measure your progress toward meeting a specification. For example, you might quickly run a test to compare three different ways to generate power for a device; the results should help you make a design decision. Once you have a fully functional works-like prototype (e.g., a MVPr), you may begin conducting verification testing. Verification testing requires careful planning of procedures and analysis techniques and is discussed in Chapters 9 and 10.

It is important only to test attributes that are expected to be present in the final device. For example, if a specification relates to the strength of a particular device made of stainless steel in the final design, but the prototype is made of plastic, it is not necessary to test the strength of the plastic components used in the prototype. An exception would be if the question being asked pertains to which *design* is the strongest; the geometry of a device can significantly change the distribution of stresses. In this case, prototypes made of the same material but with different geometries and stress-strain curves could be generated and compared because the question is about the relative strength of a design, not its absolute strength.

7.12 Prototype Review and Documentation

Prototypes are the most tangible results of your design process. For this reason, it is critical to document your prototypes, either with labeled photographs, videos, or both. Documentation of each prototype should be accompanied by decisions regarding future design directions. For example, if you show a prototype at a design expo, you should be prepared to receive and record comments. These notes can be used to help justify your next design iteration. You should include any drawings and processes used to create each prototype.

Table 7.4 illustrates one way to document the evolution of a prototype. The example is from an undergraduate design project for a small ring inserted into the nose during oral surgery. The device aims to alert a physician when sustained pressure may put a patient at risk of nasal necrosis. It is important to point out that the intention of Table 7.4 was to detail the mechanical design. Similar tables were generated to document the evolution of the electrical design and associated Arduino computer code. However, you will notice that the prototype began to incorporate the electrical circuitry during step seven. The team also downloaded and 3D printed an anatomically accurate model of a nose and created a finite element model in COMSOL to analyze the distribution of mechanical forces.

Table 7.4 A student-created summary of mechanical prototypes and design iterations resulting in a final prototype

Design Iteration # and Description	Image	Negatives	Iteration
1. Use a foam tube as the absorbent material with inner and outer nodes of aluminum foil. Resistivity would be measured across the foam (lower Ohms = less pressure).		The foam did not effectively hold the conductive gel.	Needed a more absorbent material.
2. Use a sponge as the absorbent material with inner and outer nodes of aluminum foil.		The conductive gel would dry out in the sponge over time.	Needed a material that would be conductive, cushion the nostril, and not dry out.

Table 7.4 A student-created summary of mechanical prototypes and design iterations resulting in a final prototype—cont'd

Design Iteration # and Description	Image	Negatives	Iteration
3. Copper flakes mixed into silicone to make an electrically conductive pliable polymer.		The composite did not conduct electricity consistently.	Stopped exploring the resistivity idea.
4. Two ridges of a pliable material, a conductive wire in the valley between the two ridges, and a conductive layer on top of the ridges. When the pliable material was pressed, the conductive layers would touch, completing a circuit.		Sponge was difficult to manipulate.	Used a moldable material.

Continued

Table 7.4 A student-created summary of mechanical prototypes and design iterations resulting in a final prototype—cont'd

Design Iteration # and Description	Image	Negatives	Iteration
5. Created a ring of silicone in place of the sponge. Used a twisted wire on the inside and aluminum foil on the top superglued to the silicone ring. Covered the prototype in silicone for biocompatibility.		Wire was too pliable and would bend out of its circular shape, continuously touching the outer conductive layer.	Used a more rigid inner conductive layer.
6. Same design as previously, only using a key ring in place of the twisted wire.		Worked well at the large scale.	Downsized to fit in the nostril.

Table 7.4 A student-created summary of mechanical prototypes and design iterations resulting in a final prototype—cont'd

Design Iteration # and Description	Image	Negatives	Iteration
7. Decided a clip may be easier to use at the smaller scale. Same model as previously, but the silicone is cut, and the inner wire has a loop and an extended piece to secure the two sides together when in use.		Wire clip mechanism was difficult to use.	Found a more efficient clipping method.
8. Two extended handles to pull the two ends together and secure properly in the nostril.		Did not secure well.	Found a more efficient clipping method.
9. 3D printed a small circular clip (similar to a hose clamp), coated it with copper tape, and placed it around a silicone piece.		Could not make the clip thin enough so the two copper layers would not touch when fully assembled.	Increased silicone ridge height or could use thinner clip.

Continued

Table 7.4 A student-created summary of mechanical prototypes and design iterations resulting in a final prototype—cont'd

Design Iteration # and Description	Image	Negatives	Iteration
10. Extended the height of the silicone ridges with a rubber band for testing purposes.		Worked, but with a silicone ridge at this height, it would not fit into the nostril. This was the smallest clip size to fit around the tube.	Brainstormed a better solution.
11. Used a rigid metal clip on the inside, coated in copper tape. First tested with rubber bands as the ridges to elevate the outer copper surface.		Did not work because the rubber bands were not pliable enough to allow the two copper layers to touch. The inner clip worked well.	Used silicone instead of the rubber bands.
12. Silicone ridges used instead of the rubber bands. Updated the outer copper layer to have a silicone coating for biocompatibility.		Final prototype.	Needs better soldering methods.

Key Points

- Prototyping involves a broad set of techniques for transforming an idea or sketch into a physical artifact. It is an iterative process used throughout the design process to test ideas, make progress toward meeting your specifications and solicit feedback from mentors, advisors, users, and customers.
- Looks-like prototypes are meant to convey a design concept and simulate how a device will be used. Works-like prototypes are intended to perform the functions of the device.
- Several prototyping methods are commonly used in medical device design, including crafting, traditional and CNC machining, rapid prototyping, molding and casting, microprocessor programming and circuit layout, smartphone app design, and microfluidics.
- Most design projects require mixing and matching different prototyping methods as the design evolves and the project advances from one phase to the next.
- You should document the evolution of your prototypes. Such a record is an excellent way to demonstrate how you have iterated on your design over time.

Exercises to Help Advance Your Design Project

1. Create a file or folder in your Design History File to record your prototype designs. Consider a format similar to Table 7.4.
2. Visit a technician, machinist, or lab director who you think could be a good resource as you enter the prototyping phase of your design project. Explain your project and discuss prototyping options.
3. Early in your prototyping, visit a local hardware store, department store (especially the kitchen and toy sections), or crafting store. Look for existing parts, components, tools, and materials that you might use to build your prototypes. Bring a notepad or camera with you and record ideas for design improvements that come to mind. Sometimes seeing what exists can be a very effective way to iterate on your design.
4. Explore the suppliers' websites listed in Section 7.3 for parts that could be used in your prototypes. Do the same for one or more of the other sections that most pertain to your particular project. Create a Parts and Materials document to capture what you find.
5. Discuss what you expect to be the final material and manufacturing method used to create a commercialized product, assuming that thousands of units would need to be made. Manufacturing considerations are discussed in Chapter 8, but recording your notes from a preliminary discussion can be helpful later in the design process.
6. Choose a prototyping technique that you think you may use in your design project. Spend some time learning about this technique online. Keep records of the best and most helpful resources you find and add them to your Design History File.

References and Resources

There are a wide variety of resources for prototyping, and we cannot provide a comprehensive list of resources that are relevant to your project. Many of the best and most up-to-date resources are online tutorials. Such tutorials are posted by individuals on sites such as YouTube or by companies providing resources for their customers (e.g., SolidWorks tutorials on building digital representations of 3D objects). Likewise, there are many handbooks and encyclopedias (e.g., Linden's Handbook of Batteries) that may be relevant to your project. You can likely find these online or with the help of a librarian.

Many medical devices involve electromechanical components. Some helpful resources include.

Boothroyd, G., Dewhurst, P., & Knight, W. A. (2010). *Product design for manufacturing and assembly* (3rd ed.). CRC Press.

Budynas, R., & Nisbett, K. (2020). *Shigley's mechanical engineering design*. McGraw-Hill.

Engineering Product Design. (n.d.). *What is additive manufacturing?* https://engineeringproductdesign.com/knowledge-base/additive-manufacturing-processes/.

Hallgrimsson, B. (2012). *Prototyping and modelmaking for product design*. Laurence King Publishing.

Horowitz, P., & Hill, W. (1989). *The art of electronics*. Cambridge University Press.

Hybrid Manufacturing Technologies. (n.d.). *Seven families of additive manufacturing*. https://hybridmanutech.com/resources/.

Loughborough Mott, R., Vavrek, E. M., & Wang, J. (2017). *Machine elements in mechanical design* (6th ed.). Pearson.

Scherz, P., & Monk, S. (2016). *Practical electronics for inventors* (4th ed.). McGraw-Hill Education.

Sclater, N. (2011). *Mechanisms and mechanical devices sourcebook* (5th ed.). McGraw Hill Education.

Skakoon, J. G. (2008). *The elements of mechanical design*. ASME Press.

L. University. (n.d.). (Seven categories of additive manufacturing. https://www.lboro.ac.uk/research/amrg/about/the7categoriesofadditivemanufacturing/.

In addition to these resources, there is an entire culture, "The Makers Movement," that provides online resources. One of the most prominent of these is MAKE (https://makezine.com/). In addition to posting projects on their website, MAKE also published books on prototyping, which can be ordered through their website. Some of the most relevant book titles (in no particular order) are:

How to Use a Breadboard
Tools: How They Work and How to Use Them
Making Things Smart
3D Printing: The Essential Guide
Getting Started With Arduino
Getting Started With Sensors
Getting Started With RFID
Getting Started With 3D Printing
Encyclopedia of Electronic Components
  (Volumes 1, 2, and 3)

The Makerspace Workbench: Tools, Technologies, and
  Techniques for Making
Design for CNC
Getting Started With CNC
Fabric and Fiber Inventions
Zero to Maker: A Beginners Guide to the Skills, Tools,
  and Ideas of the Maker Movement
Motors for Makers
Mechanical Engineering for Makers

Detailed Design

8

Manufacturing is more than just putting parts together. It's coming up with ideas, testing principles, and perfecting the engineering as well as final assembly.
– James Dyson, CEO, serial inventor, Dyson Ltd.

A user interface is like a joke. If you have to explain it, it's not that good.
– Martin LeBlanc, founder and CEO, Iconfinder

Chapter outline

Biomedical Engineering Design. https://doi.org/10.1016/B978-0-12-816444-0.00008-0
Copyright © 2023 Elsevier Inc. All rights reserved.

8.1 Introduction

At this point in a design project, you have generated potential solutions to the problem, selected what you expect to be the solution that best meets the customer needs and associated target product specifications, and created a prototype. As your design evolves, the goal of iterating your prototypes is to improve performance and add value. This chapter discusses several aspects of design that should be considered when converting a design solution into a more detailed and complete design solution. Depending on the type and stage of the project and context of the problem, some of these may be more important than others and will dictate the focus of your design team. For example, academic design teams working on a project that is a continuation of a previous project may need to focus more on manufacturability and less on human factor issues (that may have been addressed by previous design teams). However, if the prototype has been refined and ready for use, you may also be ready to conduct validation studies with actual users.

Design for "X" (DFX) refers to design methods that ensure that a particular characteristic, function, or quality criteria is reflected in the final design. In this chapter, we present design methods that focus on performance, functionality, and the ability to manufacture the product, specifically where "X" is usability, functionality, manufacturability, maintenance/serviceability, and the environment, because these have the greatest impact on cost, safety, market acceptance, and the commercial success of the product. These should be considered when designing any medical product.

When implementing the DFX approach, you will encounter a common phenomenon in design—changes that improve one part of the design will often impact another part. For example, when considering materials for your device, you might select a material that can only be sterilized with a particular sterilization process. This sterilization process might then require use of packaging materials that conflict with the goals involving Design for the Environment (DFE). To navigate these situations, it is helpful to consider the concept of *satisficing,* termed as such by the Nobel Laureate economist and design theorist Herbert Simon. To satisfice is to recognize that when a decision is complex and multi-dimensional, the best possible choice is to optimize such that you find a satisfactory (but not necessarily the best) solution, along as many dimensions as possible. In other words, you should not optimize your design along only one dimension (e.g., material selection) at the expense of other dimensions (e.g., sterilization).

In academic design projects, the focus is on the design of the actual device. However, in the medical device industry, the term "product" includes not only the device, but also its packaging and labeling. For this reason, this chapter includes a discussion of packaging for medical devices. Medical device labeling is discussed separately in Chapter 12.

Design teams in industry must address all the topics covered in this chapter. Developing an awareness of these topics will be extremely helpful to those involved in new product development activities for medical device companies. Time and resource limitations often make it infeasible for an academic design project to address all of these in a final design. Some of these topics, such as usability and material selection, can typically be addressed using information presented in this chapter. If possible, most DFX topics should all be at least considered during the detailed design phase.

8.2 Design for Usability

People exhibit a wide variety of abilities and constraints. Designers strive to design products that can be safely used by as many people as possible. Knowledge of the user's cognitive and physical abilities and characteristics, and potential environment in which the device will be used, is essential to the design of successful medical devices. This includes an understanding of basic human skills and abilities, anthropometry and biomechanics, accessibility requirements, and cultural design considerations. *Usability* refers to how well a product can be safely and easily used by specific users (healthcare providers, patients, and caregivers) representing populations with various abilities (e.g., geriatric and pediatric patients, people with disabilities) to perform a specific function without confusion, difficulty, or injury to the patient or user. *Design for usability* requires a focus on the user and an understanding of the various characteristics of the user, the user interface, and the environment in which the device will be used (also known as use environment).

In this section we present methods and resources to help you identify and understand the user, user interface, and use environment during device use. Interactions between these three components can determine if the device will be used safely and effectively.

8.2.1 The Importance of Considering Usability in Design

Many of the problems and complaints regarding medical devices are related to poor usability. Considering usability when designing medical devices can reduce the chance of these problems occurring. According to the Food and Drug Administration (FDA), the following situations are common causes of problems resulting from the use of medical devices:

- Device use requires physical, perceptual, or cognitive abilities that exceed the abilities of the user.

- Device use is inconsistent with the user's expectations or intuition about device operation; the use environment affects operation of the device, and this effect is not recognized or understood by the user.
- The particular use environment impairs the user's physical, perceptual, or cognitive capabilities when using the device.
- Devices are used in ways that the manufacturer could have anticipated but did not consider.
- Devices are used in ways that were anticipated but inappropriate (e.g., inappropriate user habits) and for which risk elimination or reduction could have been applied but was not.

To prevent these situations from causing problems, you need to understand how the device will be used, how it may be misused, who will be using it, in what environment it will be used, and how the user will interact with it. For example, peritoneal dialysis is a home dialysis procedure that is the most common type of dialysis used by children worldwide. During this procedure, the child or caregiver (or both) are trained to connect a sterile dialysis catheter, which is imbedded in the child's abdominal cavity, to a dialysis machine (while maintaining sterility) every night and to disconnect the catheter in the morning. This connection requires that the child or caregiver follow a specific protocol to avoid contaminating the catheter tip during the connection/disconnection procedure, referred to as *touch contamination*. Contamination during this procedure is one of the leading causes of infection in the abdomen (peritonitis) and is a leading cause of morbidity and mortality in these children.

Understanding potential sources of discomfort, user error, safety problems, ineffective use, or noncompliance is important to reducing risk and an important part of the risk management process introduced in Section 8.9 and further discussed in Chapter 9. In the situation described above, the user may be a child, and the use environment would be the child's home. The potential for incorrect handling is high and the impact of errors resulting from misuse are great. The dialysis equipment and the procedures for using it properly must be designed to reduce risk by reducing the potential for misuse and adverse outcomes.

8.2.2 Design for Usability Methods and Disciplines

Human factors engineering, usability engineering, user-centered design, ergonomics, and industrial design are all methods and disciplines that focus on the user, the user interface, and the environment in which the device will be used (use environment). They are valuable tools for achieving the goals of design for usability.

Human factors engineering (sometimes referred to as **usability engineering**) considers how humans interact with the world and applies it to the design of products and systems with the goal of increasing safety, efficiency, and comfort. The FDA defines human factors engineering as "The application of knowledge about human behavior, abilities, limitations, and other characteristics of medical device users to the design of medical devices including mechanical and software driven user interfaces, systems, tasks, user documentation, and user training to enhance and demonstrate safe and effective use." The predominant standard on human factors engineering is *AAMI HE75: Human Factors Engineering—Design of Medical Devices*. To gain a better understanding of the topic, you are encouraged to review sections of this standard on human skills and abilities, anthropometry and biomechanics, alarm design, controls, virtual displays, software user interfaces, and medical hand tool and instrument design.

User centered design (UCD) is defined by the FDA as "an iterative design process in which designers focus on the users and their needs in each phase of the design process." UCD involves user input throughout the design process to ensure that user needs are being met by design solutions.

Ergonomics refers to the application of knowledge of the interactions between humans and systems toward the design of optimized systems. This can include the design of a more comfortable handle for surgical instruments that redistributes the weight and reduces surgeon fatigue, or an improved user interface that allows a nurse to quickly and easily set and confirm the correct drug dose delivered by an intravenous (IV) pump, thereby increasing patient safety and saving nursing time.

Industrial design is a discipline with a strong emphasis on the user. It focuses on usability and the function and appearance of a product, how it is manufactured, and the value and experience it provides for users. A comparison of engineers to industrial designers reveals different approaches to problem solving and emphases on different aspects of design. Engineers focus on technical aspects such as functionality, performance requirements, analytical modeling, and design validation. They tend to be more analytical and more concerned with product performance and the design of the internal components that make the product work. Industrial designers focus on aesthetics, ergonomics, usability, safety, and the user experience. They tend to be more visual and more concerned with the interaction between users and products. Industrial designers are concerned with the psychological impact of a product's design on the user or potential customer, usability (ease of use, low potential for error), safety (no sharp edges or other potential hazards), quality of the overall product experience, and perceived value of the product. Industrial designers provide a unique, diverse perspective to a design team, and thus are often part of industry product development teams.

The methods presented in this section emphasize a focus on the user and are applicable to both academic and industry design projects. You can use the information discussed in the following sections to gain a better understanding of the user's physical and cognitive abilities, the user interface, and the environment in which the product will be used. Some of this information is contained in an FDA guidance document (*Applying Human Factors and Usability Engineering to Medical Devices*) available online, and can help you design a better device and user interface.

In industry design projects, design engineers often work with in-house human factors experts (or outside consultants) who are assigned to most of the human factors engineering tasks. However, possessing a basic understanding and awareness of design for usability will benefit your career in the design of medical devices.

8.2.3 Users

Design for usability considers the user and how they will interact with the device. Depending on the specific device and its application, users of a device might be limited to professional caregivers, such as physicians, nurses, nurse practitioners, physician assistants, physical and occupational therapists, social workers, and home care aides. For some devices, users may include nonprofessional caregivers, including patients who operate devices on themselves to provide self-care and family members or friends who serve as caregivers to people receiving care in the home. This could include parents who use devices on their children or supervise their children's use of devices. Other users might include medical technologists, radiology technologists, or laboratory professionals, as well as professionals who install and set up the devices, and those who clean, maintain, repair, or reprocess them.

The personal characteristics of the intended users need to be documented and considered. These can impact the user's ability to properly operate a medical device and include:

- patient population (e.g., pediatric, adolescent, geriatric);
- physical size, strength, and stamina;
- physical dexterity, flexibility, and coordination;
- sensory abilities (vision, hearing, tactile sensitivity);
- cognitive abilities including memory (this can be important when multiple medications are prescribed to elderly patients with diminished cognitive abilities);
- medical condition for which the device is being used;
- comorbidities (i.e., multiple conditions or diseases);
- literacy and language skills (particularly important in global healthcare settings where nonverbal instructions are likely necessary);
- general health status;
- mental and emotional state;
- level of education and health literacy relative to the medical condition involved;
- general knowledge of similar types of devices;
- knowledge of and experience with the device;
- ability to learn and adapt to a new device; and
- willingness and motivation to learn to use a new device (e.g., when healthcare providers trained to use paper charting are expected to adapt to electronic health records).

8.2.4 Use Environment

In addition to considering users, it is also important to consider the environments and conditions, both standard and extreme, in which a user may operate a device. These may include a variety of conditions that could impact optimal user interface design and thus are important for device designers to understand. Medical devices might be used in clinical environments such as hospitals, surgical suites, intensive care units, and sterile isolation areas, or nonclinical environments such as homes, offices, community settings, or moving vehicles (including emergency transport vehicles). The following situations should be considered when evaluating the environment in which your device will be used:

- Lighting levels might be low or high, making it difficult to see device displays or controls. This occurs when using a smart phone outside on a sunny day.
- Noise levels might be high, making it difficult to hear device operation feedback and audible alerts and alarms, or to distinguish one alarm from another. It is not uncommon for nurses to turn off patient alarms when the frequency of false alarms results in alarm fatigue.
- Humidity levels might be high, and temperature might be high or low, potentially affecting function of the device and impeding performance of the user. This is especially true in global health settings where most work conditions are not controlled, or with military applications in extreme temperatures.

- Strong odors might be distracting to the user or cause a physical reaction, impeding performance. For example, the use of electrocautery equipment to make incisions burns tissue, which can produce a strong odor.
- The room could contain multiple models of the same device, component, or accessory, making it difficult to identify and select the correct one. For examples, several different types of surgical instruments (e.g., forceps, of which close to a dozen types exist) are often mixed in one storage bin.
- The room might be full of equipment or clutter or busy with other people and activities, making it difficult for people to maneuver in the space and providing distractions that could confuse or overwhelm the device user.
- The device might be used in a moving vehicle (such as an ambulance), subjecting the device and the user to jostling and vibration that could make it difficult for the user to read a display or perform fine motor movements.

8.2.5 User Interfaces

User interfaces are defined by the FDA as "all points of interaction between the user and the device, including all elements of the device with which the user interacts (i.e., those parts of the device that users see, hear, touch)." This includes "all sources of information communicated by the device (including packaging and labeling), and all physical controls and display elements (including alarms and the logic of operation of each device component and of the user interface system as a whole)." As a designer, you can intentionally create interaction points based upon your understanding of the users, tasks, and environments. A user interface may be as simple as a single mechanical button or as complex as an elaborate graphical interface on a touch screen. The goal is for the user interface to be simple, flexible, able to reduce errors, and be as robust as possible across a wide range of use cases. Making a list of tasks along with a storyboard of the user interactions, such as the storyboard described in Figure 6.8, can clarify user interactions and ways to improve the design of the user interface.

8.2.5.1 User Interface Design

In the most ideal design, a user interface is so intuitive that the user does not even recognize it as an interface. Aspects to consider when designing user interfaces include:

- *The size and shape of the device.* This is particularly a concern for handheld and wearable devices. More is discussed in Section 8.2.9 on the use of anthropometric data in human factors and ergonomics.
- *Feedback to the user.* Information is usually sent to the user to indicate device status and how their actions are impacting the function of the device. These might include indicator lights, displays, auditory and visual alarms, and tactile feedback. Care should be taken not to tax the cognitive load of the user. Often this is done by carefully considering how data is presented to the user. Breakout Box 8.1 discusses uses of color, but the general advice to consider the feedback to be sensed and cognitive ability of the user can be applied more broadly to other senses and attributes of user feedback.

Breakout Box 8.1 Use of Color to Improve Usability

Color does not only impact the aesthetics of a product but can contribute to improved usability. Color-coding can be used to differentiate between components, controls, and functions. It can highlight safety features, suggest hazards, and provide warnings to the user. Figure 8.1 shows color-coded wires (A) used to communicate where to connect each wire to a device, and a color-coded control box with a red "stop" button (to be pressed in an emergency to cut power) and green "start" button (B). In the United States, red implies danger, yellow suggests caution, and green implies safe operation. Red is also used to indicate the positive (+) terminal of an electrical component, and black indicates the negative (-) terminal.

FIGURE 8.1

Color-coded wire connectors and switches used to reduce error. (A) color-coded wires used to indicate where to connect wires and (B) color-coded control box with a red "stop" button and a green "start" button.

Colors have different meanings in different cultures, so designers must ensure that the appropriate standards and conventions for the use of color in medical devices is appropriate for the specific target market. There are many standards that address issues regarding the proper use of color in medical device design. These include IEC 60601-1-8: 2006, ANSI/ NEMA Z535.1-2017, and ANSI/AAMI HE75 (Section 6: Basic Human Skills and Abilities, Section 14: Cross Cultural/ Cross National Design, and Section 21: Software User Interfaces).

- *Hardware components.* Many medical devices use physical touchpoints. These may be switches, buttons, knobs, latches, handles, thumb wheels, toggle switches, slide controls, push buttons, rotary knobs, and triggers. Choosing which is best will depend on a range of factors including the intended user, the size and shape of the device, the activation forces required, feedback for each control, tasks to be completed, and use environment.
- *Software and graphical user interfaces.* User interfaces for medical devices are increasingly software-driven. In these cases, the user interface might include controls such as a keyboard, mouse, stylus, and touchscreen. Future devices might be controlled through other means, such as gesture, eye gaze, or voice. It is important to consider the intended user; some patients (e.g., geriatric) may resist using graphical user interfaces.
- *The logic of overall user-system interaction*, including how, when, and in what form information and feedback are provided to the user. This often involves hierarchical structure and just-in-time information presentation in such a way that it does not overwhelm the cognitive or physical abilities of the user. When possible, the flow should be consistent with users' expectations, abilities, and likely behaviors at each touch point. For example, users might expect the flow rate of a liquid or gaseous substance to increase or to decrease by turning a control knob clockwise or counterclockwise, respectively based on their previous experiences. The potential for use error increases when this expectation is incorrect or counterintuitive; for example, when an electronic control dial is designed to be turned in an opposite direction to that of previously mechanical dials.
- *Error prevention and correction.* A good interface also will prevent a device from being used improperly. Mechanisms that require correct orientation of wires, connectors, or settings can be designed so that a user cannot make a mistake. Recessed or covered power buttons are sometimes used to prevent unintended actuation of a medical device.
- *Components* that the operator connects, positions, configures, or manipulates
- *Components or accessories* that are applied or connected to the patient
- *Packaging and labeling*, including operating instructions, training materials, and other materials. Although not traditionally included in user interface design, packaging and labeling can include operating instructions, training materials, warnings, and other materials that help a user operate a device. Ideally, the user interface will eliminate the need for additional information for operation.

Many design projects involve the development of user interfaces. Computer interface design is based upon an explicit understanding of users, tasks, and environments. It is an iterative process to allow refinement by user-centered evaluation and addresses the whole user experience. Involving a group of users during the design process is key to a robust user interface. For healthcare workers to accept software as part of treatment and diagnosis, the software must be valuable, useful, readily accessible, and credible. ISO 9241-210:2019, *Ergonomics of human-system interaction—Part 210: Human-centered design for interactive systems,* is a standard that addresses the planning and management of user-computer interaction.

8.2.5.2 Principles of Good User Interface Design
Jakob Nielsen, the father of user-interface design, developed ten principles associated with good interface design, which are summarized below.

Visibility of system status—the system should always keep users informed about what is going on within the system through appropriate feedback within a reasonable timeframe

Match between the system and the real world—the system should speak the users' language with words, phrases, and concepts familiar to the user, rather than system-oriented terms. It should follow accepted conventions, making information appear in a natural and logical order to the user.

User control and freedom—users often choose system functions by mistake and will need a clearly marked "emergency exit" to leave the unwanted state without having to go through an extended dialogue. The interface should allow undo and redo functions.

Consistency and standards—users should not have to wonder whether different words, situations, or actions mean the same thing. For example, the term "anterior" generally refers to the front of the body. However, in obstetrics, this term refers to the top of the pelvis (symphysis pubis) when the patient is in labor. Standards and conventions for the user community should be followed.

Error prevention—a careful design that prevents a problem from occurring in the first place is better than a good error message. The interface should either eliminate error-prone conditions or check for them and present users with a confirmation option before they commit to the action.

Recognition rather than recall—minimize the user's memory load by making objects, actions, and options visible on the user interface. The user should not have to remember information from one part of the dialogue to another. Instructions for use of the system should be visible or easily retrievable whenever appropriate.

Flexibility and efficiency of use—accelerators that are unseen by the novice user may often speed up the interaction for the expert user such that the system can cater to both experienced and inexperienced users. The interface should allow users to tailor frequent actions.

Aesthetic and minimalist design—dialogues should not contain information that is irrelevant or rarely needed. Every extra unit of information in a dialogue competes with the relevant units of information and diminishes their relative visibility.

Help users recognize, diagnose, and recover from errors—error messages should be expressed in plain language (no codes), precisely indicate the problem, and constructively suggest a solution.

Help and documentation—even though it is better if the system can be used without documentation, it may be necessary to provide help and documentation. Any such information should be easy to search, focused on the user's task, list concrete steps to be carried out, and not be too large.

The most effective design strategies for reducing or eliminating problems resulting from use of a device with computer controls or inputs involve modifications to the user interface. To the extent possible, the "look and feel" of the user interface should be logical and intuitive. A well-designed user interface will facilitate correct user actions and prevent or discourage actions that could result in use errors. Addressing these errors through device design is usually more effective than revising the labeling or training because labeling might not be accessible when needed and training depends on memory, which may not be accurate or complete.

8.2.5.3 *User Expectations and Use Errors*

An important aspect of the user interface design is the extent to which the logic of information display and control actions is consistent with users' expectations, abilities, and likely behaviors at any point during use. Users will expect devices and device components to operate in ways that are consistent with their experiences with similar devices or user interface elements. For example, confusion and misuse of medical devices can occur when the user's expectations (due to habit) conflict with a new design. Figure 8.2 shows a keyboard used for physiological monitoring with an "alarm silence" key in the upper right, indicated by the arrow in the top of the image. A replacement keyboard used by the same nurses (shown in the lower half of the image) contains a "power" button in the same location as the "alarm silence" key on the earlier version of the keyboard. Since the nurses were used to the location of the "alarm silence" button, when they tried to silence alarms, they used the key in the old location that they were used to, and instead cut the power to the primary alarm notification system, disabling the entire system.

A good design prevents a device from being used improperly. Keying mechanisms and similar features that require correct orientation of cable and wire connectors and color-coding are examples of design features that can be employed to prevent misconnection of cables, wires, or other components. Recessed or covered power buttons can be used to prevent unintended actuation of a medical device.

FIGURE 8.2

Original (top half) and newer (bottom half) keyboards used for a physiological monitoring system. *Arrows* indicate keys with different functions in the same location, sometimes resulting in unintentional loss of power to the alarm notification system.

8.2.6 User Variability, Anthropometry, and Biomechanical Data

Designers should evaluate and understand the characteristics of all intended user groups, use environments, and user interfaces that could affect the users' interactions with the device. These characteristics should be considered during the design process so that devices can accommodate the wide range in variability and limitations among users.

Ethnography (presented in Chapter 3) is a method that involves observing use of the device in the actual use environment. It is a valuable tool for gathering much of the information discussed previously and is often used to identify customer needs and opportunities for new product development. For medical devices, this could include observing a surgeon using a surgical instrument in the operating room, a disabled patient using assistive technology in the home, office, or school environment, or a patient on home peroneal dialysis connecting and disconnecting equipment required for treatment.

Once all information is gathered regarding the user, the user interface, and the use environment, you are ready to begin modifying and improving your design concept(s). Dimensions, materials, colors, surface finishes, layouts of controls, and other characteristics that can impact safe use of a device can be determined. The following example in Breakout Box 8.2 illustrates how you can use this information to design a usable medical device.

For a particular anthropometric measurement shown in Figure 8.3, the cited percentile indicates the percentage of the distribution that falls below that value. For example, a value at the 2.5 percentile means that 2.5% of the values in the distribution are less than that value. So, a value at the 50th percentile means

---

Breakout Box 8.2 The Patient as User

Sometimes the patient is the user of a medical device. Disposable, handheld medical devices, such as injection pens for delivery of insulin, growth hormone, or epinephrine, inhalers for delivery of asthma medication, and intermittent catheters used to drain urine from a patient's bladder, are often used in the home environment and thus present significant design challenges. These devices are commonly used by patients with a range of health issues, abilities, and physical attributes (size, strength, impairments, etc.), and differences in education level and training in the use of medical devices. For example, diabetic patients often exhibit limb neuropathy or loss of visual acuity or both, which can negatively affect their fingertip dexterity and vision and impact their ability to use a medical device. Rheumatoid arthritis patients can lose 5% to 10% of fingertip strength and dexterity. Patients with tremor disorders (e.g., MS patients, Parkinson's disease patients, and essential tremor patients) often deal with tremors and issues involving motion control of the arm, hand, or both. Diminished strength can also be an issue; unscrewing a bottle cap is no longer possible for some arthritic patients. Home environments can range from quiet, well-lit, and relaxing, to noisy, dark, and distracting. These types of devices must be designed to work for a range of patients with various limitations in a range of home environments.

Operating these devices requires manipulation by the user's hands (patient or caregiver). The fit of the device with the hand is an important design characteristic and refers to the overall shape and appearance of the device relative to the location of physical interaction with the user. The fit of a hand tool (such as an electric toothbrush) or other device is defined by the relationship of the device to hand size and finger reach. Anthropometry and biomechanics are used in tandem to determine proper fit, correct sizing, and shape. They both involve measurements and physical attributes of the human body. Anthropometry includes static structural measurements of individual parts of the body in fixed positions such as hand length or head mass. Biomechanics includes dynamic, functional measurements of individual parts of the body in motion such as handgrip strength or arm reach envelope, and can help determine limits for strength, endurance, speed, and accuracy.

When designing a medical device, you may not be able to meet your intended users in person to measure grip strength, arm length, and other physical attributes needed to properly design a new device for their use. In this situation, it can be helpful to make use of anthropometric and biomechanical data. These are often presented as a range of percentiles for men and women of different age groups. Figure 8.3 shows anthropometric data for adult women.

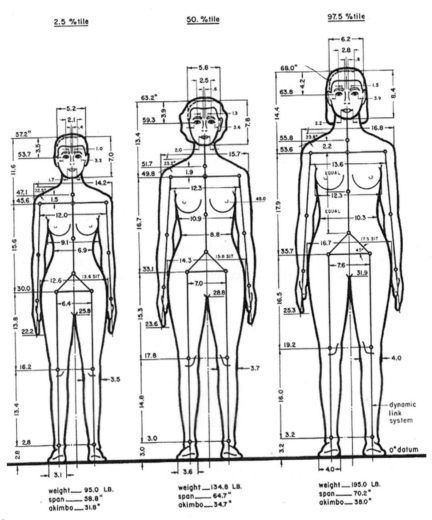

FIGURE 8.3

Anthropometric data for the 97.5, 50th, and 2.5 percentile adult female. Annis, J. (1978). Variability in human body size, in *Anthropometric source book, vol. 1*. National Aeronautics and Space Administration.

that half of the values fall below that value and half are above that value. A value at the 97.5 percentile means that only 2.5% of the distribution is higher than the value. Since measurements are for single characteristics only and there is significant variability within a single person among percentiles of their individual measurements, one cannot assume that the "average person" can be described as having all measurements in the 50th percentile. Average values apply only to the distribution of data for the characteristic being measured. Common anthropometric and biomechanics measurements and data can be found in *Human Factors and Ergonomics Design Handbook* (Tillman et al. 2016), and ANSI/AAMI HE75 Section 6: *Basic Human Skills and Abilities* and Section 7: *Anthropometry and Biomechanics*.

Several steps should be followed to determine device form (inhaler handle width or length of injector pen barrel) from anthropometric data. First, the intended user groups should be identified. The data collected for each unique user group is impacted by nationality, gender, age, and other factors. Second, a list of measurements most relevant to device usability should be established. These will have the greatest impact on device function. For example, when designing a toothbrush, hand length and finger reach are probably the two most important measurements related to usability. Third, the target percentile range should be established. Due to the variation and distribution of anthropometric and biomechanical characteristics, it is often impractical—if not impossible—to design a product that can be used by everyone. One approach would be to design using average data (50th percentile). However, this approach may leave out people with values well outside of the average. To accommodate more users, another approach would be to design for both ends of the distribution, including users between the 5th and 95th percentiles for a particular measurement. This, for example, could result in a toothbrush handle that could be used by a small woman (fifth percentile for hand length) and a large man (95th percentile for hand length). The difference between these two extremes is significant, as illustrated in Figure 8.4. A small female hand and a large male hand vary as much as 4.45 cm (1.75 in) in length. When considering the age of users, the reduction in finger strength between a 25- and a 65-year-old woman can be up to 50%. These variations pose a challenge to designers when designing a device for users in this range of hand lengths and ages. Fourth, prototypes should be created from designs based on anthropometric and biomechanical data and should be evaluated by potential users.

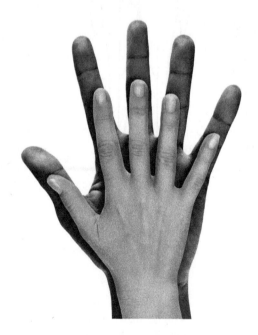

FIGURE 8.4

Fifth percentile female hand in comparison to 95th percentile male hand.

(Photo courtesy of Metaphase, ©2020 Metaphase.)

Caution should be taken when using anthropometric and biomechanical data. Generally, these data were compiled from a population (e.g., healthy subjects) that may not represent the users of your device and may not have been required to perform tasks in the same context and environments in which your device will be used. A proper design prevents the user from injuries resulting from repetitive motion, extreme postures or hand angles, contact stress, exposure to vibrations, and exertion beyond physical capabilities. By observing the use of your device by potential users, you can identify potential causes of injury as well as potential design changes that could eliminate the cause of injuries. For example, patients who inject themselves using injector pens may have diminished strength and manual dexterity in their hands. As a patient grips the pen with her hand and positions the pen over the injection site, she then will depress a knob at the end of the pen, releasing the needle into her skin to complete the injection. If the force required of her thumb to trigger the injection mechanism is too high for her to comfortably deliver, then while applying this force her hand may become unstable, allowing an undesired motion of the needle through her skin and within the underlying tissue, causing trauma to her soft tissues. Observing this during use will allow the designer to develop a solution to eliminate (or reduce the probability of) this source of injury.

Physical user interfaces typically include touchpoints and controls that enable the user to control the device. A touchpoint is an area of user interaction and a point where the user contacts the device. Examples include handles used to position monitors in the operating room and grips used to hold handheld digital thermometers. Controls allow users to change the state of a product. Examples include thumb wheels, toggle switches, slide controls, push buttons, rotary knobs, and triggers used to control the drip rate of an IV pump or flow rate of an anesthesia device. Making a list of functions that need to be controlled—and determining how they should be used and arranged—helps when selecting controls. To assist in the design of controls, a task analysis can be conducted in which a list is compiled of user tasks, the order of tasks, and difficulties associated with each task. This list can be combined with drawings to visualize the tasks, such as the storyboard described in Figure 6.8.

8.2.7 Impact of Materials on Usability

Materials can have an impact on the user experience. Designers should consider the impact of a material on the usability of a device. Other material selection criteria typically include appearance, functionality, mechanical properties, manufacturability, and cost. Later in this chapter, we will consider requirements for cleaning, sterilization, and biological performance of materials. Depending on the application, other important criteria may also include heat transfer, energy isolation, vibration reduction, light transmission and reflection, and surface roughness (friction). For example, materials with high thermal conductivity (e.g., metals) will transfer heat (or cold) more easily to a user's hands; this may be therapeutic for an injury or uncomfortable to the user depending on the user's tolerance for heat and cold. Low thermal and electrical conductivity materials (insulators, such as ceramic and polymeric materials) can prevent injuries resulting from burns or electric shock. Materials that have vibration-dampening properties and absorb vibrational energy (such as rubber, cork, and many polymers) reduce or prevent vibration-related injuries and damage to products.

Some devices, such as mobile stands with IV poles and bags (especially those carrying equipment found in the intensive care unit), are often used by patients as a means of support. These devices require a stable mechanical base to prevent tipping. Attention must be paid to the device's center of gravity when considering not only how the device is intended to be used but also how the device might be used in other ways. The weight of the materials used in these devices along with the geometry and spatial

Breakout Box 8.3 Reducing Use Errors Through Human Factors Engineering

In 1985, the Food and Drug Administration (FDA) received reports of incidents in which exposed male connector pins of electrode lead wires were accidentally inserted into either AC power cords or wall outlets instead of the patient cable that connects to the device monitor. In 1986, a death occurred when electrocardiogram (ECG) lead wires were inserted into a pulse oximeter power cord. The FDA received additional reports of similar use errors that resulted in electrical shocks, burns, and possible brain damage to patients. In 1993, the FDA issued a recall of infant breathing monitor cables which incorporated a design that had been linked to the deaths of at least three babies. In one case, a baby was electrocuted when the exposed metal ends of the lead cables were inadvertently plugged into a live electric power cord rather than into a cardiac and breathing monitor (as shown in Figure 8.5). The electrode leads intended to fit into the patient cable also fit into the extension cord.

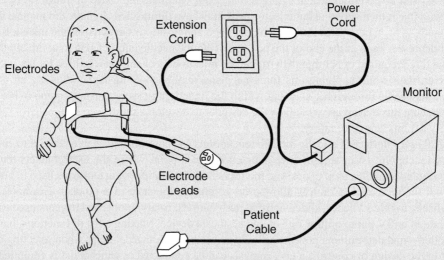

FIGURE 8.5

Hazard analysis of electrical leads and AC power sources. Electrode leads were intended to be connected to the patient cable (which was connected to the monitor) but were accidentally connected to the extension cord (which was connected to the power outlet).

After the incident in 1986, the FDA asked monitor manufacturers to assess their devices for potential electrode lead wire and patient cable connection hazards and, when necessary, to implement design changes to prevent insertion of electrode lead wire connectors into AC power cords and outlets. This included the design of "protected" leads which, through design changes to leads and monitors and the development of adapters for use with existing devices, would modify the user interface by making it impossible to insert a "protected" lead wire connector into an AC power cord or outlet, producing a safer electrode lead wire and patient cable configuration.

Standards development organizations took action to prevent electrode lead wires from being connected to electrical power sources. In March 1995, the International Electrotechnical Commission (IEC) published a second amendment to IEC 601-1(1988), the safety standard for electromedical equipment, which included a requirement that electrode lead wires be unable to contact hazardous voltages.

The Underwriters Laboratories (UL) adopted a modified version of IEC 601-1 by issuing standard 2601-1, which became effective on August 31, 1994. In adopting the IEC standard, UL included a deviation requiring that patient-connected electrodes be designed to prevent connection to electrical power sources.

The Association for the Advancement of Medical Instrumentation (AAMI) developed a standard that covers electrode lead wires and patient cables for surface ECG monitoring in cardiac monitor applications (ANSI/AAMI EC53-1995). This design standard addressed the safety and performance of electrode lead wires and patient cables with the added purpose of discouraging the availability of unprotected patient cable and lead wire configurations for ECG monitoring applications. The standard defined a safe (no exposed metal pins) common interface at the cable yoke and electrode lead wire connector. The standard was approved by ANSI on December 7, 1995.

Design solutions used to prevent a recurrence of this user error included lead wire designs to make it physically impossible to plug lead wires into wall outlets and adaptors to eliminate exposed lead wires and allow existing lead wires to be safely used with monitors. These solutions involved the application of human factors engineering to redesign a medical device and its user interface to prevent a dangerous use error from occurring.

location of the parts they are used in will have a direct impact on the center of gravity and thus can affect device function and usability.

Light transmission and reflection can also impact usability. For example, if there is a need to inspect or view internal components or check fluid levels while a device is in use, then materials with appropriate light transmission properties (e.g., glass or clear polymers) can be used to provide this functionality. Conversely, if a device contains light-sensitive materials that must be shielded from light, opaque materials can be used in the housing design to protect these materials.

Surface finish can impact the visual appearance, feel, and usability of a device. Surfaces of materials can be modified to produce the desired feel. For example, handle surfaces can be roughened to produce an increase in friction between a user's fingers and the device, improving the user's grip and control of the device. Conversely, surface roughness can be reduced to decrease friction between moving parts. This can also result in less effort required to effectively use the device. Smoother surfaces with fewer crevasses to harbor microbes may also be easier to clean or disinfect.

To conclude this section on design for usability, Breakout Box 8.3 below describes a situation where human factors were applied to improve the design and safety of pediatric cardiac monitor leads.

8.3 Material Selection

Before selecting the final materials for a specific application, review the most current version of the target product specifications discussed in Chapter 4 to ensure that materials that meet performance requirements are chosen. These are often different than the materials used in your prototype (as discussed in Section 7.3). In industry design projects, the choice of the final materials impacts the product cost and the return on investment for the new product. In academic design projects, lower-cost materials may be chosen for prototypes due to budget constraints, but you should identify the final materials to be used to best meet performance specifications. This will allow you to more accurately estimate the potential selling price and return on investment.

The previous section on Design for Usability presented several examples of how usability was enhanced by choosing appropriate materials. The following sections discuss additional information to be considered when selecting materials for and assigning dimensions to a new device design.

8.3.1 Expected Service Environment and Service Life

When selecting materials, you should first consider the expected service environment for the product (i.e., the conditions in which the product will be expected to perform). This would include the types and magnitudes of mechanical and electrical loading as well as the temperature, humidity, and air pressure to which the device will be exposed. Mechanical loading may result in tensile, compressive, bending, or shear stresses, or some combination of these. Mating parts result in wear and wear debris. Since some medical devices are intended for implantation into a patient's body, they will be exposed to an aqueous, saline environment, highly conducive to corrosion. For this reason, conditions such as temperature, pH, and the presence of proteins and ionic species must also be considered when selecting an appropriate material for implantable medical devices that will resist degradation in the physiological environment. After identifying characteristics of the service environment, you will need to determine the service life (the amount of time, number of uses, or number of cycles expected of the device). Establishing these parameters will allow you to determine the desired and/or required physical properties of the material to be used.

For example, when designing a total knee prosthesis (as shown in Figure 8.6), the service environment and service life will have a significant impact on material selection. Assuming that a typical knee prosthesis will be loaded approximately 1.5 million times per year at an average frequency of 1 Hz, the implant must be able to withstand 22.5 million cyclical compressive loads, with only negligible wear and no fatigue failures, to last 15 years (the standard warranty period provided by many orthopedic manufacturers). Per various studies of forces in the knee joint, load magnitude will range between three and five times the patient's body weight, with the average male patient weighing approximately 196 lb (88.9 kg) and average female patient weighing 168 lb (76.3 kg).

Note that the value for the weight of the average female patient is higher than what is shown in Figure 8.3 for the weight of an adult female in the 50th percentile. This may be due to the difference in populations represented in the two groups. The average age of a female patient receiving a total knee

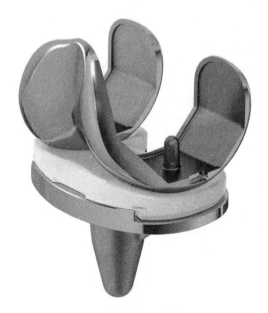

FIGURE 8.6

Total knee prosthesis comprised of a CoCrMo femoral component, a UHMWPE tibial bearing, and a CoCrMo tibial tray. The two stubs located on the inside of the femoral component are inserted into holes drilled into the distal femoral condyles (only one is visible in this photo). The larger stub located on the inferior surface of the tibial component is inserted into a hole drilled into the proximal tibial surface. These stubs provide stability to the implant.

(Photo courtesy of DePuy Synthes.)

prosthesis is higher than that of all adult females represented in Figure 8.3. The 50th percentile adult female in this older population of patients is heavier (168 lb [76.3 kg]) than the 50th percentile adult female shown in Figure 8.3 (134.8 lb [61.2 kg]).

Typical knee prostheses are made of a cobalt chrome molybdenum (CoCrMo) alloy femoral component and ultra-high molecular weight polyethylene (UHMWPE) tibial bearing surface supported by a CoCrMo tibial tray. This combination provides a low friction articulation as the femoral component slides and rotates over the polyethylene tibial component. The physiological environment contains mostly saline and includes various proteins that act as lubricants which further reduce the coefficient of friction and subsequent wear such as synovial fluid (normally present within the knee joint) that has a coefficient of friction <0.01. However, over millions of gait cycles, this relative motion generates wear debris which can cause tissue-related problems and affect implant function. In addition to mechanical loading, knee prostheses (the CoCrMo components in particular) will be subjected to a corrosive, saline environment within the knee joint.

The expected mechanical loading and potential for wear and corrosion require knee implants to be made of materials with high compressive and bending strength, high hardness (to resist wear), and corrosion resistance, such as CoCrMo alloy. The stem of a hip prosthesis, as shown in Figure 8.7, is subjected to considerable cyclical compression and bending loads, which can lead to fatigue failure over time. Therefore, hip stems are often made of materials highly resistant to fatigue and corrosion (e.g.,

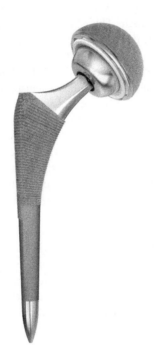

FIGURE 8.7

Modular hip prosthesis consisting of a Ti6Al4V femoral stem and neck, CoCrMo femoral head, UHMWPE acetabular cup, and CoCrMo acetabular liner.

(Photo courtesy of DePuy Synthes.)

titanium 6-aluminum 4-vanadium [Ti6Al4V] alloy). Femoral heads articulate with acetabular cups and are often made of CoCrMo alloy due to their resistance to wear and corrosion. CoCrMo and Ti6Al4V alloys are most commonly used in joint arthroplasties due to their physical properties, biological performance, and corrosion resistance. They both spontaneously form passive oxide layers which serve as barriers to corrosion, providing a significant level of corrosion resistance to the alloys.

The strength of a device component subject to mechanical loads is an important consideration when determining the geometry and dimensions of the component. In this case, strength would be an important design factor. Design factors are characteristics or considerations that impact the design of the component. Depending on the situation, these factors may include strength, thermal conditions, stress concentrations, processing, safety, stiffness, lubrication, surface finish, and other design factors. Some of these are related to dimensions, materials, and methods of processing and assembly.

Engineers often consider the *factor of safety* (also known as the *safety factor*) to ensure the safe and proper functioning of mechanical devices or components. These are commonly used to indicate the ratio of the estimated strength of the material to the calculated stress resulting from the expected loading conditions. When strength equals stress, the factor of safety = 1, and the design is said to have no safety. Calculations of stress should include the static and dynamic loading expected in the service environment as well as other applicable design factors (e.g., stress concentration, surface finish) and fluctuating loads. In some loading situations, a factor of safety = 2 indicates that a design would be marginally acceptable if the actual strength were only half of the expected strength, or if actual loads were exactly double the calculated estimate. This illustrates the importance of determining the mechanical loads (static and dynamic) that the device will be subjected to in its service environment and using these load values, along with the appropriate design factors, to estimate resulting material stresses.

Factors of safety ≥ 2.0 are often considered to represent "safe" designs, but this interpretation depends on how these are calculated, as discussed above. Strength is a statistically varying quantity, and stress is variable, too. For this reason, factors of safety > 1.0 do not guarantee that failure will not occur. Design factors for specific applications are sometimes mandated by law, policy, or industry standards. The building industry commonly uses a factor of safety = 2.0. This value is relatively low because the loads are well understood and many structures include redundancies. The aircraft and aerospace industries use factors of safety ranging from 1.2 to 3.0, depending on the application and materials. Values in the lower end of this range tend to be used for ductile, metallic materials, and values in the higher end of the range tend to be used for brittle materials. When designing medical devices, there are no established standard factors of safety; each unique medical device, application, and design may require a different factor of safety. For example, a load-bearing device such as a hip implant may require a higher factor of safety than a nonload bearing device (e.g., a Foley catheter) due to the expected mechanical loading of each device and the potential impact that device failure could have on patients (revision surgery vs. simple removal and reinsertion of a new catheter, respectively). To learn more about how to identify and use applicable design factors when calculating factors of safety, consult the mechanical engineering design references and resources listed at the end of this chapter. Factors of safety do not only apply to the design of mechanical elements and the strength of materials; they also apply to the thermal, electrical, and other aspects of a design.

8.3.2 Biological Performance

You should assess the required biological performance of the material, which involves the material's ability to peacefully coexist in its physiological environment. This consists of two components: the host response to the material (biocompatibility), and the material response to the host (biodurability). Testing for these components of biological performance is discussed in detail in Section 10.3.

8.3.2.1 Biocompatibility

Biocompatibility refers to the ability of a material to elicit a minimal response from the host when in contact with the body. It involves the local and systemic host response (other than the intended therapeutic response) to the material or its degradation products. For example, orthopedic implants often create wear debris and corrosion products (polyethylene and metallic debris, metal ions) over time. Debris collecting in the tissues surrounding the implant can cause local effects such as necrosis (cell death) in tissues surrounding the implant, which can lead to loosening of the implant. This debris can leave the implant site, travel through the body, and elicit a systemic host response. For example, elevated metal ion concentrations that can affect normal organ function have been found in blood (metallosis) and organs. Wear debris can accumulate at a remote site, resulting in chronic inflammation. When polytetrafluoroethylene (Teflon®) was used in some early orthopedic implants, deposits of Teflon particles were found in patients' abdomens (abdominal teflonoma). Other systemic effects include thrombosis, hemolysis, carcinogenesis, and allergic foreign body responses. Breakout Box 10.1 presents an example of metallosis resulting from a hip implant.

Thrombosis refers to the formation of a blood clot. Certain material surface properties can activate a coagulation cascade which will result in the formation of thrombi (clots). These thrombi can detach from the implant surface and travel through the body becoming emboli which can lead to an embolism in other parts of the body. Emboli in the bronchi may affect lung function, which can be fatal. *Hemolysis* refers to the swelling and bursting of red blood cells upon contact with an implanted material. *Carcinogenesis* refers to the formation of a tumor caused by the original implanted material or its degradation products, and any leachable polymeric components contained in the implanted material. People with hypersensitivity to certain materials such as latex, nickel, chromium, or cobalt can experience an *allergic foreign body response. Cytotoxicity* is a local effect and occurs with materials that cause cell death when in contact with cells and tissues. Cytotoxicity tests are presented in detail in Breakout Box 10.2.

Biocompatibility is evaluated through a variety of tests of the material, not the finished device. Test samples must contain the same colorants, release agents, radiopacifiers, and other additives that will be present in the final material used in the final device. ISO 10993-1:2018, *Biological evaluation of medical devices Part 1: Evaluation and testing within a risk management process,* lists test requirements for materials to be used in medical devices that come in contact with tissue or blood (or surfaces exposed to blood). This is discussed in greater detail in Section 10.3. For the knee implant example discussed earlier, the expected duration would be many years. The knee implant materials will contact blood, bone, and soft tissues. For these reasons, if a new material is under consideration, extensive implant testing would be required to determine if the materials are biocompatible and thus usable in implant design. Designers should consider materials previously used safely in similar medical device applications, or materials that have passed tests for biological performance, per ISO 10993.

All implanted materials will elicit some type of response in the host due to the body's normal immune response. Although silicone does elicit a minimal host response, the effects are clinically insignificant. For this reason, silicone is considered the "gold standard" for biocompatibility. Medical devices that contact tissue, blood, or fluid path surfaces that contact fluids that will enter the body (such as tubing used in IV pumps) must be made of materials (including leachable additives) that elicit minimal, clinically insignificant host responses.

8.3.2.2 Biodurability

Biodurability refers to the ability of a material to resist the body's attempt to destroy or reject it through corrosion, oxidation, or other forms of biological degradation. It is dependent upon the response of the material to the host. When a material is implanted in the body, a wound is created, and the implant contacts tissues and body fluids. Proteins adsorb to the implant surface and the wound triggers the inflammatory response as the body attempts to protect itself from the implant (foreign material). The body then begins the process of wound healing and elimination or isolation of the foreign material, which often includes the formation of a capsule around the injured tissue and macrophages entering the wound to remove wound debris. Next—in response to tissue damage—vasodilation, increased permeability of capillary endothelium, and increased metabolic activity occur. These responses increase blood flow, fluid volume in the wound (edema), and local temperature, which work together to remove wound debris and heal the wound.

The physiological environment and the body's inflammatory response act together to reject or destroy foreign materials. Low pH, elevated temperature (37°C), mechanical loads, chloride and other ionic species, and an isotonic saline environment can result in corrosion and oxidation of alloys and polymers, respectively. Some cells can secrete oxidizing agents and enzymes to attempt to digest foreign materials, causing polymers to swell and crack, increasing sites for further degradation reactions. The resulting degradation products can lower the local pH, further accelerating degradation. An implanted material must resist significant degradation and retain its desired physical properties to allow the medical device to function during its expected service life. This also applies to materials that are not implanted but make skin contact for a significant period of time. For example, when Ag-AgCl electrocardiogram (ECG) electrodes are attached to the body, over time the body will deplete the concentration of chloride ions contained in the electrodes, resulting in a slow drift in recorded ECG potentials. Designers can account for this drift by adding circuitry to filter out the drift effect. As a designer, you will want to consider the service environment and choose materials, surface modifications, and other adaptations that will resist degradation and keep your device functioning for its required service life.

The physiological environment can also interfere with normal device function. For example, encrustation of urological devices (e.g., ureteral stents) often occurs when minerals that have been dissolved in urine begin to form precipitates that attach themselves to these devices. This can produce a mineral shell around and inside the stent, reducing the flow of urine, and making stent removal difficult. Similarly, bacteria can grow on the surfaces of implantable devices, forming a biofilm that can interfere with normal function of the device and lead to infection.

Biodurability is evaluated through testing of finished devices or functional prototypes and is discussed in detail in Section 10.3. Test devices must contain the same colorants, release agents, and other additives that will be present in the final material used in the final device.

For devices that will never be in contact with a patient or fluid path, biological performance is not a material selection criterion. For example, the internal circuitry of an ECG monitor will not be in direct contact with patients. For this reason, the biological performance of materials inside of an ECG monitor would not be a concern nor would biological performance testing be required. On the other hand, the electrode leads of an ECG monitor will contact the patient's skin during use, and therefore must be evaluated for biological performance.

Table 8.1 lists common materials that are used in many medical devices with some level of patient contact. Materials in this list are well-known and characterized as being safe to use in many clinical applications. Many of these materials (e.g., Ti6Al4V alloy) are available in multiple grades for specific applications.

Table 8.1 Common materials used in medical device applications involving some level of patient contact

Abbreviation, chemical formula, or common name	Name	Applicable standard	Examples of medical device applications
Polymers			
PDMS	polydimethylsiloxane (silicone)		cosmetic applications for plastic surgery (breast, chin, other implants) Foley catheters, ureteral stents
PCL	polycaprolactone		bioresorbable, biodegradable applications: tissue scaffolds, sutures, ligament and tendon repair, vascular stents
PLA	polylactic acid	ASTM F2579	
PGA	polyglycolic acid	ASTM F2579	
PC	polycarbonate	ISO 11963 ASTM F997	transparent instruments and containers, tubing connectors
PDS	polydioxanone		absorbable sutures
PE	polyethylene	ASTM F639	joint replacement, sutures, tendon substitutes
PEEK	polyether etherketone	ASTM D8033	dentistry, rigid tubing
PEI	polyetherimide	ASTM D5205	skin staplers, sterilizable equipment
PES	polyethersulfone	ASTM F702	catheters, lumen tubes
PET	polyethylene-terephthalate	ASTM D5407	implantable sutures, heart valves
PMMA	polymethyl methacrylate	ASTM F451	bone cement, intraocular lenses
PP	polypropylene	ASTM 5857	heart valves, sutures, syringes
PS	polysulfone	ASTM F702	surgical clamps, artificial heart
PTFE	polytetrafluoroethylene Teflon®	ASTM F754	catheters, soft tissue augmentation, vascular grafts, sutures
PU	polyurethane	ASTM F624	sutures, catheters, heart valves, wound dressings
PVC	polyvinyl chloride	ASTM F665	tubing, bags containing blood
ABS	acrylobutadiene-styrene		device housings
Nylon	polyamide	ASTM D6779	sutures, ligament and tendon repair, catheter balloons

Continued

Table 8.1 Common materials used in medical device applications involving some level of patient contact—cont'd

Abbreviation, chemical formula, or common name	Name	Applicable standard	Examples of medical device applications
Metals			
316L SS	stainless steel	ASTM F746	fractures fixation devices (plates, screws), surgical instruments
Ti6Al4V	titanium alloy (6% Al, 4% V) (wrought/cast)	ASTM F136 ASTM F1108	surgical implants (joint replacements)
Ni-Ti	Nitinol	ASTM F2063	shape memory applications, vena cava filters, guidewires, cardiac stents
CoCrMo	cobalt alloys (wrought/cast)	ASTM F1537 ASTM F75	orthopedic implants
AE21 and WE43	magnesium alloys		biodegradable applications, biodegradable stents
AgO_2	silver oxide		antimicrobial coatings
Ceramics and glasses			
Al_2O_3	alumina	ASTM F603	dental implants
	calcium phosphates	ASTM F1609	surface coatings, bone repair
$Ca_{10}(PO_4)_6(OH)_2$	hydroxyapatite	ASTM F1185	promotes bone growth, hip implant coatings
	porcelain		dental restoration
ZrO_2	zirconia	ISO 13356 ASTM 2393	joint replacements

8.3.3 Sterilization and Device Reuse Requirements

Device reuse and sterilization requirements are additional factors that influence material choice. For example, if a product is intended for single use only, then more durable, expensive materials would not be an optimal choice. For devices intended for reuse that must be reprocessed, then more durable—and typically more expensive—materials may be indicated. Either way, if the device will be in contact with the body, then it most likely will need to be sterilized. Exceptions might include devices used in gastro-enterology, where the surfaces of the gastrointestinal (GI) tract (e.g., mouth, esophagus, stomach, intestines) are not considered sterile. These devices may require cleaning only. Choosing materials that can be safely sterilized without degradation or loss of physical properties is another consideration in material selection and is highly dependent on the sterilization method used. If the device must be resterilizable, then it also must be able to withstand the stresses of multiple sterilization cycles. Sterilization creates thermal, mechanical, and chemical stresses. This is discussed in more detail in Section 8.5.

8.3.4 Cost

Cost is a significant factor in selecting the material for a medical device. This is affected by the cost of raw material and manufacturing processes involved. For example, harder materials such as CoCrMo or Ti6Al4V alloys are typically cast or forged due to the difficulty in machining such hard materials. These processes require significant post-processing operations (e.g., grinding, polishing) that will increase the cost of using these materials. Softer materials, such as silicone or polyurethane, can be extruded into tubing for catheters or molded into other shapes with fewer additional post-processing operations, which can lower production costs.

Prototypes are most often made of materials other than the final material. Prototypes are often used for preliminary testing and to communicate size, shape, and appearance, and to demonstrate function. They may also be used for preliminary testing. A single prototype (or a small quantity of prototypes) is often fabricated for these purposes. If not intended for use in animal or human clinical studies, prototypes do not need to be sterilized or made of the final material with the same physical properties required for long-term use. For example, a prototype of a surgical instrument could be made of a polymer more easily manufactured and less expensive, since it will likely only be used for demonstration purposes to obtain feedback from potential users (surgeons). It would not need to withstand the years of use and multiple sterilization cycles required of production units, and would therefore not need to be made of stronger, more durable, or more costly materials.

Teams often choose materials that are less durable, less costly, and easier to work with when constructing the prototype in an academic design project that typically ends with a prototype that has been designed, built, and tested. Teams can also use materials that do not provide the same physical properties required in the final design and production units. Materials that do not require complex, expensive, and time-intensive manufacturing processes (e.g., casting, forging) or equipment not typically found in academic maker spaces and machine shops (e.g., injection molds) are good choices for prototypes used in academic design projects. Many of these materials are presented in Section 7.3.

In an industry design project, teams often make prototypes of materials using alternate materials that allow them to quickly construct a prototype to obtain feedback. As the design evolves, they may go through several prototype iterations and need to use materials and processes that allow for rapid production of prototypes. However, once a design is finalized, verification and other testing require samples that reflect the final design, including dimensions and materials.

Material selection can be aided by material suppliers and material selection tools such as software and books. Material suppliers are willing to share their wealth of knowledge with potential customers and can make suggestions based on specifications, required performance, and material properties, as well as cost and ease of manufacturing. There are material selection databases such as MatWeb and online resources available to assist in choosing the best material for a specific medical device design. Material costs fluctuate and depend on factors such as the quantity of material purchased, material grade and purity, ease of processing, and the economic forces of supply and demand.

8.4 Design for Manufacturability

The goal of many academic design teams is to design, construct, and test a final functional prototype to be transferred to a project sponsor or client. For some projects, the prototype is kept for a future team for further development. Unless the focus of a project is on the design of a manufacturing process (process design), teams do not typically design a process intended to produce multiple units in a production environment. In most companies, manufacturing engineers are responsible for process design and work closely with design engineers to understand critical aspects of the design. Understanding how the new product functions will help manufacturing engineers design tooling and production processes that will produce a product at a reasonable cost and high level of quality. Information regarding the project schedule, product cost, and market demand for the product help determine the manufacturing strategy used to meet required production levels. This may include manual or automated manufacturing operations, multiple production shifts, and single or multi-cavity tooling.

Considering manufacturing processes early in the project results in lower development costs, shorter time-to-market, higher quality products, and increased probability of meeting customer needs and business objectives. In this section, we share suggestions for how to make a design easier to manufacture in a cost-efficient manner, collectively known as Design for Manufacturability (DFM). Incorporating these suggestions—aside from being appreciated by industry project sponsors—will move your design closer to a commercializable product and build your awareness of how product design can impact manufacturability.

8.4.1 The Role of the Manufacturing Engineer in Design for Manufacturability

Product design has a significant impact on process design. Designing a product without considering how it will be manufactured often results in a more complicated manufacturing process. This slows the process and requires more time to produce a single unit (decreasing the production rate), increases manufacturing costs, and shrinks profit margins. Manufacturing engineers have expertise in manufacturing operations and design of manufacturing equipment such as molds, tooling, fixtures, and other equipment needed to manufacture the new product, and can advise design engineers as to how the new product can be designed to make manufacturing easier and less costly. They are responsible for evaluating a design and identifying opportunities to shorten assembly steps, reduce repetitive motion injuries among production workers, reduce waste and the number of parts needed, and reduce manufacturing costs. One of their principal roles is to optimize production processes for efficiency.

Materials and processing account for approximately 70% of a typical product's cost. Manufacturing engineers are a helpful resource to design engineers in creating a functional design that can be manufactured at a reasonable cost. As the design evolves, they can suggest design details that will maintain product function and improve manufacturability and should be consulted early in and throughout the design process. As the expert on the design of a product and how the product functions, design engineers have the final say on potential design changes suggested by manufacturing engineers. The experiences of many design engineers in industry have shown that waiting until the design is finalized before consulting with manufacturing engineers can result in required time-intensive and costly redesign efforts that could have been avoided if the design had been shared earlier.

In determining how a new product should be manufactured, manufacturing engineers will consider product- and process-related factors such as:

- the variety of manufacturing techniques (both additive and subtractive) appropriate for the product design;
- tooling, fixtures, and other equipment needed to manufacture the new product;
- the order of manufacturing operations (process design), the layout of production workers, and equipment on the manufacturing floor; and
- whether manual or automated operations or assembly steps are justified by expected production quantities.

Manufacturing engineers are also responsible for determining the steps required to fabricate a product (process design) including manufacturing and assembly operations. They will consider ways to reduce the:

- complexity and number of fabrication and assembly steps,
- potential for repetitive motion injuries among production workers,
- material and energy waste resulting from manufacturing processes,
- number of parts or tools required for fabrication and assembly operations,
- number of assembly errors and defective products,
- labor content (number of production workers needed), and
- time to manufacture a single product from start to finish.

Decisions regarding each of these process design considerations will often have an impact on product design. For example, a manufacturing engineer might suggest slightly altering the wiring so that a battery housing can be placed later in the manufacturing process. Such a product redesign is relatively simple—yet it may have a significant impact on the production line by allowing other housings in the unit to be placed at the same time by one robotic arm in one operation instead of several, thereby streamlining the process. Such simple product design changes can have a significant impact on production time and labor, resulting in lower manufacturing costs. Section 15.8 discusses how manufacturing processes can be optimized to reduce time, cost, and waste.

In an academic setting, you should consult with university technical staff (e.g., machine shop personnel, lab and electronic technicians) to discuss your design prior to submitting a request for prototyping of your parts. Machine shop personnel can provide helpful feedback regarding design changes that can make it easier to produce the parts. Experienced personnel may offer similar guidance to what manufacturing engineers offer industry design teams. Lab technicians can improve circuit design layouts and help create complex 3D parts.

8.4.2 The Role of the Design Engineer in Design for Manufacturability

Process design is typically the domain of the manufacturing engineer. However, design engineers need to be familiar with basic manufacturing operations such as machining, casting, molding, forming, joining, and additive manufacturing; as they design a new product, they can consider ways in which it could be made, helping to create a design that is manufacturable. An understanding of manufacturing processes allows engineers to design products that can be manufactured more easily and at a reasonable cost. Figure 8.8 shows the essential components of all production processes, a combination of materials, energy, mechanical forces, chemical reactions, and time.

Basic manufacturing processes were presented in Chapter 7. Familiarity with these processes and thinking about them during the design process can help you create designs that will be manufacturable. When designing a new medical device, you should consider how materials will be handled and flow through a manufacturing process, and how the product can be designed to increase production rates and reduce required labor.

To reduce manufacturing and assembly costs, consider the following:

- Design for the least amount of waste (unused material).
- Reduce the number of parts required in an assembly. Often, multiple parts can be combined into one part (or fewer parts), reducing part cost and assembly time and cost.

FIGURE 8.8

The essential components of all manufacturing processes: a combination of materials, energy, mechanical forces, chemical reactions, and time.

(Used with permission from rccProduct Design.)

- Use standard hole sizes and milled radii so that standard drill and milling bits can be used. Custom cutting tools are required for nonstandard hole sizes and add cost and time.
- Use appropriate draft angles to allow for easier assembly, molding, and removal of molded parts from a mold.
- Avoid designs that require operations hidden from the assembler's view (blind operations).
- Avoid designs that obstruct access to parts and tools.
- Avoid designs that entail operations requiring assemblers to twist or contort their hands (or body) into uncomfortable positions, and designs that can lead to repetitive motion injuries (e.g., carpal tunnel syndrome).

8.4.3 Optimizing Production Processes and Reducing Errors

Several methods for preventing production errors and optimizing production processes exist. In this section, we present three of these methods.

8.4.3.1 Poke-Yoke

Poke-Yoke is a Japanese concept developed to prevent mistakes in manufacturing. It is often used as a design methodology to "mistake proof" a design to prevent parts from being assembled incorrectly. This method includes the use of fewer types of parts, symmetrical parts that cannot be installed incorrectly, asymmetrical parts that can only be installed in the correct orientation, alignment pins and tabs, unique geometries and shapes, and markings to indicate correct assembly. For more information on the Poke-Yoke methodology, refer to "Poke-Yoke: Improving Product Quality by Preventing Defects," listed in the resources section at the end of this chapter.

8.4.3.2 Lean Manufacturing Methods

Familiarity with manufacturing process design and lean methods allows engineers to understand how their designs can impact the production line. Breakout Box 8.4 provides guidance on how to design products that are more manufacturable. Lean manufacturing methods seek out process waste and try to eliminate it, thereby reducing production costs and unit costs for the new product. This is important to the design engineers as well as production personnel and upper management. Forms of process waste include:

- *Transport*: moving products when not actually required to perform a process
- *Inventory*: all components, work-in-process, and finished product not being fully processed (labor, overhead, and other costs allocated to unfinished products that cannot yet be sold)
- *Motion*: people or equipment moving or walking more than is required to complete an operation
- *Waiting*: waiting for the next production step
- *Overproduction*: production ahead of demand that increases inventory and work-in-process preventing one-piece flow
- *Over-processing*: additional steps in a process resulting from poor process, tool, or product design
- *Defects*: effort involved in inspecting for and correcting defects
- *Unused capacity*: unused human talent or equipment capacity

Each of these forms of process waste is evaluated to identify sources of waste and opportunities to reduce it. This may result in process design changes as well as requests to change the design of the product.

8.4.3.3 Six Sigma

Another approach to reducing errors and defects is Six Sigma, a set of quality management techniques used to improve business processes by significantly reducing the probability that an error or defect will occur. The term "sigma" refers to one standard deviation of a normal distribution. Six Sigma (or six standard deviations) includes 99.99966% of a normally distributed population. The Six Sigma approach is designed to prevent errors and defects in 99.99966% of products in a normally distributed population. It is a statistically based, data-driven, continuous improvement methodology that involves five steps: define, measure, analyze, improve, and control. Six Sigma applies to any industry and has been used by many companies to reduce errors and defects, lower costs, and improve customer satisfaction.

Breakout Box 8.4 Make it Makable… From the Beginning

Theories, free body diagrams, equations, and computer models are all important, but there comes a point in time when something must be made. Now the questions change to the practical matters of turning the idea, concept, or design into a real thing. This includes questions such as:

- What material(s) should it be made of?
- What processes should be chosen using those materials?
- Do we have the ability to make it, or do we need help?
- Even if we know how to make it, do we have the production capacity to do so?
- How many will we need to make?
- How much will it cost to produce?
- What kinds of tools, equipment, skills, and facilities are needed?
- How can we get it to the customer?
- Does it have a shelf life?
- Are there risks to workers in its production and mitigations that we need to take to address these risks?
  …and more like these.

If questions like those above were not considered from the very beginning, the new product engineer will likely be in for a rude awakening when they bring their prized design to the manufacturing plant or supplier. This could mean costly redesigns with significant delays. It could be that the *process* design could force the *product* design back into a development loop to confirm the changes still support all the other product requirements.

Product AND process. Product AND process. "How are you going to make it?" must be a question always in the mind of the product engineer or designer.

But there is more. Just because you did a fine job taking manufacturing processes into account, new products entering a plant are disruptive to current production. This is true whether the plant is internal to the company or somewhere in the supply base.

Think about it. How are production operations measured and rewarded? The following are just a few ways:

- meeting production schedule quantities
- meeting production cost targets
- meeting quality goals
- continuously improving processes and removing waste from the system

Those in manufacturing are interested in keeping a well-oiled machine humming along smoothly. Anything new will be a disruption. To minimize disruption, the new product engineer should coordinate with the plant(s) and/or supplier(s) to plan new product trials and startups into the production schedule. Good communication is key and even better if production has been alerted as early as possible regarding how the new product will impact them. Whenever possible, demonstrate that their needs as a stakeholder have been considered from the very beginning. Where compromises are needed, explain the business rationale that caused the choices that were taken. The particular manufacturing challenge may not be solved, but recognizing that a concerted effort was made goes a long way.

Simply put, think about the needs of your counterparts in production operations.

David Rank
Consultant
rccProduct Design

8.5 Design for Sterilization

During the manufacturing process, microorganisms in the air and on surfaces of production equipment can contaminate the surfaces of medical devices. The level of microbial contamination is referred to as *bioburden*. If used with patients, these surface microorganisms can cause infections. To prevent infection, most medical devices intended to contact a patient's body must be sterilized. Implanted devices, such as pacemakers, cardiac stents, or total joint prostheses, must be sterilized. Generally, all parts of a device or accessory that touch the patient must be sterilized. For example, an ECG monitor does not need to be sterile because it does not contact the patient, but the ECG leads that contact the patient must be sterile.

In this section, we address the various levels and the common methods of sterilization to improve your general awareness of sterilization and how the design of a medical device can impact its ability to be sterilized.

8.5.1 Sterilization, Disinfection, Cleaning, and Decontamination

In this section, we review some basic terminology as it applies primarily to sterilization of products by medical device manufacturers (industrial sterilization processes), and more specifically for single-use products. Clinical sterilization methods are considered in Section 8.5.3 and focus primarily on reusable products sterilized in the clinical setting.

Sterilization is the process used to render a device sterile. Ideally, sterilization would destroy all pathogens and microorganisms. However, current sterilization methods do not destroy or inactivate prions or endotoxins. Absolute sterility cannot be determined; instead, a statistical definition, known as the sterility assurance level (SAL), is used to define sterility. The SAL is defined as "the probability of a single viable microorganism occurring in or on a product after sterilization." The internationally accepted definition of sterility of medical devices allows for the maximum probability of finding one viable organism on a sterilized medical device to be 1/1,000,000, or a SAL of 1×10^{-6}. Medical devices that will contact sterile tissues (e.g., implants, surgical drains, surgical instruments) are required to be sterilized to this level. The FDA recommends a SAL of 1×10^{-3} for devices that are intended for contact with intact skin only.

Disinfection reduces the number of microorganisms but not to the same SAL achieved with sterilization methods, and it does not inactivate all forms of microorganisms. Disinfection methods are used in situations where the risk of infection is low and the design of the device makes sterilization difficult. Medical devices that will contact sterile tissues, such as implants, surgical drains, and surgical instruments, are required to be sterilized. High levels of disinfection are acceptable for devices that will contact mucosal tissue or nonintact skin, such as respiratory therapy and anesthesia equipment. Flexible endoscopes (laparoscopes, cystoscopes, gastroscopes, hysteroscopes) are challenging to sterilize due to their long, narrow lumens and their overall fragility. Ideally, they should be sterilized between uses; however, due to these difficulties, current practice for most flexible endoscopes involves cleaning and high-level disinfection instead of sterilization (see below).

Cleaning refers to the removal of foreign material such as soil and dust from the internal and external surfaces of a medical device. This is difficult to accomplish in cavities and long lumens found in endoscopic instrumentation. Cleaning is important to decrease the bioburden of devices prior to sterilization, which can improve the efficiency of a sterilization process. It is also necessary for removing contaminants used in production processes such as mold release agents, cutting lubricants, polishing compounds, and other substances. Failure to remove these contaminants can impact biocompatibility and processes such as surface modifications and adhesive bonding of parts. Once used, medical devices become contaminated. Devices intended for reuse must be cleaned and resterilized. For this reason, all reusable surgical equipment is cleaned prior to sterilization.

Decontamination refers to reducing the quantity of microorganisms from a device so that it is safe to handle and discard. This is often used to prepare a device for safe handling and disposal by hospital workers.

8.5.2 Industrial Sterilization Methods

In this section, we present the three most common methods used by manufacturers to sterilize packaged medical devices; pressurized steam, ethylene oxide (EtO), and radiation (gamma and electron beam). Less common methods include gas plasma, vaporized hydrogen peroxide, dry heat, and vaporized per-acetic acid. We focus on industrial processes used for single-use devices. Most medical devices are sterilized inside their final packaging, a practice referred to as *terminal sterilization*. Packaging is cho-sen based on the intended sterilization method and is discussed in more detail in Section 8.8. The condi-tions, mechanisms, advantages, and limitations of these methods are summarized in Table 8.2 below.

8.5.2.1 Steam

Steam sterilization uses thermal energy from high-pressure saturated steam to denature DNA and destroy enzymes to kill microbes on the surfaces of devices. A packaged device is placed into a chamber (autoclave) and pressurized steam is introduced at 121°C for a specific amount of time. This process involves high temperature and humidity and is incompatible with polymeric materials with melting or softening points, or glass transition temperatures below or near the temperatures involved. Glass, most metals, paper, and some polymers such as silicones, polyurethanes, and polycarbonates are compatible (i.e., physical properties are not affected). Steam sterilization can result in corrosion of some metals.

For steam sterilization to be effective, moisture and heat must be able to penetrate packaging mate-rials and contact and transfer thermal energy to all surfaces of a medical device. This requires the use of porous packaging (made from paper, Tyvek®, or similar materials) that allows water vapor to enter and exit but prevents the same for microorganisms. Barriers to contact (e.g., air pockets within the device), surface contaminants (e.g., dirt, grease, lubricants, other manufacturing residues), and design features (e.g., O-rings that seal surfaces, tortuous paths that impede the diffusion of steam to parts of the device, nooks and crannies that hide and protect microorganisms) can interfere with effective steam sterilization. These barriers can be overcome through proper design of the device. Devices with higher masses require more time to reach effective sterilization temperatures. However, materials with high thermal conductivities will accelerate heat transfer.

8.5.2.2 Ethylene Oxide

EtO gas is the most common sterilization method used for medical devices. It is an effective sterilant due to its ability to diffuse through solid materials and its compatibility with many materials, many of which are sensitive to heat, high humidity, and radiation. EtO gas can kill bacteria, viruses, and fungi, and produces chemical reactions that interfere with normal cellular metabolism and reproduction, ren-dering microbes nonviable. The parameters that affect the efficacy of EtO sterilization are gas concen-tration, temperature, humidity, and time.

The EtO process begins with preconditioning, during which the product (typically one or more pal-lets of product) is exposed to a warm, humid environment until a uniform internal temperature and rela-tive humidity (RH) of 52°C and 55°C to 65°C are reached, respectively. The product is then moved into

Table 8.2 Summary of the conditions, mechanisms, advantages, and limitations of common industrial sterilization methods used for medical devices

Technique	Mechanisms	Advantages	Limitations
Steam (autoclave)	Steam sterilization uses thermal energy (121°C for a specific time) from high-pressure saturated steam to denature DNA and destroy enzymes to kill microbes on the surfaces of devices. Typical time: 15 min	Simple, safe, rapid, and efficient, good penetrability, easy to monitor, no toxic residues, low cost, works with glass, most metals, paper, and some polymers such as silicones	High temperature and moisture are incompatible with many thermosensitive polymers; requires breathable packaging, can cause corrosion of some metals
Ethylene oxide (EtO)	Gas produces chemical reactions that interfere with normal cellular metabolism and reproduction, rendering microbes (bacteria, viruses, fungi) nonviable. Most common sterilization method used for medical devices. Typical time: 2.5 hours (not including aeration time)	Compatible with materials that are sensitive to heat, high humidity, and radiation; can diffuse through solid materials. Good choice for sterilizing temperature and/or humidity sensitive materials. Efficacy of EtO controlled by gas concentration, temperature, humidity, and time.	Requires porous packaging. Requires mild pretreatment in warm humid environment, as well as post-processing diffusion of EtO out of device and packaging. Not appropriate for biodegradable materials (e.g., sutures). May take as long as two weeks for residual EtO gas to reach safe levels in device and packaging
Radiation (gamma)	Gama radiation from X-ray or cobalt 60 kills microbes by causing damage to DNA. Typical time: less than 24 hours depending on required dose	Excellent penetrability (e.g., metal/foil pouches), efficient and reliable even for nonporous packaging, dosimetric release, cost effective, compatible with many materials, leaves no residue. Works on wide range of materials and on many devices at once.	Isotope containment requires costly facilities; Polymer damage (through chain scission), can result in both functional, and color changes that, increase with dosage. Potential for decreased shelf life.
Radiation (e-beam)	High-energy electrons from electron accelerators kills microbes by causing damage to DNA. Typical time: 5–7 min	Same as gamma radiation, but exposure for seconds	Similar to gamma radiation

a sealed chamber where a slight negative pressure is used to evacuate air and EtO gas is introduced. After a period of time (validated to confirm effectiveness) negative pressure is used again to evacuate the EtO gas from the chamber and the product, and the product is then removed. After removal, the product will remain in a quarantined area of a warehouse while residual EtO gas (absorbed by the product) is allowed to diffuse out of the sterilized product (off-gassing). The product can be shipped to customers once the EtO residual level is determined to be safe. The sterilization process itself can take several days to complete. Off-gassing can take 1 to 2 weeks.

To be effective, EtO gas must interact with water molecules on the surface of a medical device and with the internal surfaces of the packaging. Humidification is used to deliver water molecules to these surfaces, followed by an inflow of EtO gas that reacts with the water molecules to begin the chemical reactions that eventually sterilize the surfaces. Porous, breathable packaging, often made of Tyvek®, must be used to allow penetration of water vapor and EtO gas into and out of the package. Like steam sterilization, device designs that do not hinder the diffusion of EtO gas into a device and onto all device surfaces will enhance the effectiveness of the EtO sterilization process.

Most materials are compatible with EtO sterilization. Slightly elevated temperatures well below the melting or softening points of many polymers, and slightly elevated humidity levels make this a good choice for sterilizing temperature and/or humidity sensitive materials. Devices made of biodegradable polymers, such as dissolvable sutures, are sensitive to humidity. The degradation process for these materials can be initiated when they come in contact with water molecules; as a result, they may not be good candidates for EtO sterilization. To prevent water vapor from contacting these devices while sitting on the shelf, they must be packaged in nonporous packaging, which cannot be used in EtO sterilization.

8.5.2.3 Radiation

Radiation sterilization methods use ionizing radiation in the form of gamma radiation and electron beams. Ionizing radiation kills microbes by causing damage to DNA. The main sources of ionizing radiation are high-energy photon sources such as X-ray machines and cobalt 60, and high-energy electrons from electron accelerators. Ionizing radiation can penetrate all medical device materials and does not require porous packaging. It is not hindered by long, narrow lumens, or nooks and crannies that hide microbes and act as barriers to gas-based sterilization methods.

Gamma radiation involves packaged product loaded into totes. The totes are transported by a conveyor system around a central cobalt 60 source until the appropriate radiation dose has been delivered to the product. Exposure can last for hours. In electron beam sterilization, electrons are accelerated through an electric field and focused onto the product. Exposure lasts for seconds.

Unlike gas-based methods, radiation sterilization can change the physical properties of materials in medical devices. Metals are unaffected but the physical properties of polymers may be altered. For example, chain scission is a common side effect of ionizing radiation and results in the breaking of molecular chains within the polymer. This lowers the molecular weight of the polymer, often reducing strength, ductility, and other physical properties, making the polymer more brittle. Other changes include color and opacity changes (yellowing and clouding of clear polymer tubing). Ionizing radiation methods accelerate the aging of polymers and can affect the shelf life of products, a topic further discussed in Chapter 9. After exposure to radiation, the molecular weight of a polymer can continue to decrease over time while on the shelf. These changes in physical properties can impact the function of a medical device and require testing of sterilized samples to determine the effect of radiation sterilization on device function.

The ability to penetrate all medical devices and packaging materials allows the use of nonporous packaging such as metal/foil pouches and other vapor barrier materials. This flexibility allows device designers to protect sensitive materials while on the shelf and extend a product's shelf life.

Design for sterilization involves making design choices that will make it easier to sterilize a medical device. When designing a device with sterilization in mind, designers should consider the most likely sterilization method that will be used, keeping in mind that the final design may require a different method of sterilization. For any sterilization method, device geometry and material composition should be considered; furthermore, barriers that can prevent gas from reaching, contacting, and interacting with surfaces should be eliminated. These barriers include air pockets within the device, surface contaminants such as dirt, grease, lubricants, and other manufacturing residues, O-rings that seal surfaces, tortuous paths, or long lumens that impede the diffusion of steam or EtO gas to parts of the device. For example, EtO gas may have difficulty penetrating into the center of the long, narrow lumens, present in many catheters and other tubular medical devices. Radiation may damage sensitive electronics present in electronic medical devices. Steam can soften or melt polymeric devices. Designers should choose sterilization methods that will not significantly reduce material strength or other physical properties, damage the device, or adversely affect device function, and that are able to penetrate materials and reach small nooks and crannies within the device that can harbor and protect microorganisms. In an industry design project, functional testing discussed in Chapter 9 is often completed on devices that have been produced using all final manufacturing processes, including sterilization.

8.5.3 Cleaning and Sterilization in the Clinical Environment

Devices that contact patients and are intended for reuse must be cleaned and sterilized or disinfected prior to reuse. These devices must be designed to allow for the removal of debris and contaminants from device surfaces (external and internal) and compartments. The primary sterilization method used in the clinical setting is steam (autoclave), as discussed in Section 8.5.2.1. Suggestions for designers that will allow for easier cleaning, disinfection, and sterilization are listed below:

- Avoid using textured surfaces, hinges, springs, dead-end lumens, and inaccessible cracks and crevices that may harbor organisms and organic material and make cleaning difficult.
- To assist in removing debris, consider adding flush ports to provide better access to difficult to clean surfaces. Flush ports allow liquid cleaning solutions to be flushed into and through spaces in the device to improve access to areas that may harbor organisms or organic material.
- Cleaning should not require specialized brushes, detergents, or sponges. Devices should be designed to work with standard chemical solutions and materials normally available in hospitals.

Designing for device sterilization in the clinic is most often impacted by material choice. Generally higher grade and more durable materials are required to withstand the stresses of multiple sterilization cycles yet retain their functionality. The choice of material is highly dependent on the sterilization method used. In designing a device for sterilization, you would first determine the form of sterilization that would be most appropriate for your device. Second, you would determine the thermal, mechanical, and chemical stresses that your device would be subjected to through repeated sterilization cycles. Third, you would select materials and designs that you would then test (as described in Chapter 9), redesigning as necessary.

8.6 Design for Maintenance and Service

Many medical devices, such as ECG monitors and endoscopes, are reusable and intended to be used many times. Reusable devices will require calibration, service, and maintenance. In the hospital environment, these activities are often the responsibility of **clinical engineers**, who are typically trained in biomedical engineering. Some medical devices designed to be used in the home environment will require minor maintenance such as battery or filter replacement. Just as it is important to design products that are easy to assemble, it is important to design products that are easy to service and maintain. This section presents several design practices aimed at making the service and maintenance of reusable medical devices easier. Breakout Box 8.5 provides a clinical engineer's perspective on medical technology and device design.

Breakout Box 8.5 The Clinical Engineering Perspective on Medical Technology and Device Design

Design engineers and clinical engineers have different yet complementary perspectives on medical device design. Whereas a design team translates medical science into a useful clinical technology, a clinical engineer provides technical and managerial oversight of this technology in the healthcare setting. Both disciplines emphasize the delivery of safe, effective, and reliable care, but clinical engineers are focused on these priorities in the context of how these devices are maintained, utilized, and integrated into the overall environment of care. Developing internal technical competency and expertise allows a healthcare organization to provide service, ensure compliance, and control costs.

The work of a clinical engineer can be impeded if the devices are designed with physical or software features that limit serviceability. A highly compact device in a sealed case limits the ability to resolve issues on-site and can create a logistical challenge in shipping equipment to an offsite repair depot. Similarly, creating designs with major components or highly integrated subsystems may simplify the architecture, but from a clinical engineering perspective, this may not be cost-effective if something goes wrong. Often these components need to be replaced as a complete unit which may carry a significant expense. For example, replacing an entire control console for $50,000 may resolve an issue, but replacing a hard drive within a console for $100 would be much more economical.

Some design specifications may impose operational and financial burdens on clinical engineering operations. Creating designs that simplify and minimize routine maintenance activities is desirable as it requires fewer technical resources. Small design variations can determine whether these routine activities are to be performed by technical personnel or by clinical staff, which may have regulatory or accreditation implications. Designs that require nonstandard supplies or consumables may pose a challenge or introduce risk. For example, a design that incorporates materials that require specialized cleaning wipes not commonly used by customers will introduce a nonstandard item into their supply chain. If the specialized wipe is not adequately stocked or users inadvertently use the familiar standard cleaning wipes, it could cause discoloration, weaken the material, or even break, thereby creating a service issue.

Discrete devices that remain isolated from other technology are becoming less common as the desire grows to integrate, automate, and communicate through IT networks. In some cases, it is difficult to draw the distinction between the device and the IT infrastructure. Historically, manufacturers and clinical engineers could install, modify, and update devices with minimal risk of impacting other systems. The current healthcare environment requires clinical engineers to carefully coordinate and plan for any device changes that may impact dependent systems, disrupt data interfaces, or trigger network security alerts. The ability of clinical engineers to implement or deploy these updates needs to be incorporated into the design of the device.

The utilization of a medical device is greatly influenced by human factors, such as visual aspects, control interfaces, and general ergonomics, which can often impact the successful adoption of the device regardless of how sound the engineering is. User experience varying drastically from one model to the next can cause a steep learning curve, cause confusion, or—in extreme cases—pose a safety risk. Improper use or lack of familiarity with normal operating behavior of the device can appear to the user to be a device malfunction at which point the clinical engineering team will be engaged. It is not always feasible for a healthcare organization to replace an entire fleet of equipment so it is possible that clinicians may be simultaneously using two or more generations of a device or may be transitioning from a competitor's product. Therefore, significant design innovations must be balanced with the ability of the clinical practice and technical support teams to adapt to the changes.

Matthew Dummert, MS, CHTM
Director of Healthcare Technology Management
Froedtert Memorial Lutheran Hospital and the Medical College of Wisconsin

Over the life of a reusable medical device, blades, filters, batteries, fuses, light bulbs, and other components will need to be replaced. These components must be easily accessed by users or service personnel and should not require considerable time to access and replace. Suggestions for improving access and simplifying part replacement include:

- Design battery compartments with covers that snap into place, and do not require tools to remove screws or other fasteners holding the battery compartment cover in place. There should also be sufficient space to easily grasp the batteries, remove them, and replace them by hand.
- Choose and locate filters and light bulbs that are easily accessed, removed, and replaced without specialized tools. Removal of a filter or light bulb should not require removal of other parts to provide access.
- Simplify replacement of other components such as circuit boards, fuses, and other components by including access panels that, once removed, allow for easy and quick access to the old parts. Replacing these parts should not require the removal of other parts or assemblies to gain access.

The same guidelines apply to other commonly replaced parts and components. In addition to a well-written user manual, many medical devices come with a service manual that provides detailed instructions for component replacement. The documents reduce confusion and the time required to maintain and service reusable medical devices.

8.7 Design for the Environment

All products have some impact on the environment. The most obvious sources of environmental impacts are the energy used to extract raw materials, refining and manufacturing, shipping and packaging, waste generated during manufacturing and use, and the disposal of a product. There are additional hazardous (e.g., chemicals or radioactive materials) and biological wastes generated during the use of a medical device that must be processed using additional energy and chemical agents and then transported to a specialized medical waste facility (requiring energy for transportation). Most waste generated from manufacturing or product use will ultimately be disposed of in a landfill, which may pollute soil and groundwater sources, or be incinerated, which reduces air quality.

In the United States, hospital waste alone ranks 13th in the world when compared to total waste volume of all other countries. Concerns involving the environmental effects of leached plasticizers in landfills and emissions resulting from incineration of disposed-of medical devices have resulted in significant initiatives among medical device companies to design environmentally friendly and sustainable products. Many companies are incorporating the Triple Bottom Line (discussed in Breakout Box 3.8 as balancing concern for People, Profits, and Planet), and the Triple Aim of Health Care (also discussed in Breakout Box 3.8, a component of which is Improving the Overall Health of Populations) into their corporate culture. Some have hired experts in DFE to work with design engineers (who may not be experts in this area). Regulatory bodies are also issuing standards that guide sustainable design. European Union regulations and standards such as *Restriction on Hazardous Substances* (RoHS), *Waste From Electrical and Electronic Equipment* (WEEE), *Registration, Evaluation, and Authorization of Chemicals* (REACH), and ISO 14001 *Environmental Management Systems – Requirements with Guidance for Use* provide guidance for manufacturers to identify and recognize the effects of their products and processes on the environment.

In this section, we consider DFE to reduce the environmental impacts resulting from the production, use, and disposal of a device. Basic knowledge of DFE and awareness of this topic are very helpful to engineers involved in new product development for medical device companies.

8.7.1 Product Life Cycles and Environmental Impacts

Product life cycles typically last for months to years depending on the product. They start with extraction and processing of raw materials from natural resources. These raw materials are then used to manufacture the product, after which the product is distributed and used. When the product is no longer useful, it might be remanufactured, components might be reused, materials might be recycled, or the product might be disposed of by incineration or deposition in a landfill. Each of these stages in the product life cycle may consume energy and generate emissions and waste, resulting in environmental impacts. These impacts include liquid discharges, gaseous emissions, energy consumption, depletion of nonrenewable natural resources (such as helium and metals), and generation of solid waste. Discharges, emissions, and solid waste may be comprised of toxic substances that can poison or damage the environment and contribute to pollution of air, soil, and water, as well as land degradation and ozone depletion. Products that release relatively large amounts of carbon-based gases (such as carbon dioxide and methane) into the atmosphere have large carbon footprints, which contribute to climate change.

8.7.2 Sustainability

Most plant and animal-based organic materials quickly degrade into nutrients and materials needed to grow new plants. This natural lifecycle is relatively short and can repeat indefinitely, and represents a balanced, sustainable, closed-loop system. Inorganic materials, such as minerals, are created over a much longer time and are nonrenewable resources. Once these resources are depleted, they will not replace themselves. For a product to achieve conditions of environmental sustainability the material used in the product must be balanced in a sustainable, closed-loop system, like plant and animal-based organic materials. The three challenges associated with sustainable product design are to eliminate the use of nonrenewable resources (including nonrenewable sources of energy), disposal of synthetic or inorganic materials that do not decay quickly, and creation of toxic wastes that are not part of natural life cycles.

8.7.3 Sustainable Thinking and Life Cycle Assessment

Sustainable thinking requires an engineer to consider in detail all steps in the life cycle of a product by answering questions such as:

- What is the source of your materials and components? Does it involve depletion of a nonrenewable natural resource? What processing was involved in refining and creating these materials and parts?
- How much energy was required for the transportation of raw materials and supplies, distribution to the user, and shipping to waste facilities? How much energy was involved in manufacturing the product?

- What types of pollution (e.g., soil, groundwater, air) might be involved in the solid disposal or incineration of waste generated? Over what timeline do synthetic or inorganic materials used in the product decay? Are any degradation byproducts toxic to natural life cycles?

Life cycle assessment (LCA) is used to evaluate the environmental impacts of a product during all stages of its lifecycle, identify problem areas, and make improvements that will reduce the impact. There are several software-based tools used for LCA (SimaPro, Eco-it), and references such as *Okala Practitioner: Integrating Ecological Design* (White, 2013), a guide to help manufacturers conduct an LCA of their products.

Once a product life cycle has been assessed, a key to successful sustainable product design is to develop a product with an improved product lifecycle. This involves considering design characteristics that can impact various phases of design (e.g., concept generation, material selection, manufacturing) when planning projects. Package design, transportation, use, and disposal are also considered. Each of these is proactively evaluated considering energy efficiency, material usage, cost, and environmental impact.

8.7.4 Barriers to Sustainable Design

Significant barriers to sustainable design exist within the healthcare ecosystem because 90% of medical devices are disposable, single-use products. Two of these barriers have historically been difficult to overcome. First, the business models used by many medical device manufacturers include a significant portion of revenue from sales of disposable devices. A device that is reused and purchased infrequently may not generate as much revenue as a single-use device purchased frequently. Second, there are functional, safety, and sterility considerations that often require disposability.

8.7.5 Improving Sustainable Design

There are several ways to improve the sustainability of a design that should be considered during the design process. Some of these include:

- specifying a reusable material when it is a reasonable alternative (e.g., specifying a reusable OR towel as opposed to a disposable one);
- eliminating the use of materials containing toxic or hazardous substances (such as lead or polyvinylchloride) and replacing them with more acceptable, safer, environmentally friendly materials;
- reducing product size and weight to reduce shipping costs;
- creating less bulky, more compact, stackable packaging to reduce fuel consumption during shipping;
- choosing materials that reduce toxic air emissions during disposal and incineration to limit environmental damage;
- making the disposable component smaller to minimize waste in products with a reusable component;
- using a significant amount of recycled materials where possible;
- selecting materials and components that can be recycled at the end of the product life cycle, such as lithium-ion batteries;
- if possible, manufacturing products using clean renewable energies;

- choosing materials and processes that do not release harmful air or water emissions during production;
- when possible, choosing recyclable or biodegradable packaging materials to reduce the amount of hospital waste; and
- designing reusable devices such that they can be repaired to avoid becoming waste at the first sign of dysfunction (see Section 8.6).

Additional advice and guidance for designing products for the environment can be found in the suggested reading list at the end of this chapter.

8.8 Package Design

Packaging broadly refers to the container that holds the finished product as well as the product labels, printed materials, and package inserts that accompany the finished product. Academic design projects tend to focus on the design of the medical device and only consider packaging as an afterthought, if at all. This is not the case in industry; many medical device companies employ packaging engineers or consult with packaging vendors to design and supply packaging for their medical devices. Packaging is designed early in the design process once the size and shape of the device are known and materials have been chosen. The major requirements of medical device packaging are presented in this section.

8.8.1 Sterilization and Maintenance of Sterility

As discussed in Section 8.5, sterilization methods must be considered when designing the packaging for a medical device. This includes choosing materials that are compatible with the sterilization process. For example, porous materials such as paper or Tyvek® are used for devices sterilized using steam or EtO gas to allow penetration of these sterilants into the package. Materials that can withstand high temperatures must be used for devices sterilized by high temperature steam. Once a device is sterilized, its packaging must maintain sterility during shipping and storage by providing a sterile barrier to microorganisms.

8.8.2 Protection of Product

Packaging must contain and protect the product during sterilization, shipping, and storage. Some medical device materials will degrade if exposed to airborne contaminants such as air, ozone, moisture, and light. Packaging must prevent these contaminants as well as dust and dirt from contacting the device.

Packaging must also prevent damage to the product, the packaging itself, and the sterile seal from external sources, and any internal components or accessories included in the package. Devices, components, and accessories may have sharp corners and edges that can rub against packaging materials during shipping, resulting in punctures, tears, and rips to the packaging materials, thus compromising the sterile seal and the sterility of the product and package. For example, an implantable device was packaged in a thermoformed tray that was similar to the tray shown in Figure 8.9. The implant was contained in its own compartment to keep it separated and protected from contact with other components

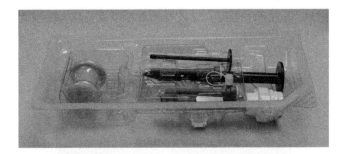

FIGURE 8.9

Example of a thermoformed polymer tray with compartments to store and secure various components of a Single Incision Laparoscopic Surgery (SILS™) port. The heat-sealed Tyvek® lid has been removed from the top of the tray to reveal the components stored inside.

in the package. Due to vibration of the package during shipping, one of these components, a special needle used in the surgical implantation procedure, migrated out of its compartment and touched the outer silicone sheath of the implant. The needle could have easily punctured the soft, outer layer of silicone, which would have resulted in leakage of the fluid-filled device during implantation. As a short-term solution, short lengths of silicone tubing were cut and slid over the end of the needle to cover the sharp end and eliminate the potential for the needle to puncture the implant. The thermoformed tray was later redesigned to provide a slight press fit between the tray compartment and needle, preventing migration of the needle within the tray during shipping. The tray in Figure 8.9 contains compartments to store and secure components of a Single Incision Laparoscopic Surgery (SILS™) port during shipping and storage.

8.8.3 Ease of Entry and Use

Medical device packaging must be easily opened in the operating room and other clinical environments to allow for immediate access to the device during emergencies or situations when the device is needed immediately. It should consider the users and environment in which the package will be opened. For home use, an easy-to-open package will enable a wider range of users to access the device. In the clinical setting, there are emergency situations where a device or tool is needed immediately, and seconds can make a difference in patient outcomes. In these cases, the packaging designer can enhance usability by considering the spacing, location, and orientation of components. Figure 8.9 shows a well-designed thermoformed tray that contains each component securely in its own easy-to-access location. In nonurgent situations, an easy-to-open package can reduce labor and time. One exception to this design requirement is childproof packaging designed to prevent ease of entry. This type of packaging is used to prevent children from opening medication containers and protect them from accidental ingestion of potentially harmful medications.

When opened in a clinical setting, packaging materials must not generate debris which could then travel through the air and contaminate a patient's wound. For example, materials such as Tyvek® provide strong seals to help contain a device in its packaging but are easily pulled away from a thermoformed tray without creating debris.

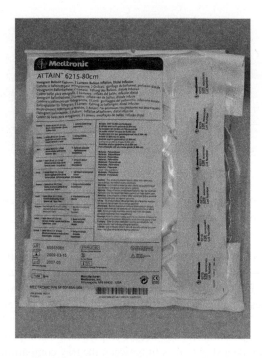

FIGURE 8.10

Medical packaging that includes legible labels and clear materials. The device is visible through the clear material between the two labels when the package is on the shelf of a storeroom, allowing healthcare providers to confirm the contents listed in the labels.

8.8.4 Convenient, Efficient Storage

Packaging must allow the product to be easily identified during storage so that users or medical personnel can easily locate the product. This requires legible labels, clear packaging, and orientation of the product that does not block the view of the device when the package is stored. Requirements for medical device labeling are presented in Chapter 12. Figure 8.10 shows packaging that meets these requirements. Packaging should make efficient use of space; it should lay flat to allow several packages to be stacked. Packaging configurations that allow for stacking of devices on a floor or shelf, and the use of nesting packages allow for neat, efficient use of storage space.

8.9 Design Risk Management

Everything we do has risks associated with it. For example, when crossing the street there is a chance that you might trip on uneven pavement, turn your ankle in a pothole, or be hit by a passing vehicle. Each hazard (uneven pavement, potholes, passing vehicles) could result in a related harm (falling, ankle turning, impact to body) and each harm has a level of severity (minor injury, serious injury, or death) and a nonzero probability that the hazard could cause the harm to occur (probability of occurrence). The severity and probability associated with these events determine risk. Careful design and maintenance of the street, proper traffic controls and warnings, and attention paid when crossing the street by pedestrians all contribute to reducing the probability of occurrence, thereby reducing residual risk but not eliminating it. It is often impossible to remove all risk; the goal of designers and the risk management process is to reduce residual risk to an acceptable level.

In this chapter, we present several aspects of design to consider as you generate iterations of your design solution and prototype. Prior to finalizing your design, it is important to determine how your device could fail (failure modes) and the effects of these failures. This information can then be used to determine the causes of these failures and develop plans to reduce the potential for these failures to occur. This process is known as *risk management* and is an important design tool that should be used throughout the design process, not just at the end of a project. It should be used when any design changes are made to confirm that these changes do not introduce new failure modes with new effects on device function. For example, replacing a power supply in an anesthesia pump, surgical power tool, or another electromechanical device with a newer, less expensive alternative could result in increased heating of the device. This could lead to device malfunction or injury to the patient or user. A larger fan could be added to dissipate the heat generated from the new power supply; however, this may require additional design changes which could in turn impact other aspects of the design.

Design risk differs from project risk discussed in Section 2.6.3. Project risk involves events that can threaten the planned completion date of a project. Design risk refers to the potential for a design to result in harm and is the focus of this section. Once you have established a preliminary design, you should begin the risk management process by identifying modes of failure and determining the effects of these failures. After conducting some of the tests presented in Chapters 9 and 10, test data and other information can be used to estimate risk and determine how unacceptable risks can be eliminated or reduced to acceptable levels.

ISO 14971:2019, *Medical Devices – application of risk management to medical devices*, is the main standard for risk management of medical devices. Per this standard, risk management involves three steps—risk analysis, risk evaluation, and risk control. The standard defines *hazard* as a potential source of harm (such as an open flame), *hazardous situation* as a circumstance in which people, property, or the environment are exposed to one or more hazards (such as exposure to an open flame), and *harm* as injury or damage to the health of people, or damage to property or the environment (such as a burn injury caused by the open flame). *Benefit* is defined as positive impact or desirable outcome of the use of a medical device on the health of an individual, or a positive impact on patient management or public health. *Risk* is defined as the combination of the probability of occurrence of harm and the severity of that harm. In mathematical terms, $risk = probability \times severity$. *Risk analysis* is the systematic use of available information to identify hazards and estimate the risk. This standard is also discussed in Chapter 11.

8.9.1 Risk Analysis

Risk analysis consists of these five steps:

1. Identify potential hazards associated with the design, considering the environment in which the device will be used and how it might be misused.
2. Determine the potential harm that can result from each hazard.
3. Determine the severity of each potential harm (nuisance/annoyance, minor injury, serious injury, death).

4. Estimate the probability of occurrence of each hazard.
5. Calculate risk by multiplying probability and severity.

Two risk analysis tools recommended by the FDA are Failure Modes and Effects Analysis (FMEA) and Fault Tree Analysis (FTA). FMEA is commonly used in the medical device industry and can be used with subjective or experiential data. Another method is Failure Mode Effects Criticality Analysis (FMECA). It requires quantitative or historical data, which is often not available for preliminary designs that have not been tested and thus might not be appropriate for academic design projects. It also requires a value for detectability in the calculation of risk (which is not required by ISO 14971:2019) that some experts consider inappropriate when estimating risks associated with the design of a medical device. In this chapter, we focus on the use of FMEA to analyze risk.

When your preliminary design has been determined, you will most often have little if any data available to you for estimating the probability of occurrence needed to calculate risk values. For this reason, you should use FMEA as a risk analysis tool at this stage in the design process to identify potential hazards and failure modes. The following example demonstrates how FMEA can be applied to the design of an IV pump as shown in Figure 8.11, the components of which are shown in Figure 8.12.

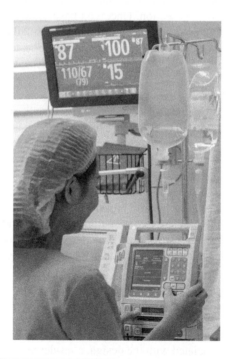

FIGURE 8.11

Example of an intravenous (IV) pump used in hospitals to deliver medication to patients. Medication from a hanging IV bag (*top right* of image) flows via gravity feed through tubing and is delivered to the patient through a needle placed into the patient's vein. A healthcare provider is shown controlling the medication delivery rate through the user interface of the IV pump.

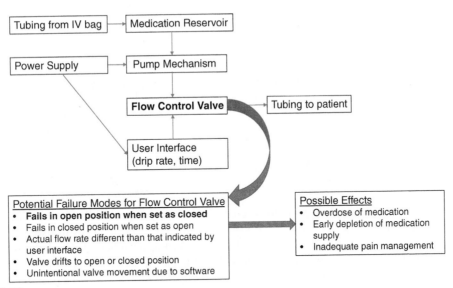

FIGURE 8.12

Schematic diagram of intravenous pump. Potential failure modes of the flow control valve and possible result-ing effects are shown in boxes at the bottom of figure.

Applying FMEA to the design of an IV pump involves five main steps. For this example, we focus on the flow control valve only. To complete the analysis, these steps would be repeated for each com-ponent in the device (step #5).

1. List all components of the device.

   (See Figure 8.12 for schematic diagram of the IV pump.)

2. Describe the function of each component.

   The flow control valve is used to control the flow of medication from the pump into the tubing that delivers the medication to the patient's bloodstream. Flow rate is set through the user interface.

3. Identify failure modes (How could each component fail?).

   For the flow control valve, several modes of failure are possible as shown in Figure 8.12. It can fail in the open or closed positions, there can be a mismatch between the flow rate indicated by the user interface and the actual flow rate, the valve might drift into the open (if originally closed) or closed (if originally opened) position, or an un-commanded, unintentional (software directed) movement of the valve could occur.

4. If the failure occurs, what effects would it have on device function?

Table 8.3 Example of criteria for severity based on a 10-point scale of potential harm associated with a medical device. Severity scores for harms that fall between the descriptions and criteria shown can be interpolated.

Severity score	Description	Criteria
1	Negligible	Nuisance or annoyance only; would not cause injury
3	Minor	Minor injury
5	Serious	Moderate injury
9	Critical	Serious injury or death
10	Catastrophic	Multiple deaths or serious injuries

For purposes of this example, we will focus on one failure mode, specifically the valve failing in the open position (where the valve opens after it was set to the closed position). If this were to occur, a possible overdose of medication could result. This could lead to depletion of the medication supply available to the patient. Depending on the condition of the patient and the type of medication being delivered (pain medication, antibiotic) a valve stuck in the open position creates the potential for a drug overdose with serious consequences to the patient.

5. Repeat the process for all remaining components in the device.

Severity and probability of occurrence values are needed to complete the risk calculation. Severity can be determined subjectively and can be reasonably determined without test data. For example, the criteria shown in Table 8.3 might be used as a guide to quantify severity. The probability of occurrence values can be estimated by evaluating field complaints of similar devices (historical data) and reviewing product recall data (described in Section 12.7), through laboratory testing presented in Chapter 9, with analytical or simulation methods, or with expert judgment.

8.9.2 Risk Evaluation

Once risk values have been calculated, it must be determined whether or not they are acceptable.

Figure 8.13 provides guidance for determining if risks are acceptable. For example, a heart valve design with a high probability of occurrence for strut fractures resulting in catastrophic failure and death (high severity) would represent an unacceptable risk. This risk is represented in the light gray (unacceptable) region of Figure 8.13. A motor used for an IV pump that infrequently (low probability of occurrence) became noisy during operation due to inadequate lubrication but continued to pump fluid at the required rate (low severity) would represent an acceptable risk. This risk is represented in the dark gray region in Figure 8.13.

Risks in these two regions are easily classified as acceptable or unacceptable. However, many risks will fall into the white region labeled as low as reasonably practicable (ALARP) in Figure 8.13. ALARP means that after considering the costs and benefits of reducing or accepting the risk, it has been reduced to the lowest level possible without using unreasonable amounts of money, time, or effort in proportion to the level of risk reduction. In this case, "reasonably practicable" steps have been taken to reduce residual risk. Practicability refers to the availability and feasibility of solutions

Severity	Negligible	Minor	Serious	Critical	Catastrophic
Probability				.	
Frequent	ALARP	ALARP			
Likely		ALARP			
Occasional		ALARP	ALARP		
Remote			ALARP		
Unlikely			ALARP	ALARP	

FIGURE 8.13

Risk acceptability chart indicating regions where risk values are unacceptable (light gray; risk reduction required), broadly acceptable (dark gray; no further action required), and as low as reasonably practicable (ALARP - white; investigation of reasonable options to further reduce risk required).

that can reduce risk, as well as the ability to reduce risks without making the use of the device financially infeasible. Risks that fall into the broadly acceptable region are acceptable with no further risk reduction efforts needed. Ideally, risks in the ALARP region will be reduced and moved into the broadly acceptable region. Risks that fall into the ALARP region are generally acceptable if the benefits justify the residual risks. Risks that fall into the unacceptable region must be reduced to the level of ALARP or broadly acceptable risks. If it is determined that this is not possible, then the design must be abandoned.

8.9.3 Risk Control

Once it is determined that there are unacceptable risks, risk mitigation plans are established to reduce risk to acceptable levels. These plans focus on reducing severity levels and/or probabilities of occurrence. Risk reduction options include:

- design changes (more robust design, improved factor of safety, improved usability to reduce confusion, additional alarms);
- procedure changes (improved maintenance procedures, safety precautions);
- warnings (improved instructions for proper device setup, use, and maintenance included on devices and/or packaging and in user manuals and other labeling); and
- manufacturing changes (to reduce defective assemblies and potential for component failures).

Once a risk mitigation plan is created, it must be decided which of these risk-reduction options will be implemented. These changes are then implemented and the preliminary design is reevaluated (or the prototype is retested) to determine the effect of these changes on risk levels. It must also be determined whether new hazards and risks have been created because of these changes.

To summarize risk data and calculate risk values, you may want to use the form shown in Table 8.4 to begin thinking about risk and where problems might occur with your design. At this (preliminary

Table 8.4 One tabular method for documenting risk data, calculating risk values, and recording risk mitigation plans

Potential failure mode	Harm	Severity	Probability	Risk	Risk mitigation plan

design) stage of your project, information regarding failure mode and harm is available from an FMEA, and severity can be estimated. At this time, you can estimate probabilities using your judgment of where potential problems might exist in your design. You can then fine-tune these probabilities later after testing your prototype. For academic design projects, probabilities are typically not available due to a lack of adequate test data. You will revisit and complete Table 8.4 in Chapter 9 when test methods are presented that you can use to generate test data needed to determine probabilities, which will allow you to calculate more accurate risk values and develop risk mitigation plans. An example of a completed table is provided in Section 9.11.

In an industry design project, resources are available to produce a statistically significant sample of prototype devices for testing. This helps in determining failure rates and probabilities of failure modes. In academic design projects, time and budget constraints may only allow construction and testing of a single prototype that will be transferred to a client or sponsor at the end of the project. For this reason, nondestructive testing is not an option. The small sample size and limitations on the type of testing may prevent determining accurate probabilities for use in calculating risk. Academic design teams may be limited to reasonable estimates of probabilities.

8.10 Documentation and Design Reviews

As you consider and apply the information presented in this chapter and the design concept advances toward a more detailed design, all decisions regarding the evolution of your design should be included in your project's Design History File. This will document your rationale for all design choices and will remind you (and your sponsor and mentor) why these choices were made. This information will also help you answer questions during formal design reviews and presentations and justify your design decisions. Preliminary FMEA results should be included. Documenting design decisions is a habit to develop and is part of good professional design practice. In industry, this information will help an auditor (internal or from the FDA) follow your design decisions, and help you defend your design choices if necessary.

Key Points

- When designing a product, engineers should consider the user, use environment, and user interface.
- Manufacturing engineers should be consulted early in a design project to consider design characteristics that will make the product easier and less costly to manufacture. In an academic design project, appropriate technical staff should be consulted prior to producing prototype parts.

- Material selection should be based on the expected service environment, cost requirements, sterilization methods, and production processes.
- The most common sterilization methods used for medical devices include steam, ethylene oxide, and radiation. Sterilization methods are chosen for compatibility with the material used to manufacture the device. Steam sterilization using an autoclave is the most common sterilization method used in hospitals.
- Packaging requirements include compatibility with sterilization processes, protection of the product during shipping and storage, maintenance of a sterile barrier, ease of entry, efficient storage, and quick identification of the product in the package.
- Life Cycle Assessment involves determining the total environmental impact of the product, including the energy and wastes associated with creating raw materials, production processes, product use, and product end-of-life.
- Risk management is part of detailed design. It involves identifying potential modes of failure, consequences of these failures, and mitigation plans to reduce risk. FMEA is a widely accepted and commonly used method for risk management.
- Risk is the product of probability of occurrence and severity of a particular harmful event or failure. Risk can be reduced through design changes, procedure changes, manufacturing changes, and warnings.

Exercises to Help Advance Your Design Project

1. Consider your design project and determine which of the "Design for X" approaches discussed in this chapter your project would benefit from most.
2. Apply the approach you selected in Exercise 1 to your design. Generate a list of potential design changes that will make your design better with respect to "X." Remember that there are often tradeoffs between improvements such that improving one aspect of the design might adversely affect another (e.g., making a product more usable by reducing its size and weight might require a higher density of parts and components which could make it more difficult to maintain and service). Consider your design and identify any serviceable components such as filters, batteries, fuses, or other components. Can they be easily replaced or serviced? If not, what design changes can you make to improve the serviceability of your design?
3. Sketch and then iterate upon the ideal user interface for your device. Annotate this drawing with your rationale for your design decisions. How are you using the various principles of user interface design outlined in the chapter?
4. Assume that your product will be shipped to customers around the country. Mockup the kind of packaging that would be most effective for your device (either as a drawing or a physical prototype). Think about what kind of packaging you would use to protect it during shipping and storage. Will it need to be sterilized? If so, what method might you use, and how would this impact your choice of packaging materials and specific package design?
5. List all the cognitive and physical demands that you expect a user to experience in interacting with your device. Are there revisions to your design that would reduce these demands? You may want to consider both the internal functional aspects and the user interface.
6. Create a user's manual for your product. Consider including process steps, diagrams, assembly instructions, and a repair or maintenance section.

7. Perform a preliminary FMEA of your device. Estimate the severity of each potential effect. As you consider failure modes, be sure to build scenarios that cover your range of users and expected use environments. Save your work in a file that you can update.

8. When you have refined your design solution by applying what you learned in this chapter, revisit your project statement. Does it still accurately capture the problem being solved by your prototype? If someone read your project statement and then used your prototype, would the connection between the problem and your solution be clear? Should your project statement be updated?

References and Resources

AIAA S-110. (2005). Space systems-structures, structural components, and structural assemblies, section 4.2. AIAA.

Anderson, D. M. (2014). *Design for manufacturability: How to use concurrent engineering to rapidly develop low cost, high-quality products for lean production.* CRC Press.

Ashby, M., & Johnson, K. (2013). *Materials and design* (3rd ed.). Butterworth-Heinemann.

Association for the Advancement of Medical Instrumentation. (2013). *Human factors engineering—design of medical devices. (ANSI/AAMI HE75: 2009/(R)2013).* AAMI. https://www.aami.org/.

Beer, F., & Johnson, R. (1992). *Mechanics of materials* (2nd ed.). McGraw-Hill.

Buchanan, G. (1988). *Mechanics of materials.* Holt, Reinhart, and Watson, 55.

Burr, A., & Cheatham, J. (1995). *Mechanical design and analysis* section 5.2 (2nd ed). Prentice-Hall.

Center for Devices and Radiological Health. (2016, February 3). *Applying human factors and usability engineering to medical devices: Guidance for industry and Food and Drug Administration staff. (FDA-2011-D-0469).* FDA. https://www.fda.gov/.

Eckelman, M. J., & Sherman, J. (2016). Environmental impacts of the US health care system and effects on public health. *PLoS One, 11*(6), e0157014.

Evans, B. (2008, August). *The greening of medical device design* (pp. 40–48). MDDI, Canon Communications

"Factor of safety. Wikipedia. https://en.wikipedia.org/wiki/Factor_of_safety#. [Accessed 27 April 2021].

Goldberg, J., & Malassigne, P. (2017). Lessons learned from a 10-year collaboration between biomedical engineering and industrial design students in Capstone design projects). *International Journal of Engineering Education, 33*(5), 1513–1520.

Gratteau, H., & Millenson, M. (1993). *FDA requesting a recall on infant monitoring cables* Chicago Tribune. https://www.chicagotribune.com/news/ct-xpm-1993-08-28-9308280097-story.html.

Interaction Design Foundation. https://www.interaction-design.org/literature/topics/user-centered-design.

International Organization for Standardization. (2019). *Medical Devices—Application of risk management to medical devices (ISO 14971:2019).* https://www.iso.org/standard/72704.html.

Juvinall, R. (1967). *Stress, strain, and strength.* McGraw-Hill.

Kadamus, C. (2008). *Sustainability in medical device design* (pp. 42–47). MDDI, Canon Communications.

Lerouge, S., & Simmons, A. (2012). *Sterilisation of biomaterials and medical devices.* Woodhead Publishing.

Maddox, M. E. (1997, May). *Designing medical devices to minimize human error,* MDDI, Canon Communications.

Marshall, J., Hinton, M., Wrobel, L., & Troisi, G. (2009). Designing sustainable medical devices (pp. 56–60). MDDI, Canon Communications

Medical Devices: Establishment of a performance standard for electrode lead wires and patient cables. 62 fed. Reg. 25477 (proposed May 19, 1994) (to be codified at 21 CFR pt. 898).

NASA-STD-5001. (2008). *Structural design and test factors for spaceflight hardware, section 3.* NASA.

Nielsen, J. (1993). *Usability engineering.* AP Professional.

Package Engineering for Sterilization. (2018). *Medical design briefs*. https:// www.medicaldesignbriefs.com/ shiftentercomponent/content/article/mdb/features/technology-leaders/28352.

Patel, A., Pope, J., & Neilson, M. (2021). *Design considerations for medical device manufacturers. AAMI Horizons*. Springer, 73–75.

Privitera, M. B. (Ed.). (2019). *Applied human factors in medical device design*. Elsevier Academic Press.

Research-Based Web. (2006). *Design & usability guidelines*. US Department of Health and Human Services.

Rutter, B. (2018, December). *Key human factors drive design for disposables, test kits, and IFUs*. Medical Design Briefs.

Samaras, G. M. (2013). *Use, misuse, and Abuse of the device failure modes effects analysis*. Medical device and Diagnostic industry. https://www.mddionline.com/software/use-misuse-and-abuse-device-failure-modes-effects-analysis.

Sapiente, J. (2013, March 4). *Reduce medical device costs the automotive way*. Medical device and Diagnostic industry. https://www.mddionline.com/design-engineering/reduce-medical-device-costs-automotive-way.

Sherman, M. (1990). *Medical device packaging handbook* (2nd ed.). Marcel Dekker, Inc.

Shigley, J. (1976). *Mechanical engineering design* (3rd ed.). McGraw-Hill.

Shigley, J., & Mischke, C. (1986). *Standard handbook of machine design*. McGraw-Hill, 2–15.

Shumbun, N. K. (1989). *Poka-Yoke: Improving product quality by preventing defects*. Productivity Press.

Snow, A. (2001, March 1). *Integrating risk management into the design and development process*. Medical Device and Diagnostic Industry. https://www.mddionline.com/news/integrating-risk-shiftentermanagement-design-and-development-process.

Tilley, A. R. (2002). *The measure of a man and woman: Human factors in design*. Henry Dreyfuss Associates, John Wiley and Sons, Inc.

Tillman, B., Fitts, D. J., Woodson, W. E., Rose-Sundholm, R., & Tillman, P. (2016). *Human factors and ergonomics design handbook* (3rd ed.). McGraw-Hill.

Ullman, D. (2017). *The mechanical design process* (6th ed.). David Ullman, LLC.

Ulrich, K. T., Eppinger, S. D., & Yang, M. C. (2020). *Product design and development* (7th ed.). McGraw-Hill.

White, P. (2013). *Okala practitioner: Integrating ecological design*. Okala Team.

Whitmore, E. (2012). *Development of FDA-regulated medical products* (2nd ed.). ASQ Quality Press.

Young, W. (1989). *Roark's formulas for stress and strain* (6th ed). McGraw-Hill.

VALIDATION AND VERIFICATION TESTING

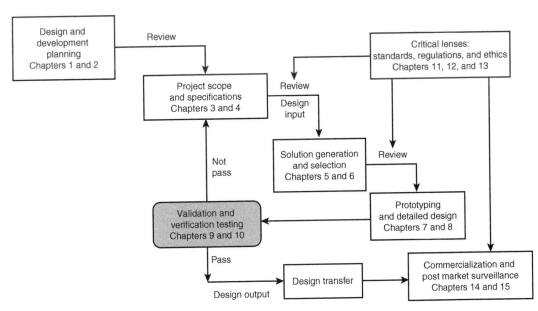

5

In the next two chapters you will learn how to plan and conduct tests appropriate for your medical device and analyze test results. Although you may have engaged in feasibility tests earlier in the design process, testing takes center stage during verification (are technical specifications met?) and validation (are customer/user needs met?). These tests aim to demonstrate that a new product is both safe (not harmful) and efficacious (performs as required). Some of the most impactful academic design projects are simple; still, the design and execution of the tests and the analysis of test results present significant technical challenges.

Chapter 9 is focused on benchtop and laboratory testing in nonliving systems. You will learn how testing is critical in verifying specifications, validating your design with customers, determining the probabilities and severity of known failure modes, performing a risk analysis, and discovering new failure modes. Guidelines are provided on electrical tests, static and dynamic mechanical tests, testing to failure, and using testing standards. In addition, you will learn how to simulate biological fluids, tissues, and organs in nonliving systems, as well as the role of aging and shelf life studies.

Chapter 10 builds upon the guidance for testing in nonliving systems but focuses on the unique challenges in conducting tests *in vitro* (cell and tissue) and *in vivo* (biocompatibility, animal, and human clinical trials). Testing in living systems is often required for verification and validation of certain specifications and user needs, respectively. Furthermore, data from living systems is often necessary in obtaining regulatory clearance, convincing health insurance companies to reimburse for use of the device, and creating promotional materials. Working with living systems brings with it ethical and legal protections for test subjects. You will therefore learn how you can obtain the proper approval and safety training to perform tests in living systems.

By the end of Chapters 9 and 10 you will be able to:

- Design and conduct the appropriate tests in nonliving and living systems to verify each of your specifications.
- Design tests that validate that your solution will meet the needs of users.
- Select and execute appropriate data analysis methods for design verification and validation.
- Present test results in a way that will convey the efficacy and safety of your design.
- Become qualified in engaging in living system testing and in drafting proposals to university committees for *in vitro* testing and *in vivo* testing.
- Complete a risk analysis of your design solution and create mitigation plans for reducing risk.

Testing for Design Verification and Validation

If you truly have faith in your convictions, then your convictions should be able to stand criticism and testing.
— **DaShanne Stokes, author, sociologist, and activist**

Assumptions aren't facts; they're opportunities for research and testing.
— **Laurie Buchanan, PhD, author and life coach**

Chapter outline

Biomedical Engineering Design. https://doi.org/10.1016/B978-0-12-816444-0.00009-2

9.1 Introduction

This chapter presents common laboratory tests (bench tests) for proving that a new product meets all performance specifications (design verification) as well as customer needs (design validation). These tests are used to confirm that design input = design output. They are often used to determine technical feasibility, quantify the performance of a design solution, and identify potential design improvements.

    This chapter focuses on bench tests that involve feasibility, prototype, and product testing in nonliving systems. We also present package tests and shelf-life studies. Testing in living systems (biocompatibility tests and animal and clinical studies) are discussed in Chapter 10. We first present the different types of tests that engineers might use during the design process. We then present verification tests that can be used to confirm that the design solution meets the technical specifications identified in Chapter 4. Validation tests are then discussed as a way to determine if customer needs (discussed in Chapter 3) have been met. This chapter concludes with advice on how to report test results in writing and oral presentations. The goals of this chapter are for you to (1) understand and appreciate the value of testing and the role it plays in the design process and (2) plan and implement the appropriate test procedures and data analysis methods to demonstrate that your device is safe and efficacious. Meeting these goals will help prepare you for a career in medical device design.

9.2 Reasons for Testing

Medical device design aims to create a safe (not harmful) and efficacious (does what is claimed) solution. Following a robust design process, complying with design controls and regulatory requirements, and testing the design solution will ensure these goals are met. There are several reasons why testing is an integral part of the design process. It is an important requirement of design controls, provides support for regulatory and reimbursement submissions and sales of a new product, helps improve the product and prevent failures and recalls, and can save time and money when used to detect design defects before they reach the customer. These reasons are presented in greater detail in this section.

9.2.1 Testing to Meet Design Controls Requirements

Testing is an essential component of design controls, as shown in Figure 1.3. It is an iterative process that often guides the design process. Test results determine how close a design solution is to meeting all technical and nontechnical requirements and helps identify any changes needed for the next design iteration. It often indicates a need to change the design to improve performance, increase safety, and reduce residual risk. Once design changes have been implemented, testing is repeated to confirm that the revised design performs as expected and does not introduce new problems. Test results often determine if design iterations and additional testing are needed and should be added to the project schedule.

9.2.2 Testing to Prevent Failures and Product Recalls

Testing allows engineers to discover potential design defects in a new product before it reaches the customer. Simple design changes may unknowingly introduce a new variable that results in decreased performance or safety. Failure to repeat testing of the new design iteration can lead to product recalls caused by design changes. Testing can help identify hidden problems, improve safety, reduce risk, and reduce the potential for product recalls. Generally, the extent of testing before market introduction will increase with higher potential risk.

Most new medical devices will be tested to determine how they could fail. In Section 8.9, Failure Modes and Effects Analysis (FMEA) is introduced to identify potential failure modes and their effects (impact). To perform a risk analysis of a design (without actual test results), the probability of occurrence for each failure mode and the severity of the failure's impact are estimated, and an estimated risk value is calculated. Testing on final (or nearly final) devices can determine actual probabilities, allowing for a more accurate risk analysis to be conducted in which actual probabilities are used instead of estimates. Methods for determining more accurate probabilities are presented later in this chapter.

9.2.3 Legal and Ethical Reasons for Testing

There are ethical and legal considerations associated with the testing of a medical device. If a device failure causes harm to a user, is the designer legally or ethically responsible? Such cases are considered in more detail in Chapter 13. Proper documentation of rigorous failure testing not only protects end users from harm but may also provide designers and companies with some level of legal protection while reducing their liability.

There are several ways in which testing and ethics intersect. First, in the context of a medical device, using a market introduction of a commercialized medical device is not an acceptable alternative to premarket testing as a way to find defects in medical devices. Some popular innovation methods advocate introducing a Minimally Viable Product (MVP) into the market as soon as possible with the intent of quickly identifying failure modes reported by their customers.

Improvements are then implemented, and field testing is repeated. First, while this method may be acceptable in some industries, it is ethically and legally unacceptable for medical devices. Second, simple design changes are often assumed to have no impact on product function or safety, and additional testing is not conducted. These changes, however, may unknowingly introduce a new variable that results in decreased performance or safety. Failure to repeat testing of the resulting design iteration has often resulted in product recalls caused by design changes. Third, the engineering profession, through various engineering societies (e.g., National Society of Professional Engineers, Biomedical Engineering Society, Institute of Electrical and Electronics Engineers), and most companies have strict ethical codes (some of which are presented in Section 13.3) that include testing as a pathway to protect the public from harm.

9.2.4 **Testing to Prevent Financial Loss and Support Sales**

Regular and iterative testing can help reduce project risks by identifying defects early in the design process. The 1-10-100 rule shown in Figure 9.1 illustrates the benefits of detecting and correcting defects as early in the design process as possible. If detected before a design is finalized and transferred to production, assume that the relative cost of detection and correction would be $1. If detected after the design is finalized and production has begun, the relative cost would be $10. This increase often includes the cost of revised or new tooling needed to correct the design defect. If detected after market release, the relative cost could climb to $100 depending on how many units have been shipped to

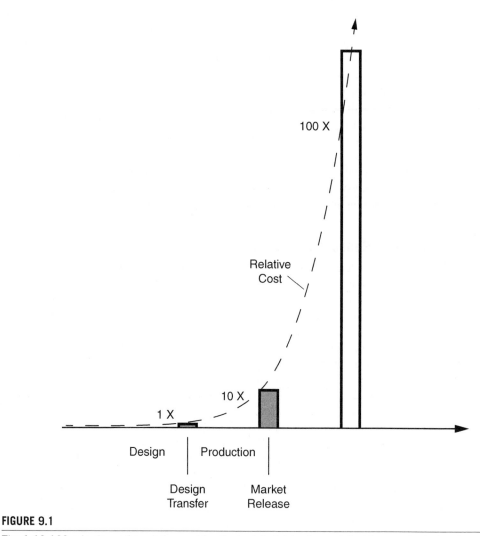

FIGURE 9.1

The 1-10-100 rule shows the relative cost (vertical axis) of detecting a defect in a product at various stages in the design and commercialization process.

customers. This increase is due to the costs associated with retrieving product from the field (e.g., shipping product back to the manufacturer, replacing returned product) and any legal actions that could result. From a financial perspective, this illustrates the importance of detecting a design defect as early as possible in the design process. Early detection not only protects the end user of the device (and complies with a company's ethical responsibility to produce a safe and efficacious product) but also makes good business sense.

The adage "time is money" applies to the design process. As you navigate your design process, perhaps your most precious resource is time. In this regard, the detection of a design flaw or defect can seem like a setback. However, by detecting the problems and analyzing risks early in the design process through testing, you will avoid wasting time and money on concepts that will not meet customer needs or solve the customer's problem.

Another reason for testing is to support sales. Test results regarding device performance and capabilities are often cited in marketing and sale literature (as discussed in Section 14.4) and publications in medical journals. If results are favorable, sales personnel may use them to persuade customers to purchase the new device. If testing includes evaluation of competitive devices, results are often used to compare the new device's performance with that of the competitive devices (competitive benchmarking).

9.2.5 Testing for Regulatory Approval and Reimbursement

Two significant barriers to the commercialization of a new medical device are obtaining regulatory approval (discussed in Chapter 12) and convincing health insurance providers to reimburse for use of the new product (discussed in Sections 6.3.2 and 15.7). Without reimbursement from health insurance providers, the patient would pay the often-high price of a medical device, making it very costly and thus significantly limiting the product's potential market. A significant portion of most medical device regulatory submissions will include test results to demonstrate the safety and efficacy of the device, as well as evidence that design controls have been followed and risk analyses have been conducted. Reimbursement decisions are based on evidence that the new medical device is reasonable and necessary.

9.3 Stages and Forms of Testing

Testing appears in many forms throughout the design process. It determines how close a design solution is to meeting all performance and other requirements and helps identify what changes may be needed for the next design iteration. In this respect, the purpose of the testing presented in this section is to measure some aspects of performance. Various forms of performance testing are conducted at different stages in the design process, starting with feasibility tests involving rough prototypes or analytical models, moving on to tests of more detailed and final prototypes, and ending with tests of the final product. When testing indicates the need to revise the design, some of these tests will need to be repeated. This section focuses on the forms of performance testing and when they are typically conducted, to help you consider the most appropriate tests for your design solution.

9.3.1 Feasibility Testing

It is important to identify concepts that are not technically feasible as early as possible in the design process to avoid wasting time and money on those that will not meet (or have the potential to meet) customer needs or solve the original problem. *Feasibility testing*, often called *proof-of-concept testing*, occurs early in a design process to determine if a design concept has the potential to meet performance requirements and technical specifications. It helps you answer the question, "could our design concept possibly work?" These tests often involve rough prototypes or analytical models, as presented in Section 6.2. Feasibility tests of physical prototypes are often conducted to determine if one critical specification needed to meet performance requirements can be (or has the potential to be) met or if a particular function or subsystem of the design has the potential to function as required.

Analytical models can be used to determine the technical feasibility of a design concept. The use of basic mathematical models such as free body diagrams, finite element, circuit, and stress analyses, as well as computer simulations, are helpful tools for identifying feasible concepts worth further development and exploration. For example, to determine the feasibility of a new syringe design, a manual calculation or computer simulation can be used to calculate the minimum pressure needed to produce the required flow rate of liquid out of the attached needle and into the patient's body. This will depend on several factors such as the needle and barrel diameters, lengths of the barrel, syringe tip, and needle, viscosity of the liquid, and resistance of the body tissue. Once this minimum pressure has been calculated, the diameter of the plunger can be used to calculate the minimum plunger force required to produce the desired flow rate. This can then be compared to anthropometric and biomechanical data of the typical forces that physicians, nurses, and other healthcare providers can apply to a syringe plunger to determine if the new design could function as required.

9.3.2 Prototype Testing

Prototype testing is typically conducted on the first fully functional prototype in a laboratory setting to determine if a design solution meets specifications and performance requirements. At this stage of a project, it is often found that only some requirements have been met. Performance testing of these prototypes helps identify what, if any, changes are needed to improve the design. The detailed design approaches presented in Chapter 8 are then applied, a new prototype is constructed, and testing of the improved prototype is repeated. Some animal testing might begin, as presented in Chapter 10, on more advanced prototypes. Breakout Box 9.1 provides an example of bench tests conducted on prototypes of a new female incontinence device.

9.3.3 Product Testing

When testing advanced prototypes confirms that the latest design iteration is acceptable, the design is temporarily frozen. Depending on the device and its clinical application, it may be transferred to production to begin scaling up for manufacturing. Additional prototypes may be built to support animal or clinical studies prior to manufacturing. *Product testing* is conducted to determine if the product, manufactured using actual production processes and final materials, meets performance requirements and is safe and efficacious. These tests may be performed at different stages during the manufacturing scale-up process and are used to determine the potential effects of any changes introduced by the manufacturing process on the performance or safety of the new device. These effects may be caused by new variables that were not present in the previously tested prototypes (and thus were not tested) such as mold release agents, lubricants, new production tooling, new assembly procedures, sterilization processes, or packaging. Ideally, testing of final production units (manufactured per the final design and materials) using the final production, sterilization, and packaging processes would be included in the testing plan for a new medical device.

The decision as to when to test a product may be affected by financial considerations. For example, there would be a high financial risk to investing in expensive multi-cavity injection molds before receiving acceptable results from a clinical study. If the study indicated that design changes were needed, depending on the complexity of the changes, the tooling might need to be scrapped or revised at considerable cost. To reduce financial risk, many companies will produce devices for clinical studies using less expensive temporary tooling. They will wait to receive positive clinical study results before freezing the design and committing to building expensive production tooling, after which product testing of final production units would be conducted. However, other companies might be willing to take this risk if they feel that it is outweighed by the potential benefits of saving time and allowing for a quicker product introduction.

9.4 Testing in Industry and Academic Design Projects

The goals of testing in academic and industry design projects include verifying that specifications, performance requirements, and customer needs have been met, and identifying design defects and failure modes and determining risk values. Test results are used to create risk mitigation plans and propose design changes to minimize risk. This section highlights the similarities and differences in how testing is conducted in these two environments.

9.4.1 Testing in an Industry Design Project

Figure 9.2 shows the various types of testing required for industry design projects prior to market introduction, including preclinical and clinical testing. These tests include bench, animal, and clinical testing, as well as package and shelf-life testing and process validations. This testing aims to confirm that the new product will provide an acceptable return on investment and generate data to support regulatory and reimbursement approvals as the product moves closer to commercialization. Finalizing—or freezing—a detailed design solution cannot be completed until all testing has been completed to prove that the device meets all performance claims, is safe, and can be manufactured at a reasonable cost using specified manufacturing processes. Test results will often determine if a project will be allowed to advance in an industry design project. It is not uncommon for a project to be abandoned or put on hold if a problem is identified that will require significant time and resources to correct. Industry design teams often include personnel responsible for conducting many of the tests presented in this chapter.

9.4.2 Testing in an Academic Design Project

An academic design project may include many of the tests used in an industry design project. This typically involves feasibility and prototype testing using some form of bench testing in a laboratory setting.

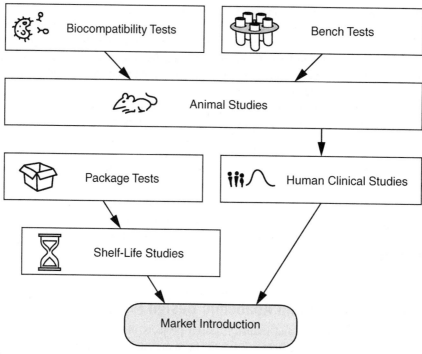

FIGURE 9.2

Typical flow of testing required in industry design projects prior to market introduction. Testing in academic design projects is typically limited to bench testing and possibly some biocompatibility and other testing in living systems.

Due to limited time and resources, you may only be able to conduct some of the tests presented in this chapter. For example, it would be very rare for an academic design project to include animal or clinical studies, package tests, or shelf-life studies involving production units.

Testing may reveal a design flaw that would require design changes or prompt you to pivot to a different design solution. In this situation, the core project statement and many specifications (identified in Chapters 3 and 4) will remain the same; however, you will be tasked with pursuing one of the alternate design solutions you created using the methods presented in Chapters 5 and 6.

Because resources in academic settings must serve a range of needs and projects, universities will often invest in general testing equipment. For this reason, you may sometimes not have a testing device that meets your exact requirements. However, with some modifications, these general devices can be adapted to meet the needs of your project. For example, the standard grips on a mechanical testing device may need to be modified to work with your device. As mentioned in Chapter 7, you may be able to design and machine a supplementary grip that will work with your device. In this situation, ask your mentors and advisors if they can suggest resources (people and equipment) that may be helpful when you design and conduct tests of your device.

Suppose the appropriate test equipment is not available nor readily obtainable. In that case, it may be worth outsourcing your device testing to an external testing laboratory that has the appropriate test equipment and specializes in the type of testing that you need. Before doing so, you will want to consider your project budget, schedule for testing, and the need for a nondisclosure agreement (NDA) to address confidentiality issues and protect your potential intellectual property. Some testing companies and laboratories have established relationships with universities for various types of testing. Ask project mentors, advisors, and faculty about these relationships and the possibility of outsourcing your testing to one of these external testing resources.

9.4.3 Guidance on Planning and Conducting Tests

You will be responsible for planning, conducting, and analyzing tests for your device. There are several reasons why you should adopt a consistent and coherent format for your tests. First, doing so will make it easier for others to understand your tests at design reviews and in written reports. Second, suppose the appropriate test equipment is not available, and you need to outsource testing to an organization that specializes in the type of testing you need. In that case, that organization will need detailed protocols. Third, a common format may be required by the company (if an industry design project), sponsor, mentor, or client who may want to continue your project after your experience has concluded. The details of a test report should be such that an engineer with similar training to yours would be able to understand and repeat your test procedures.

Most test protocol formats contain the following information:

- basic information (test name, date, and location, and team members);
- an introduction conveying the purpose of the test and desired information to be obtained. Specifications or needs are often referenced by number, as shown in Table 4.3;
- a methods section detailing the materials, equipment, devices, algorithms, procedures, and data analysis techniques used, including images of test setups and test samples;
- a results section providing a summary of the analyzed data;
- an interpretation section indicating whether or not the test provided the desired information, as well as any other pertinent findings that may impact design decisions;
- a recommendation or action item that resulted from the test; and
- a reference to raw data, either as a link to a digital repository or appendix.

If it has not already been provided for you, it would be helpful to create a general template that includes the sections listed above. Each test that you plan, conduct, and analyze would then use this same template.

9.5 Design Verification

Design verification answers the question, "did we make the product right?" It is typically accomplished by testing prototypes to experimentally confirm that the design meets the most recent version of the specifications. In the language of design controls (presented in Chapters 1 and 4 and Section 11.4), this means that design input = design output, and design specifications are met. Verification testing also helps you identify specifications that still need to be met and provides a level of assurance that the product will function as required.

Typically, a test is designed and conducted, and results are compared to ideal and marginal values for each specification. For example, consider a specification for a handheld device that states that it must not weigh more than 5 lb (2.27 kg) and ideally would weigh 3 lb (1.36 kg) or less. A test to weigh the device would be included and conducted as part of verification testing. If the device weighed 5 lb or less, it would pass this requirement, and the specification would be met. If the device fails this test, you would then determine how to reduce the weight or decide if a slightly higher weight would be acceptable. Prototypes are refined until all specifications are met through the iterative design process shown in Figure 1.3.

In this section, we provide guidance on designing and conducting verification tests applicable to your device. Included are the use of testing standards, bench testing in laboratory settings, and packaging, distribution, and shelf-life testing. Each test should be carefully planned per the guidelines presented in Section 9.4.3. The test protocol for your device should include only those tests needed to verify that each specification has been met. For example, tests to determine if a product is waterproof are not necessary if being waterproof is not a requirement included in the list of specifications.

9.5.1 Use of Standards in Verification Testing

Standards (presented in detail in Chapter 11) play an important role in design verification. They include test procedures that have been created, revised, validated, and approved by experts and practitioners. They provide you with industry-accepted and endorsed test procedures in one place.

Standards can prevent your team from "reinventing the wheel" and creating custom test procedures that may exist, saving time and effort and ensuring that the appropriate test methods will be used. Some standards include performance requirements (minimally acceptable values) that you can use to determine if your design performs acceptably. Ideally, these requirements have been incorporated into the list of target specifications established in Chapter 4.

Testing per standards allows medical device companies to advertise that their products comply with a particular standard (or standards). This is often a customer requirement and is also important to regulatory and reimbursement agencies.

The use of standards benefits customers. For example, when considering which brand of ureteral stent to purchase, a urologist may compare the loads required to straighten the proximal curls (located in the kidney) between stents from different manufacturers. This performance characteristic is associated with resistance to migration of the stent. If all companies test per their own standard, then direct "apples to apples" comparisons of their products would not be possible. Using the same standard test procedures avoids confusion and allows for direct comparison between stents from different manufacturers, thereby allowing the urologist to make a better, more informed purchasing decision.

9.5.2 Bench Testing

The term "bench testing" refers to tests done on a laboratory bench or in a laboratory setting outside of living systems. Bench tests may be conducted during any phase of the design process. They can be used to evaluate one particular performance specification early in the design process to test a new feature or functionality or verify the entire detailed design solution prior to moving on to animal and human clinical studies. In this section, we discuss some common forms of bench testing.

The goal of bench testing is to simulate, in a laboratory setting, the actual use, service environment, and function of the device. Bench tests should attempt to simulate the physiological service environment (as discussed in Chapter 8), including mechanical and electrical loading, lubrication and wear conditions, fluid composition, concentration, temperature, and pH, magnetic fields, and other energy sources (radio frequency, gamma radiation), as appropriate. Prior to testing, test samples and test equipment should be inspected and calibrated, respectively. During testing, it is helpful to monitor changes in performance characteristics over time such as flow rates (tubular devices where fluid is transported, such as catheters and stents), volume of wear debris generated (in designs where parts contact each other, such as joint replacements), or voltage output and leakage current of electronic devices (such as ECG monitors). Upon completion of all testing, test devices should be inspected for signs of degradation and changes in physical properties over time. This would include electronic devices with decreased sensitivity and accuracy and values that are out of calibration, implantable devices showing signs of corrosion and wear, product failures (product no longer functions as intended), loss of physical properties such as strength and flexibility of polymeric devices, discoloration of materials, obstruction and kinking of tubular devices, and fracture of devices.

9.5.2.1 Nondestructive Versus Destructive Testing

Single-use, disposable medical devices, such as ECG electrodes, are designed to be used once and then discarded. Other devices such as ECG monitors and some surgical instruments are designed for multiple uses. It is important to confirm that all devices perform as required for the expected service life of the device (single or multiple uses). For multiple-use devices, it is critical to know if and how their performance changes over time. These devices could be tested for a minimum number of individual uses or hours of operation, based on the expected service life of the device, or they could be tested until failure.

Nondestructive tests leave the device intact, functional, and typically undamaged after testing. These tests are often conducted when there is a limited number of test samples needed for other purposes such as demonstrations to customers, sponsors, and clients. Data from these nondestructive tests can confirm that the product will meet performance requirements but will not yield information regarding how long the device can function until failure.

Destructive testing results in damage to the device. It is conducted to confirm device performance and determine how and when a device will fail. Results of destructive testing can be used to determine more accurate values for probabilities of occurrence used in the design FMEA and risk analysis process (presented in Section 8.9), and the need for product warnings. Comparisons of results from nondestructive and destructive tests are necessary to confirm factors of safety. For example, assume that a device will be subjected to a maximum compressive load of 45.4 kg (100 lb) during normal use, and a test demonstrates that the device can withstand this load with no damage. To determine the factor of safety, the device must be tested to failure. If the load at failure is 90.8 kg (200 lb), then it can be concluded that a factor of safety = 2 exists for the maximum compressive load.

In an academic design project, you may be required to deliver a functional prototype to a sponsor or client. If you have enough funds to create only one prototype, then destructive testing would not be appropriate because it would destroy your only prototype. For this reason, you may only want to perform tests that do not damage, weaken, or decrease the device's service life. In many situations, destructive testing is not necessary. You can demonstrate that your device meets specifications (design verification) through nondestructive testing. However, you should be aware that rigorous testing (both destructive and nondestructive) is often required before a device is allowed to be used in a healthcare setting.

9.5.2.2 Static Versus Dynamic Mechanical Testing

Many medical devices will be subjected to various forms of mechanical stress. This includes devices that do not have a primary mechanical function. For example, a pacemaker's electronics, wires, and sensors will experience mechanical loading while in the body. Most mechanical tests can be classified as either static or dynamic.

Static testing, in which a mechanical component or device is loaded with a single static mechanical load (sometimes until failure) is used to determine mechanical properties and the maximum load to failure. This type of testing is most appropriate for single-use medical devices that will be loaded once and then discarded. Knowing the maximum load to failure and the expected load during use of the device allows design engineers to calculate a factor of safety for a single-use device.

For load-bearing, implantable devices such as joint replacements, static testing may have less value. These devices will be repeatedly loaded and unloaded (cyclical loading) during each gait cycle when in the body. For this reason, cyclical loading better simulates the service environment and actual loading conditions of these implants. It typically involves mechanical loads that produce fluctuating tensile and compressive loads and can lead to fatigue failure due to the formation and propagation of surface cracks. This type of testing using dynamic loads is often referred to as fatigue testing. When designing a dynamic fatigue test, load magnitude, load frequency, number of cycles, range of motion, and environmental variables such as temperature, pH, and composition of test fluids are chosen to simulate the expected service environment most accurately, as presented in Section 8.3.1.

To determine the time or number of cycles to failure, testing must continue until the product fails (no longer functions as required). This form of destructive testing, if conducted with an appropriate sample size, can yield a distribution of time-to-failure for the design. It allows design engineers to predict the percentage of devices that will exceed the required service life and determine the typical failure modes. This can be useful information when exploring ways to extend the device's longevity and reduce the failure rate.

Testing until failure of devices at different mechanical loads allows construction of an S-N (stress vs. number of cycles) curve that can be used to understand the relationship between the number of loading cycles until failure and the stress that is applied to a component, device, or material sample during mechanical loading. Knowledge of this behavior allows the determination of the *fatigue limit* of a device or material, which indicates the stress below which it will not fail due to mechanical loading and resulting fatigue stress. It also allows determination of *fatigue life*, which is the maximum number of loading cycles until a fatigue failure occurs. *Fatigue strength* (or *endurance limit*) is the stress that will produce a fracture after the fatigue life is reached. Materials such as some steels and titanium alloys will not fracture if stresses are below the fatigue limit.

Most materials do not exhibit a distinct fatigue limit. Decreasing stress in these materials increases fatigue life, and increasing stress decreases fatigue life. In other words, materials can withstand a much higher single, static load than they can withstand a fluctuating load applied through many cycles. This relationship implies that fatigue strength (involving many loading cycles) will be much less than a material's ultimate tensile strength (involving one loading cycle). For example, the ultimate tensile strength of Ti6Al4V alloy is 930 MPa (135 kpsi), but its fatigue strength is only 420 MPa (61 kpsi) at 10^8 cycles, which is 45% of the ultimate tensile strength. Many implantable materials have a fatigue strength between 30% and 50% of their ultimate tensile strength.

9.5.2.3 Electrical Testing

There are a variety of tests required for electronic medical devices that evaluate device performance and safety for the patient or user. Safety concerns include electrical shock, burns, and electrocution hazards. For example, some electrical components can produce significant amounts of heat, which can present burn hazards to the patient or create a fire hazard in a clinical setting. Tests for these and other potential hazards may include thermal and leakage current measurements and determining the potential for electrical shorts if a wire fails or becomes exposed. Standards for testing electronic medical devices provide approved and recommended test procedures and define the maximum safe levels of leakage current and other related electrical hazards. IEC 60601-1 *Medical electrical equipment – Part 1: General requirements for basic safety and essential performance* is an international standard that covers electrical equipment used in medical practice, emphasizing basic safety and performance requirements.

Electrical testing is also conducted to evaluate device performance in the expected service environment. For example, there are tests to determine the immunity of an electronic device to electromagnetic interference (EMI) that can occur when electromagnetic fields produced by one electronic device interfere with the normal function of a nearby electronic device. ISO 14117: 2019 *Active implantable medical devices—Electromagnetic compatibility—EMC test protocols for implantable cardiac pacemakers, implantable cardioverter defibrillators and cardiac resynchronization devices* is an example of an international standard that includes tests for electromagnetic compatibility of cardiovascular devices.

9.5.2.4 Testing in Simulated Biological Environments

It is often possible to test medical devices in a biological service environment without testing in living systems. It is important to know which aspects of the biological environment are critical to simulate in these situations. This section discusses two common methods for simulating this environment, including use of biological fluids and duplication of human anatomy.

Some medical devices will be in contact with biological fluids during use. To simulate the service environment, devices should be exposed to these fluids to determine their potential effects on the device. However, it can be difficult (and potentially unsafe) to use actual body fluids during testing. In many situations, it is possible to test the device in a simulated fluid that exhibits similar physical and chemical properties to the actual fluid. For example, it is possible to simulate the non-Newtonian rheological properties of blood. Companies such as Nasco Life/Form sell blood-like substances and animal blood (www.innov-research.com). For many applications, it is possible to simulate blood using materials that can be easily found in a grocery store. For example, glycerin is colorless and nontoxic and used in many medical and cosmetic applications. Xanthan gum is a powder polysaccharide food additive used in cooking as a thickening and binding agent. Two simple blood-like fluids can be made with mixtures of ~75/25% (water/xanthan gum by weight) or ~60/40% (water/glycerin by weight). More sophisticated formulations use a combination of glycerin and xanthan gum. The percentages can be changed to roughly estimate other biological fluids (e.g., saliva, lymph, semen). Standard test solutions for simulated testing in urine (see Breakout Box 9.1) and other bodily fluids also exist.

Some device testing requires access to a tissue model that simulates some aspect of human anatomy and physiology. The model used will depend upon the test required. For example, it may be helpful to have an accurate anatomical model of a particular organ or organ system. Databases of 3D printable geometries were presented in Section 7.3. Anatomic models can also be purchased through websites listed at the end of this chapter.

Other device testing may require a physical model that mimics a critical property of tissue. For example, in situations that involve ultrasound, it is important to simulate the appropriate acoustic properties of the bulk tissue. Phantoms with acoustic properties that mimic different tissue or bone might also be needed. Similarly, the mechanical homogeneity and consistency of tissue are often required; in this case, BFX ballistics gelatin can be used with a plaster mold (first sprayed with cooking spray) to replicate parts of the body. The mixtures can be altered to mimic various densities. Thin-walled tubes within the gel can be used to mimic veins and arteries.

At times the simplest way to test a device is to use real tissue. Chapter 10 discusses testing in living systems, but several models exist that can very closely mimic human tissue and do not require approvals to use. For example, you might try chicken skin and muscle (readily available at a grocery store or butcher shop) as a surrogate for living biological tissue. You might also check a local butcher shop for animal organs. If there are ongoing animal tests at your institution, the researcher conducting those tests might also be a source for tissue and organs. Approval is not required to perform experiments on animals that are no longer alive.

9.5.2.5 Advanced Testing

Not all medical devices are subjected to significant mechanical loads or present serious electrical hazards. These devices may not require the same kind of mechanical and electrical testing as other devices. However, they may be subjected to other conditions that can result in failure modes that would degrade performance or pose a hazard, including environmental variables such as temperature, pH, the composition of biological or nonbiological fluids, or light exposure. These devices might require destructive and nondestructive testing as well as static or dynamic testing. Furthermore, more advanced testing might combine two or more variables. For example, one might measure wear debris generated during cyclic mechanical loading of an orthopedic implant in service environments with a range of protein concentrations and mechanical loads. This may be important in understanding the effects of proteins and mechanical loading on the rate of wear debris generated.

During testing, changes over time are often monitored through observation or direct measurement of performance. Observations may include visual inspection for signs of wear, corrosion, changes in physical properties, or other signs of degradation. In some cases, these changes could be detected using imaging technology (e.g., video camera). Performance measurements could include flow rates (in tubular devices where fluid is transported, such as catheters and stents), volume of wear debris generated (in designs where parts contact each other, such as joint replacements), electrical sensitivity, accuracy (in a pacemaker lead), or mechanical strength and flexibility (in components of an implantable joint replacement). These measurements are often directly related to a critical specification that must continue to be met over time under normal, or perhaps even extreme, situations.

We conclude this section with an example of bench testing presented in Breakout Box 9.1, which describes the testing required for a female incontinence catheter used to control the flow of urine in women who cannot control the flow of urine from their bladder.

9.5.3 Cadaver Testing

In some situations, you may want to test the usability of your device and determine how well it will interface with the human body. In these situations, it can often be helpful to conduct tests in a human cadaver. For example, a team wanted to test the usability and accuracy of their handheld patellar tendon graft knife. Bench testing on a laboratory knee model with simulated patellar tendon material (to

Breakout Box 9.1 Bench Testing of a Female Incontinence Catheter

Female incontinence is a condition in which a woman cannot control urine flow out of her bladder. This can be due to a spastic bladder, where sudden, unpredictable contractions of the bladder muscles occur. It may also be due to a urethral sphincter that does not function properly (due to a spinal cord injury, trauma, or disease) and fails to close off the urethra at the bladder neck, allowing urine to flow out of the bladder and into the urethra. The uncontrollable flow of urine can result in odor, skin irritation, and infection, as well as embarrassment. Female incontinence is currently managed with diapers, catheters (short term), and other devices. It can also be treated through implantation of an artificial urinary sphincter. Each of these solutions presents tradeoffs between cost, ease of use, potential complications, and level of invasiveness.

Surgitek, a urological device company, developed an incontinence catheter, as shown in Figure 9.3, for use by women to control their urine flow. The device consisted of a silicone tube with an inflatable bladder balloon and a magnetic valve that could be actuated (opened and closed) manually by the patient through the use of an external magnetic actuator containing a neodymium-iron-boron (NdFeB) magnet. The catheter was placed into the urethra, and the bladder balloon was inflated to prevent catheter migration out of the bladder. The distal end of the catheter was located between the labia and was visible to the patient. The patient sat upright on a toilet seat to use the device, using gravity to create fluid pressure within the bladder to assist with urine flow. The handheld external actuator was then held close to the distal end of the catheter. This attracted a NdFeB magnet located in the catheter valve, disturbed the seal of the valve mechanism, and allowed urine to flow through the valve and out of the patient's urethra. Once the bladder was emptied, the actuator was pulled away from the catheter, and the spring-loaded valve reseated itself to block the flow of urine through the catheter.

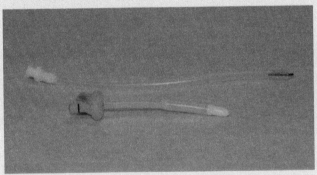

FIGURE 9.3

Female incontinence device (foreground) used to control the flow of urine from the bladder. An inflation device used to inflate the catheter balloon appears in the background.

To test the efficacy of the device, a bench test was conducted to determine if (1) the catheter valve could prevent urine flow through the catheter and allow flow when actuated by the external actuator, (2) the inflatable balloon could seal the urethra and prevent leakage past the balloon, (3) the inflatable balloon would not leak, resulting in shrinkage of the balloon and subsequent urine leakage around the balloon, and (4) the catheter would allow an acceptable flow rate of urine. Test results helped predict how long a catheter could be used in a patient before it must be removed and replaced.

This test consisted of 10 catheters mounted into a test fixture designed to simulate the service environment (chemical, physiological, thermal, and mechanical characteristics) in which the catheter would reside in the body. This involved mounting the catheters into the ends of vertically secured glass funnels used to simulate the bladder and urethra while in an upright, sitting posture. The catheters were mounted vertically with the balloon inflated at the base of the funnel and the magnetic valve accessible just below the bottom of the glass tube exiting the funnel. This orientation allowed gravity to assist urine flow and allowed contact between the magnetic valve and the external actuator. All test setups were housed in an incubator to maintain a temperature of 37°C (body temperature).

At the start of the test, the funnels were filled with an artificial urine solution formulated per *ASTM F623 Standard Performance Specification for Foley Catheter*. The funnels were filled with 0.7 liters of artificial urine (approximately one-third of the 2-liter average daily urine output of a healthy woman). To simulate typical voiding intervals for patients

continued

Breakout Box 9.1 Bench Testing of a Female Incontinence Catheter—cont'd

with intermittent catheters, the catheter valves were actuated every 8 hours (3 times per day) to empty the funnels. During each emptying of the funnels, the volume of artificial urine passing through the catheter and out of the magnetic valve was measured and the time needed to empty the funnels was recorded. This allowed calculation of the flow rate in mL/sec. The device's service life was expected to be close to 28 days; thus, the test continued for this amount of time. At the end of the 28-day test period, the test was stopped and the catheters were removed from the test apparatus and inspected for signs of leakage, encrustation, and corrosion of the metallic components. Any leakage observed during the test period was noted. The last volume of artificial urine emptied from the funnels was collected, and a sample was analyzed for the presence of Nd ions.

The artificial urine solution contained similar concentrations of chemical components (e.g., sodium chloride, sodium citrate, magnesium sulfate, potassium chloride, calcium chloride, sodium oxalate, and others) typically found in human urine that often form small mineral precipitates. These can attach themselves to the surfaces of implanted (or indwelling devices). The resulting *encrustation* can prevent the valve from properly closing and the inflated balloon from adequately sealing the urethra, resulting in urine leakage through the catheter or around it, respectively. Using artificial (or human) urine instead of distilled water allowed testing of device longevity (i.e., how long a catheter can remain in the patient before it begins to leak, due to either malfunction of the internal valve or improper sealing of the bladder neck by the inflatable balloon).

---

emulate the physical properties of tendon tissue) yielded some useful information; however, they wanted to determine accuracy and ease of use in an actual human knee. Testing in a live human model would require a clinical study (presented in Chapter 10), but cadavers are not subject to the same regulatory restrictions. For this reason, the team decided to test their device in a cadaveric knee which allowed them to use the device and gain firsthand knowledge of the difficulties experienced by orthopedic surgeons as they attempt to resect the middle third of the patellar tendon for use as a graft to replace the anterior cruciate ligament. As a result, they were able to generate new ideas for improving the accuracy and usability of the graft knife.

Cadaver testing may not be appropriate for all types of devices. For example, nonliving tissue will degrade over time, resulting in changes in physical properties such as stiffness, strength, electrical conductivity, and thermal properties. Thus, cadaver testing does not simulate all physical properties of live human tissue and may not be a good option for quantitative testing. For example, the graft resection procedure described above required minutes to complete, and the stiffness of the patellar tendon tissue was adequate to test the knife properly. However, due to biochemical and physical changes to cadaver tissue over time, cadaver testing would not have helped determine the long-term stability, function, and biological performance of an implanted orthopedic device.

If cadaver testing would be helpful in testing your device, you should ask a mentor or advisor about the possibility of conducting these types of tests. Cadaver laboratories are often found in academic institutions associated with medical schools, nursing schools, and biomedical science programs that offer gross anatomy courses.

9.5.4 Package Testing and Distribution Simulation

Section 8.8 discusses packaging design as a critical consideration in designing a medical product. As a reminder, medical packaging is just as much a part of the product as the device itself and packaging design considers sterilization, product protection, efficient storage, distribution stresses, and usability. It is unlikely you will have the time or resources to conduct package and distribution tests as part of an academic design project. However, considering the package testing requirements for your device will make you aware of this common form of testing conducted by medical device companies.

As presented in Section 8.8, medical device packaging needs to protect the product during shipping, on the shelf in a warehouse, and in a hospital storeroom. As shown in Figure 9.4, the product will be exposed to various distribution stresses associated with shipping and storage, from the time it is manufactured until it is opened and used. Package testing aims to ensure that the device will meet performance requirements when the customer uses it. This requires testing of the device and its packaging, distribution simulation, and aging studies. Package testing is required by regulatory bodies; documentation of these tests are included in the Design History File (DHF).

After the product is manufactured, it is often shipped to an off-site sterilizer. If shipped using ground transportation, the product (including its packaging) will be subject to various distribution stresses, including mechanical stresses such as vibrational and handling stresses (packages dropped, tossed onto surfaces, and potentially crushed by boxes stored on top of them), and environmental stresses such as heat, cold, or humidity. The packaged devices are then shipped back to the manufacturer's (or a distributor's) warehouse and subject to additional distribution stresses. When an order is placed, the product is shipped to the customer (subjecting it to additional distribution stresses) and then stored in a hospital storeroom until it is ready for use.

Product stored in a manufacturer's warehouse or hospital storeroom will often be stored in a temperature and humidity-controlled environment. However, storage conditions can vary during shipping. Product shipped via ground transportation across a long distance is often stored in a truck or temporary warehouse along the way. Depending on the time of year and location, these warehouse temperatures can climb to as high as 140°F (60°C) or drop below freezing. For devices using temperature-sensitive polymeric materials, these conditions could soften the materials (high temperatures) or cause the material to crack (low temperatures), affecting product performance. Humidity-sensitive materials (such as nylon) can absorb water in the air and cause them to swell, thus resulting in dimensional changes to the product stored in high humidity environments. Shipping products via air transportation can result in similar effects. Not all freight cabins are pressurized, nor are they temperature- or humidity-controlled.

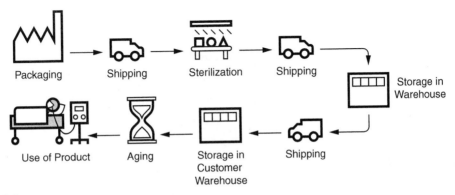

FIGURE 9.4

Distribution history of a typical sterilized medical device.

9.5.5 Distribution Simulation

Packaging tests should be conducted on the final, sterilized product, in the actual packaging intended to be used with the product. Environmental variables such as temperature, humidity, and pressure (when appropriate) are simulated to represent the potential extremes that the product will be exposed to. Through testing, the distribution history is also simulated to represent the mechanical and handling stresses that the product will be subjected to during shipping and storage. These include manual handling, stacking, and vehicle vibration tests per ASTM D999-08 (2015) *Standard Test Methods for Vibration Testing of Shipping Containers.* Each test involves testing an entire shipping container that contains many individually packaged devices.

As shown in Figure 9.5, the manual handling test includes a series of drop tests designed to simulate the rough handling of packages during shipping. The tests involve a shipping container filled with product dropped on all four sides and then on all six corners from a height of 3 feet.

As shown in Figure 9.6, the stacking test involves a compressive load applied to the top of the shipping container. It is designed to simulate the expected top loading of a container due to stacking during shipment in a truck trailer or storage in a warehouse.

As shown in Figure 9.7, the vehicle vibration test involves the application of vibratory loads in trailers during shipping in a truck. This long-term dynamic test is designed to simulate the loose load vibrations that a shipping container will be subjected to during ground transportation.

Upon completion of all package tests, the outside packaging of the shipping container is inspected for signs of damage such as crushing, flattened corners, cuts or holes in side panels, and other damage.

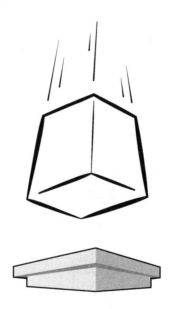

Drop Test

FIGURE 9.5

Manual handling test.

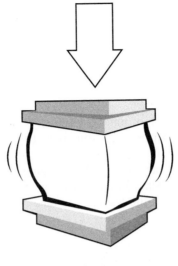

CrushTest

FIGURE 9.6

Stacking test.

Vibration Test

FIGURE 9.7

Vehicle vibration test. A full shipping container is placed onto a vibration table to simulate vibratory loads that occur when products are shipped via ground transportation.

Next, the shipping container is opened and the individual device packaging is inspected for signs of damage such as burst seals (compromising sterility of the product), crushed packaging materials, and cracked or torn trays or pouches. Inspection for migration of components out of their tray cavities resulting in damage to the components themselves or damage to other components, the product, or the packaging materials is also conducted. Finally, the product is inspected for damage and may be tested for functionality to determine if it has been damaged. Package tests often yield valuable information to product and packaging designers regarding redesigning the packaging to ensure protection during shipping. Similar inspections would be conducted on products stored in a high or low temperature or high humidity environment to determine the effect of these environmental extremes on product function and packaging performance.

9.5.6 Shelf-Life and Aging Studies

The term *shelf life* refers to the length of time that a product can sit on a shelf with no significant changes in physical properties or performance. This period typically begins after sterilization and applies to the device and its sterile packaging. Some polymeric materials can degrade (especially if sterilized with radiation) while in storage, and sterile packaging will not maintain sterility indefinitely. Changes in physical properties are often referred to as aging.

There are two approaches to shelf-life testing: real-time and accelerated aging studies, both of which are discussed below.

9.5.6.1 Real-Time Aging Studies

Real-time testing involves first conducting a distribution simulation. Instead of inspecting the product immediately after packaging tests, the packaged product is stored in the expected storage conditions (simulated storage environment). The product may be stored on the shelf for several years, with samples being tested periodically to evaluate the product and packaging for changes in properties and function. If a company wants to claim a shelf life of 2 years, then real-time testing must be performed for at least 2 years.

9.5.6.2 Accelerated Aging Studies

Accelerated aging studies are designed to accelerate the aging process. They are typically conducted similar to real-time studies but at higher temperatures which can accelerate the aging process. By testing at a higher temperature, changes in material properties occur sooner. By correlating these changes with real-time studies, the shelf life can be extrapolated from accelerated aging data. For example, suppose a significant change in material properties (i.e., a 20% reduction in tensile strength of a polymeric catheter material) occurs after 2 years of real-time testing at ambient temperatures and a similar reduction occurs after 6 months of accelerated testing at a higher temperature. In that case, it may be possible to conclude that 6 months of accelerated testing at higher temperatures would be equivalent to 2 years of real-time testing at ambient temperatures. This correlation could be used to determine the shelf life of similar products tested using accelerated aging studies. The advantage of this approach is that it takes much less time and would not delay a product introduction while real-time shelf-life studies are being conducted.

9.6 Design Validation

Design validation involves determining if a new design meets customer needs and answers the question: "did we make the right product?" It focuses on the customer and relies on feedback representing the user's perspective to determine if customer needs have been met. In industry, it is common for engineers to be involved in design validation which ideally would involve direct contact with intended users. In an academic design project, such direct contact may or may not be possible. You may need to accept feedback from a client, sponsor, or mentor as a substitute in this situation.

Chapter 3 introduces you to interviews, surveys, focus groups, and other social science tools to understand the problem from the perspective of the customers, clients, caregivers, and other users. You can use these same tools to obtain validation feedback from these same stakeholders as well as project sponsors, faculty advisors, and technical and clinical mentors. Design validation can be as simple as giving the final prototype to stakeholders and asking if the design meets their needs. If the answer is positive, then the design has been validated by those asked. If not, then you will have information to guide you in improving the design. Once improvements have been made, the validation process is repeated until stakeholders are satisfied with the final design. Remember that stakeholders may not be aware of the technical constraints of your project and may request design changes that are not possible. Design validation is presented in greater detail in Section 10.6.

9.7 Analysis and Interpretation of Test Data

The goals of testing discussed earlier in this chapter are rarely achieved with raw data. For this reason, the methods you use to analyze and interpret test results are just as important as the design and execution of the test itself. In this section, we provide guidance on the analysis and interpretation of test data.

9.7.1 Choosing Data Analysis Methods

A rigorous plan and well-executed set of test procedures must be coupled with some form of data analysis. The nature of this analysis will depend upon the tests, and there are some essential aspects of testing and analysis that you will want to consider. First, the test protocol and analysis methods should work together; therefore, it is good practice to determine both simultaneously. A common mistake made by design teams is to generate data and then realize—during data analysis, when it is too late—that the incorrect tests were run. This mistake can be costly, require additional testing, and illustrates the importance of identifying and conducting the correct tests. Second, when designing a test, the number of trials must be determined. In some cases, the answer may simply be one trial (e.g., weighing the handheld device described earlier in this chapter). In other cases, a more elaborate power analysis should be conducted to determine the smallest sample size needed to achieve a given statistical significance. Third, the nature of your analysis will vary depending upon your tests and the data generated. You will need to choose the appropriate statistical methods, recognizing the strengths and limitations of each. Fourth, some analysis methods process raw data to extract a parameter of interest. Matlab, MiniTab, Python, R, and other languages have many extensions or toolboxes designed to perform statistical analyses of the data collected. Some tests will require image capture and analysis. For example, a tool such as imagej.net may help you determine the size and motion of bubbles in a fluid captured by a video recording. In other situations, your data may be used as the input to a model or calculation, with the intent of indirectly determining values for a variable that is difficult to measure directly (e.g., an internal force, flow, temperature, or concentration). These should be planned before testing begins.

9.7.2 Interpretation of Test Results

Engineers are comfortable reporting results based on numerical, graphical, and tabulated data. Design decisions are sometimes based on more nuanced and qualitative interpretations of engineering results. For example, assume a mechanical part of your device includes a metal cantilever that you analyzed and found the maximum stress to be below the metal's yield strength. That result is straightforward to report. However, if you are asked if the device is mechanically safe for all loading conditions at a design review, that is a more nuanced question and will be difficult to answer. Learn to express your professional opinion in a way that conveys your level of confidence in that opinion.

As the interpretation of results is built more on a combination of context and intuition, there are a few important rules to remember. We offer two tried and true methods to help you improve your intuition skills. First, a good mentor or advisor will help you interpret your data and explain the rationale behind their interpretation. Listen carefully as you will gain insights to help you build your own intuition for future designs. Second, all results can be viewed through two different lenses: (1) a generous lens whereby the results suggest that you are on the right track and you can advance in the design process, and (2) a critical lens whereby the results are not entirely trusted, making further testing necessary. Many problems with navigating the design process during testing stem from not applying one of these lenses. Being too generous can lead one to speed through testing and find a critical flaw when it is too late. Being too critical can cause the design process to stall. It is essential to try to balance the views from these two lenses.

9.8 Engineering Competency and Test Design

Some impactful medical devices appear to be very simple. Often, much of the technical rigor is hidden in the engineering design process, including testing of the device. Even with a simple device, testing requires several technical skills and provides engineers with an opportunity to demonstrate these skills. Consider an improved attachment between a feeding tube and a pump. As a small piece of plastic, it may appear to have been easily designed and manufactured to meet a few critical specifications. However, testing such a device often requires detailed knowledge of its design, applicable test methods, appropriate data analysis methods, and how to simulate the appropriate service environment. Designing and conducting the appropriate tests, choosing the correct data analysis methods and using them appropriately, making sound judgments from your analysis, and then applying what you learned to guide design changes are all ways to demonstrate your engineering competency.

9.9 Testing and Risk Management

In Section 8.9, we recommend that soon after a preliminary design is established, the risk management process should be initiated to identify potential failure modes, their effects, and design changes to reduce the probability of these failures from occurring. The severity of the effects could be determined, but the probability of occurrence could only be estimated due to the lack of bench test results (typical for this early stage). Data from some of the tests described in this chapter may allow you to estimate these probability values more accurately. If appropriate sample sizes are used, it is possible to determine a failure rate (and probability of occurrence) for a particular failure mode. In industry design projects, field complaints about similar devices (historical data) can also be used to estimate probabilities of occurrence more accurately. However, these data must be considered carefully because customers do not always report problems, leading designers to potentially underestimate failure rates and probabilities of occurrence. Once accurate probability values have been determined, they can be entered into Table 8.4 along with values for severity (as determined in section 8.9); risk values can then be calculated. As discussed in Section 8.9, failure modes with significant risk values will require mitigation plans to reduce the risk to an acceptable level. Breakout Box 9.2 demonstrates how the risk management process was applied to a medication pump to reduce the potential for failure.

9.10 Documenting and Communicating Test Results

As discussed in previous chapters, it is important to fully document all aspects of the design process in the DHF. This includes documentation of all verification testing activities discussed in this chapter with descriptions of the design, execution, analysis, and interpretation of all tests and test data. How test results are communicated to stakeholders will vary greatly depending upon the phase of testing, the audience (e.g., design panel, project manager, upper-level management, regulatory body), and format (e.g., written report, presentation, video). In this section, we provide some guidance on sharing your test results with others.

9.10.1 Connecting Tests to Critical Design Data

There are often many tests that will be completed during a design process. In documenting tests, two points become critical for effective communication. First, every test should have a purpose with a strong connection to some unanswered question that, once answered, will advance an idea to reality. As such, testing documentation should clearly indicate or state why a particular test was run in the context of the design process. For example, a common connection in verification testing would be to the technical specifications defined in Chapter 4 and refined throughout the design process. Second, when many tests are completed, having some organizing framework can help a reader see the big picture. In this regard, a table such as the templates shown in Tables 4.2 and 4.3 can serve as an organizing framework for a verification test. Some teams have found it helpful to add a column to Table 4.2 to indicate whether or not the specification was met or not. In this way, a glance at the table will quickly reveal progress toward meeting all specifications. The same two principles can be applied to many other aspects of design. For example, FMEA could be used as an organizing framework, with the test results determining various probabilities and potential outcomes of failure.

9.10.2 Reporting Test Results to Stakeholders

There are often many stakeholders who will want to see your test results. During the early phases of a project, this could include an immediate supervisor or a project manager, who may often be an engineer. As a project progresses to a more detailed design and rigorous testing, requiring more resources, the range of stakeholders will expand, and the level of scrutiny may increase. During later phases of a design process, test results often make up the majority of a design review and may be used to support marketing and sales of the new device. When regulatory and reimbursement reviewers evaluate a new product, they will base part of their evaluations on test results. When a problem occurs, such as a product recall, test results may be reviewed at the start of an investigation.

Breakout Box 9.2 Design Risk Management for a Medication Pump

Risk management can be used to improve the design of a device and reduce risk. In this breakout box, we discuss how it can be applied to a medication pump.

The pain and discomfort experienced during injections of some medications can occur when a bolus of the medication is injected too quickly using a conventional syringe. Fear of these injections may prevent some people from seeking medical care. To solve this problem, a medication pump could be designed to provide a slow, controlled rate of medication injected into body tissues. This could be a hand-held, battery-operated pump. For convenience, it could be used with a glass cartridge that is purchased prefilled with the proper dose of the desired medication and inserted into the device. When activated, the pump would drive a piston (as shown in Figure 9.8) which would push on a rubber stopper located in the cartridge, forcing the medication out of the glass cartridge and into a short length of clear tubing connected to a needle inserted into the patient. To eliminate painful injections, the pump would provide a slow, steady flow of medication.

Before the medication can be injected, air must be purged from the tubing and needle to prevent it from being injected into the patient. To do this, a small volume of medication can be pumped until a drop of liquid appears at the end of the needle, indicating that all air has been evacuated from the tubing and needle and that it is safe to insert the needle into the patient. Before the dose of medication could safely be injected, the user would need to ensure that the needle was in tissue and not in a vein. To confirm proper location of the needle, the device could create a negative pressure to pull fluid from the tissue surrounding the inserted needle (aspiration) into the clear tubing for visual inspection. An O-ring could be used to provide a tight seal between the piston and inside of the glass cartridge shown in Fig. 9.8 so that retraction of the

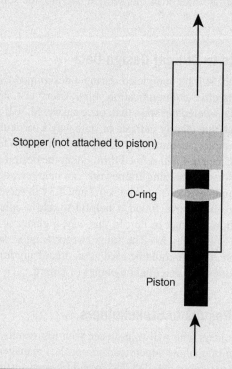

FIGURE 9.8

Schematic diagram showing piston pushing on the rubber stopper in the cartridge containing medication. The O-ring provides a tight seal needed to create a slight suction when the piston retracts during aspiration of fluid. Arrows indicate the direction of travel of the piston and flow of medication out of the cartridge and into the short tubing and needle (not shown).

Breakout Box 9.2 Design Risk Management for a Medication Pump—cont'd

plunger produces a slight suction, aspirating fluid from around the needle, and into the clear tubing. The color of this fluid would then be visually inspected to confirm placement of the needle into a vein if the fluid was red, and muscle, fat, or other tissue if the fluid was clear. If located in a vein, the needle would then be repositioned, and aspiration repeated until a clear colored fluid was observed in the clear tubing. Injection of the full volume of medication could then proceed.

A risk analysis of this device could be conducted by considering the results of bench testing performed on the device shown in Table 9.1. These values would indicate failure rates of production units tested in a laboratory which could then be used as accurate estimates of the probability of occurrence for each of the failure modes listed. If an additional risk analysis were to be conducted after a product was introduced, then customer complaint data might be available and could be included in the risk analysis. For purposes of this example, we will assume that customers have used the product and customer complaint data is available. Table 9.1 includes customer complaint data related to each of the failure modes listed. These percentages would be based on the number of complaints received compared to the total number of units in the field. This information would also be used to determine accurate estimates of probabilities of occurrence for each failure mode listed. Note that some problems reported by customers were not observed during bench testing.

Table 9.2 includes a sample of the failure modes identified along with the effects of these failures, the severity of each effect, and the probability of occurrence for each failure mode. Risk mitigation plans developed to address these failures are also included.

Table 9.2 indicates high risk values for failure modes #2 and #3. These values are high enough to require action to reduce risk. The risk mitigation plans for each of these potential failure modes appear in Table 9.2 and would be implemented to eliminate or reduce risk to an acceptable level (as low as reasonably practicable), as discussed in section 8.9. They include warnings, procedure changes, and design changes.

Table 9.1 Results of bench tests and customer complaint data for the medication pump

Problem	Bench Test Observations	Customer Complaint
Loud motor during operation	–	3%
Cannot determine if aspirating properly	–	10%
Purge cycle not working	–	5%
Cracked, worn, O-ring	30%	5%
Broken glass cartridge	20%	10%

Table 9.2 Risk analysis data for a medication pump. In this example, severity values range from negligible (1) to critical (9), and probabilities range from 0.05 to .0.3

Failure Mode/ Hazard	Effect/Harm	Severity	Probability	Risk	Mitigation Plan
1. Leak in system due to cracked tubing or defective adhesive seals, resulting in failure to purge air in tubing	Could inject air into vein creating air embolism, resulting in potential medical complications for patient	Serious (7)	Unlikely (0.05) (per customer complaint data)	0.35	Inform users of importance of purging air from tubing. (Warning) Instruct users to look for liquid exiting needle to confirm proper purging of air in tubing (Procedure change)

continued

Table 9.2 Risk analysis data for a medication pump. In this example, severity values range from negligible (1) to critical (9), and probabilities range from 0.05 to .0.3—cont'd

Failure Mode/ Hazard	Effect/Harm	Severity	Probability	Risk	Mitigation Plan
2. Worn O-rings prevent suction on stopper when piston retracts, preventing proper aspiration of fluid from around needle when inserted into tissues	Could inject medication into vein, causing a severe reaction in the patient	Critical (9)	Occasional (.3) (per bench test data)	2.7	Warn users of the potential for worn O-rings to prevent proper aspiration (Warning) Instruct users to periodically inspect O-rings for signs of wear. Provide maintenance kit with new O-rings and silicone lubricant and instructions for replacement and lubrication of O-rings (Procedure change)
3. Broken glass cartridge causing leakage of medication into pump housing, resulting in potential damage to electrical components	Possible sparking and overheating could cause minor injury to patient and user	Minor (5)	Occasional (0.2) (per bench test data)	1.0	Warn users to stop procedure if a glass cartridge breaks and use blow dryer to dry pump housing. (Warning, Procedure Change) Build all new production units with new hermetically sealed motor to prevent shorting due to liquid spill. Replace all units in field with newer units with new motor. (Design Change)
4. Loud motor	Could annoy user and patients	Negligible (1)	Unlikely (.03) (per customer complaint data)	.03	None required (risk is acceptable) Active noise cancellation could be implemented if a significant number of customer complaints are received.

Due to the potentially diverse audience for test results, it is worth considering how much data to display and how to display it. As in all communications, this will depend upon your target audience and the nature of your data. Sometimes presenting one key number will help initiate a conversation. If using this approach, be prepared to respond to questions with additional data and graphical representations. On the other hand, if the purpose of a meeting is to discuss a single test, you should be prepared to share more detailed test results and information.

Key Points

- Design verification answers the question, "did we make the product right?" Design validation answers the question, "did we make the right product?" Answering both questions relies on testing.
- Testing is an iterative process that determines how close a product is to meeting all requirements.
- Testing is important to a design team to guide the design process and demonstrate to external stakeholders that a new product is safe, efficacious, and provides value above and beyond available solutions.
- Verification testing can be significantly aided by testing per industry standards. This will save time and ensure that accepted and approved test procedures are used.
- When testing prototypes or the final product, it should be conducted in a test environment that mimics the expected service environment.
- A medical product includes the device and its labeling and packaging. Distribution simulations mimic the distribution stresses that the product will be subject to during sterilization, shipping, and storage of the product.
- Shelf-life studies are used to determine how long a product can be stored with no significant changes in physical properties or performance.
- It is critical to document all testing procedures and results and present them in a way that is appropriate for the target audience.

Exercises to Help Advance Your Design Project

1. Create a test protocol that you can use to determine if each of the specifications for the device that you are developing has been met. Test your device per this list.
2. For the specifications that were not met in Exercise 1 above, what design or other changes would you recommend so that they can be met?
3. Think about the expected distribution history of your device if it were to be manufactured. Might it be exposed to extreme cold or heat? If so, could it survive these extremes? If not, what special handling procedures would be needed to ensure that it will function properly when delivered to a hospital or user?
4. Will destructive testing be helpful and/or necessary when testing your device? What are some reasons why it might not be?
5. Complete a risk analysis for your device using the table in Table 9.2. If you are not able to test a large sample size, estimate the probability of occurrence and calculate risk.

6. Which risks calculated in Exercise #5 do you think are high enough to warrant developing a risk mitigation plan? Develop mitigation plans to reduce these risks. What types of mitigations are represented (e.g., warnings, procedure changes, design changes, manufacturing changes)?

General Exercises

7. What are the benefits of cadaver testing, and for what types of tests is it inappropriate?
8. What are the reasons for testing medical devices? Why is it important to test them prior to commercialization?

References and Resources

Anatomy Warehouse. (n.d.). https://www.anatomywarehouse.com.

ASTM International. (2015). *Standard test methods for vibration testing of shipping containers.* ASTM D999-08). https://www.astm.org/Standards/D999.htm.

ASTM International. (2019). *Standard performance specification for Foley catheter.* ASTM F623). https://www.astm.org/Standards/F623.htm.

Budynas, R., & Nisbett, J. (2008). In *Shigley's mechanical engineering design.* McGraw Hill.

Global Technologies. (n.d.). *GT Simulators by Global Technologies.* https://www.gtsimulators.com.

Innovative Research, Inc. (n.d.). *Innovative Research.* https://www.innov-research.com.

International Organization for Standardization. (2015). *Medical electrical equipment—Part 1: General requirements for basic safety and essential performance (IEC 60601-1).* https://www.iso.org/standard/65529.html.

International Organization for Standardization. (2019). *Active implantable medical devices—electromagnetic compatibility—EMC test protocols for implantable cardiac pacemakers, implantable cardioverter defibrillators and cardiac resynchronization devices.* ISO 14117 https://www.iso.org/standard/73915.html.

US Food and Drug Administration. (n.d.). *Product classification.* https://www.accessdata.fda.gov/scripts/cdrh/shiftentercfdocs/cfpcd/classification.cfm.

Swanson, S., & Freeman, M. (1977). *The scientific basis of joint replacement.* John Wiley and Sons.

3B Scientific. (n.d.). https://www.a3bs.com.

Whitmore, E. (2012). In *Development of FDA-regulated medical products: A translational approach.* ASQ Quality Press.

Testing in Living Systems

Science and research do not compel us to tolerate...the careless and callous handling of animals in some of our laboratories.
— Lyndon Baines Johnson, 36th President of the United States

The history of cancer research has been a history of curing cancer in the mouse.
— Ronald Klausner, former director, National Cancer Institute

Chapter outline

Biomedical Engineering Design. https://doi.org/10.1016/B978-0-12-816444-0.00010-9

10.1 Introduction

Understanding the interactions between medical devices and living systems is critical in determining safety and efficacy, a step that is necessary before regulatory bodies can clear a medical device. As biological systems respond in complex ways and often adapt over time, tests in nonliving systems cannot capture the variability or responsiveness of a physiological environment. Some adverse effects on a patient, and failure modes of a device, only arise after an extended period and thus, can only be revealed with testing in living systems.

Tests in living systems are conducted in an analogous way to bench testing described in Chapter 9. As in bench tests, you should have a clear purpose, a detailed protocol, the proper approvals, and appropriate analysis and documentation methods before conducting tests. However, the significant difference is that you are testing in living systems and have a duty to protect the safety and well-being of your test subject(s), whether they are cells, animals, or humans. Working with living systems can expose you and others to risks, such as physical harm. For these reasons and others, you must have a sponsor (e.g., faculty member, physician) obtain approval of a study proposal by an institutional committee. If you are participating in studies, you will need to get the proper training and certification. Another difference is that, unlike bench tests that typically assess the impact of one variable at a time, testing in living systems allows you to determine the device's effect on the whole living system.

This chapter introduces the types of biological and clinical tests conducted on living systems, including biocompatibility and biodurability studies, animal studies, validation studies, and clinical device testing and clinical trials. The chapter also details the process of obtaining approval to perform tests in living systems from the appropriate university committees. Familiarity with these tests and applicable terminology, procedures, and considerations will help you in your current project and as you advance in your career.

10.2 The Purposes and Types of Testing in Living Systems

To commercialize most medical devices, it is required by law to demonstrate safety and efficacy, which generally involves testing in living systems. In the context of a design project, some specifications can only be tested in a living system. For example, animal testing will likely be needed if a specification dictates the time to perform a surgical procedure. Some specifications target a clinical outcome (e.g., reducing the incidence of bedsores), requiring human testing to determine if the specification is met. Some devices do not interact directly with patients (e.g., medical simulators, test equipment, service robots), and they do not need regulatory clearance. However, they do require validation testing with target users. These validation tests are considered a form of human subject research. For these reasons, nearly every medical device will undergo testing in living systems before commercialization. In this section, we discuss in detail the types and purposes of testing in living systems.

10.2.1 Types of Tests in Living Systems

There are three types of tests in living systems that are most relevant to medical device testing. First, tests in cells and tissues—the most basic living structures—are known as in vitro tests, from the Greek "in glass." They are similar to bench tests in that they are performed in a laboratory. Second, tests in whole multicellular organisms (e.g., zebrafish, rats) are known as in vivo ("within the living") tests. A best practice in animal testing is to begin testing with the lowest phylogenetic form (e.g., using a mouse instead of a pig) if such testing will provide you with the desired information. Third, in medical device design, human subjects are often used for human factors testing as presented in Section 8.2 (e.g., validation of device use) and performance testing (e.g., verification of a specification timing how long it takes to use a device). As indicated earlier, tests on organs (Sections 5.7.5 and 6.5.3) or cadavers (Section 9.6.3) are not living system tests. They only need the same level of preparation as a bench test, with accommodation for handling biohazardous waste.

10.2.2 Testing as a Way to Guide the Design Process

Testing in living systems can help guide your design decisions as you navigate the design process. For example, the same specification might be tested in increasingly rigorous ways as your device is refined. An early benchtop performance test (as discussed in Chapter 9) may determine if a silicone part meets a particular mechanical specification. However, a biocompatibility test or a later biodurability test in an animal model would verify that the physiological environment (e.g., body pH, temperature, biochemistry) does not degrade the mechanical performance of a device over a specified period. More rigorous and long-term testing will likely require a clinical study in humans. On the other hand, some testing in living systems is prohibited in humans; examples include tests that involve deliberately causing harm (e.g., nerve injury, fracture), stopping treatment of a life-threatening event (e.g., hemorrhage), or intentionally not killing cancer cell. The only way to test these interventions may be in an animal model.

Consider using a new grade of silicone in a tool that will be used in open surgery. Although this material will only temporarily contact parts of the anatomy, it is essential to know how this new material will respond mechanically and chemically to changes in the physical environment (e.g., body pH, blood). You might initially address this question in a bench test with an environment that mimics the service environment. Later verification tests might use an appropriate animal model. More rigorous and long-term testing may require a clinical study. An equally important question would be, "how will the contact with the silicone part affect the patient long-term (e.g., causing adhesions that take days to form and longer to reach final status)?" The results of these successive tests may lead to rethinking design decisions about material selection and perhaps lead to a material selection change.

Testing in living systems can serve dual purposes in validation testing. Assume you plan to make a series of refinements in your surgical tool design and print a few 3D prototypes, each one successively closer to the final design. After each prototype, you seek feedback from a user, or a group of them, using the tool during a surgical procedure and intend to use that feedback to make each subsequent refinement.

10.2.3 Testing for Failure Modes

As introduced in Section 8.9 on design risk management, a Failure Mode and Effects Analysis (FMEA) begins by listing failure modes and then estimating the probability of failures and the severity of the effects. Section 9.9 provides guidance on conducting benchtop tests to verify these estimates, and Breakout Box 9.2 provides an example with a medical device. Testing in living systems can allow more accurate quantification of the probable rate of occurrence of known failure modes and, because living systems are responsive, discover previously unknown failure modes.

When a device is first tested in its actual biological service environment, the range of possible failure modes expands. The reason for this is a device that performs well in a controlled setting may fail when exposed to the unique service environments found in biological organisms. Some failure modes can only be detected in a living system. For example, cytotoxicity tests require living cells. Devices may also lose performance when subjected to biological environments. For example, if corrosion due to low pH were tested in a controlled laboratory experiment, the compounding effects of chloride ions, proteins, and other organic molecules would still be unknown.

Living systems adapt over time, revealing new failure modes that do not present in short-term studies. Long-term tests of living systems aim to determine the types of adaptive responses that may be unexpected as a device is used over long periods. Often long-term tests are conducted after a product has been commercialized and is already being used in patients. Breakout Box 10.1 documents the case of an approved metal-on-metal (MoM) hip implant being used in patients for more than ten years before it was discovered that blood poisoning from cobalt, which was known to cause minor medical issues, caused metallosis, with severe systemic medical problems in some patients. Examples of other long-term effects realized only after device implantation are presented in Section 12.9.3 on product recalls.

Breakout Box 10.1 Cobalt Chromium Molybdenum (CoCrMo) Hip Implants and Metallosis in Human Subjects

The first hip implant to provide a low friction arthroplasty was used in 1962 and was made of a cobalt-chromium alloy and ultrahigh molecular weight polyethylene. For years, these alloys had been safely used in many joint replacements, so it was assumed that when metal-on-metal (MoM) hip implants made of CoCrMo articulating components were introduced in the 1990s, the alloy would perform acceptably. However, after months or years of wearing the metallic articulating surfaces, some patients experienced symptoms including cardiac myopathy, visual loss, hearing loss, tinnitus, depression, or memory loss, and often required revision surgery. In 2005, these implants were acknowledged to be the cause of the symptoms observed in patients. Excessive wear due to the design of the articulating surfaces produced unexpectedly high levels of metallic wear debris, leading to elevated levels of cobalt ions in a patient's blood and wear debris in tissues surrounding the implant site (metallosis). Normal serum cobalt levels are <0.5 g/L, toxic levels are >5 g/L, and patients with MoM hip implants had levels as high as 122 g/L. Although cobalt-chromium alloys had been safely used for decades in many metal-on-polyethylene implants, the use of CoCrMo alloy in MoM implants produced many severe adverse reactions. For this reason, MoM total hip replacements are no longer approved for use in the United States. A surgical image of metallosis in the tissues surrounding the implant site, obtained during revision surgery after a failed metal-on-polyethylene implant, is shown in Figure 10.1. Note the black deposits of metallic wear debris collected in the tissue and fat surrounding the implant site.

Breakout Box 10.1 CoCrMo Hip Implants and Metallosis in Human Subjects—cont'd

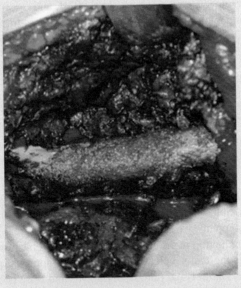

FIGURE 10.1

Severe metallosis around the hip joint. Microscopic view showed histiocytic infiltration with abundant metallic debris.

(From Kwak, H. S., Yoo, J. J., & Lee, Y. K., et al. (2015, March). The results of revision total hip arthroplasty in patients with metallosis following a catastrophic failure of a polyethylene line. CiOS Clinics in Orthopedic Surgery, 7[1], 46–53.)

Testing in living systems requires careful consideration of subjects and controls to detect new failure modes and estimate failure probabilities. Disease can be studied using an animal with a gene knocked out. For example, if the insulin gene is knocked out in a mouse, murine offspring will all suffer from diabetes. A diabetic population of mice can be used to conduct experiments on various treatments for diabetes. If using a homogeneous population (e.g., diabetic mice), realize that you may inadvertently miss essential failure modes that would only appear in a more heterogeneous population. In a design project, the analog would be testing your technology with several healthcare providers (or even across several healthcare systems) instead of only one. Before commercializing a device, tests should be completed on a heterogeneous population to reduce the probability of unanticipated failure modes.

10.2.4 Testing for Regulatory Approval and Reimbursement

To commercialize a novel technology, two types of organizations need to be convinced of safety and efficacy. First, the regulatory agencies (e.g., Food and Drug Administration [FDA], EMA, and PDMA) must clear or approve a medical device before being sold. Regulatory agencies often require clinical trials (investigational studies) to demonstrate safety and efficacy. These are presented in more detail in Section 10.5 on Human Testing. Second, health insurance providers will determine if a newly commercialized device warrants reimbursement, a significant factor in whether or not healthcare providers will purchase and use a new device.

10.2.5 Financial Aspects of Testing in Living Systems

With successive testing in living systems, the cost, time, effort, and rigor increase. For this reason, the results of initial tests are often used to determine if a device will pass from one stage-gate phase to the next. Companies carefully consider whether or not a test is justified. Consider that patient testing is costly for many reasons, including:

- The equipment, space, and training involved often are beyond those required for nonbiological tests.
- Clinical staff is needed for patient recruitment, randomization, data collection, and analysis.
- Companies may be held liable for adverse events during a clinical study of a new medical device.
- In industry, for device commercialization, 150 to 300 test subjects are required to achieve results with sufficient statistical significance for regulatory agencies and ultimately receive regulatory approval.

There are, however, many tangible benefits for patient testing. It can generate trust in the product among customers and users. Statistics generated from clinical test data are often used in promotional materials. For example, a company would likely publicize that clinical testing of their implantable device demonstrated a 23% reduction in the development of an adverse biofilm. Carefully designed and conducted living systems testing can accelerate the regulatory approval process and increase acceptance by healthcare providers and hospital purchasing agents and from health insurance organizations, who determine purchasing and reimbursement for the device.

10.3 In Vitro Testing

In vitro tests involve live cells or organs and are conducted in a laboratory setting. Cell tests may be undertaken in preparations involving isolated cells or intact tissues and range from "blank" cells (e.g., oocytes) to more specialized cells (e.g., rat liver cells in culture). Tests in intact tissue from an organism (e.g., dissected muscle from a rabbit) or even artificially created tissue (e.g., tissue-engineered cardiac muscle) are also performed in a laboratory setting.

For medical devices, the most common reason for in vitro testing is to determine the biological performance of a new material to be used in a device. From Section 8.3, biological performance includes biocompatibility and biodurability. This section provides guidance on testing these host and material responses by testing material interactions with living cells and tissues.

10.3.1 Biocompatibility Testing

A device is biocompatible if it evokes the intended host response without creating a toxic or harmful reaction. Although defined in Section 8.3.2, biocompatibility in practice refers to the short- and long-impact(s) of a medical device in patients. You may have selected a material that is considered to be biocompatible because it is already in use in another medical device, thinking that no biocompatibility testing will be required. However, that is only true if the same material formulation, including chemical composition, additives, colorants, release agents, sterilization technique, and radiopacifiers, has been tested previously. There have been metals, polymers, glass, and ceramics that have been extensively tested and were determined to be biocompatible for a specific medical device application. Table 8.1 provides a list of materials that have already been tested and used in other medical devices. However, slight changes in the chemical composition, manufacturing processes, nature of patient contact, or sterilization methods would require confirmation by additional testing for biocompatibility. For this reason, virtually all device materials must undergo biocompatibility testing.

As presented in Sections 8.3.2 and 11.3, the guiding standard for in vitro testing of biocompatibility is *ISO 10993:2018: Biological evaluation of medical devices*. Regulatory bodies accept the ISO 10993 series of standards to determine the biocompatibility of a testing program. The standard contains 22 parts, each referred to as ISO 10993-1, ISO 1009-2, etc. For example, Part 1 of the standard titled "Evaluation and testing within a risk management process" deals with categorizing medical devices based on the nature and duration of contact with the body. ISO 10993-5 covers testing for cytotoxicity, and ISO 10993-10 covers testing for irritation and skin sensitization. A list of all parts of the ISO 10993 series can be found on the ISO website. Although ISO 10933 does not mandate specific tests, it does provide testing requirements and guidance.

10.3.1.1 Criteria for Determining Required Biocompatibility Tests

The three major criteria used to determine the required biocompatibility tests are (1) contact duration, (2) type of body contact, and (3) device type. The types of tests required will depend on these three factors for your particular medical device:

Contact duration is divided into three categories:

Limited Duration <24 hours (e.g., home CPAP mask, surgical tools)
Prolonged Duration 24 hours to 30 days (e.g., indwelling ureteral stents, most bandages)
Permanent >30 days (e.g., pacemakers, artificial joints)
Transient Contact is defined as less than 1 minute (e.g., hypodermic needles, tongue depressor) of contact, but this category is not used in biocompatibility testing.
The eight types of body contact include skin, mucosal membranes, breached surfaces, tissue, bone, dentin, blood, and circulating blood.
The three device types are surface, external communicating, and implant devices. These represent the different degrees of invasiveness of devices. Although the category of device is often self-evident, sometimes it is not so clear. One example is a feeding tube, which is surgically placed through the abdominal wall and is affixed within the stomach. For feeding, it is an external communicating device (ECD). However, the stomach is a mucosal membrane. Is it an implant or an ECD? The FDA considers it to be an ECD in part because of clinical problems associated with accidental disconnection and clogging of the tube.
Intravenous (IV) tubing and IV bags do not directly contact the patient, and you might suppose that they do not require biocompatibility tests. However, IV fluids contact the surfaces of the tubing as they are transported from the IV bag, through the tubing, into the catheter, and into the patient. Once in the patient, the fluid contacts blood and travels through the bloodstream to other organs and tissues. Substances in the IV tubing can leach out from the tubing and into the IV fluid, potentially causing an adverse reaction in the body. Conversely, substances within the fluid may bind polymers to the tubing and thus be prevented from reaching the patient. Therefore, materials used in fluid path contact devices must be tested for biocompatibility, even if they do not directly contact the patient.

10.3.1.2 Selecting the Right Biocompatibility Tests

ISO 10993-1 outlines the types of biocompatibility tests that may apply to your device. Some tests involve cell cultures, while others involve animals. Some tests require a sample of material, while others require an extract of the material. Biocompatibility tests never involve humans and generally do not require an entire intact device. A brief overview of the 13 types of tests is provided below.

1. *Cytotoxicity.* Evaluate the material's effect on cell growth, cell depth, and other effects. All 72 combinations must be tested for cytotoxicity.
2. *Sensitization.* Determines the potential for contact sensitization as small amounts of material can produce allergic or sensitization reactions when in contact with skin and other tissues. All permutations must be tested for sensitization.
3. *Irritation.* Determines the potential for irritation in a suitable model using sites such as skin, eye, and mucous membranes. Except for long-term indirect ECDs, all other permutations must be tested for irritation.
4. *Intracutaneous reactivity.* Assesses the localized reaction of tissue to a material. Applicable to devices that can breach the skin and contact circulating blood and other tissues.
5. *Acute systemic toxicity.* In an animal model, determines the potentially harmful effects of single or multiple exposures of less than 24 hours each.
6. *Subacute and sub-chronic toxicity.* Determines the effects of exposure for a period not less than 24 hours and not greater than 10% of the animal's life span used for testing.
7. *Genotoxicity.* Uses cell cultures or other techniques to determine whether a material induces gene mutations, changes in chromosome structure or number, or other DNA or gene toxicities resulting from exposure to the test material. One genotoxic material is chromium, as it can interact with DNA and can be carcinogenetic.
8. *Pyrogenicity.* Assesses a material's potential to induce fever-producing reactions in test animals. Fever-inducing is limited to devices causing biological effects.
9. *Implantation.* Evaluates local pathological effects on tissue at both gross and microscopic levels of a material specimen (not a medical device itself) that is surgically implanted in the tissue of an appropriate test animal.
10. *Hemocompatibility.* Evaluates the effects of blood contact with the test material. Tests assess the potential for hemolysis (lysing of red blood cells) and thrombogenicity (the ability of a material to form blood clots).
11. *Chronic toxicity.* Determines the effects of single or multiple exposures to the material during at least 10 % of the life span of the test animal.
12. *Carcinogenicity.* Assesses the tumorigenic potential during the significant portion of a test animal's life span. Carcinogenicity tests should be performed only if other sources suggest that the test material may induce malignancies.
13. *Reproductive toxicity.* Assesses genotoxicity gene mutations, chromosomal abnormalities, DNA effects. Also assesses endocrine toxicity, which in turn can impact reproductive functions such as ovulation and spermatogenesis.

Counting all of the permutations of three durations (eight types of body contact and three device types), 72 possible classifications will help you determine the biocompatibility tests that will apply to your device. To know which biocompatible tests are relevant, you must first determine which of the 72 device classifications apply to your device. A helpful summary is contained in the Biocompatibility

Matrix in Table 10.1. In general, the longer the contact duration and the more invasive the device, the more testing will be required. All 72 combinations of devices, contact, and contract duration require tests for cytotoxicity, sensitization, and irritation or intracutaneous reactivity. Degradation and reproductive biological effects must also be considered for novel materials involving degradation (resorbable sutures, surgical films) or reproduction (intrauterine device).

You may, for example, conclude that your device requires a cytotoxicity test (ISO 10933-5) using cell monolayers. To claim a noncytotoxic response, the cell response to your experimental material should be no worse than the cell response to the negative control. The goal of a cellular biocompatibility test is to grade a material on its level of cytoxicity on a scale from 0 to 3 (i.e., noncytotoxic to highly cytotoxic). The noncytotoxic response is the negative control and the known cytotoxic response is the positive control. A detailed example of such a test is outlined in Breakout Boxes 10.2 and 10.3.

10.3.2 Biodurability Tests in Living Systems

Biodurability is concerned with the effect of the host (living tissue) on the material and is defined as the ability of materials to resist alteration *in vivo*. While device durability can be evaluated and verified through bench tests (in nonliving systems), biodurability requires testing in living systems to determine if specifications and performance requirements can be met in the context of a living system. As with many of the tests in this chapter, biodurability tests may start with in vitro and then progress to more rigorous animal and human tests which will be discussed in more depth in Sections 10.4 and 10.5.

Consider that medical grade silicone is biocompatible and biodurable, and thus has been widely used in extracorporeal applications due to its material properties. Specific properties include high molecular weight and surface tension, hydrophobic nature and hemocompatibility. It is considered a biodurable material as well because of its chemical stability and thermal stability and virtually no thermal degradation below 150°C. There are many medical-grade types of silicone and selecting the optimal type is one of many design decisions; it is usually a balance between optimal mechanical properties and potential degradation when in use. As noted previously, use of a material in one situation does not guarantee that it will be biodurable and biocompatible in another. Bench and in vitro testing may be required for verification and validation.

Indwelling devices, such as ureteral stents, Foley catheters, left ventricular assist devices, and some orthopedic screws and plates, are not intended to remain in a patient permanently. A first step in assessing the biodurability of these devices may be short-term implants in animal models. The devices can then be retrieved for inspection and an evaluation of biodurability. Evaluation tests include those for wear (as in the hip implant presented Breakout Box 10.1), oxidation (e.g., of some polymeric insulation wires), corrosion (e.g., of bare metal stents), and fiber and particle release, (e.g., OR towels used in surgery). Fiber and particles resist biochemical and physiological clearance mechanisms in the body. They also cause scarring adhesions, which may inhibit normal function and may cause chronic pain.

Implantable devices such as heart valves and joint replacements are intended to remain in a patient's body for decades, sometimes permanently. Assessing the biodurability of these implantable devices likely requires retrieval, inspection, and testing of the device after a period of time. For ethical and practical reasons, premature retrieval of these devices is not possible in a human study due to the trauma and risk associated with the early removal of an implanted device. In general, implant retrievals are only possible in long-term animal studies to resist unwanted changes such as material degradation and loss of physical properties. Some devices, such as resorbable stents or absorbable sutures, are intended to degrade within the body. In these situations, degradation is the desired change.

Table 10.1 Summary of biocompatibility endpoint testing. Based upon ISO 10993-1 and FDA Guidance.

Medical device categorization by						
Nature of body contact		Contact duration				
Category	Contact	A—limited (<24 h) B—prolonged (>24 h to 30 d) C—permanent (> 30 d)	Cytotoxicity	Sensitization	Irritation or intracutaneous reactivity	Acute systemic toxicity
Surface device	Intact skin	A	X	X	X	
		B	X	X	X	
		C	X	X	X	
	Mucosal membrane	A	X	X	X	
		B	X	X	X	O
		C	X	X	X	O
	Breached or com-promised surface	A	X	X	X	O
		B	X	X	X	O
		C	X	X	X	O
External commu-nicating device	Blood path, indirect	A	X	X	X	X
		B	X	X	X	X
		C	X	X	O	X
	Tissue/bone/ dentin	A	X	X	X	O
		B	X	X	X	X
		C	X	X	X	X
	Circulating blood	A	X	X	X	X
		B	X	X	X	X
		C	X	X	X	X
Implant device	Tissue/bone	A	X	X	X	O
		B	X	X	X	X
		C	X	X	X	X
	Blood	A	X	X	X	X
		B	X	X	X	X
		C	X	X	X	X

X = ISO 10993-1:2009 recommended endpoints for consideration.∗
O = Additional FDA recommended endpoints for consideration.∗
∗All Xs and Os should be addressed in the biological safety evaluation either through the use of existing data, additional endpoint-specific testing, or a rationale as to why the endpoint does not require additional assessment.

Table 10.1 Summary of biocompatibility endpoint testing. Based upon ISO 10993-1 and FDA Guidance—cont'd

Biological effect

Material-mediated pyrogenicity	Subacute/subchronic toxicity	Genotoxicity	Implantation	Hemo-compatibility	Chronic toxicity	Carcinogenicity
O	O		O			
O	X	X	O		O	
O						
O	O		O			
O	X	X	O		O	O
O				X		
O	O			X		
O	X	X	O	X	O	O
O						
O	X	X	X			
O	X	X	X		O	O
O		O		X		
O	X	X	X	X		
O	X	X	X	X	O	O
O						
O	X	X	X			
O	X	X	X		O	O
O		O	X	X		
O	X	X	X	X		
O	X	X	X	X	O	O

Breakout Box 10.2 Designing a Cytotoxicity Testing Protocol

Imagine that you work for a medical device company that aims to gain FDA approval to 3D print custom implantable devices, for example, hybrid hierarchical polyurethane. The first step toward that goal is to design and conduct *in vitro* cytotoxicity assays on your 3D printed test material. Your company will use these data internally to determine whether your tested material would be suitable for use in the device. If successful, the test data might also be included in your submission to the FDA for approval. You have been tasked with designing the test protocols that other employees will execute. After some research, you use the following standards to design your test protocol:

- ISO 10993-5: Biological evaluation of medical devices—Part 5: tests for *in vitro* cytotoxicity
- ISO 10993-12: Biological evaluation of medical devices—Part 12: sample preparation and reference materials
- ASTM F748-16: Standard Practice for Selecting Generic Biological Test Methods for Materials and Devices
- ASTM F 813-20: Standard Practice for Direct Contact Cell Culture Evaluation of Materials for Medical Devices

Preliminary work: Discuss your need for at least 12 flasks of an established cell line (per ISO 10993 with the cell culture lab). Because they are readily available and only require a BL-1-level laboratory, Madin-Darby Canine Kidney (MDCK) epithelium cells are chosen. Be sure you have completed your Biosafety Level 1 Training.

1. Prepare 60% to 80% confluent cell monolayers in 12 flasks following standard cell passage protocol. Microscopically examine each cell culture. Reject any flasks in which the monolayer is not of correct confluency or cells are not well attached to the substrate. Identify 12 flasks that are suitable and of roughly equivalent confluency, label each flask with the material that will be inserted into the flask with a unique name (e.g., A1, A2, A3, B1, B2, B3, …) and randomly assign flasks to the following four groups:

   A. 3 flasks for the untreated controls (if any flasks are unsuitable, use fewer flasks in this group)
   B. 3 flasks for the negative controls (HDPE—high-density polyethylene, no or mild cytotoxic response)
   C. 3 flasks for the positive controls (copper, which has been shown to elicit a cytotoxic response)
   D. 3 flasks for the test specimens of the experimental material to be tested

2. For all flasks (test, controls, and untreated cells), remove the old media and replace it with 3 mL of fresh media. The flasks containing untreated cells to which no material needs to be applied (group A) can be placed in the incubator and loosely capped.

3. Clean your control and test materials thoroughly without letting them touch each other. Since these materials will be placed in direct contact with the cells, you want to be particularly careful to clean them well and not allow cross-contamination. Place each type of material on its own clean absorbent sheet in the hood and allow them to dry. Clean forceps thoroughly. Place on a clean absorbent sheet in the hood and allow to dry. Make sure that both the forceps and the material are completely clean and dry before moving on to the next step, as alcohol can kill the cells.

4. Place a single test or control specimen gently in each of your experimental flasks using the alcohol-cleaned forceps. The material specimen should be placed in direct contact with the cell monolayer as close as possible to the center of the flask. Be careful to avoid scraping or agitating the cells as you place the material specimen in each flask. To avoid cross-contamination, clean the forceps between different materials. First, place all the negative controls (HDPE) into their assigned flasks, then all the test materials, and then all the positive controls. This will avoid any possibility of contaminating the HDPE or test material flasks with residue from the copper.

5. Once the flasks have been loaded with test or control specimens, return all flasks to the incubator, making sure they are loosely capped. Try to ensure that test material remains at the center of the flask.

6. Incubate the flasks with the materials in them for 24 ± 2 hours.

7. After 24 ± 2 hours, microscopically examine each cell culture and evaluate the degree of cytotoxicity. Each flask will be ranked on a 0 to 3 scale, with 0 representing no cytotoxic response, 3 indicating a severely cytotoxic response, with 1 (mildly) and 2 (moderate) indicating different levels of cytotoxic response. As these rankings are qualitative, begin by examining the negative (polyethylene) and positive (copper) controls to establish your ranges. Be sure to examine the entire flask to ensure that you have accounted for any variability in how you may have handled the test preparations. The justification of your ranking should be noted in your log and be based upon your observations of:

Breakout Box 10.2 Designing a Cytotoxicity Testing Protocol—cont'd

- *Cell Shape.* If you scan around the flask, do the shapes deviate from the negative and positive controls? Unusual shapes may indicate cytotoxicity.
- *Viability and Cell Density.* If cells are floating or low density, it may be due to cytotoxicity.
- *Estimated % Confluence.* Cytotoxicity can inhibit cell growth without killing cells.

  Take a high-quality, focused picture for each flask using consistent microscope magnification (20× recommended). Be sure at least ten representative cells are visible in each picture. Each image should be given a unique filename that includes the flask name (e.g., A1), magnification (e.g., 20×), and date.

8. When you have fully documented your flasks with pictures and written descriptions, you may dispose of the liquid into the waste beaker and the flasks into the biohazardous waste containers.

9. If no test specimens elicit a cytotoxic response, the material is considered to have passed the assay. If all three test specimens elicit a cytotoxic response, the material is considered to have failed the assay. If only some of the specimens elicit a cytotoxic response, the ASTM standard specifies that the tests should be repeated.

Breakout Box 10.3 Assessing Cell Shape

Breakout Box 10.2 suggested a qualitative assessment of cell shape as one measure of a cytotoxic response to a material. As an engineer, you may be able to perform more rigorous quantitative assessments. *Image J* (a program available at no charge from the National Institutes of Health) contains several helpful image analysis and measurement algorithms. For example, given an image of cells and some threshold, the program can identify unique cells and determine the Area (A) and Perimeter (p) of each cell. These measurements can then be used to calculate the dimensionless parameter called SI.

$$SI = \frac{4\pi A}{p^2}$$

SI = 1 is for a perfect circle and SI = 0 is for a perfect line. In other words, SI can be used to quantify the degree of elongation of a cell.

10.4 Animal Testing

Animal models have been used for nearly all of recorded history to gain new knowledge about anatomy and physiology and test healthcare interventions. In the context of medical device design, testing in animals generally occurs after in vitro testing but before testing in humans. In some cases, animal tests are used to detect potential negative outcomes, either to a patient or to the device in a physiological environment. In other cases, a test can only be conducted in an animal model. In Section 10.3.1.2, for example, biocompatibility tests 5, 6, 9, 11, and 12 can only be performed using live animals instead of cells or tissues. Test #13 requires the use of pregnant animals. Likewise, some biodurability tests require a device to be implanted in an animal for some period of time, subsequently explanted, and then subjected to additional bench verification testing.

This section discusses the ethical, legal, and logistical elements of animal testing. To reinforce the importance of animal testing, Breakout Box 10.4 gives a snapshot of two cases where if animal tests had been performed, it would have prevented more than 100 pediatric deaths and prevented more than 10,000 congenital disabilities in humans.

10.4.1 The Animal Welfare Act and Replace, Refine, Reuse

President Johnson's quote that appeared at the beginning of this chapter was from a speech given at the passing of the 1966 Animal Welfare Act (AWA). The act (along with amendments over the years) ensures that test animals will be treated humanely when used as research subjects. The AWA, which is

Breakout Box 10.4 The Importance of Animal Models in Assessing Failure Modes

Two past human tragedies illustrate that harm or death to humans could have been avoided if proper drug tests on animals had been conducted prior to market introduction. The liquid form of elixir sulfanilamide, known in 1937 as an antibiotic, was not tested for toxicity; the FDA did not require a toxicity test at the time. However, the liquid form contained diethylene glycol (used in anti-freeze), a then well-known deadly chemical when ingested. The manufacturer sold it as a liquid antibiotic in 1937 that caused over 100 deaths. This hastened the 1938 enactment of the Federal Food, Drug and Cosmetic Act, which increased the FDA's authority to regulate drugs. Frances Oldham Kelly, MD, was a new toxicology expert hired by the FDA and helped the process of preventing further sales of the drug.

Another example occurred when Thalidomide, a tranquilizer, has been used due to its anti-vomiting effect to treat nausea symptoms such as those caused by morning sickness in pregnant women. In the late 1950s and early 1960s, thalidomide was prescribed and used ubiquitously during that time, although not in the United States (except in sponsored trials), as it was not an FDA-approved drug for use in pregnancy. Only after 10,000 babies were born around the world (only ~40 in the US) with phocomelia (i.e., missing or malformed limbs as depicted in Figure 10.2), was the relationship that thalidomide caused phocomelia realized; thalidomide is teratogenic (causing a specific type of birth defect). A fetus demonstrating phocomelia in utero is shown on the left in Figure 10.2; the newborn cadaver showing the upper extremity phocomelia is shown on right in Figure 10.2. The reason it was not sold in the United States is one physician employee of the FDA, the same Dr. Frances O. Kelly who help prevent clearance of elixir sulfanilamide, did not clear it for sale in the United States, despite considerable pressure to do so, because she wanted to see more data that the drug was not teratogenic, did not cause congenital birth defects.

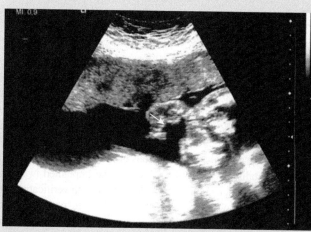

FIGURE 10.2

On left, a transverse ultrasound image of a 19-week-old fetus with missing left forearm is diagnostic for left phocomelia. On right, a newborn cadaver with upper-extremity phocomelia.

(From Travessa, A. M., Dias, P., Santos A, et al. (2020, March). Upper limb phocomelia: A prenatal case of thrombocytopenia-absent radius (TAR) syndrome illustrating the importance of chromosomal microarray in limb reduction defects. Taiwanese Journal of Obstetrics and Gynecology, 59[2], 318–322.)

under the auspices of the United States Department of Agriculture (USDA), requires that basic standards of care and treatment be provided to animals used in biomedical research. Institutions that operate animal testing facilities must provide their animals with adequate care and treatment in housing, handling, sanitation, nutrition, water intake, veterinary care, and protection from extreme weather and temperatures. Most countries have laws ensuring that animals used for testing are treated humanely.

The Guide for The Care and Use of Laboratory Animals (available online) is intended to help investigators plan and conduct animal experiments with the "highest scientific, humane, and ethical principles." There are three primary guiding principles in animal testing: replacement, refinement, and reduction—known as the three Rs.

Replacement refers to methods that avoid using animals. The term includes absolute replacements (i.e., replacing animals with inanimate systems such as computer programs) as well as relative replacements (i.e., replacing vertebrate animals with animals that are lower on the phylogenetic scale). In other words, animals should not be used for research unless the desired effect or outcome is unattainable in another way. One example is studying cardiac death. Instead of inducing cardiac death in pigs, a researcher can now develop a set of computer models that are able to accurately portray symptoms and etiologies in a specific patient. The medical goal is to help cardiologists predict sudden cardiac death and possibly develop technology to prevent it from occurring in patients.

Refinement refers to modifications of experimental procedures for animal husbandry (the science and breeding of animals) to enhance animal well-being and minimize or eliminate pain and distress. Institutions and investigators should take all reasonable measures to eliminate pain and distress—such as not depriving animals of regular feeding, warmth, or sleep—through refinement. It is possible that pain is sometimes unavoidable and therefore may not be eliminated based on the goals of the study.

Reduction involves strategies for obtaining comparable levels of information with the use of fewer animals or for maximizing the information obtained from a given number of animals (without increasing pain or distress) so that in the long term, fewer animals are needed to acquire the same scientific information. One example is to use both upper extremities of the same animal—one for the procedure and one for the control. This approach relies on an analysis of experimental design, applications of newer technologies, the use of appropriate statistical methods, and control of environmentally related variability in animal housing and study areas. One example of reduction is when using animals that are already part of ongoing experiments by others.

An important first step is in deciding whether animal testing is appropriate for your project. In other words, ask the question, "Why do need to conduct this experiment?" The answer lies in the project's specifications. Examples of the type of specifications that would be best tested on an animal include the following three: reduce the volume of blood loss, limit scarring, and reduce healing time.

10.4.2 Institutional Animal Care and Use Committees

The various laws are oversight bodies require some practical oversight. In the United States, institutional oversight is provided by an Institutional Animal Care and Use Committee (IACUC or ACUC). This committee is unique to a particular institution and is responsible for ensuring that all animal studies at that institution are conducted in compliance with the AWA.

Each ACUC evaluation committee must be composed of at least three members, including one administrator, one veterinarian, and one layperson who is not affiliated with the facility in any way. If needed for a specific application, external experts will serve as *de facto* committee members. For example, if a proposal is for developing technology to treat cataracts and the test proposes the use of horses, the ACUC committee may engage an equine ophthalmologist.

ACUC members perform two related duties. First, they ensure that shared animal handling facilities are following basic standards of care and treatment. Second, they review and approve proposals for animal studies. ACUC members will check that the rationale for your study is sound, relevant, and that it has considered minimizing discomfort, distress, and pain. It is not uncommon for a proposal to be returned with suggestions on how to improve it before being ultimately approved.

An application for animal testing protocol submitted to the ACUC has many standard components that must address the rationale for the proposed test, objective, and relevance, and the proposal must provide a detailed protocol. Many good templates are available online from academic medical centers. Each university's or facility's ACUC is unique to its institution; however, certain parts are common to all.

There are often many questions that must be addressed, some of which (e.g., recombinant, biohazard) will not apply to a particular animal test. Animal tests involving invasive, painful, or repeated surgeries generally will receive the highest level of scrutiny. Most proposals will need to address:

- personnel—principal investigator (PI) (must be a faculty member) and team members
- training and qualifications—demonstrate all study team members are ACUC trained and certified
- animal requirements—species, gender, age, and weight ranges
- location(s)—specific laboratory where testing will be set-up
- objectives—overall purpose of the project
- importance—relevance of work to human health or for the good of society
- rationale for animal use—justify using live animals, as opposed to other models
- species—choice and explain rationale
- number of animals and rationale—how many animals are needed and why
- procedure description—test protocol
- anesthesia—what type and how it will be administered
- pain/distress—categorize and explain how it will be minimized including euthanasia for terminal studies

There are many factors to consider when designing animal tests. For example, animal tests require resources to house, feed and care for animals to be tested. If surgery is involved, anesthesia and veterinary time are also needed. However, the costs for testing on animals can be minimized in some cases, especially if there are other animal studies at your university. If there are other nonsurvival animal studies at your institution, you may request access to those animals before they are euthanized. In some medical schools, for example, laparoscopic surgery is sometimes initially taught by operating on female pigs. After a teaching session for surgeons, the pigs may be available for animal testing before they are euthanized. If you are considering animal testing, you should look for these types of opportunities. The following sections detail the type of decisions you must make in conducting animal testing.

10.4.3 Choosing an Animal Model

Particular animal species have been found to mimic aspects of the normal or disease physiology in humans. In the context of medical device design, these species can serve as a model to study the intended and unintended consequences of a human-device interface.

In choosing an animal model you should determine which particular characteristics are most important to simulate. If you suspect your device may have unintended adverse consequences, you have an ethical obligation to demonstrate the safety of your device by testing it on an animal. In this section we present the various considerations in choosing and justifying the choice of an animal model.

The test objective and the information to be gained should guide the selection of an appropriate animal model. For example, if your device aims to measure bone density by taking small biopsies during surgery, this will require (1) using a vertebrate animal model and (2) a similar anatomy and scale to those of a human. This example reveals an additional important consideration in choosing an animal model—similarity to humans may be one of two types. First are anatomical/physiological analogs—similar structures that perform similar functions. For example, the sheep is used extensively to model pregnancy in humans for several reasons, including singleton births, a gestation length ~60% that of humans, and similar placental properties. Second are genetic homologue models; shared genetic makeup means that cellular functions are similar. Homologous animals will usually have similar genetic diseases, symptoms, and reactions to interventions at the cellular and tissue levels. For example, insulin is often tested in rabbits because the relationship between insulin and decreased blood glucose is similar in humans and rabbits.

You should also recognize that there are some diseases that only occur in humans. For example, pre-eclampsia (high blood pressure) is a disease of pregnancy unique to humans; there are no animals that experience it. Another disease example for which there is no animal model is multiple sclerosis. Likewise, some aspects of cancer are unique to humans, as suggested by Dr. Klausner's quote at the beginning of this chapter; cancer researchers have cured even advanced cancer in mice many times over, but not yet in humans.

Other considerations include the availability and cost of animal subjects and local expertise upon which you can draw. Many institutions house animals but must have their facilities approved for particular species. The result is that they only house those species that are in use by members of their institution. Likewise, staff veterinarians will be most familiar with animals that are under their care. You should consult with these experts and members of the ACUC, as they can help you plan your experiments and perhaps point you to previously approved protocols for you to append your tests.

Established animal and organism models balance availability, cost, similarity to humans, and ethical and legal protections. Table 10.2 summarizes the rationale for choosing nine of the most commonly used species.

10.4.4 Designing an Animal Test Protocol

A test protocol must be developed once the objective, personnel, and an appropriate animal model have been identified. All guidelines outlined in Chapter 9 still apply, including recommendations for equipment, materials, procedures, data collection, and analysis methods.

To gain more insight into the kinds of language and details required, we consider a student-prepared proposal for testing a novel device to improve the surgical procedure of bowel packing, which is a standard surgical procedure needed for many abdominal and pelvic open surgical procedures. During abdominopelvic surgery, bowel packing is necessary to provide adequate surgical exposure and protect the intestines from injury. A surgeon performing pelvic surgery will first use OR sponges and towels to sweep the intestines cephalad and keep the bowel retracted out of the surgical field behind metal retractors. This

Table 10.2 Nine common animal/organism models with typical applications and rationale for use as an experimental model

Animal or organism	Typical applications	Rationale for use
Brown rat	Cardiovascular disease, psychiatric stress	HR 300–600 bpm, similar cardiac responses, behavioral response to stress and fear similar, predictive response to therapy
House mouse	Genetic diseases, spinal cord injury, neuronal disease, Huntington's disease, bone and joint disease, cancer avatar	98% of some genes comparable to human genes, genetically modifiable, similar nervous and reproductive systems
Primates	Neuroscience, learning, neurosurgery	Highest cognitive capability, responds to host of disturbances, similar anatomy
Rabbit	Intraperitoneal cavity, pregnancy, implants, diabetes	Human hCG hormone triggers leporine pregnancy response, similar anatomical structures, adhesion formation similar, as is the effects of insulin
Roundworms	Aging	Gene mutation can extend life to 150–250 days (10 times longer than normal)
Sheep	Cardiology, pregnancy, fetal development	Fetal hypertension, preterm labor, cystic fibrosis, cardiac output in pregnancy
Swine	Hemorrhage, trauma, cardiac function, urology	Comparable cardiac size and output, similar blood clotting mechanisms, similar cardiovascular and hemodynamic responses to hemorrhagic shock, similar urinary tract
Yeast	Cell processes, cancer, anemia, aging	Similar biological properties, human cell counterparts, telomeres linked to cancer
Zebrafish	Organ development, movement disorders, Alzheimer's disease, heart disease, ophthalmologic diseases	70% genetically similar, many genes homologous, vertebrate, transparent embryos, similar CNS connectivity and pathfinding, similar visual anatomy

is the standard medical procedure for pelvic surgery. Figure 10.3 shows a completed surgical preparation during an open laparotomy, which involves making a surgical incision through the abdominal wall.

The procedure, which needs to be repeated every 1 to 2 hours due to leakage induced by peristalsis, requires about 5 minutes of OR time. Peristalsis also causes loops of bowel to escape around the towels. Additionally, OR towels leave fibers on the intestines after surgery is completed. Rarely, a sponge (hidden behind the OR towels) is inadvertently left inside the patient after the surgical wound is closed, and a second surgical procedure is needed to retrieve it.

In this example, gynecologic cancer surgeons, who served as project sponsors, identified three clinical goals for the design project:

1. Reduce the typical time it takes to pack the bowel to below 5 minutes.
2. Reduce the need to repack the bowel every 1 to 2 hours during surgery.
3. Reduce adhesions caused by bowel packing during abdominal surgery.

Shown in Figure 10.4 is the design solution, consisting of a soft and flexible molded silicon device that replaces the need for OR towels. The device's side flaps and contoured bottom press against the

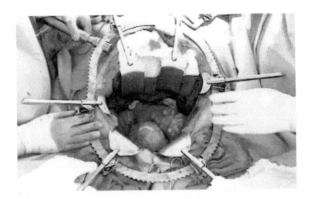

FIGURE 10.3

A surgeon's view of a packed bowel. The white and blue OR towels are held in place by the upper retractor blades. These, along with OR cotton pads and sponges (already packed, and hence not visible) attempt to retain the position throughout surgery.

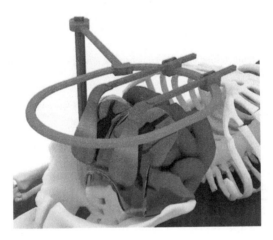

FIGURE 10.4

Image from computer animation of a bowel packing device. In lieu of OR towels, OR pads, and sponges (hidden behind the exposed OR towel), intestines are held in place by a clear silicone barrier with flaps shaped to conform with the bowel. As with OR towels, the retractor blades secure the bowel packing device in place.

abdominal walls to prevent having to repack the bowel due to peristalsis-induce leakage of the intestines from behind and through the towels (retention failure). The dual top flaps shield the bowels from the surgery itself.

After creating this design solution, the team wished to test the prototype beyond bench testing for material selection, geometry, and strength. The first design goal could be tested by measuring how long it takes to pack the bowel with the device and the traditional method in one large animal already scheduled for euthanasia as part of a separate study. The second design goal could be tested in the laboratory

---

Breakout Box 10.5 Partial Proposal for Study of Adhesion Formation During Bowel Packing (prepared by the design team)

Objective

We propose to quantify and compare the extent of intraperitoneal postoperative adhesion formation during bowel packing, a necessary procedure for abdominal and pelvic laparotomies. The standard of care for bowel packing includes using sponges and OR towels to retract the intestines out of the surgical field. We plan to compare adhesion formation following standard of care cotton laparotomy packs and use a silicone-based elastomer for bowel-packing.

Importance of Research

Nearly 10% of inpatient surgical procedures in the United States require bowel packing. Cotton laparotomy packs currently in use are abrasive and leave behind foreign particles, which result in intraperitoneal postoperative adhesions. Adhesions can cause complications, including infertility, small bowel obstruction, and pelvic pain. A second procedure known as adhesiolysis is needed for some patients to divide or remove adhesions. More than $2B was spent on ~260K adhesiolysis procedures in 2016. Using a nonfibrous, less adhesiogenic material for bowel packing can mitigate and possibly eliminate the occurrence of postoperative adhesions. This experiment will quantify the adhesiogenic nature of silicon-based technology relative to cotton laparotomy packs currently used to pack the bowels during surgery.

References

1. Sakari, T., Christersson, M., & Karlbom, U. (2020, December). Mechanisms of adhesive small bowel obstruction and outcome of surgery; a population-based study. *BMC Surgery*, 20(1), 1–8.
2. ten Broek, R. P., Bakkum, E. A., Laarhoven, C. J., & van Goor, H. (2016, January 1). Epidemiology and prevention of postsurgical adhesions revisited. *Annals of Surgery*, 263(1), 12–19.

Rationale for Animal Use

All bench testing and biocompatibility testing has been completed. It is not possible to simulate the physiological process of adhesion formation (a form of scarring) in a biofidelic way. Animal testing is therefore required.

Species Selection

We choose the rabbit in part because it is a well-characterized animal for the human peritoneal cavity. Rabbits are also readily available through the Principal Investigator (PI).

Animal Number and Rationale

This is a pilot study to collect preliminary results. We have determined that performing the procedure on 30 animals, once per animal, is sufficient for small-sample statistically significant results. Three groups of 10 rabbits each will be divided into sham surgery, bowel packing with towels, which is the standard of care, and bowel packing with our technology. After 14 days, the rabbits will be euthanized and reoperated to assess postsurgical adhesion formation.

---

by designing and building a bowel simulator that pulsates and bleeds (e.g., cooked noodles in water on a vibrating surface). Testing the third goal required a live animal model.

The first tests were performed in two female pigs undergoing a laparotomy. Both test animals were already under anesthesia and about to be euthanized as part of another terminal experimental protocol. Bowel packing was performed in the traditional way as a control using one animal and then again using the team's prototype on the other pig. The time for each procedure was measured and compared. Gaining approval through an amendment to the original ACUC-approved protocol is required to perform these studies.

The third tests were performed in three sets of rabbits. One set would be a sham surgery (a surgery where no procedure is actually performed), one set would undergo a conventional laparotomy, and one set undergoes a laparotomy using the teams' device. Required for this test would be 30 rabbit-sized prototypes of the device. Breakout Boxes 10.5 (prepared by the student team) and 10.6 (prepared by a veterinarian) contain select sections from their proposal to the ACUC.

Breakout Box 10.6 Partial Proposal for Study of Adhesion Formation During Bowel Packing (prepared by a veterinarian)

Survival surgery will be performed on the rabbits to apply cotton laparotomy packs or our technology to the abdominal cavity. Rabbits will be held off pellets the night before the surgery but will be allowed free access to hay and water. The following morning, the rabbits will be sedated with an intramuscular (IM) injection of 1–2 mg/kg pharmaceutical-grade acepromazine, 20 mg/kg pharmaceutical-grade ketamine, and 0.005–0.010 mg/kg IM pharmaceutical-grade buprenorphine. After the rabbit is sedated, a butterfly catheter is placed in one of the marginal ear veins, an endotracheal tube placed, and the rabbit is given Pentothal (brand name of thiopental) IV 40 mg/kg to maintain a surgical plane of anesthesia throughout the procedure.

The rabbit's abdomen will be shaved and disinfected with chlorhexidine and alcohol. The surgical field will be aseptically draped, and sterile instruments and warmed fluids will be used. The surgeon will be aseptically gowned and gloved and wearing a cap and mask. All assistants in the room will be wearing surgical caps and masks. Throughout anesthesia, the rabbit's respiratory rate, heart rate, carbon dioxide level, reflexes, and temperature will be monitored and recorded every 15–20 minutes. The rabbit will be maintained in a surgical plane of anesthesia with repeated boluses of thiopental or isoflurane inhalation anesthesia. Once fully anesthetized, an 8-cm ventral midline incision is made to expose the abdominal contents. A cotton laparotomy pack soaked in Lactated Ringer's or 0.9% NaCl, or a silicone elastomer sample will be placed inside the abdominal cavity, ensuring contact with the bowels for a 6-hour period. During this period, the abdominal cavity will remain open to simulate the exposure that results from a surgical procedure. A circulating warm water blanket drapes over incisions, and extra heat support (e.g., heat lamp, hot water bottles) will be used to maintain body temperature over this time. If necessary, moisture in the abdominal cavity will be maintained via the addition of Lactated Ringer's or 0.9% NaCl.

Upon completion of the 6 hours, the cotton laparotomy pack and silicone elastomer samples will be removed from the abdominal cavity and the cavity will be closed in a two-layer closure using absorbable sutures. No material will be introduced into the abdominal cavity of the animals receiving a sham surgery (control group). The abdominal muscles (linea alba) will be closed with 3-0 braided sutures in a simple continuous pattern. The skin will then be closed using the same suture in a subcuticular pattern. There will be no skin sutures or staples.

All animals will survive for 14 days with no restrictions on food or water intake. Normal housing will be utilized. After the 14-day period, each rabbit will once again be sedated with ketamine and acepromazine as described above. After sedation, each rabbit will be euthanized with an IV injection of >100 mg/kg pentobarbital. Death will be confirmed by a lack of an auscultable heartbeat and no corneal reflexes. After euthanasia, postmortem evaluation of the animals will be conducted to quantify the nature and extent of intraperitoneal adhesion.

10.5 Human Testing

The goal of all medical device design projects is to introduce a safe and effective new device to the healthcare ecosystem. Benchtop and animal testing can verify many of your specifications, however, some specifications can only be verified through testing in humans. In addition, formal validation of your device can only be gained through surveys, focus groups, and ethnographic observations with the intended users. Both clinical studies and validation studies are a form of human subject research. Even if you do not plan to engage in human studies as a part of your current design project, it is important to understand the factors that must be considered in conducting tests with human subjects as it is almost always an important part of a medical device design process. One challenge with performing tests on patients is that you are not allowed to communicate with them without specific permission to do so (more on this in Section 10.5.2.4.)

The justification and oversight of a proposed study are required to reduce the risk of an adverse event occurring with a test subject. The effort involved in a human subject study is significant. Beyond the testing of your technology, you need to be concerned about recruiting patients, ensuring patient privacy, and ensuring safety. One possible exception is if you have an industry or clinical mentor who has an existing approved clinical study. For example, if you are developing a new walker for disabled children, your physician advisor may already have an ongoing clinical study for the same patient population. A validation study may require less effort if the healthcare providers, who would be the test subjects, are easy to recruit through a clinical or industry project sponsor.

This section introduces you to the policies and procedures of testing in humans, including Institutional Review Board approval, registering clinical trials with the appropriate regulatory bodies, and informed consent. Although our focus will be on the United States, most other countries have similar protects and policies regarding human subject research. The ethical dimensions of human subject research are presented in more detail in Chapter 13.

10.5.1 Institutional Review Board Proposals and Approval

Nearly all countries have laws that protect the safety and dignity of humans, including the use of human subjects in experiments. Much of the oversight of compliance with United States laws falls to an Institutional Review Board (IRB), a committee that oversees all human subject research at its institution. In this regard, an IRB performs a similar function to the ACUC in that it reviews and approves all applications that involve testing using human subjects. The IRB is responsible for ensuring the welfare and safety of all human test subjects, in part by following a set of regulations for ethical treatment of human subjects. IRB oversight is mandated by the Office of Human Research Protections (OHRP), a division of the US Department of Health and Human Services, which provides advice and educational materials on ethical and regulatory issues in biomedical and behavioral research. In addition to reviewing proposals, the IRB also conducts annual or semi-annual reviews reported to OHRP. Should a subject be harmed or a protocol or procedure not followed after approval, it must be reported to the IRB contemporaneously. The IRB has the authority to suspend or even terminate a human subject study.

Although each IRB operates differently, an IRB committee is legally required to consist of at least five members. The members as a group must be professionally competent, diverse, and ethical—they must also be knowledgeable about applicable standards, professional conduct, regulations, and institutional policies. At least one member must be primarily concerned with scientific matters; another must be principally concerned with nonscientific issues (often a clergy member, ethicist, or patient's rights advocate). One member must be a person not affiliated with the institution. Meeting times are often posted on a website, along with guidance and forms required for submitting a proposal.

The number of IRB committees, and the frequency with which they meet, are institution-dependent and reflect the volume of human subject research being performed at that institution. You are allowed to consult with the office of the IRB before submitting a proposal. This can be very helpful, especially if you are submitting a proposal for the first time or for a type of study that is new to you. It is not unusual for a proposal to be sent back with comments, to be reviewed again at a later date. For this reason, the more well-thought-out the proposal, the fewer number of resubmissions needed to obtain approval. As some IRBs charge for their services, you should be sure that you have included these fees in your project budget.

10.5.1.1 IRB Reviews

In preparing an IRB proposal, it is helpful to understand how they are reviewed. When a proposal is submitted, the IRB will ask a series of questions that will help gauge the level of risk to the subjects. A member of the IRB (often the Chair) will ensure that the answers to these questions align with the proposed work and then classify your proposal into one of three types. The first and lowest level of risk is IRB-exempt, which means that the study can be approved without a full review of the committee. In general, exempt studies are those in which each patient's or subject's identity is unknown. An anonymous survey of a public database may qualify for exempt status. The second is an expedited review, which the IRB would consider to be of minimal risk. Examples of minimal risk studies include noninvasive measurement of blood hemoglobin (analogous to a pulse oximeter measurement) and a survey of cardiac patients to learn about causes of noncompliance using a prescribed cardiac vest. The third and highest level of risk requires a review by the full committee at a formal meeting. A proposal to measure the electrical conductivity of the liver during surgery is an example of a study that a full committee would typically review. Ultimately, the IRB decides what level of risk each study would be classified as on a case-by-case basis.

You should not assume your proposal will qualify for a particular category. In some situations, this may be to your benefit. For example, the IRB in the liver measurement example may have already approved a similar study (with similar protocols and team participants), and you are only submitting an addendum. In this case, they may decide to expedite the review.

IRB submission is a detailed process and may seem daunting. However, if approached systematically utilizing the categories of questions on the application, in some cases, you can prepare an IRB application early in the design process and have it approved by the time testing is planned to begin.

10.5.1.2 IRB Proposals

The most common set of questions asked on a standard IRB application can be divided into three categories: general information, protocol, purpose, and recruitment. They typically include the following:

- What is the study title?
- Who are the PIs and other team members, and what are their roles?
- Have all team members passed the required compliance tests?
- What is the purpose of study and a synopsis of protocol?
- Is this a clinical trial?
- How is it supported?
- Where will it be conducted?
- What is the sample size?
- What is the study protocol?
- Who is being studied? Are they healthy?
- How will you recruit participants and ensure that no subjects have been coerced into participating?
- What type of consent is needed?
- If a device is to be used, what is it?
- How will you ensure privacy?

The first three questions are generally straightforward but do require some thought. For example, the title should be phrased to have meaning to a physician. One example is "Patient Survey to Determine Wearable Cardioverter Defibrillator Compliance"; another is "A Novel Method for Monitoring Uterine Contractions." The PI must be a faculty member or an affiliated physician and is the one who ultimately submits the proposal to the IRB portal and communicates with the IRB. The study team includes those involved in the research. Design team members can be on the study team as long as each member has passed the required certifying exams. These often include online tests on Basic Human Subjects Research and Health Privacy Issues for Researchers, Conflict of Interest (COI), and Responsibilities and Duties for Researchers. It is best to determine who will participate well in advance to ensure each study team member obtained the necessary training before submitting the proposal.

As with animal studies, the objective must be stated in the context of the information to be gained and the anticipated clinical benefit. One example of an objective statement is "To measure the effectiveness of a new orthotic device to reduce gait disturbances, which would be a clinical benefit to Multiple Sclerosis (MS) children." The objective of a study should be one sentence. A rationale must be provided for using the number and target population of human subjects. Published literature and previous bench, *in vitro,* or animal studies are often used as part of the rationale. You will likely be able to reuse text you have already written about from your project statement and from your Design History File (DHF) for the study rationale.

The protocol should include step-by-step details about the conduct of the study before (e.g., recruitment of participants, equipment procurement, and environment preparation), during (e.g., actions taken by study participants and subjects), and after (e.g., post-procedures or follow-up with the subjects, data analysis) the actual experiment. It should be made clear the parts of your device with which subjects will come into contact and each part's intended function.

There are many other types of questions, but many of these are not related to most design projects. For example, you may conduct tests in a clinical environment, but this is not designated as a clinical trial, so some questions related to clinical trials would not apply. Likewise, you may be able to skip questions about drugs, genetic testing, or radiation exposure if these are not part of your protocol. Although we provide basic guidance below, you should check with your host institution's IRB website before beginning your proposal.

The protections in place for the study subjects must be included in the protocol. Each of the following issues should be explicitly addressed and aim to protect subjects and ensure that the data collected will achieve the study aim without introducing bias.

- inclusion and exclusion criteria for the target population;
- recruitment of a target population, ensuring that no subjects have been coerced into participating;
- Informed Consent written in language understandable to a middle-schooler to be signed by the subjects or their guardians;
- explanation of the risks and benefits of participation, including any gifts or payment;
- assurances of maintaining patient dignity before, during, and after the study;
- means of assuring privacy of sensitive health data;
- consideration of vulnerable populations (e.g., children, prisoners, elderly, disabled participants); and
- grounds for study termination (either voluntarily by a subject or by the researchers).

The following section presents the practical implications of these protections.

10.5.2.3 Subject Protection

A fundamental principle of biomedical research is that it should not increase the risk of harm or injury nor leave patients deliberately untreated for a disease. This is the result of the Belmont Report

Breakout Box 10.7 Example IRB Proposal for a Contraction Monitor

Abstract: Management of labor in the developing world includes midwives manually measuring contraction frequency and duration, manually screening for abnormal labors, and having high-risk patients moved to an acute-care facility. The problem is with few midwives and too many patients, contraction frequency is often not measured, and many times abnormal labors are not detected. This leads to about 200,000 preventable maternal deaths as a result of neglected abnormalities of labor. In resource-poor environments, midwives use the paper partograph to monitor the progress of labor, which is the standard of care. Important components of the partograph include contraction frequency and duration. Abnormal contraction patterns, such as tachysystole (increased contraction frequency without progressive cervical dilation), if left untreated, increase risks for many complications, including uterine hemorrhage, fetal asphyxia, neonatal seizures, and maternal and neonatal death. To address this problem, we have developed a contraction monitor to continuously measure contractions which is self-powered, untethered, and provides a real-time digital readout of contraction frequency and duration. A midwife can glance at it to make a more informed clinical decision in a timely manner.

Objective: To demonstrate that our proposed mechanism for uterine contraction monitoring can display sufficiently accurate data that is consistent with that required by protocol in many developing countries.

Background: In low-resources settings, midwives take care of far too many labor patients at the same time to provide proper care to each one. In practice, use of a paper partograph is cumbersome and time-consuming. In a crowded labor ward with few providers, filling in the partograph out is often abandoned. As a result, partographs are often not kept current and abnormal labors are often not realized in time to intervene. As an example, a continuously measured hypertonic contraction strip is presented in Figure 10.5; a partograph representation is shown in Figure 10.6.

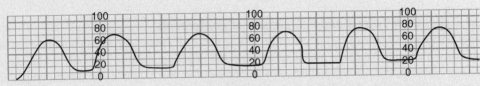

FIGURE 10.5

A sample contraction strip (in mm Hg of uterine pressure) over 10 minutes (horizontal axis) where each peak represents one contraction. Without corresponding cervical dilation is an abnormal contraction pattern (tachysystole), which may require intervention (e.g., cesarean section) to deliver the baby sooner than that one naturally occurring.

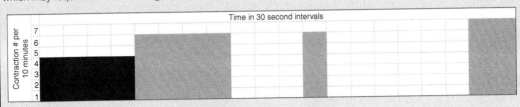

FIGURE 10.6

A partograph representation of six contractions for 10 minutes, with portions not filled in. This limits midwives' ability to make an abnormal labor diagnosis. If more of the partograph was filled out, this would enable a midwife to diagnose tachysystole.

presented in Section 13.5. Many national and international laws have been put in place as safeguards, and an IRB will want to see evidence of the following in your protocol:

- communicating with subjects about what to expect,
- communicating to subjects their rights, and
- providing subjects as much autonomy as possible.

Breakout Box 10.8 **Sample Email to Patients From Their Healthcare Provider**

Name/Address/Date

Dear [Patient],

You may be interested in participating in a research study entitled "Patient Survey to .... " under my direction. The purpose of the research is to study factors that affect to the usage patterns of *name of product*.

I am contacting you on behalf of a study team from *name of university* that is interested in studying the *name of product* with the ultimate goal of making it better for future patients. One team member would like to ask you a series of questions about the *name of product* in a phone call or virtual meeting that will last approximately 10 minutes.

If you wish to participate in the study, please reply to this email with a copy to *email address of study team member*. Your decision to participate or not participate will not affect the medical care you receive. If you do not wish to participate, please ignore this email. There is no need to opt out.

Thank you for your consideration.

Sincerely,

Name of Provider, degree

Title

To demonstrate how these concepts can be applied, Breakout Box 10.7 provides a portion of an IRB proposal to test a self-contained contraction monitor intended for pregnant women for use in resource-poor settings.

10.5.2.4 Inclusion Criteria and Recruitment of Participants

An important aspect of protecting human subjects is to ensure that the appropriate population is being studied. Most medical devices are intended for a specific target population. Typically, a clinical study would target subjects that are representative of that population. For example, the target population for the contraction monitor would only be pregnant patients being evaluated for labor contractions (this is called antenatal testing). For practical reasons, it cannot include patients in labor. An IRB will assess whether or not the study inclusion criteria match the study objective.

Once the target population has been clarified, a recruitment plan is needed. The IRB wants to know your recruitment strategies before they are implemented. There are many possible ways to reach a population, ranging from emails and posters to physician recommendations. For example, in the contraction monitor study, recruitment of test subjects could take place when patients register at the fetal assessment center for contraction monitoring. However, recruitment methods cannot violate a participant's right to privacy nor interfere with their healthcare treatment. An IRB will want to see that you are using an appropriate channel to reach your target population.

Recall from the introduction to Section 10.5 that you cannot directly contact patients without permission. The initial contact must be made by the treating healthcare provider. Direct contact with patients is possible once permission is granted. For example, you may interview patients in a waiting room after they have registered and given permission to partake in a design or research study.

Email has become a common way of recruiting subjects. Another way to contact a patient is to create a draft of an email that would then be sent from the provider to patients under their care that fit the study recruitment plan. A sample email is presented in Breakout Box 10.8, with key terms italicized.

As part of your recruitment plan, an IRB will also want evidence that you can recruit enough subjects in the target population such that statistically significant results can be achieved. As the sample size depends on the nature of the study, advice on determining sample size is beyond the scope of this chapter. Formulae can be found on reliable websites; you may also want to consult with a biostatistician.

Breakout Box 10.9 Informed Consent Authorization

1. Why is this research being done?

   To assess the performance of an investigational contraction monitor, called TOCO, that has been developed to monitor uterine contractions for possible use in low-resource settings around the world. The contraction monitor system that is currently used at this hospital is too expensive and complicated for use in many regions of the world. Therefore, we have developed a self-contained contraction monitor. We hope that information from this study will help to develop a system that can make safe and effective labor monitoring accessible in low-resource settings around the world.

   The use of our self-contained uterine contraction monitor in this research study is investigational. The word "investigational" means that the monitor is not approved for marketing by the Food and Drug Administration (FDA) and is still being tested in research studies.

2. What will happen if you join this study?

   If you agree to be in this study, we will ask you to do the following two things:
   - Allow a healthcare provider to apply the TOCO monitor to your skin around your uterus for a portion of your fetal assessment (up to 1/2 hour). The study monitor will be attached to the same belt that is used to attach the current standard of care contraction monitoring device. The study monitor will record the frequency and duration of your contractions during that time. The accuracy of the readings will be compared to the output of the current standard of care contraction monitor normally used.
   - Answer a few brief questions about your experience while wearing the device.
   - Your participation in this study will not affect your standard clinical care.

3. What are your options if you do not want to be in the study?

   You do not have to join this study. If you do not join, your care will not be affected.

4. Can you leave the study early?
   - You can agree to be in the study now and change your mind later.
   - If you wish to stop, simply notify the study team member or clinician with whom you are in contact.
   - Leaving this study early will have no effect on your regular medical care.

   If you leave the study early, the hospital may use or release your anonymized health information that it has already collected if the information is needed for this study or any follow-up.

   Patient Signature   Date

10.5.2.5 Free and Informed Consent

Before participating in a study, subjects must provide their *free and informed consent*. The term "free" in this context means voluntary and without coercion. Examples of coercion include offering money or food to someone living in poverty, a reduced sentence to a prison inmate, or money to a poor college student, in exchange for participation in a study. The desire for food, money, or a reduced prison sentence may influence a potential study participant's willingness to participate in a study, causing them to ignore the potential risks of participation and focus on the financial or other benefits instead. The term "informed" means that the subject understands what they will experience, their rights as human subjects, and any potential benefits or harms that may come from participating. These are outlined in writing and reviewed with subjects prior to their giving written consent. Consent is indicated by a signature and date; however, in some cases (e.g., a phone interview) consent may be given orally and documented in writing by a member of the study team.

Breakout Box 10.9 contains an example of a partial consent form for the contraction monitor study. Three points are notable. First, unlike the technical tone used in the ACUC or IRB proposal, the consent form has been simplified, both in detail and in language. The IRB will in fact want to see that the consent materials are appropriate for the target audience. For adults, this is middle-school-level language. Other subjects (e.g., children, mentally disabled patients) require different accommodations. Second, consent forms should emphasize that the study is voluntary and that subjects can choose not to participate or withdraw from the study at any point without penalty. Third, to avoid confusion, second-person language is required. Specifically, instead of language such as "the research subject will perform," use "you will perform." The consent form in Breakout Box 10.9 was drafted by a student design team in conjunction with their medical advisor. Some IRB websites will also contain examples, templates, and guidelines to help you design a consent form that is acceptable to your particular institution.

10.5.2.6 Subject Privacy

An important consideration in every medical procedure, including human subject research, is the privacy of the patient. This includes all data associated with the subject's medical record. In most situations, a medical record belongs to the patient's healthcare facility, even if it is a private medical office. The record is only accessible by treating providers and to the patient's representatives when requested. Without explicit permission, nontreating providers or family members are not permitted to discuss a person's medical records with anyone, even with insurance companies. Laws, such as the Health Insurance Portability and Accountability Act (HIPAA) in the United States, which is discussed in detail in Chapter 13, were created to protect confidential patient information. Violations of patient privacy by an individual healthcare provider can be grounds for termination and legal action. Institutional violations can result in severe fines and the revoking of state and federal licenses. For this reason, protecting patient information is a high priority for hospital systems. Some healthcare systems have additional staff who monitor clinical studies for potential HIPAA violations.

Some studies on human subjects require access to patient data or may result in data that will be included in their medical records. For example, consider a validation study that aims to have patients fill out a survey in a cardiology waiting room. Before such a study can be conducted, you must gain permission from the clinic and obtain permission from each individual patient. You would only be able to survey those patients who answer affirmatively.

Regarding protection of subject privacy, an IRB looks for a distinct section of a proposal. explaining how subject privacy will be maintained. At a minimum, most IRB applications will make clear that the results of the study will only be shared outside of the study team in aggregate form—no individual data will be reported. As a general rule, studies should be designed so that only the personnel who need access to the data are allowed to use it.

One common technique to protect the privacy of a test subject is to separate identifying information (e.g., name, address, medical record number) from the rest of the study and assign each participant a unique identifier. All data collected is then attached to this identifier rather than a name. In this way, it is possible for the study to move forward without anyone being able to match healthcare data to a specific individual. In addition, the IRB will want to know how data will be secured (often through encryption and password protection on local servers) and assurance that it will be destroyed after the study is over (often by a specific date).

The IRB may require that a separate privacy form be signed by participants as part of Informed Consent. An example of one can be found in Breakout Box 10.10 this was also prepared by the student team in collaboration with its medical advisor. As with the informed consent template, you should check the website of your local IRB for possible required templates.

Breakout Box 10.10 Example Subject Privacy Form

We have rules to protect information about you. In the United States, the federal *Standards for Privacy of Individually Identifiable Health Information* (known as the Privacy Rule) protect your privacy. By signing this form, you provide your permission, called your "authorization," for the use and disclosure of information protected by the Privacy Rule to members of our research team.

The research team working on the study will collect information about you. This includes things learned from the procedures described in the consent form (Breakout Box 10.9). They may also collect other information including your name, date of birth, and information from your medical records. This could include information about HIV and genetic testing, treatment for drug or alcohol abuse, or mental health problems. For research, all data collected will be anonymously collected; your medical information will be given a code name.

Only a few members of the research team may know your identity and that you are in the research study.

People outside of the hospital may need to see or receive your information for this study. Examples include government agencies (such as the Food and Drug Administration), external members of the Institutional Review Board, and safety monitors.

We cannot conduct this study without your authorization to use and give out your information. You do not have to give us this authorization. If you do not, then you may not join this study.

We will use and disclose your information only as described in this form and in our Notice of Privacy Practices; however, people outside the hospital who receive your information may not be covered by this promise or by the federal Privacy Rule. We try to make sure that everyone who needs to see your information keeps it confidential, but we cannot guarantee that your information will not be redisclosed.

The use and disclosure of your information have no time limit. You may revoke (cancel) your permission to use and disclose your information at any time by notifying the Principal Investigator (PI) of this study by phone or in writing. If you contact the PI by phone, you must follow up with a written request that includes the study number and your contact information. The PI's name, address, phone, and fax information are on page one of this consent form.

If you do cancel your authorization to use and disclose your information, your part in this study will end and no further information about you will be collected. Your revocation (cancellation) would not affect information already collected in the study, or information we disclosed before you wrote to the PI to cancel your authorization.

Signature   Date

10.5.3 Clinical Trials

A clinical trial, also known as an investigational study, is a study in which test subjects are assigned to different groups, and each group receives one or more interventions/treatments—or no intervention at all—so that researchers can evaluate the safety and effectiveness of the intervention(s) on human health outcomes. A successful clinical trial is the last and most rigorous test of patient safety and effectiveness of a new or updated medical device or drug. For most medical devices, clinical trials are a required step on the way to becoming a commercialized product.

Sponsored clinical trials are legally required to be registered on a public website (clinicaltrials.gov), which is a database of privately and publicly funded clinical studies conducted around the world (220 countries) and is maintained by the US National Library of Medicine. The website is also a resource for finding studies in a medical specialty or specific area, including publications, and study results (there are over 2700 completed medical device clinical trials with study results available, and typically 2500 ongoing medical device trials). Depending on the technology being studied, clinical trials may be inpatient or outpatient investigations. In addition to their use for regulatory approval, the data generated from clinical trials may be used in other ways. For example, data can be used in marketing materials or to convince hospitals and health insurance providers of the relative benefits and lower risks of new technology compared to existing solutions.

There are two primary types of clinical trials; one for drugs and the other for devices. The remainder of this section will discuss these two pathways.

Significant risk devices require a clinical trial for FDA approval. Significant risk devices include all implants, invasive devices, devices that introduce energy into the patient, technology that supports or sustains human life, and devices having substantial importance in diagnosing or treating disease. To be able to conduct clinical trials for significant risk devices, an investigator must apply for an investigational device exemption (IDE). An IDE is a license to allow a manufacturer to ship unapproved medical devices across state lines for the purposes of a clinical study. If the FDA approves the device for initial testing, an IDE should be granted, and the clinical trial may begin.

Most devices are not required to undergo a clinical trial because they are considered a nonsignificant risk (FDA Class I and most Class II devices). There are three levels of clinical trials for devices, each with a successive increase in the number of subjects and level of risk. The first level is a *pilot or feasibility study*, which typically consists of 10 to 40 patients, where the focus is on safety and whether the potential benefits or value of the data justifies the risk. Statistics are not relied upon and endpoints—an FDA-defined term indicating when to finish the study (e.g., after n patients)—are not as important. Results are sometimes used to answer basic research questions. Preliminary safety and efficacy data generated from a pilot study is often used to justify the expense of moving to the second-level clinical trial.

The second level clinical trial is a *pivotal study* that determines if a medical device or technology could be cleared or approved for commercialization. It generally involves 150 to 300 patients and requires defined endpoints and statistical analyses. A clinical trial endpoint may be either an objective measure (e.g., volume oxygen [VO_2] max, ventilatory threshold [VT1], hospitalization rate) or a subjective measure (quality of life, pain level, hemorrhage volume estimation), but statistical analysis is required in both cases. Regulatory approval for high-risk devices is based on a risk-benefit analysis, which is assessed in part by the outcomes of primary safety and effectiveness endpoints. The pivotal study results should convince the regulatory body that the new device presents a "reasonable assurance of safety and effectiveness."

The third level device clinical trial is a *post-approval study*. This study is conducted after the product is on the market and involves many patients. Its main goals are to gain fuller understanding of the long-term benefits and to learn of possible adverse events. Section 12.9 delves into post-approval study more deeply.

Clinical trials for drugs follow a different pathway but are relevant to medical device design because some devices include the release of a drug (e.g., a drug eluting stent), Drug clinical trials are divided into numbered phases, each with larger and more diverse populations and increasing levels of rigor. In the United States, there are up to four phases. The first three must be completed successfully for the FDA to approve a drug for sale. Below is a summary of the clinical trial phases for a drug in the United States:

Phase I Trials are generally conducted on 20 to 100 subjects and take a few months to conduct, depending upon the rate of test subject enrollments. The primary aim is to determine dosing and to determine therapeutic thresholds and toxicity thresholds.

Phase II Trials are generally conducted on 50 to 300 subjects and may take months or years to complete. Whereas Phase I focused on dosing and safety, Phase II trials focus on confirming dosing, efficacy and ruling out longer-term adverse reactions. Most often subjects are chosen randomly to diversify the test population. Such randomization is critical to limit bias.

Phase III Trials include hundreds to thousands of patients, usually at multiple sites, and would follow much of the same protocols as a Phase II trial. The primary goal is to identify drug-drug interactions and possible minor adverse events. In many cases, these trials would be randomized and double-blinded, meaning that neither subjects nor healthcare providers would know who is receiving which treatment. Once these studies are completed, the data are to be submitted to the FDA for evaluation and approval.

Phase IV Trials are part of post market surveillance and occur after a drug is commercially available and being used in medical practice. It typically consists of follow-up medical exams, tests, or questionnaires. The aim is to collect long-term data and to track adverse events; if frequent and severe enough, these may trigger a recall, as presented in Section 12.9.

There are some similarities and differences between device and drug clinical trials. Each has a control group and treatment to compare against. Some differences include:

- Most drug trials are randomized and blinded. Many device trials are difficult to blind and there is no device equivalent to a placebo drug; performing a sham surgery on a patient who needs a knee implant is unethical.
- Most drug trials involve more than 1000 patients; most device trials involve fewer than 400 patients.
- Many device trials require some training of healthcare providers; no such specialized training is required for providers who administer drugs for a drug trial.
- Imaging may be used in device trials to evaluate device performance *in vivo*. This would not normally be used for drug trials.
- Healthcare provider technique can play a significant role in device trials.
- Medical devices and drug trial endpoints are primarily interested in outcomes. Specifically, there should be no (or clinically insignificant) adverse reactions for safety, and specific clinical benefit for effectiveness.

10.6 Validating Your Device

Design validation involves determining if your target users or customers agree that your solution meets their needs, thereby providing value over existing solutions. Validation asks the question, "did we make the right product?" It is based on the opinions of the users and customers regarding your device and can only be achieved through human testing. For example, validation of a reduced-size surgical instrument would involve asking surgeons to use the tool, evaluate it, and say whether or not the reduced size makes the surgical instrument easier to use during surgery. This is contrasted with a verification test which would test the specification to reduce the size of the instrument by some percentage.

In this section we discuss some of the methods for designing and conducting validation tests. As validation tests are meant to gather information on human perceptions, many of the methods build upon qualitative research methods that were developed by social scientists and introduced in Sections 3.6 and 3.7. Despite the term "qualitative," these are rigorous testing procedures that should match the careful design and analysis of the engineering tests presented in Chapter 9.

10.6.1 Internal and External Validation of a Medical Device

Internal validation involves asking advisors, mentors, and perhaps a few users or customers their opinion of your design and prototypes. The aim is to help you navigate the design process and give you that you are moving in the right direction. You have likely been engaging in this form of internal validation throughout your design process, as you sought out instruction and constructive criticism from others. For example, a design review is a type of internal validation. Although internal validation involves humans, it does not require IRB approval. A design review in academia would be considered part of the educational mission and can proceed without IRB approval.

External validation involves users and customers who typically are outside of your institution. Regulatory and reimbursement groups will want to see some evidence of external validation. In the context of a design process, external validation is often a last step before the Design Transfer, and therefore requires a final prototype where the design has been frozen. Design changes after this point would only be due to feedback from validation activities or field complaints. In industry, external validation of a device is often used to create marketing materials (e.g., "4 out of 5 physicians surveyed said Device X was preferred over Device Y") including testimonials from physicians or users. Furthermore, management may want to see some external validation of prototypes before moving a product to the next stage of the stage-gate process.

10.6.2 Choosing Validation Methods

The essence of external validation is to put a product into the hands of a customer or user with the intent of getting their honest feedback. The most common forms of external validation are surveys, interviews, focus groups, and observations. Much of the guidance presented in Section 3.7 remains relevant:

- Think carefully about the types of questions you ask to be sure you avoid biasing the responses you receive.
- Organize questions in a way that makes sense to your subjects. These may be in order of importance, from broad in scope to narrower, or by categories. The seven questions in Chapter 3, Table 3.3 start broad and then become specific.
- Medical professionals are busy. Limit the length of a survey or interview to a maximum of 10 minutes unless there are special arrangements for a longer period. Table 3.3 would normally take 10 to 15 minutes to complete.
- Be sure you are asking the right population. If a device is most likely to be used by surgical nurses, then that group of users should be asked. It is not uncommon for a device to be designed with input from physicians when they have little interaction with the device. It is possible that patients would be the best population for validating a device.
- You may mix multiple-choice and single-choice (yes/no, slider, and Likert) with open-ended (free response) questions.
- Consider the best way to reach your target population. Options may be in-person (e.g., paper-based, oral interview), by phone, or by virtual online tools (e.g., Google forms, Survey Monkey, Qualtrics). These methods can also be used to capture raw data in a common format.
- Sometimes you will need to use a mixed-methods approach (combining qualitative and quantitative methods) because no one population or methodology can get you all the information that you need.

There are some critical differences between where you were in your design process in Chapter 3, and when you have a nearly final prototype ready for validation. First, the types of questions you ask to validate a device will be very different from the questions you asked to better understand a clinical problem.

Second, whereas IRB approval may not have been necessary to gather information regarding a problem, external validation of a device is a form of human subject testing and will usually require IRB approval. You will need to create a proposal, as previously presented, that will detail inclusion criteria, the number of participants needed to reach statistical significance, a recruitment plan, data analysis methods, informed consent, and assurances of privacy. In addition, you would need to provide the questions that will be part of the study.

Third, most medical devices must be used to be validated. Determining how a subject might use your device in a way that will allow give you meaningful feedback, yet still be safe, can be challenging. For example, consider how you would determine a surgeon's perspective on a new implantable device. You may have asked a surgeon to inspect your prototypes during the design process, but one validation test would require many surgeons to use the device in surgery as part of a clinical test. There are other options for validation testing, such as simulating surgical implantation into ballistic gel (as presented in Section 9.6) or using an animal model (as done for the bowel packer presented in Breakout Boxes 10.5 and 10.6).

Fourth, you may have designed your device for a particular user. However, there are many stakeholders and sometimes multiple users for the same device. For example, a comprehensive validation for an implant might require input from surgeons, surgical nurses, patients, and perhaps even the patient's postoperative caretakers. All would have different perspectives on the value of the device.

Fifth, the level of rigor often increases for external validation when the design and analysis methods of social scientists are used. We introduce the ideas of validity, reliability, and coding in Section 10.6.4.

10.6.3 Design Validation Questions

The basis of most design validation studies whether they involve surveys, interviews, or focus groups, involves asking users questions. Strictly speaking, validation of a design could be as simple as asking each user (yes or no) if your device solves their problem. However, most robust validation studies will go well beyond this single question.

The exact scope and phrasing of your questions will vary depending upon the study participants, and which aspects of a device you aim to validate. You will need to develop your own questions. Common areas to ask about include:

- a critique of the usability of existing solutions
- a comparison of new technology to existing devices
- changes to actions or processes as a result of your device
- potential drawbacks (or least favorite aspects) of your solution
- potential strengths (or most favorite aspects) of your solution to others
- safety concerns they may have encountered or foresee in the future
- barriers to other users/customers adopting your new solution
- value that might be added in the future
- interest in evaluating future iterations of your solution
- whether the device solves the problem it was designed to solve

These are offered only as a guide, but you can use them to generate your own questions. For example, you may make the second point more specific by asking, "how does the liver retraction device compare with the current method of having a surgical nurse hold back the liver manually?" Likewise, you might also combine the ideas above into one question. For example, you may combine points 5 and 6 to ask, "what safety concerns, if any, you anticipate due to the increased speed of closing a wound using our device?"

10.6.4 Validity, Reliability, and Coding

The open-ended questions that make up most surveys, interviews, and focus groups often generate qualitative data in the form of textual answers from many individuals. Social scientists who work with these types of data have developed a range of qualitative research methods and analysis techniques. Three important aspects of conducting a rigorous qualitative study include validity, reliability, and coding. They are presented to serve as a starting point in case you are asked to conduct a robust validation study. Consult the resources listed at the end of this chapter for a more comprehensive treatment.

In the context of social science research, validity is how accurately a method measures some desired information. This includes the entire methodology and can be broadly based upon theoretical constructs, statistical methods, or expert judgment. In the context of open-ended questions in a survey, validity most often means gaining some assurances that when participants read a question that they interpret it in the way you intended. These assurances are gained through validation of the questions, typically by asking a representative population how they interpret the questions. In this case, your representative population is serving as experts. Here the term validation is used in a similar way to engineering validation, with the only exception being that human subjects are asked about their interpretation of a question rather than their opinion of a device. These representatives can help you gain confidence that the questions are asking what you intended to ask. Furthermore, these experts may be able to suggest improvements to the questions, directions and flow, or the length of time needed to complete the survey or interview.

Reliability is how consistent and repeatable the results are when the same methods are repeated. For the purposes of question-based studies, reliability is based on the execution of the survey and on the analysis of the data. A reliable survey (or interview or focus group) is one that accurately captures the views of the target population in such a way that if a different group of people from the target population were sampled, the results would not change significantly. It may be tempting to simply increase the population size. However, social scientists have a few alternatives to increase reliability without increasing the sample size. This might be a test-retest protocol in which the participants answer the same questions at different times so their answers (before and after) can be evaluated for consistency. Another technique is to ask for similar information but in two questions that are phrased differently.

The second aspect of reliability is the consistency of the analysis methods. For quantitative data this is simple, as applying the same statistical methods to the same data would yield the same results. There is a degree of interpretation in qualitative data. To systematize the analysis of qualitative data, social scientists have created a variety of tools. The most commonly used tool with regard to textual data is known as *coding*. This is different than computer coding and involves developing themes that characterize the data. For example, if several study participants mention topics related to "timing," this might become a theme. When coding answers, one would then note every instance where "timing" is referenced. This may expand beyond the exact words, for example, if an answer includes the comment, "…it took forever" or "… it took way too long." Such a statement would be assigned to the theme "timing" even though the actual word "time" was not used. This is an example of concept coding; however, a similar idea can also be extended to emotional expressions. Themes could be determined before data is gathered (often using a theoretical construct). In the context of device validation, it is common to develop the theme after data collection to get a sense of the answer themes.

A powerful way to gain reliability is through a method known as interrater reliability. Different coders read the same data with the same themes and independently assign their own scoring. The coders then compare how they scored the text to determine where they agree and where they do not. Frequent agreement among scores (which can be quantified) would be considered a high degree of interrater reliability. The more coders involved, and the more agreement between them, the higher the reliability.

10.6.5 A Practical Approach to Validation

While amassing data for validation can be fascinating, it should be limited. If taken too far it can distract you from other aspects of the design process. Unless you already have someone on your team who is well-versed in qualitative methods, you will likely perform a less rigorous external validation of your device. In the context of an academic design project, you might only ask a few users or customers, or only ask for numerical responses. You may not need to perform a full coding of the responses from textual prompts. Below is a practical example of validation in the context of a design project.

Consider a project where a video chat feature was designed and implemented for a patient to show a wound or skin condition to a healthcare provider at a distant location to solicit an assessment of whether or not to seek treatment. The design team developed a smartphone app that included the video chat feature and wanted to validate it by asking wound-care patients about their comfort level in using the technology. After obtaining IRB approval to conduct the survey, team members interviewed 35 patients waiting in a wound care clinic. The main question was, "how comfortable would you feel using this app?" The answers were on a three-point scale (not comfortable, indifferent, or comfortable) along with a representative textual response associated with each scale response. The results are summarized in Table 10.3.

The textual answers from individuals varied. Rather than coding every answer, the team simply found representative text to illustrate the three categories of uncomfortable, indifferent, and comfortable. The results after 35 patients were encouraging (71% rating as comfortable) and these data could be used as validation of the video chat feature of the app.

Table 10.3 Survey results for comfort level using video chat feature of an app.

Comfort level	n	%	Representative text
Uncomfortable	4	11	Do not really want to use. Perhaps only with a family member.
Indifferent	6	18	Would use, but not at work.
Comfortable	25	71*	I love to video chat, even with healthcare providers I do not know (e.g., nurses).

*$p<.05$, t-test

Breakout Box 10.11 Example Test Data From Bowel Packing Study

For the bowel packing problem (see Section 10.4.4) to have statistical significance, subsequent design teams collected and measured frequency and density of adhesions in 30 rabbits; one-third treated with our technology, one-third treated with standard of care technology (OR towels and sponges), and one-third treated with no technology (sham surgery). The team used the nonparametric Mann-Whitney-Wilcoxon test (independent measures, small sample) to compare the three groups; $p<0.05$ was considered significant. Results are presented in Figures 10.7 and 10.8.

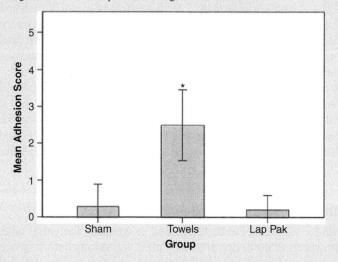

FIGURE 10.7

Frequency of adhesions per rabbit for the sham surgery, OR towel, and silicon elastomer prototype. Adhesions were manually counted after 14 days. The towel frequency was significantly greater than the other two surgeries, $p<0.05$.

(From Liu, B. G., Ruben, D. S., Renz, W., Santillan, A., Kubisen, S. J., & Harmon, J. W. (2011). Comparison of peritoneal adhesion formation in bowel retraction by cotton towels versus the silicone lap pak device in a rabbit model. Eplasty, 11, e42. Used with permission.)

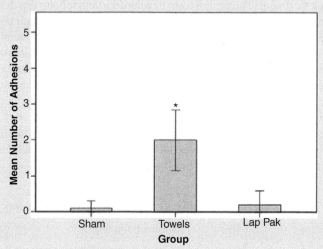

FIGURE 10.8

Adhesion grades. Mean adhesions score per rabbit for the three surgeries. Adhesion scores were determined by frequency, tenacity, density, and ease of dissection. The frequency scale is 0 to 4, where 0 means no adhesions and 4 means more than 3 adhesions, tenacity >1000, opaque and dense, and sharp dissection needed. As in Figure 10.7, the towel score was significantly greater (it had more severe adhesions) than the other two surgeries, $p<0.05$.

(From Liu, B. G., Ruben, D. S., Renz, W., Santillan, A., Kubisen, S. J., & Harmon, J. W. (2011). Comparison of peritoneal adhesion formation in bowel retraction by cotton towels versus the silicone lap pak device in a rabbit model. Eplasty, 11, e42. Used with permission.)

10.7 Documenting Living Systems Testing and Results

As in other phases of the design process, all protocols, analyses, results, conclusions, and interpretations should be documented. The same guidance given for documenting testing presented in Chapter 9 applies to living systems testing. Living system documentation includes additional considerations such as approved proposals explaining recruiting subjects, maintaining the privacy of individuals, documenting free and informed consent, as well as where and how data will be stored. Experimental results can be represented in tabular form (e.g., Table 10.3) or graphically (e.g., Breakout Box 10.11)

You may or may not be able to draw statistically significant conclusions from a small sample size. Even if your results show promise, you should acknowledge the limitations of small sample statistical significance in your interpretation. In industry, portions of a DHF would be used in submitting regulatory applications. Results of testing in living systems will also be of interest to marketing and sales personnel. Claiming that a specification (e.g., decreasing hemorrhage by 30%) was met in a clinical study is far more convincing than claiming the same result in a computer simulation or bench test.

We conclude this chapter with Breakout Box 10.11 (above) showing the results of the bowel packing device study described in Breakout Boxes 10.5 and 10.6. These are the type of results that should go in a DHF, along with the experimental protocol for a living systems test. In this case, copies of the approved ACUC application from Breakout Boxes 10.5 and 10.6 should also appear in a DHF.

Key Points

- Living systems are complex in ways that nonliving systems are not. New and unexpected reactions may occur when a new medical technology interacts with a living system.
- Living systems tests are broadly classified as in vitro (cells and tissues) and in vivo (animal and human) tests.
- Testing in living systems typically occurs after bench testing and aims to further demonstrate the safety and efficacy of a device in animals and in humans under physiologic conditions.
- Biocompatibility is a measure of the effects of a material or a device on the living system. There are up to 13 biological effects outlined in ISO 10993 that need to be considered when testing for biocompatibility. A material that is biocompatible in one medical application may not be compatible in a different application.
- Biodurability is a measure of the effects of the living system on the device material and its performance.
- Testing with animals must be approved by the Institutional Animal Care and Use Committee. Testing in humans must first be approved by the Institutional Review Board.
- Clinical trials for significant risk medical devices are required by law to demonstrate safety and efficacy, only after which can a new device become a salable product.
- Planning, gaining approval, and conducting living systems tests usually take months. Design teams that plan a living system test must factor this time into their workplan early in the project.

Project Related Exercises

1. Navigate to the Office of Human Subjects (OHRP) website and click on the Education and Outreach section. Navigate to the Online Education link and watch the less than 13-minute mini-tutorial on IRB Review Criteria. What are the top three concepts learned as a result of watching the video? For exercises 2 to 4, you may need an institutional license to sign up for the CITI Program courses.
2. Take the CITI (Collaborative Institutional Training Initiative) Human Subject Research course on Biomedical Basics at www.citiprogram.org. You will receive a certificate that you can add to your resume. It is also required by some IRBs to conduct human studies.
3. Take the CITI Human Subject Research course on Socio-Behavioral-Educational Basics at www.citiprogram.org. You will receive a certificate that you can add to your resume. It is also required by some IRBs to conduct human studies.
4. Navigate to the CITI website at www.citiprogram.org and select a course of your choosing. Some courses relevant to biomedical designers would be Biosafety, Responsibilities and Duties for Researchers, and Good Laboratory Practice. After completing the course remember to add it to your resume.
5. Search the clinical trials database at clinicaltrials.gov for two devices that are closely related to your project. Extract information that may be helpful in designing your own studies and prepare a summary of your findings.
6. Choose a project specification that can only be tested in an animal model. Outline an objective, protocol, and any data analysis methods. Justify your choice of animal model and number of animals needed. Use your institution's ACUC resources as a guide.
7. Choose a project specification that can only be rigorously tested in a human. Outline an objective, protocol, and any data analysis methods. Justify the number of test subjects needed. Use your institution's IRB resources as a guide.

General Exercises

1. Poly Trimethylene Carbonate (PMC) polymers are biocompatible materials used to make bioabsorbable staples. If the desired absorption is within 1 month, propose an *in vivo* and an *in vitro* test protocol, along with the expected results, to demonstrate efficacy.
2. The hip implant presented Breakout Box 10.1 became less biodurable over time due to wear. This degradation caused trace amounts of cobalt to be released into the bloodstream. The clinical effect was metallosis and systemic clinical complications (e.g., hearing loss, cardiac myopathy, depression) in some patients. Your project is to design an improved elbow implant that involves chromium cobalt contact. Discuss if biodurability is a concern. If it is, how would you demonstrate that the elbow implant will continue to perform as expected over time? If not, how do you demonstrate the durability loss over time is clinically insignificant?
3. Modular hip implants allow for a separate femoral head to be placed onto a femoral stem that is implanted into the femur. The head articulates with an implanted acetabular cup, lined with a polyethylene bearing, to create a low friction bearing surface. As the head is loaded during the gait cycle, small-scale relative micromotion (fretting) can occur, causing damage to the surface of the titanium alloy neck and cobalt alloy head. When in the presence of saline, this often leads to corrosion. How would you evaluate corrosion in test animals and in human clinical study subjects?

4. In a BME instrumentation class, you are asked to develop technology to make a "smart" cane to assist a visually impaired person (i.e., a cane that would audibly and/or tactilely warn a visually impaired user of an off-the-ground obstruction, such as an overhanging tree branch). The final test is to walk inside, downstairs, and outside while blindfolded to demonstrate that the technology is working as expected.
   a. Is IRB approval needed for this test?
   b. Do participants need to be consented? If so, write the consent form.
5. Your team is working on a project to develop a walker for Medical Intensive Care Unit (MICU) patients. Having finished bench testing, you now wish to test the device on MICU patients, and the team submits an IRB application.
   a. Research the primary reason for having MICU patients become peripatetic while on an ICU unit. How many MICU healthcare workers are required to accompany the patient on a walk?
   b. Draft a one-page outline of a protocol for the IRB application.
   c. Draft parts of the consent form. Specifically, answer the questions below in language that would be understandable to a middle-schooler.
   What are the risks and benefits to the test subjects of this study?
   Can one leave the study early?
   Will participants derive benefit?
6. Your team has developed technology to help obstetricians diagnose preterm labor early enough to delay labor onset. The current standard of care is that once a pregnant patient goes into preterm labor it cannot be stopped for more than a few days. You wish to demonstrate that the technology can be useful. What animal model would you use? What test would you plan to demonstrate the efficacy of your device?

References and Resources

Australian Bureau of Statistics. (n.d.). Sample size calculator. Retrieved November 3, 2021, from https://www.abs.gov.au/websitedbs/d3310114.nsf/home/sample+size+calculator.

Ayhan, E. R., Yönetken, A., & Kuloğlu, O. A. (2018). Teeth of implant production and characteristics by using Ti-Cr-Co powders. *International Journal of Scientific Engineering and Research*, 9(8), 60–65.

Balnaves, M., & Caputi, P. (2001). *Introduction to qualitative research methods: An investigative approach*. SAGE Publications.

Bren, L. (2001). Frances Oldham Kelsey. FDA medical reviewer leaves her mark on history. *FDA Consumer*, 35(2), 24–29.

Chittester, B. (2020). How to conduct a well-controlled clinical trial. *MasterControl*. https://www.mastercontrol.com/gxp-lifeline/medical-device-clinical-trials-how-do-they-compare-shiftenterwith-drug-trials-/.

Committee for the Update of the Guide for the Care and Use of Laboratory Animals, Institute for Laboratory Animal Research, Division on Earth and Life Studies, & National Research Council. (2010). *Guide for the care and use of laboratory animals* (8th ed.). National Academies Press.

Davidson, M. K., Lindsey, J. R., & Davis, J. K. (1987). Requirements and selection of an animal model. *Israel Journal of Medical Sciences*, 23(6), 551–555.

de Aguilar-Nascimento, J. E. (2005). Fundamental steps in experimental design for animal studies. *Acta Cirurgica Brasileira*, 20(1), 2–8. https://doi.org/10.1590/s0102-86502005000100002.

Drummond, J., Tran, P., & Fary, C. (2015). Metal-on-metal hip arthroplasty: A review of adverse reactions and patient management. *Journal of Functional Biomaterials*, *6*(3), 486–499.

Fishman, G.I., Chugh, S.S., DiMarco, J.P., Albert, C.M., Anderson, M.E., & Bonow, R.O., et al. (2010). Sudden cardiac death prediction and prevention: Report from a national heart, lung, and blood Institute and heart rhythm society workshop. *Circulation, 30*;122(22):2335–2348.

Fujita, S., Pitaktong, I., Steller, G. V., Dadfar, V., Huang, Q., Banerjee, S., et al. (2018). Pilot study of a smartphone application designed to socially motivate cardiovascular disease patients to improve medication adherence. *mHealth*, *4*, 1.

Gessner, B. D., Steck, T., Woelber, E., & Tower, S. S. (2019). A systematic review of systemic cobaltism after wear or corrosion of chrome-cobalt hip implants. *Journal of Patient Safety*, *15*(2), 97–104.

Geyer, M. A., & Markou, A. (1995). Animal models of psychiatric disorders. In F. E. Bloom, & D. Kupfer (Eds.), *Psychopharmacology: The fourth generation of progress* (pp. 787–798). Raven Press.

Hillisch, A., Pineda, L. F., & Hilgenfeld, R. (2004). Utility of homology models in the drug discovery process. *Drug Discovery Today*, *9*(15), 659–669.

Ianetti, T., Morales-Medina, J. C., Merighi, A., Boarino, V., Laurino, C., Vadala, M., et al. (2018). A hyaluronic acid- and chondroitin sulfate-based medical device improves gastritis pain, discomfort, and endoscopic features. *Drug Delivery and Translational Research*, *8*(5), 994–999.

Kadam, P., & Bhalerao, S. (2010). Sample size calculation. *International Journal of Ayurveda Research*, *1*(1), 55–57. https://doi.org/10.4103/0974-7788.59946.

Knight, S. R., Aujla, R., & Biswas, S. P. (2011). Total hip arthroplasty—Over 100 years of operative history. *Orthopedic Reviews*, *3*(2), e16.

Kumar, V., Dhabalia, J. V., Nelivigi, G. G., Punia, M. S., & Suryavanshi, M. (2009). Age, gender, and voided volume dependency of peak urinary flow rate and uroflowmetry nomogram in the Indian population. *Indian Journal of Urology: IJU: Journal of the Urological Society of India*, *25*(4), 461–466.

Liu, B. G., Ruben, D. S., Renz, W., Santillan, A., Kubisen, S. J., & Harmon, J. W. (2011). Comparison of peritoneal adhesion formation in bowel retraction by cotton towels versus the silicone lap pak device in a rabbit model. *Eplasty*, *11*, e42.

MacCracken, C., Dutta, P. K., & Waldman, J. C. (2016). Critical assessment of toxicological effects of ingested nanoparticles. *Environmental Science: Nano*, *3*, 215–282.

Menkinoff, J., Kaneshiro, J., & Pritchard, R. (2017). The common rule, updated. *New England Journal of Medicine*, *376*(7), 613–615.

Ratner, B. D. (2019). Biomaterials: Been there, done that, and evolving into the future. *Annual Review of Biomedical Engineering*, *21*, 171–191.

Ratner, B. D., Hoffman, A. S., Schoen, F. J., & Lemmons, J. E. (2012). *Biomaterials science: An introduction to materials in medicine* (3rd ed.). Academic Press.

Rawal, B. R., Yadav, A., & Pare, V. (2016). Life estimation of knee joint prosthesis by combined effect of fatigue and wear. *Procedia Technology*, *23*, 60–67.

Reeves, T. D., & Marbach-Ad, G. (2016). Contemporary test validity in theory and practice: A primer for discipline-based education researchers. *CBE-Life Sciences Education*, *15*(1), rm1.

Rollin, B. E. (1981). *Animal rights and human morality*. Prometheus Books.

Saldana, J. (2015). The Coding Manual for Qualitative Researchers (3rd ed.). SAGE.

Shi, H., Vorvolakos, K., Dreher, M., Walsh, D., & Duraiswamy, N. (2017). *In vitro evaluation of coating performance of guidewire surrogates*. 2017 Design of Medical Devices Conference.

Society for Biomaterials. (n.d.). (*Giving life to a world of materials: Resource guide*. https://www.biomaterials.org/about/resource-guide.

Sohal, I. S., Cho, Y. K., O'Fallon, K. S., Gaines, P., Demokritou, P., & Bello, D. (2018). Dissolution behavior and biodurability of ingested engineered nanomaterials in the gastrointestinal environment. *ACS Nano*, *12*(8), 8115–8128.

Steens, W., Von Foerster, G., & Katzer, A. (2006). Severe cobalt poisoning with loss of sight after ceramic-metal pairing in a hip—a case report. *Acta Orthopaedica, 77*(5), 830–832.

Storer, R. A. (1995). Standard test method for determination of total knee replacement constraint. *Annual Book of ASTM Standards, 13*(1) 544–550.

Swearengen, J. R. (2018). Choosing the right animal model for infectious disease research. *Animal Models and Experimental Medicine, 1*(2), 100–108.

Trayanova, N. A. (2011). Whole-heart modeling: applications to cardiac electrophysiology and electromechanics. *Circulation Research, 108*(1), 113–128.

Tower, S. S. (2010). Arthroprosthetic cobaltism: Neurological and cardiac manifestations in two patients with metal-on-metal arthroplasty: A case report. *Journal of Bone and Joint Surgery American Volume, 92*(17), 2847–2851.

Tower, S. S., Bridges, R., Cho, C., & Gessner, B. (2019, November 13). Arthroprosthetic cobaltism is common: A screening study of 241 patients with joint replacement with Urine-cobalt and cobaltism-symptom-inventory (CSI). *Food and Drug Administration Expert Panel.* https://www.fda.gov/media/132881/download.

Ulery, B. D., Nair, L. S., & Laurencin, C. T. (2011). Biomedical applications of biodegradable polymers. *Journal of Polymer Science Part B: Polymer Physics, 49*(12), 832–864.

US Food and Drug Administration. (2019). *Concerns about metal-on-metal hip implants.* https://www.fda.gov/medical-devices/metal-metal-hip-implants/concerns-about-metal-metal-hip-implants.

US Food and Drug Administration. (2020). *Use of international standard ISO 10993-1, "biological evaluation of medical devices—Part 1: Evaluation and testing within a risk management process," guidance for industry and food and drug administration staff.* https://www.fda.gov/media/85865/download.

US National Library of Medicine. (n.d.). ClinicalTrials.gov. https://clinicaltrials.gov.

Utembe, W., Potgieter, K., Stafaniak, A. B., & Gulumian, M. (2015). Dissolution and biodurability: Important parameters needed for risk assessment of nanomaterials. *Particle and Fibre Toxicology, 12*(11).

Vanderpool, D. (2012). HIPAA—Should I be worried? *Innovations in Clinical Neuroscience, 9*(11–12), 51–55.

Vigmond, E., Vadakkumpadan, F., Gurev, V., Arevalo, H., Deo, M., Plank, G., et al. (2009). Towards predictive modelling of the electrophysiology of the heart. *Experimental Physiology, 94*(5), 563–577.

Wallin, R. F., & Arscott, E. F. (1998). A practical guide to ISO 10993-5: Cytotoxicity. *Medical Device and Diagnostic Industry, 20*(4), 96–98.

Wang, X. (2013). Overview on biocompatibilities of implantable biomaterials. In R. Lazinica (Ed.), *Advances in biomaterials science and biomedical applications in biomedicine, 27* (pp. 111–155).

CRITICAL LENSES: STANDARDS, REGULATIONS, AND ETHICS

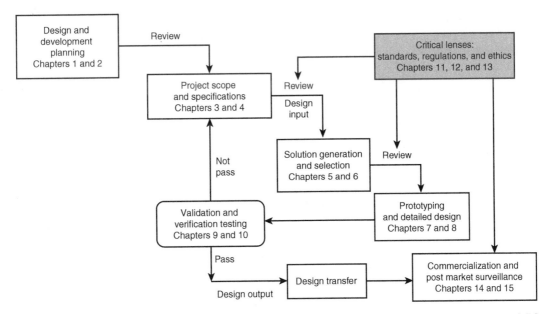

6

In the next three chapters you will learn about three perspectives that are important in all design processes but are especially critical in the design of a medical device. Each of these perspectives present unique requirements that impact the design process.

Chapter 11 describes standards as widely accepted agreements that ensure medical products and processes are safe, effective, and of high quality. There are standards for how to perform tests, determine the interactions between materials and living systems, and guide the entire design process (known as Design Controls). Standards have been mentioned in other chapters, but you will learn how standards are created, accelerate innovation, encourage acceptance of new products, help gain regulatory clearance, and save you time and effort.

Chapter 12 describes the complex regulatory review processes, mandated by law in most countries, for protecting the well-being of the public. Although regulatory review is often viewed as a barrier to commercializing a medical device, it provides a helpful framework for thinking more broadly about the design process. You will learn about device classifications, the clearance and approval processes, proper labeling and warnings, and recalls.

Chapter 13 describes the ethical implications of bringing new healthcare technologies into the world. The focus is on applied ethics as a lens through which you can view design decisions. Case studies are used to illustrate ethical principles and demonstrate how many laws and policies arise due to product failures that resulted from lack of oversight, conflicts of interest, violations of confidentiality, negligence, and even malicious intent. The ultimate aim is for you to develop habits of mind that will allow you recognize, dissect, and navigate ethical dilemmas that will inevitably arise during your career.

By the end of Chapters 11, 12, and 13 you will be able to:

- Create a list of standards that are relevant to your design project and will help you navigate the various phases of your design process.
- Use one or more standards to design tests relevant to design verification.
- Create a plan for both domestic and international regulatory clearance.
- Perform a thorough assessment of ethical implications of your solution and how they might intersect, and potentially conflict with, legal, social, political, and economic systems.

Medical Device Standards and Design Controls

11

You shall have just scales, just weights, just ephat [dry measures] and just hin [liquid measures] ...
— Leviticus 19:36 (Sheman and Zlotowitz, 1993)

Chapter outline

11.1 Introduction

As indicated by the biblical quote, standards have existed since at least ancient times. That standards have played a role in every other chapter and is a clue to the significance of current standards in biomedical engineering design. The general purpose of an internationally or regionally accepted standard is to establish a universally or regionally accepted measurement, material composition, method, procedure, or process. Standards also guide design activities including material selection and composition, the design of experiments, test procedures and acceptance criteria, and packaging and design documentation. In addition, standards can help reduce costs, increase productivity, generate economic growth, improve quality, and promote fair comparisons with competitive products. When a medical device is certified as having met a standard, it accelerates ultimate regulatory approval. Many stakeholders only accept or pay for a medical device if it can be demonstrated that it complies with applicable standards.

The formal definition of a medical device standard is provided by the International Organization for Standardization (ISO) and the International Electrotechnical Commission (IEC):

> A document, established by consensus and approved by a recognized body, that provides, for common and repeated use, rules, guidelines or characteristics for activities or their results, aimed at the achievement of the optimum degree of order in a given context.

In brief, for engineering applications, a standard is a technical guideline to ensure that materials, processes, representations, and services are fit for their purpose. Standards are prominent in medical device design to ensure safety, quality, and effectiveness. In this chapter, we present more about how standards are created, how they are used in making design decisions, and how to control the design process. Breakout Box 11.1 describes one biomedical engineering graduate's perceptions on the importance of standards to academic design students as well as to practicing biomedical engineering designers.

11.2 Need For and Types of Standards

Understanding the need for standards in medical device design is demonstrated well by the following medical example. The medical challenges of administering hypodermic drugs or vaccines or drawing blood (vascular application) are that the procedure (1) must protect the healthcare worker and (2) must be leak-proof to avoid incorrect dosing and medication waste. Independent of the system used to administer a drug or a vaccine, at least one connection is necessary. Safety requires that connection(s) be leak-proof between the drug or vaccine and the patient.

For medications or vaccines, an injection needle into the body is a common delivery method. There are close to 50 gauge sizes, including half-sizes of needles, ranging from gauge 6 (5 mm outside diameter [OD]) to gauge 34 (0.159 mm OD). A particular size gauge needle is chosen for the medical application (e.g., diabetes management, bolus injection, pediatric blood draw). Multiple manufacturers distribute needles. Similarly, there are multiple types and manufacturers of syringes. A standard type of connection is necessary to ensure compatibility between syringes and needles manufactured by competing companies. The standard intravascular or hypodermic connector is called a Luer connector, the lock-type of which is shown in Figure 11.1.

There are different types of Luer connectors (lock, slip, eccentric slip), but each one ensures a leak-proof transfer of fluid. One characteristic of all Luer connectors for syringes and needles is a conical fitting with a 6% taper, as specified in ISO 80369-7, *Small-bore connectors for liquids and gases in healthcare applications*. This standard also specifies other dimensions and requirements for small-bore connectors to be used in intravascular applications or hypodermic connections. For example, the male Luer cone length must be <7.5 mm, but is 4 mm at the tip and 4.45 mm at the end. The female Luer piece is specified to be 9 mm deep (to have enough room for the 7.5 mm cone). Typical drawings of male and female components are shown in Figure 11.2.

Breakout Box 11.1 Biomedical Engineering Design Team Alumna's Perspective on Standards

In an industrial setting, as a design moves from the concept phase to the development phase, the regulatory team is tasked with going through a list of common standards and selects the standards that they believe to be relevant to the product. Even within academic design resulting in a prototype, one can readily determine if some standards are relevant based on whether the device is implanted, enters the body temporarily, uses batteries, lasers, electricity, gas, Radio Frequency Identifications (RFIDs), is controlled by software, or has parts subject to corrosion.

Design engineers are typically tasked with adding the list of standards to the product's Design History File (DHF) and ensuring that the design complies with these standards. In a large company, many of the standards will be built into procedures that the design engineer must follow as part of design controls. Although design engineers in industry do not need to read the actual standard, they work in accordance with the company procedures that are compliant to the standards.

The tricky part comes when there are standards that apply to the design, but were not known to exist, not on the list of common standards that were checked, or that were not thought to apply to the project based on the design in its early stages. A medical device company likely has procedures implemented that help you design in accordance with all the standards you would need to follow to comply with applicable agency regulations and apply for clearance or approval. This can also be done by design teams. However, there are other standards that may simply help your device attach to/interact with other common medical equipment; if you were not aware of the standard, you could start to sell your product and then receive complaints that it is working in one hospital but not in another. This can lead to the need for a design change after prototyping or an industry product launch, which will likely be frustrating because it requires redesign, testing, and further expense. You may even have complaints coming in from unsatisfied users or customers. If there is a safety risk as a result of your design, a recall of the product you have already sold is possible.

Even if there is not a safety risk as a result of noncompliance with a given standard, there can potentially be a project-related or business-related risk. For example, say your team project's goal is to decrease the incidence of falls from hospital surgical tables, and the team is designing a device that attaches to the side rails of a surgical table. You go into the field and measure the rail size on 10 surgical tables, unaware that surgical tables have a standard for the size of side rails. Your measurements are identical, you assume that all side rails are the same size based on your measurements in the field, and you move forward with your design. Once you start to design and prototype your device, you find out the team's measurements are from room tables made by three of the five most common surgical bed companies in the US (Skytron, Steris, Maquet, Stryker, and Schaerer Medical). The team was unaware of ISO11.140/IEC 60601-2-52, *Particular Requirements for The Basic Safety and Essential Performance of Medical Beds*. You also did not realize that tables made for Europe by those same companies actually follow a different standard and have different rail sizes.

There isn't much of a safety risk here, but this could lead to an unsatisfied project advisor or a lot of returned products as well as a bad reputation for your company. If this was an industry design project, you would have decreased your market significantly by not designing the product to be compatible with side rails of all surgical tables. Are you still able to sell enough product to make up the cost of the design time and labor and turn a profit? Do you now need to invest more time and money in a design change? Could you have saved a lot of time and money if you had come across the standard for rail size of surgical tables at the beginning of the project?

From a design team standpoint, it is beneficial to search for standards that apply to your project. If able to obtain access to one or more in the planning or even the prototyping stage, the information may affect your design or testing, or both. Doing so will become part of your DHF and will be useful for subsequent teams if the project is pursued.

Mary O'Grady, MS (2011)
Stryker Corporation

FIGURE 11.1

A syringe and needle connection with a male Luer lock fitting and a needle with female Luer lock fitting (purple, 21-gauge, 0.83 mm OD) which screws into it. Each needle gauge has a distinct color associated with it so that healthcare providers can discern among the different sizes and prevent connection errors.

(From William Rafti, William Rafti Institute. https://en.wikipedia.org/wiki/File:Syringe_and_hypodermic.jpg.)

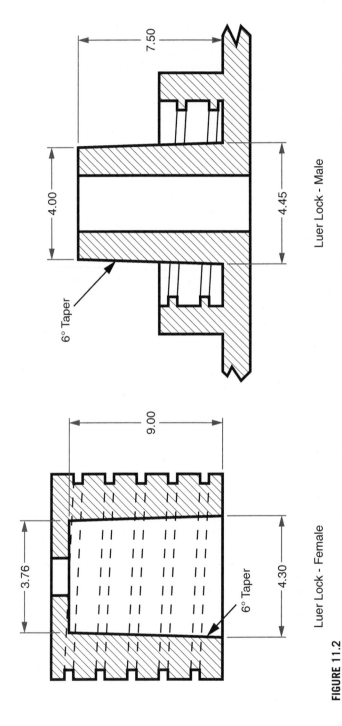

FIGURE 11.2

Dimensioned sketches of Luer locks. On the left is a partially dimensioned sketch of female Luer lock; the right image is a sketch of the male connector. All distances are in mm; tolerances are + <3%.

There are a number of other medical procedures (e.g., respiratory) that require specialized connectors. There is a set of ISO standards, one of which is ISO 80369-1, *Small-bore connectors for liquids and gases in healthcare applications—Part 1,* which includes general requirements and contains an overview of the groups of standards associated with connectors, as shown in Figure 11.3.

Parts 2 through 7 are for specific medical applications (with Part 7 being exclusively for vascular and hypodermic connections). Part 2 is for common test methods for all connectors. It is not uncommon for standards to build upon and reference one another. For example, additional guidance on Luer connectors can be found in ISO 594, *Conical fittings with 6% (Luer) taper for syringes, needles and certain other medical equipment.*

As indicated in Chapter 1, standards are used to guide design activities including material selection and composition, experimentation, design characteristics, packaging, and design documentation. They define standard terminology to prevent confusion when communicating technical information. These standards also address quality and safety; for that reason, most regulatory bodies, including the US Food and Drug Administration (FDA), the European Medicines Agency (EMA), and the Pharmaceuticals and Medical Devices Agency (PMDA) in Japan more readily approve healthcare products for clinical use that satisfy a specific set of standards related to developing technology.

Standards build confidence and trust among various stakeholders in the healthcare ecosystem, including users and customers, regulatory bodies, manufacturers, insurers, and healthcare providers. Regulatory agencies only clear a medical device for sale if it can be proven it met applicable standards. Doing so is one way to ensure safety and effectiveness. In this way, compliance with standards eases the pathway to regulatory approval, commercialization, and adoption of a product by the wider healthcare ecosystem. In addition, standards can help reduce costs, increase productivity, generate economic growth, improve quality, and promote fair comparisons with competitive products. One example is the Luer lock presented above, which creates a leak-free connection between a male-taper connection and a female counterpart. This design of a locking mechanism is ubiquitous in medical applications. ISO 594 is a standard that describes characteristics of Luer taper connectors while ISO 80369-1 specifies dimensions for all small-bore connectors to be used in intravascular or hypodermic applications. Because virtually all healthcare product manufacturers comply with these standards, all of their products are compatible with one another, thus enhancing safety, saving time, and promoting commerce across brands and borders.

11.2.1 Innovation and Standards

Standards are sometimes viewed as constraints rather than helpful guidelines. As presented in Section 3.3.2, creativity and innovation can thrive from having well-articulated constraints. In this respect, standards can play two roles that roughly map to two different pathways to innovation. A *fundamental innovation* is one that is new to the world whereas an *adaptive innovation* is one that generates the need for further innovation. Below, we present two cases that detail these two roles' standards play in innovation.

The principle of Nuclear Magnetic Resonance (NMR) as applied to medicine is an example of a fundamental innovation that gave rise to Magnetic Resonance Imaging (MRI) as we know it today. The developers of full body scanning instrumentation developed safety guidelines for its use to accelerate the widespread diffusion of this new imaging modality. Furthermore, new standards and updates to existing standards helped keep the technology relevant. Examples of recent MRI safety standards

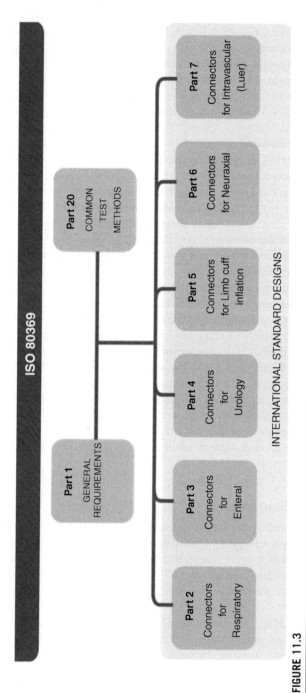

FIGURE 11.3

Eight parts of the ISO 80369 standard. General and testing requirements as well as six medical applications, with Part 7 focused exclusively on hypodermic and intravascular connections.

include: (1) The Intersocietal Accreditation Commission's *Standards and Guidelines for MRI Accreditation*, (2) ISO standard 10974, *Assessment of the safety of magnetic resonance imaging for patients with an active implantable medical device*, and (3) ASTM F2503, *Standard practice for marking medical devices and other items for safety in the magnetic resonance environment*.

It is not uncommon that a fundamental innovation takes a relatively long time for widespread medical use. The car was invented in 1885 and it was not mass-marketed until 23 years later. Nuclear Magnetic Radiation (NMR) was first discovered as a physics research tool in the late 1930s when it was found that atomic nuclei behaved differently when exposed to a sufficiently strong magnetic field. This technology became the basis of spectroscopy to determine molecular structure. In biomedical engineering, however, the first human scan using similar technology that produced an image did not occur until 1977, newly termed Magnetic Resonance Imaging (MRI). Approval for medical use first occurred in 1984, with insurance reimbursement beginning in 1985. Over the next decade, MRI imaging grew to such a volume that there was one MRI scanner per 100K people. MRI imaging continues to grow, with about one MRI scanner per 28K people in the United States.

An adaptive innovation is one that is need generated. A quintessential example of that is the need to create a prototype from Computer-Aided Design, CAD software (e.g., AutoCAD, Creo, Solidworks) files. To be able to prototype, data exchange between CAD output and input to CAM (3D printing, as described in Section 7.4) or CAE—computer-aided engineering (e.g., finite element modeling)—needed to be standardized as each CAD system represented geometry data differently. This need led to the creation of ISO 10303, *Automation systems and integration—Product data representation and exchange*, known as the Standard for the Exchange of Product model data, or STEP standard. In addition, a *de facto* standard was the software template library, which had been developed for converting most of CAD data into stereolithography input files for 3D printing. (As indicated in Section 7.4, all CAD software can export its data to Standard Triangle Language [STL], also known as Standard Tessellation Language, in STL file format).

As indicated by its name, an STL file approximates outer surfaces using multiple triangles (or other polygons) to provide a reasonable representation of surface geometry. An STL file cannot represent color, texture, material, substructure, or other properties of the fabricated object. Not uncommonly, an STL file has problems that occur at boundary edges and intersecting faces. Because exported CAD files usually lose detail, specifically some features in the CAD data, they need to be refined using a variety of techniques before manufacturing a prototype of sufficient quality. CAD software can now save data in multiple formats, as can be seen in Figure 11.4

Additive manufacturing technology can now produce geometries in full color with functionally defined gradations of materials, microstructures, and textures. As such, the need for a newer standard has led to a 2016 standard ISO/ASTM 52915, *Specification for additive manufacturing file format (AMF)*. Adherence to this standard reduces the effort needed to refine output files for prototyping.

11.2.2 Types of Standards

There are a number of different classes of standards; seven primary classifications are summarized in Table 11.1. A specific standard may contain a mix of several types. For example, a prescriptive testing method may be combined with performance requirements and guidance for analysis and interpretation while requiring use of accepted scientific units (standard measures). In addition, some standards may only apply to a local jurisdiction, such as a region or single state. For example, one prescriptive Illinois

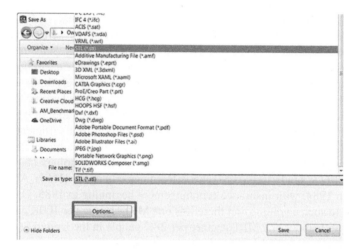

FIGURE 11.4

Representative screen shot of saving a CAD file in .stl format, one of more than two dozen other types of formats. The options choice includes specifying the resolution. As expected, the finer the resolution, the more accurate the part. A file size of <200 KB for a part will produce an approximate spherical exterior that looks like a golf ball. A file size >1 MB for a part will produce a visually smooth spherical surface.

standard for a minimum hospital door frame width for wheelchairs is 864 mm (34 inches). Although prescriptive, it is also performance-based because virtually all standard wheelchairs can fit through an opening that size. It also ensures interoperability because it specifies a standard-sized door width to fit the width of a standard door frame.

Most standards are voluntary. However, a standard may be mandated by industry, government, trade agreement, a professional society, or an individual company itself. Those mandated by government or an international trade agreement may eventually become legally required. Although regulations sometimes do not mandate adherence to specific standards, medical products are still expected to meet relevant standards.

11.3 Standards Organizations and Standards Generation

Standards originate through a variety of means that vary across the globe. Often a technical need or new technology warrants a new standard. At times, standards build upon and supplement one another. In this section, we explore the various types of standards-generating bodies and how they generate and encourage adoption of a standard.

11.3.1 Standards Development Organization

Standards Development Organizations (SDOs) create and publish standards (Table 11.2). Some SDOs are independent and not-for-profit, such as the ISO and ASTM International (ASTM is the acronym for the former American Society of Testing Materials). Other standards are a part

Table 11.1 A table of seven types of standards and examples of each

Standard Type	Description	Examples
de facto	Broadly recognized and used but lack formal approval by a standards organization. These are standards that are in place by common use and are unlikely to change in the foreseeable future because of widespread acceptance and market forces.	QWERTY keyboard, STL format, HTML (before 2001, when a standard was published)
Regulatory	Created by a Standards Development Organization (SDO) or a governmental agency with the intent of enhancing public welfare and safety.	Standards are generated by the Environmental Protection Agency (EPA) and Occupational Safety and Health Administration (OSHA)
Standard measure	Defined and agreed-upon measurements necessary for product comparisons, and frequent trade across international boundaries.	Kilogram, meter, liter, second
Prescriptive	Guidelines and best practices for processes. These include the design process (as discussed later in this chapter in the context of Design Controls) and testing (as presented in Chapters 9 and 10).	Testing methodology for feeding tubes, orthopedic implants, tissue adhesive strength, and friction
Performance	Definitions of proper function and performance, including sizes, weights, concentration levels, material composition, flow rates, angles, normal operating temperature ranges, and levels of sterility. May define acceptable and unacceptable performance levels.	Specifications for resins and metal alloys for surgical implants, antimicrobial susceptibility tests, performance test for glucose monitoring
Interoperability	Specify a fixed format or design with the goal of ensuring compatibility between systems using the same physical entity or data.	CAD file format, electronic health records, Luer fittings
Management	Requirements for the business processes and procedures used by companies (e.g., quality assurance for manufacturing or environmental management systems).	Design controls and risk management. ISO 13485 is prime example.

of a professional or trade society, such as the Institute of Electrical and Electronics Engineers (IEEE), the American Society of Mechanical Engineers (ASME), the Association for the Advancement of Medical Instrumentation (AAMI), or the National Electrical Manufacturers Association (NEMA). Each of these organizations draft, adopt, and disseminate its own standards, sometimes cross listing them with one or two other SDOs. Unlike the ASME or IEEE professional organizations, the Biomedical Engineering Society (BMES) does not produce its own standards.

De facto standards sometimes end up being sponsored by an SDO. For example, the Hypertext Markup Language (HTML) used for webpages was first developed in 1991. After its fourth revision, it became an internationally recognized standard nearly a decade later (ISO/IEC 15445 *Information technology—Document description and processing languages—Hypertext Markup Language [HTML]*). There are also organizations that attempt to bring together other SDOs. For example, the American National Standards Institute (ANSI) does not generate its own standards but is a private nonprofit organization that brings together other SDOs to reach consensus (e.g.,

Table 11.2 Some FDA-recognized medical standards organizations

Abbreviation	Standards Development Organizations (SDOs) Name	Example
ISO	International Organization for Standardization	ISO 13485 is the primary standard for quality management systems for the healthcare industry. Many other ISO standards apply to medical devices.
ASTM International	ASTM International (formerly American Society for Testing and Materials)	ASTM publishes hundreds of standards in specialties ranging from biocompatibility to urology with a focus on performance, safety, and testing.
IEEE	Institute of Electrical and Electronics Engineers	ISO/IEEE 11073 is a family of standards for point-of-care (POC) and personal health devices (PHD)/medical device communication (MDC)
ASME	American Society of Mechanical Engineers	ASME 18.1 provides design guidance and safety standards for platform lift and stairway chairlifts.
CLSI	Clinical and Laboratory Standards Institute	GP17-A3 specifies guidelines and procedures to ensure laboratory safety.
IEC	International Electrotechnical Commission	IEC/EN 60601 is a series of technical standards to ensure basic safety and essential performance of electrical medical equipment.
ITU-T	International Telecommunication Union—Telecommunication Standardization Sector	H.323 is a family of standards for multimedia and VoIP.
AAMI	Association for the Advancement of Medical Instrumentation	AAMI publishes standards in diverse areas such as medical devices, biological evaluation, dialysis, electromedical equipment, healthcare technology management, sterilization, and human factors.

ANSI/AAMI/ISO 5840-1:2015). ANSI is the only United States representative that is permitted to serve as a member of ISO. The vast majority of standards in the United States are voluntary consensus standards, meaning that compliance is not mandatory. However, customers or companies will often choose not to purchase a product that does not comply with specific standards. This potential impact on sales often provides an incentive for companies to comply with relevant standards.

Some governmental agencies promote creating standards. For example, the National Institute of Standards and Technology (NIST, formerly known as the National Bureau of Standards) was created over 100 years ago to promote economic growth by working with industry to establish common standards. NIST offers a breadth and depth of technical expertise, a reputation as an unbiased and neutral party, and a long history of working collaboratively with the private sector. NIST also publishes best practices as a way to promote consistency and growth across the healthcare industry. One publication, for example, is NIST SP 1800, *Securing Electronic Health Records on Mobile Devices,* which describes what NIST considers best practices on securely protecting health records on wireless devices. Other biomedical engineering-related publications are NIST SP 500-288 *Specification for WS Biometric Devices (WS-BD)* and, in software, *Healthcare— Standards and Testing.* These documents are available online; websites can be found in the resource section.

11.3.2 The Creation and Dissemination of Standards

There are a variety of factors that initiate the development of a standard. Governmental SDOs may be prompted by the legislature to develop standards. Professional and trade organizations may see a new technology on the horizon that requires some consistency to grow into widespread application and safe practice. For example, the need for new and updated standards regarding safety and effectiveness of publicly used masks to prevent COVID-19 infection spread in 2020 caused ASTM to put a call out for committee members with expertise to develop three proposed standards. One committee was for medical face mask materials, another for barrier face coverings, and a third for an air-purifying respirators for first responders. Some standards already existed (e.g., ASTM F2100 *Standard Specification for Performance of Materials Used in Medical Face Masks*), but needed to be revised or replaced due to technical improvements, changes in use, or changes in policy.

Many AAMI standards are applicable to rehabilitation and assistive device projects. AAMI standards are developed by the Rehabilitation Engineering and Assistive Technology Society of North America (RESNA), a nonprofit corporation whose mission is to promote the health and wellbeing of people with disabilities and the aging population. They generate standards that are divided into 13 medical categories (e.g., wheelchairs, vision and hearing impairments, cognitive deficits).

In recognizing a standard, a regulatory agency is not formally endorsing it nor officially requiring compliance. Rather, recognition means that regulatory evaluators are familiar with the standard and trust that products submitted for regulatory review have been tested per the standard and are fully compliant. Practically speaking, complying with recognized standards accelerates the regulatory review process in the United States and can be mandatory for approval in other nations.

11.3.4 Finding Medical Device Standards

There are thousands of standards that have been created specifically for medical devices and many more cited in these standards. Due to the dynamic nature of health care, there are new standards being released almost every week. For example, recent technical advances in computer-assisted and robotic surgery, implantable "smart" sensors, nanotechnology, personal protective equipment, and improved healing of injury through tissue engineering are prompting new standards to be created. It is not always easy to find the standards that are relevant to your specific design project. We suggest using a combination of search methods as described below.

A powerful and robust way to find standards is to search the FDA 510(k) database for similar products. As shown in Figure 11.5, a good place to start is with the 510(k) database and enter the product name.

For example, a search for relevant standards for endosseous dental implants takes you to the premarket notification database that lists all the companies that have received 510(k) clearance for similar products. Selecting a company takes you to the company's premarket notification, where the associated regulation number and product code are listed. The regulation number indicates the regulation that lists test requirements for the product and the product code indicates the standards that apply. This is the most specific type of search when developing medical devices. Another method is to use the FDA searchable database of consensus standards. You can reach this site by going to the FDA website and searching for "consensus standards."

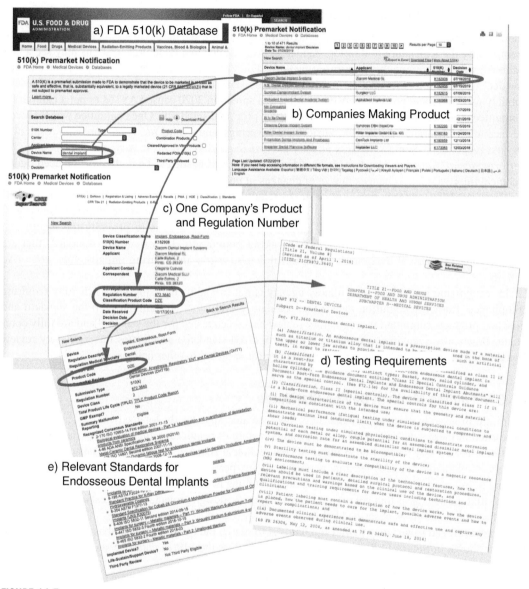

FIGURE 11.5

Screenshots of FDA's websites to determine tests and standards associated with a product. Using endosseous dental implants in this case leads to (a) FDA's 510(k) database, (b) companies that make the product, (c) a product code and registration number, (d) testing requirements, and (e) relevant standards.

There also are online general standards search engines. The National Institute of Standards and Technology (NIST) has a website, standards.gov, that is useful for information on standards and conformity assessment, and has a limited search capability. Document-center.com, a commercial site, is a useful standard searching site. The limitation of using either of these sites is that multiple classifications of standards (e.g., clothing, electronics environment, food technology, metallurgy, packaging) along with healthcare technology are included with the 40 other classes of standards. Some medical device standards that apply to your device and need to be complied with are likely from classes other than health care (e.g., ISO 9001).

Although not recommended, one option for finding relevant standards is to use a generic search engine. For example, you may be looking for testing methods that apply to an initial search term such as "endosseous dental implant standards"; ISO standard 14801: *Dentistry—Implants—Dynamic loading test for endosseous dental implants* appears on the first page. A limitation from a generic search is that results do not indicate whether or not the standards found are recognized by regulatory agencies.

11.3.5 Examples of Medical Device Standards

We conclude this section with some examples of medical device standards summarized in Table 11.3. Some are for devices themselves—one example is ANSI/AAMI/ISO 5840-1, *Cardiovascular*

Table 11.3 Sample medical products and selected associated standards. The publication years are current as of October 2020. Future standards will have a later publication date.

Product	Standard #	Standard Title
Prosthetic heart valves	ANSI/AAMI/ISO 5840-1:2015	Cardiovascular Implants—Cardiac Valve Prostheses
General catheterization system	ISO 10555-1:2013	Sterile and Single Use Catheters, Intravascular Catheters—Part I: General Requirements
Vascular stents—elastic recoil	ASTM F 2079-09(2017)	Standard Test Method for Measuring Intrinsic Elastic Recoil of Balloon-Expandable Stents
Tubular vascular stents	ANSI/AAMI/ISO 7198:2016	Cardiovascular Implants and Extracorporeal Systems—Vascular Prostheses—Tubular Vascular Grafts and Vascular Patches
Hemodialysis equipment	ANSI/AAMI RD16:2007	Cardiovascular Implants and Artificial Organs—Hemodialyzers, Hemodiafilters, Hemofilters and Hemoconcentrators
	ANSI/AAMI/IEC 60601-2-16:2008	Hemodialysis Systems
Mechanical ventilators	IEC 60601-2-12(2001)	Medical Electrical Equipment: Particular Requirements for the Safety of Lung Ventilators—Critical Care Ventilators
Oxygen monitor	ISO 21647:2004 (IEC 80601-2-55)	Medical Electrical Equipment—Particular Requirements for the Basic Safety and Essential Performance of Respiratory Gas Monitors
Pulse oximeter	ISO 80601-2-61:2017	Medical Electrical Equipment—Part 2-61: Particular Requirements for Basic Safety and Essential Performance of Pulse Oximeter Equipment

Continued

Table 11.3 Sample medical products and selected associated standards. The publication years are current as of October 2020. Future standards will have a later publication date.—cont'd

Product	Standard #	Standard Title
Electronic/infrared thermometers	ASTM E1112-00(2018) & ASTM E1965-98(2016)	Standard Specification for Electronic Thermometer for Intermittent Determination of Patient Temperature
Thermometer probe covers	ASTM E1104-98(2016)	Standard Specification for Clinical Thermometer Probe Covers and Sheaths
Electro-encephalography	IEC 80601-2-26:2019	Medical Electrical Equipment—Part 2-26: Particular Requirements for Basic Safety and Essential Performance of Electroencephalographs
Neurostimulators	AAMI/ISO 14708-3:2017	Implants for Surgery—Active Implantable Medical Devices—Part 3: Implantable Neurostimulators
	ANSI/AAMI NS4:2013(R)2017	Transcutaneous Electrical Nerve Stimulators
Cochlear implant	ANSI/AAMI C186:2017	Cochlear Implant Systems: Requirements for Safety, Functional Verification, Labeling and Reliability Reporting
Ophthalmic implants	ISO 11979-2:2014	Ophthalmic Implants—Intraocular Lenses—Part 2: Optical Properties and Test Methods
	ISO 11979-3:2012	Ophthalmic Implants—Intraocular Lenses—Part 3: Mechanical Properties and Test Methods
	ISO 11979-5:2006	Ophthalmic Implants—Intraocular Lenses—Part 5: Biocompatibility
	ISO 11979-8:2017	Ophthalmic Implants—Intraocular Lenses—Part 8 Fundamental Requirements
Hip prostheses	ASTM 1440-92:2008	Standard Practice for Cyclic Fatigue Testing of Metallic Stemmed Hip Arthroplasty Femoral Components Without Torsion
	ASTM F1875-98(2014)	Standard Practice for Fretting Corrosion Testing of Modular Implant Interfaces: Hip Femoral Head-Bore and Cone Taper Interface
	ISO 14242-1:2012	Implants for Surgery—Wear of Total Hip-Joint Prostheses— Part 1: Loading and Displacement Parameters for Wear-Testing Machines and Corresponding Environmental Conditions for Test
	ISO 14242-2:2016	Implants for Surgery—Wear of Total Hip-Joint Prostheses— Part 2: Methods of Measurement Parameters for Wear-Testing Machines and Corresponding Environmental Conditions for Test
Stents	ASTM F2514:-08(2021)	Standard Guide for Finite Element Analysis (FEA) of Metallic Vascular Stents Subjected to Uniform Radial Loading
	ASTM 2081-06(2017)	Standard Guide for Characterization and Presentation of the Dimensional Attributes of Vascular Stents
	ISO 25539-1 (2017)-02	Cardiovascular Implants—Endovascular Devices— Part 1: Endovascular Prostheses

Table 11.3 Sample medical products and selected associated standards. The publication years are current as of October 2020. Future standards will have a later publication date.—cont'd

Product	Standard #	Standard Title
	ISO /TS 17137 2019	Cardiovascular Implants and Extracorporeal Systems—Cardiovascular Absorbable Implants
	ASTM F1828-2017	Standard Specification for Ureteral Stents
Glucose monitoring	CLSI POCT05-A-2008	Performance Metrics for Continuous Interstitial Glucose Monitoring
	IEEE 11073-10417-2017	Health Informatics—Personal Health Device Communication—Part 10417: Device Specialization—Glucose Meter
	IEEE/ISO 11073-10417 (2017)	Health Informatics—Personal Health Device Communication—Part 10417: Device Specialization—Glucose Meter
Foley Catheters	ASTM F623 (2019)	Standard Performance Specification for Foley Catheters

implants—Cardiac valve prostheses—Part 1: General requirements. This standard defines operational conditions for heart valve substitutes such as laboratory testing to assess flow, leak, strength, and durability. Other standards listed are for testing medical devices, such as ASTM F1798-97 *Standard Guide for Evaluating the Static and Fatigue Properties of Interconnection Mechanisms and Subassemblies Used in Spinal Arthrodesis Implants.* These examples, which include general requirements and experimental guidance, demonstrate the diversity of medical device standards.

During your career, you will become familiar with specific standards that are most relevant to the types of products your company develops and the segment of the medical device industry in which the company competes. However, there are a few standards that are ubiquitous and transcend product classifications and apply to almost all medical device types. The primary example is ISO 13485, *Medical devices—Quality management systems—Requirements for regulatory purposes.* This standard defines the design process that each medical device company or team must follow. Two examples relating to risk management include ISO 10993-1:2018, *Biological evaluation of medical devices—Part 1: Evaluation and testing within a risk management process,* and ISO 14971:2019, *Medical devices—Application of risk management to medical devices.* These standards are addressed in greater detail in Section 8.9, with a testing example presented in Section 9.11.

11.4 Design Controls and ISO 13485

Standards exist not only for guidance on design decisions (e.g., material choice, testing procedures) but also for guidance of the design process itself. A design process that allows for detection and correction of technical and nontechnical errors throughout the process results in less time needed for design, lower manufacturing costs, and higher quality parts. One way to achieve that is through a Quality Management System (QMS).

A QMS is defined as being a formalized system that documents processes, procedures, and responsibilities for achieving quality policies and objectives. In practical terms for the design process, the purpose of a QMS is to ensure each time a process is completed, the same type of information, methods, skills, and controls are applied in a consistent manner. In design, these are called Design Controls.

Design controls are part of part of ISO 9001, *Quality Management Systems—Requirements*, and part of ISO 13485 *Medical devices—Quality management systems—Requirements for regulatory purposes*. This set of practices and requirements aims to reduce risk and increase the probability that a design meets customer needs. ISO 13485 includes requirements that help companies provide medical devices and services that consistently meet customer and regulatory requirements. These are not product standards; rather, they are quality management standards focused on the customer. As shown in Figure 12.2, more than 60% of recent FDA recalls were due to design defects or material problems. Many of these could have been prevented if effective design controls had been implemented, thereby leading to better design decisions. In this section we outline the components of design controls.

11.4.1 Design Control Requirements

Design control requirements do not specify the exact methods to be used, but rather outline a generic process. Each company and design team determines how they meet these requirements. At the most general level, ISO 9001 describes the seven principles of a QMS and the basic rationale for each. In alphabetic order, these are:

- **Continuous Improvement**—to maintain current levels of performance, to react to changes in internal and external conditions, and to create new opportunities
- **Customer Focus**—to meet customer requirements and strive to exceed customer expectations
- **Engagement of People**—to empower competent people to enhance their capability to create and deliver value
- **Evidence-Based Decision Making**—to understand cause-and-effect relationships and potential unintended consequences
- **Leadership**—to establish unity of purpose and direction, and to create conditions in which people are engaged in achieving the organization's quality objectives
- **Process Approach**—to consistently achieve effective and efficient results when activities are understood and managed as interrelated processes functioning as a cohesive system
- **Relationship Management**—to achieve sustained success, the organization must manage relationships with all interested parties (employees, partners, and suppliers) to optimize their impact on the organization's performance.

In the context of biomedical engineering design, ISO 13485 specifies requirements for a QMS that consistently meets customer and applicable regulatory requirements. ISO 13485 includes sections on quality management systems, management responsibility, resource management, product realization, and measurement, analysis, and improvement. Section 7.3 of the standard describes nine parts of design controls. The FDA's Design Control sections are included in the Quality System Regulation (QSR), which is modeled after the ISO standard. A comparison of the titles in each are found in Table 11.4.

A graphical overview of design controls in the context of medical device design is represented in the textbook model as depicted in Figure 1.3, which enhances the design control requirements of the ISO 9001 and ISO 13485 standards and FDA's Waterfall Model. As before, procedures for controlling design changes and documentation are not indicated on the drawing but are an integral part of the design process.

Table 11.4 The similarities and nuanced differences between the Design Control components of ISO 13485 and the FDA Quality System Regulation

Section	ISO 13485	FDA Quality System Regulation (QSR)
1	General	General Requirements
2	Design and Development Planning	Design and Development Planning
3	Design and Development Inputs	Design Input
4	Design and Development Outputs	Design Output
5	Design and Development Review	Design Review
6	Design and Development Verification	Design Verification
7	Design and Development Validation	Design Validation
8	Design and Development Transfer	Design Transfer
9	Control of Design and Development Changes	Design Changes
	Design and Development File	Design History File

A controlled design process does not proceed linearly through the nine elements of design controls. For example, design and development planning must happen at the beginning of a project and should then be revisited throughout the process (e.g., as updates are made to project specifications, schedule, or budget). In the remainder of this section, we briefly describe each of the elements of design controls that lead to a robust design process.

11.4.2 Design and Development Planning

Design and development planning requires design teams and manufacturers to establish and maintain plans that describe all required design and development activities and identifies the person or group responsible for their implementation. The plan includes who will be involved and which activities are needed to provide input into the design and development process (e.g., needs-finding activities described in Chapter 3 or deliverables such as project schedules and budgets, as shown Section 2.6). Specific methods for design and development planning are not part of the design control requirements; each design team or company develops its own plan.

11.4.3 Design Input

Design inputs include performance and technical requirements, specifications, applicable standards, and regulatory requirements. These serve as the basis for the design of a device. Generally, these make up the list of quantified and testable specifications as outlined in Section 4.3. These specifications were based upon the methods of discovery outlined in Chapter 3 that incorporate user and customer needs and performance of competitive products. Design inputs may also be derived from considering the many Design for X elements (e.g., packaging, human factors, manufacturability, biocompatibility) presented in Chapter 8. As such, the list of design inputs often evolves over time, usually as a result of activities such as design reviews. The goal is to develop a list of design specifications and performance requirements that can be verified through testing. As presented in Section 11.4.5, Design Reviews at the input stage are necessary to ensure that customer needs are met.

Reviews uncover specifications and requirements that are unnecessary, inadequate, incomplete, ambiguous, or in conflict.

11.4.4 Design Output

Design output consists of the deliverables of a design process that include product and process documentation needed to transform a design into a prototype or finished product (final specifications, drawings, and assembly instructions). They also include test procedures, test results, and risk analyses to verify that a product meets the design input requirements. Breakout Box 11.2 presents an example relating customer needs to design inputs and design outputs.

Breakout Box 11.2 Connecting Design Inputs and Outputs

As a design input and output example, one medical problem obstetricians face is providing proper counsel on safe exercise limits during pregnancy with patients who are athletes. To address this, it is important to know the relationship between maternal heart rate and fetal heart rate (FHR) during exercise; FHR is normally between 110 and 160 bpm. A typical exercising patient's heart rate is close to the low end of that range. What the obstetric project sponsors wanted in new technology is shown in the first column of Table 11.5. Sample design input is presented in the middle column; corresponding design output is shown in the last column.

Table 11.5 Abbreviated design input and design output based on wants of project sponsor

What Obstetric Providers Wanted	Sample Design Input	Sample Design Output
Must be accurate, wearable, and able to provide data to providers	Record fetal heart rate (FHR) to within 5 BPM Capable of 1.5 hours of transmitting FHR data	Printed Circuit Board (PCB) with filtering circuit Wireless specification to IEEE 802.11 Signal filtering code Error handling code Power specification to IEEE/ISO/IEC 8802-15-4
Must be portable and robust	Weight <250 grams Survive a meter drop Reusable after cleaning Sweat resistant	Weight 180 grams Housing survived ASTM D5276 drop test Exposed housing made of ABS Passed Ingress Protection (IP-7) waterproof test
Must minimally interfere with patients' activities	Limit surface area to <3 cm Limit height to <1 cm	1.70 cm × 1.70 cm × 1 cm

---

11.4.5 Design Reviews

Design reviews are a required component of design control and are common in industry and academic design projects. ISO 13485 Section 7.3.5 states, in part, the following requirements of design reviews:

At suitable stages, systematic reviews of design and development shall be performed in accordance with planned and documented arrangements to:

a) evaluate the ability of the results of design and development to meet requirements, and

b) to identify and propose corrective actions (if any) or improvements or both.

In academic design projects, design reviews enable a panel of experts outside the design team to review and critique both the process and progress of a design project. Participants in such reviews usually include faculty and advisors; they may also include representatives of the government or a company comprised of technical experts on a product or prototype under consideration. In academia, design reviews are usually part of a course grade.

Records of the results of the reviews and any necessary actions shall be maintained and include the identification of the design under review, the participants involved, and the date of review. Most importantly, the review leads to a summary of actions with which the project team needs to follow up during subsequent design progress. The feedback from these reviews is invaluable. One template for a design review summary is presented in Breakout Box 3.12.

11.4.6 Design Verification

Design verification answers the question, "did we make the prototype right?" Verification involves creating a formal testing plan (guidelines can be found in Section 9.6) demonstrating that design output meets the design input requirements. Examples of verification testing standards include biocompatibility testing (ISO 10993; Section 10.3), software testing (ISO 62304), and electrical safety testing (ISO/IEC 60601).

ISO/IEC 60601 is a series of technical standards (sometimes a more comprehensive standard is bundled as a group of smaller standards) relating to the safety and performance of electrical medical equipment. The first 11 standards in the series are collateral standards that define the requirements for aspects of electrical safety and performance. These aspects include electromagnetic compatibility, environmentally conscious design, and medical systems for use in the home health environment. There are more than 50 other standards related to specific medical equipment, such as operating tables, computed tomography scanners, and high-intensity ultrasound devices. Following the 60601 series is fundamental to manufacturers of electronic medical equipment.

Each verification test has a testing plan that typically includes the reason for the test, materials, equipment and facilities needed, an experimental protocol, and results and conclusions. In both industry and academic design projects, it is recommended to create a generic format that can be used for all testing. Section 9.9 presents more on design documentation of verification activities.

Some verification testing is specified in detail in standards. For example, the current standard recognized by the FDA for cardiac monitoring on a hospital floor is ANSI/AAMI/IEC 60601, *Medical electrical equipment—Part 2-27: Particular requirements for the basic safety and essential performance of electrocardiographic monitoring equipment*. One requirement in the standard for a hospital-based system is to detect inadequate connections to the patient. The standard specifies that the test for an unsuitable connection is to discharge a small current (<1 µA) to pairs of electrodes. If an open circuit is detected in any of them, the cardiac monitor should sound an alarm to the healthcare workers' central station on the hospital floor. Using this and other criteria as a guide to create verification bench tests would ensure compliance with the standard.

11.4.7 Design Validation

As indicated in Section 9.7, design validation answers the question, "did we make the right product?" It aims to determine if the solution meets the users' needs, even if it meets all technical specifications. While design verification generally starts with the testing of specifications using laboratory tests, at the heart of validation testing is the perspective of the user.

The activities described in Section 3.7 are very early forms of validation to determine if there is need for a new solution, even if it is not yet clear what that solution will be. Validation continues at other stages as well. For example, seeking feedback on sketches of potential solutions or prototypes (described in Section 6.5) is a form of validation testing. Design for usability, as discussed in Section 8.2 (e.g., human factors), includes perspectives of users. In industry, ultimate successful validation takes the form of a product being adopted by the healthcare ecosystem.

The QSR defines design validation as "establishing by objective evidence that device specifications conform with user needs and intended use(s). The most meaningful validation testing is performed with a prototype device and takes the form of soliciting feedback from users on their experience using the device. The most general format of such tests is to provide potential users with the device and other materials (e.g., instructions) and ask each one to use the device. Use of the device could be followed up by a survey or focus group, as outlined in Section 3.7, but is now more targeted toward the user experience.

11.4.8 Design Changes

As indicated throughout this text, design is an iterative process in which new information is gained and the project and design solutions are reevaluated on an ongoing basis. Design control requirements call for a robust method of tracking these changes to ensure that the evolution of the design is transparent. This includes documents, drawings, and prototypes. When more than one prototype can be made, it is critical to capture the advantages and drawbacks of each iteration. The graphical technique shown in Table 7.3 can be used to document the evolution of prototypes including design changes and their rationale.

In industry, where several groups may be working on the same design, tracking and communicating design changes is critical; a change by one group may have a subtle or significant impact on the design changes of another. For this reason, companies have robust processes in which proposed design changes are requested by engineering, manufacturing, purchasing, marketing, or other personnel, along with justification for the changes. This request is then approved by personnel (i.e., design engineers, manufacturing engineers, regulatory affairs specialists) familiar with its impact on product function, manufacturability, product cost, regulatory compliance, and other areas before they can be implemented. The changes are then made to the applicable documentation, including drawings, assembly and work instructions, Bills of Materials, purchasing specifications, and inspection instructions. Auditors from regulatory agencies want to see evidence that a design change process has been put in place and evidence that the process is being followed.

11.4.9 Design Transfer

Design transfer is the handoff of a completed design and tested prototype to manufacturing. This is considered to be the last formal step in the engineering design process. Design transfer is most often done through a Device Master Record (DMR) that contains all documentation needed to produce a finished medical device. The DMR is divided into five categories of information:

- device specifications,
- product process specifications,
- quality assurance procedures,
- packaging and labeling specifications, and
- installation, maintenance, and servicing procedures.

Although some of this information is also in the DHF, merely referencing relevant sections of the DHF meets the requirements of the DMR (i.e., they do not need to be reproduced). On a more detailed level, the DMR includes instructions for all manufacturing and assembly processes, Bills of Materials, product and tooling drawings, specifications, and sterilization processes (if applicable). The specific role of the design engineer in design transfer in presented in Section 14.3.

In the context of design controls, the DMR is cross referenced in ISO 13485, Section 4.2.3(c), titled "Medical Device File." As in other parts of design controls, there are no specific recommendations about how this must occur. Companies have found a variety of successful ways to transfer their designs to production.

11.4.10 Design History File

The DHF was introduced in Section 2.9 and presented in every other chapter throughout this text. In the context of design controls, the DHF is a compilation of the entire design history of a finished medical device. The design controls used, and standards employed, must be documented. Included in the DHF are reference records necessary to demonstrate that the design was developed in accordance with proper design controls. This includes user needs and design inputs, design outputs, design verification and validation results, design reviews, and design transfer materials related to manufacturing. For commercialized products, additional components of the DHF likely include approved labeling, sterilization procedures, design transfer documentation, and design change procedures.

11.5 Example in Applying Standards

Imagine that you are working to solve a pediatrician's challenge to measure core body temperature in fidgety and uncooperative children under 6 months old. During your medical research you find that rectal temperature is the gold standard because it is most accurate, but that measurement is at least 10 seconds in duration. Based on a large sample of children, the less accurate temporal measurement of near core temperature—taken wirelessly from the forehead—is not currently recommended, especially among babies aged up to 6 months; the rectal method is thus still recommended for the measurement of core body temperature in young babies.

You decide to focus on two specifications: (1) to measure temperature to within <1% error and (2) to do so in 5 seconds or less.

You have designed your system. The basic principle is to use a sensor that converts thermal energy into electrical energy (a thermopile sensor, which includes an infrared absorber.) The absorber's voltage change is on the order of microvolts. Because the voltage change is low and noise is present, data

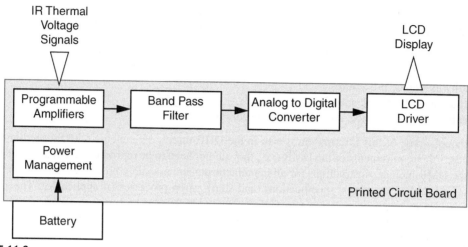

FIGURE 11.6

Block diagram for an Infrared (IR) thermometer designed to convert the measurement of two thermal sensors to an electrical signal that can be displayed on an Liquid Crystal Display (LCD) screen.

must first be amplified and then filtered before passing to the analog, the digital converter, and the LCD display. There is also software to infer core body temperature based on initial transient temperature readings. To communicate your solution, create the block diagram shown in Figure 11.6.

Before building your first prototype, are there standards that help you verify your specifications? In your search, you discover a number of associated standards, two of which appear immediately to be relevant testing standards: (1) ASTME1965 (2016), *Standard Specification for Infrared Thermometers for Intermittent Determination of Patient Temperature*, and (2) ISO 80601-2-56:2009 *Medical electrical equipment—Part 2-56: Particular requirements for basic safety and essential performance of clinical thermometers for body temperature measurement*. Both standards are recognized by multinational regulatory agencies. This is important because it means that any tests included in the standards can be used later in the design process when you are ready for verification testing of your prototype. Furthermore, ASTM 1965 specifies the maximum permissible laboratory error for skin IR thermometer to be no more than 0.3°C (0.54°F). This is less than your specification (for clinically relevant temperatures above 30°C (86°F), so meeting the standard ensures that you meet the <1% error specification.

The thermal test suggested in the standard can be performed in a benchtop setting and involves setting up a blackbody radiant thermal emitter. You learn that a blackbody radiates thermal energy characteristic of its contact temperature, with an emissivity close to one. Practically, this means you can have very fine control over the exact temperature of the cavity walls of a blackbody radiator. With a precisely known temperature you can then test the accuracy of the IR sensors and electronics in the prototype. After more research, you find a helpful paper that describes how to construct a blackbody radiator, as shown in Figure 11.7.

To conduct the test, the blackbody is secured in place in a liquid bath (with a volume of at least 2 L) requiring a minimum diameter of 10 cm and a minimum depth of 15 cm (to accommodate the

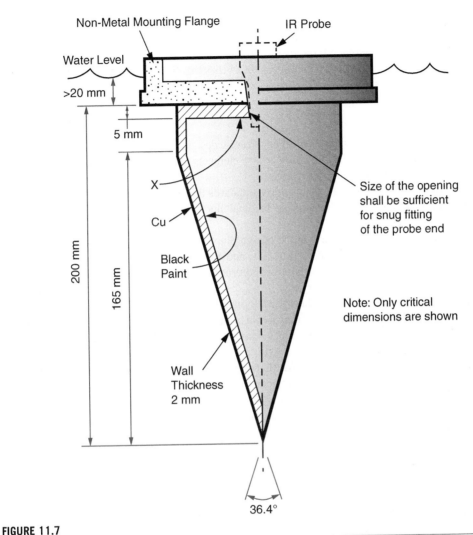

Non-Metal Mounting Flange

IR Probe

Water Level

>20 mm

5 mm

X

Cu

Black
Paint

200 mm

165 mm

Wall
Thickness
2 mm

Size of the opening
shall be sufficient
for snug fitting
of the probe end

Note: Only critical
dimensions are shown

36.4°

FIGURE 11.7

Blackbody Radiator. "Snug fitting" means tolerances must be <0.01 mm at the opening for the probe end.
(Adapted from Irani, 2001).

blackbody's maximum diameter and length). The temperature of the bath should be within 0.03°C of a target temperature (~37°C) measured by an immersed contact thermometer which should be traceable to a calibrated standard. Develop a detailed step-by-step protocol using this as a guide.

While reviewing ASTM 1965 and ISO 80601-2-56, you discover more helpful information that will be relevant in testing later iterations of your device. For example, there is guidance on other possible tests, including: (1) a drop test (one meter fall), (2) an electromagnetic susceptibility test, (3) an electrostatic discharge test, and (4) a storage test (i.e., for possible degradation due to temperature and humidity). There is also guidance on delays between recalibration and reverification and on how to present documentation about the product (e.g., displayed temperature range, maximum laboratory error, body site[s] used as a reference for adjusting the displayed temperature value).

From the example above, ASTM 1965 and ISO 80601-2-56 are useful resources as a guide to which tests to perform. The same is true for testing many other medical devices.

Key Points

- Standards play an essential role in biomedical engineering design and in establishing technical guidelines for designers and manufacturers for critical elements such as selecting materials, guiding experiments, and formalizing the design process.
- For medical device design controls, the primary standards that apply to the design process are ISO 13485 and ISO 9001. For manufacturing medical electronic devices, the most important standard is ISO/IEC 60601.
- Knowledge of relevant standards early in the design phase ensures quality, guides verification testing, helps guide design, saves time and money, and enhances safety.
- Standards can sometimes accelerate the innovation process by helping technological discoveries become widely accepted across industry.
- There are more than one hundred standards development organizations that range from not-for-profit to professional technical organizations to governmental agencies. Specific standards are typically created by organizational committees with significant technical knowledge about the technology being considered.
- A good place to start searching for standards that apply to a specific medical device is in regulatory agency website resources for similar devices (e.g., FDA.gov).
- Regulatory bodies require that medical products are developed per established design controls specified in relevant standards. Teams and companies are allowed to determine their own design processes, but then must follow design controls and show evidence that processes and controls have been followed.
- The Design History File is a critical component of design controls and should be maintained throughout the design process.
- The transition from product design and prototyping to manufacturing is design transfer, the last formal step in the design process. The Device Master Record (DMR) includes specifications and instructions for manufacturing, production process specifications, labeling, and packaging requirements and installation, maintenance, and servicing procedures and methods to commercialize a final medical product.

Exercises to Help Advance Your Design Project

1. As demonstrated in Figure 11.5 determine the standard type(s) recognized by regulatory bodies associated with your project.
2. What are the fundamental differences among the standards associated with your project (if more than one)? If accessible, read the most relevant standard and document how portions would relate to and impact/direct the design process.

3. What are the differences in your project between the testing specified (Figure 11.5e) and standard required to be met (Figure 11.5d)?

General Exercises

1. There are only six products in FDA class I that require QMS controls and demonstrate substantial regulatory equivalence. Five of these devices protect or treat patients, but one of the six—surgical gloves—primarily protects providers. Discuss why it is one of the Class I devices that require QMS controls.

2. Alarm fatigue is a considerable problem in ICU nursing. This occurs because of the interfacing equipment with each patient (e.g., breathing monitor, heart rate monitor, pulse oximeter), each with a built-in alarm system governed by ANSI/AAMI/IEC 60601. This causes many alarms to be set off due to positional changes that are not medically relevant. Some nurses ignore the alarms for a long time. Your team has been asked to address this problem. As a possible solution, consider a change to the standard test procedures to address this problem.

3. Assume your project involves development of a wound simulator. Determine what, if any, standards must be followed.

Reference and Resources

Allen, R. H., & Sriram, R. D. (2000). The role of standards in innovation. *Technological Forecasting and Social Change, 64*(2–3), 171–181.

American Society of Mechanical Engineers (ASME). (1997). *ASME position paper: Standards and technical barriers to trade.* https://www.thefreelibrary.com/ASME+position+paper+on+standards+and+technical+barriers+to+trade-a019642359.

Association for the Advancement of Medical Instrumentation (AAMI). (1994). *Blood pressure transducers (ANSI/AAMI BP22:1994/(R)2016).* https://www.aami.org/.

Association for the Advancement of Medical Instrumentation (AAMI). (2008). *Manual, electronic, or automated sphygmomanomometers (ANSI/AAMI SP10:2008).* https://www.aami.org/.

Association for the Advancement of Medical Instrumentation (AAMI). (2011). *Medical electrical equipment— Part 2-27: Particular requirements for the basic safety and essential performance of electrocardiographic monitoring equipment* (ANSI/AAMI/IEC 60601-2-27:2011/(R)2016). https://www.aami.org/.

ASTM International. (2016). Standard specification for infrared thermometers for intermittent determination of patient temperature (ASTM e1965–e1998(R)2016). https://www.astm.org/.

Baura, G. D. (2012). *Medical device technologies: A systems based overview using engineering standards.* Academic Press.

Chhaya, M. P., Poh, P. S., Balmayor, E. R., van Griensven, M., Schantz, J. T., & Hutmacher, D. W. (2015). Additive manufacturing in biomedical sciences and the need for definitions and norms. *Expert Review of Medical Devices, 12*(5), 537–543.

Damadian, R., Goldsmith, M., & Minkoff, L. (1977). NMR in cancer: XVI. FONAR image of the live human body. *Physiological Chemistry and Physics, 9*(1), 97–108.

Document Center. (n.d.). https://www.document-center.com/

Drashti, P., Kothari, C. S., Shantanu, S., & Manan, S. (2019). In-depth review on "innovation and regulatory challenges of the drug delivering medical devices". *Journal of Generic Medicines, 15*(1), 18–28.

Dunphy, S. M., Herbig, P. R., & Howes, M. E. (1996). The innovation funnel. *Technological Forecasting and Social Change, 53*, 279–292.

Graz, J. C., & Hauert, C. (2014). Beyond the transatlantic divide: The multiple authorities of standards in the global political economy of services. *Business and Politics, 16*(1), 113–150.

Gross, R., Hanna, R., Gambhir, A., Heptonstall, P., & Speirs, J. (2018). How long does innovation and commercialisation in the energy sectors take? Historical case studies of the timescale from invention to widespread commercialisation in energy supply and end use technology. *Energy Policy, 123*, 682–699.

Grunkemeier, G. L., Jin, R., & Starr, A. Prosthetic heart valves: Objective performance criteria versus randomized clinical trial. The Annals of Thoracic Surgery, 82(3), 776–780.

International Organization for Standardization. (2009). *Medical electrical equipment—Part 2-56: Particular requirements for basic safety and essential performance of clinical thermometers for body temperature measurement* (ISO 80601-2-56:2009). https://www.iso.org/standard/44106.html

Irani, K. (2001). Theory and construction of blackbody calibration sources. In A. E. Rozlosnik, & R. B. Dinwiddie (Eds.), *Thermosense XXIII (vol. 4360). International Society for Optics and Photonics.*

Jakobs, K. (2017). Corporate standardization management and innovation. In R. Hawkins, K. Blind, & R. Page (Eds.), *Handbook of innovation and standards*. Edward Elgar Publishing.

Judson, L. V. H. (1976). Weight and measure standards of the United States: A brief history (special publication 447). *Department of Commerce, National Bureau of Standards.*

Mallett, R. L. (1998). Why standards matter. *Issues in Science & Technology, 15*(2), 63–66.

Mansoor, K., Shahnawaz, S., Rasool, M., Chaudhry, H., Ahuja, G., & Shahnawaz, S. (2016). Automated versus manual blood pressure measurement: A randomized crossover trial in the emergency department of a tertiary care hospital in Karachi, Pakistan: Are third world countries ready for the change? *Open Access Macedonian Journal of Medical Sciences, 4*(3), 404–409.

Mattli, W., & Büthe, T. (2003). Setting international standards: Technological rationality or primacy of power? *World Politics, 56*(1), 1–42.

National Institute of Standards and Technology, US Department of Commerce. (2021). *Healthcare—standards and testing.* https://www.nist.gov/itl/products-and-services/healthcare-standards-testing.

O'Brien, E., Waeber, B., Parati, G., Staessen, J., & Myers, M. G. (2001). Blood pressure measuring devices: Recommendations of the European Society of Hypertension. *British Medical Journal, 322*(7285), 531–536.

O'Brien, G., Lesser, N., Wang, S., Zheng, K., Bowers, C., & Kamke, K. Securing electronic health records on mobile devices. National Institute of Standards and Technology, US Department of Commerce. https://doi.org/10.6028/NIST.SP.1800-1https://nvlpubs.nist.gov/nistpubs/SpecialPublications/NIST.SP.1800-1.pdf

Providing Regulatory Submissions in Electronic Format Standardized Study Data. (2014). https://www.fda.gov/downloads/Drugs/GuidanceComplianceRegulatoryInformation/shiftenterGuidances/UCM292334.pdf. [Accessed 11 April 2019].

Quality Management Principles. (2015). https://www.iso.org/publication/PUB100080.html. [Accessed 13 May 2019].

Sherman, N., & Zlotowitz (Eds.). (1993). *Stone Edition The Chumash*. New York, NY: Mesorah Publications.

Standards.gov. (n.d.). https://www.nist.gov/standardsgov

Slickers, K. A. (2003). Quality Standards: Harmonizing standards: Revisions to ISO 9000 prompts changes to ISO 13485, the standard specific to medical device manufacturers. *Quality Magazine, 42*(1).

Tyrvainen, P., Silvennoinen, M., Talvitie-Lamberg, K., Ala-Kitula, A., & Kuoremaki, R. (2018). Identifying opportunities for AI applications in healthcare—renewing the national healthcare and social services. In *2018 IEEE 6th International Conference on Serious Games and Applications for Health (SeGAH).*

US Food and Drug Administration. (n.d.). *Recognized Consensus Standards Database Search.* https://www.accessdata.fda.gov/scripts/cdrh/cfdocs/cfStandards/search.cfm

US Food and Drug Administration. (1990). Device recalls: A study of quality problems. US Department of Health and Human Services, Public Health Service, Food and Drug Administration, Center for Devices and Radiological Health.

US Food and Drug Administration. (2021). *Providing regulatory submissions in electronic format—standardized study data.* https://www.fda.gov/downloads/Drugs/GuidanceComplianceRegulatoryshiftenterInformation/Guidances/UCM292334.pdf.

Van Oeck, R. (2008). A whack on the side of the head. *Grand Central Publishing.*

Zuckerman, A. (1999). *Standards battles heat up between United States and European Union, quality progress. 39–42.*

Regulatory Requirements

12

If a State has reliable scientific information that demonstrates that a warning is needed for a particular food [or medicine], then in the interest of public health, it should share that information with the FDA and petition for a new national standard.
— Nathan Deal, 82nd Governor of Georgia

Chapter outline

Biomedical Engineering Design. https://doi.org/10.1016/B978-0-12-816444-0.00012-2

12.1 Introduction

Medical device regulations and regulatory bodies exist to protect the health and safety of the public. Medical devices sold around the world typically require review and approval by a governmental or sanctioning organization (regulatory body) confirming the devices to be safe (will not cause harm) and effective (do what they are claimed to do). The type and level of review depend on the classification of a medical device, which is based on its level of risk or potential to cause harm (death or severe injury) to a patient. Classifying a medical device is the first step in determining its regulatory pathway, which will impact the planned market introduction date and design of the device. Regulatory review requires manufacturers to submit different types of information and supporting data, typically generated during the design process, to demonstrate the safety and efficacy of the device. This includes data from verification and validation tests described in Chapters 9 and 10. Regulatory bodies also require evidence that procedures are in place to ensure that quality is maintained throughout various aspects of the business (such as design, manufacturing, and distribution) and that procedures exist for monitoring safety and efficacy of the device after market introduction. If problems are reported by users or customers, regulatory bodies often oversee corrective actions that may include a product recall.

In this chapter, you will become familiar with regulatory terminology and requirements, device classifications, pathways to market, post market surveillance activities, and reasons and procedures for product recalls.

12.2 Regulatory Considerations in Academic and Industry Design Projects

In an established medical device company, dedicated regulatory affairs personnel will be part of your project team. They will be responsible for determining the regulatory strategy for a new product, managing submissions to regulatory bodies, preparing for and guiding the company through audits and inspections, and managing post market surveillance of safety and efficacy.

Design engineers, although not responsible for regulatory oversight, are often in close communication with regulatory affairs personnel as they consider the impact of regulatory requirements on design decisions and project timelines. For example, regulatory requirements are considered when narrowing and refining design solutions (discussed in Section 6.3.3), analyzing risk (introduced in Section 8.9 and continued in Section 9.11), designing and executing tests (discussed in Sections 9.2.1 and 10.2.3), and conducting clinical trials (presented in Section 10.5.3). The Design History File (DHF) maintained throughout your project will be part of regulatory submissions. As discussed in this chapter and Section 14.5.2, design engineers may be involved in post market surveillance and product recalls. For these reasons, you will need to understand regulatory terminology and requirements, approval processes and pathways to market, risk management, and ethical obligations to the public.

In an academic design project, you are unlikely to have the time and resources needed to submit documentation to regulatory bodies. An understanding of regulatory requirements and approval processes can serve as a framework for guiding you through the design process and will help prepare you for a career in the medical device industry.

12.3 History of FDA Legislation and Regulation

The United States represents the largest market in the world for medical devices. Many countries model their regulatory approval processes after those of the US Food and Drug Administration (FDA). For this reason, much of this chapter discusses FDA processes and classifications. The European Union (EU), Japan, and other countries also represent large medical device markets. Section 12.6 discusses regulatory bodies in these regions to highlight differences in definitions and processes.

The FDA was established in 1906 to regulate food and drugs. FDA regulation of medical devices began in 1938. Figure 12.1 depicts a history of significant FDA legislation establishing and expanding the regulation of medical devices, and the adverse outcomes that prompted their creation and passage. The regulations that apply to medical devices may present hurdles to medical device companies, but they were established to require proof of safety and efficacy and protect the health and safety of the public. The first three legislation examples shown in Figure 12.1 were motivated by serious injuries or death caused by unsafe drugs or devices.

Table 12.1 describes the impact of significant US medical device legislation and regulations since 1976, when the Medical Device Amendments were passed, providing the FDA with greater authority to regulate medical devices. It demonstrates the constantly changing regulatory environment for medical devices. The regulatory environment is always changing and will continue to change throughout your career. As part of the need for lifelong learning, medical device designers need to stay up to date with the latest regulatory requirements and their impact on the design and commercialization process.

12.4 Product Classifications: Device, Drug, Biologic, or Combination Product?

A new medical product can be classified by the FDA as a device, drug, biologic, or combination product. Some products are classified as combination products with characteristics of more than one classification. For example, a drug-eluting stent is classified as a combination product because it is comprised of both a drug and device. Its primary intended use is to restore the diameter of an artery, which is considered a medical device function. As a secondary intended use, the stent acts as a drug delivery product because it releases a controlled dose of medication. This medication is classified as a

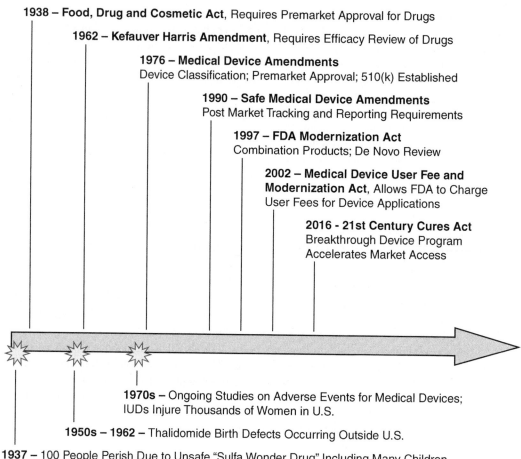

1938 – Food, Drug and Cosmetic Act, Requires Premarket Approval for Drugs

1962 – Kefauver Harris Amendment, Requires Efficacy Review of Drugs

1976 – Medical Device Amendments
Device Classification; Premarket Approval; 510(k) Established

1990 – Safe Medical Device Amendments
Post Market Tracking and Reporting Requirements

1997 – FDA Modernization Act
Combination Products; De Novo Review

2002 – Medical Device User Fee and Modernization Act, Allows FDA to Charge User Fees for Device Applications

2016 - 21st Century Cures Act
Breakthrough Device Program
Accelerates Market Access

1970s – Ongoing Studies on Adverse Events for Medical Devices; IUDs Injure Thousands of Women in U.S.

1950s – 1962 – Thalidomide Birth Defects Occurring Outside U.S.

1937 – 100 People Perish Due to Unsafe "Sulfa Wonder Drug" Including Many Children

FIGURE 12.1

History of significant FDA legislation providing for the regulation of medical devices and the adverse outcomes that prompted their passage.

drug. Thus, this type of device is a combination product. Classifications are guided by the definitions presented in the following sections and are used to determine the regulatory pathway for a new medical device.

12.4.1 Medical Devices

Medical devices cover a range of applications, medical specialties, risk, complexity, invasiveness, and cost. Examples include tongue depressors, CT scanners, heart valve replacements, and endoscopes. Medical devices also include *in vitro* diagnostic products, (e.g., general-purpose lab equipment, reagents, test kits), combination products (e.g., drug-eluting stents, birth control implants), companion

Table 12.1 Summary of significant medical device regulations from the last 50 years and their impact

Year	Title	Impact
1976	Medical Device Amendments	• Medical devices explicitly regulated by FDA • Ensures safety and efficacy of medical devices; requires registration with FDA • FDA authorized to regulate devices during development, testing, production, distribution, and use
1990	Safe Medical Device Amendments	• Required design validation, increased recall authority, tracking requirements, civil penalties
1996	FDA Quality Systems Regulations	• Emphasis on design control and post market surveillance
1997	FDA Modernization Act	• Added Office of Combination Products • Early collaboration • Set the "least burdensome" provision • Established De Novo review
2002	Medical Device Fee and Modernization Act	• Established User Fees for medical device reviews • Increased surveillance of medical devices
2016	21st Century Cures Act	• Deregulation of certain Clinical Decision Support software • Created Breakthrough Device program • Designed to accelerate market access in a time of significant innovation/technology advances

diagnostics (tests used with a therapeutic drug to determine its applicability to a specific patient), software (depending on the specific application), mobile medical apps, and some accessories. Products emitting electronic radiation, such as diagnostic ultrasound products, x-ray machines, and medical lasers, may also meet the definition of a medical device.

In the United States, if a product is labeled, promoted, or used in a manner that meets the following definition found in section 201(h) of the Federal Food, Drug, and Cosmetic (FD&C) Act, it will be regulated by the FDA as a medical device and will be subject to premarketing and post marketing regulatory controls:

A medical device is "an instrument, apparatus, implement, machine, contrivance, implant, in vitro reagent, or other similar or related article, including a component part, or accessory which is (1) recognized in the official National Formulary, or the United States Pharmacopoeia, or any supplement to them, (2) intended for use in the diagnosis of disease or other conditions, or in the cure, mitigation, treatment, or prevention of disease, in man or other animals, or (3) intended to affect the structure or any function of the body of man or other animals, and which does not achieve its primary intended purposes through chemical action within or on the body of man or other animals, and which is not dependent upon being metabolized for the achievement of its primary intended purposes.

This definition helps distinguish between medical devices and other FDA regulated products such as drugs and biologics. Software functions that meet this definition may be regulated as medical devices.

Medical devices are regulated by the Center for Devices and Radiological Health (CDRH). Other centers include the Center for Veterinary Medicine, Center for Tobacco Products, and Center for Food Safety and Applied Nutrition. Our discussion will be limited to centers most relevant to medical devices that impact human health.

12.4.2 Drugs

If the primary intended use of a product is achieved through chemical action or metabolism, the product is typically classified as a drug. According to the FD&C Act, a drug is:

- a substance recognized by an official pharmacopoeia or formulary;
- a substance intended for use in the diagnosis, cure, mitigation, treatment, or prevention of disease;
- a substance (other than food) intended to affect the structure or any function of the body; or
- a substance intended for use as a component of a medicine but not as a device or a component, part, or accessory of a device.

Human drugs are regulated by FDA's Center for Drug Evaluation and Research (CDER).

12.4.3 Biological Products

Per the FDA, a biological product is defined as:

> ...a virus, therapeutic serum, toxin, antitoxin, vaccine, blood, blood component or derivative, allergenic product, protein (except any chemically synthesized polypeptide), or analogous product, or arsphenamine or derivative of arsphenamine (or any other trivalent organic arsenic compound), applicable to the prevention, treatment, or cure of a disease or condition of human beings.

Biological products, including blood and blood products, and blood banking equipment, are regulated by FDA's Center for Biologics Evaluation and Research (CBER).

12.4.4 Combination Products

According to the FDA, combination products include:

- a product comprised of two or more regulated components (i.e., drug/device, biologic/device, drug/biologic, or drug/device/biologic) that are physically, chemically, or otherwise combined or mixed and produced as a single entity;
- two or more separate products packaged together in a single package or as a unit and comprised of drug and device products, device and biological products, or biological and drug products;
- a drug, device, or biological product packaged separately that, according to its investigational plan or proposed labeling, is intended for use only with an approved individually specified drug, device, or biological product where both are required to achieve the intended use, indication, or effect, and where upon approval of the proposed product, the labeling of the approved product would need to be changed (e.g., to reflect a change in intended use, dosage form, strength, route of administration, or significant change in dose); or

- any investigational drug, device, or biological product packaged separately that, according to its proposed labeling, is for use only with another individually specified investigational drug, device, or biological product where both are required to achieve the intended use, indication, or effect.

Combination products are regulated depending on their primary intended use. For the drug eluting stent example presented earlier, the primary intended use is to mechanically restore the diameter of an artery, which is the function of a medical device. For this reason, it would be regulated as a medical device—not a drug—and would follow one of the regulatory pathways presented in Section 12.6. Other examples of combination products include hormone releasing intrauterine devices, prefilled syringes containing medication, and dermal patches filled with drugs.

12.4.5 Software

The FDA regulates certain mobile medical apps, software embedded in medical devices, and certain clinical decision support software. This section focuses on mobile medical apps. According to the FDA website:

Mobile apps are software programs that run on smartphones and other mobile communication devices. They can also be accessories that attach to a smartphone or other mobile communication devices, or a combination of accessories and software. Mobile apps span a wide range of health functions. Many mobile apps carry minimal risk; those that can pose a greater risk to patients (if the device were to not function as intended) require FDA review.

Mobile medical apps are mobile apps that meet the definition of a medical device and are an accessory to a regulated medical device or transform a mobile platform into a regulated medical device. The FDA is taking a tailored, risk-based approach that focuses on the small subset of mobile apps that meet the regulatory definition of "device" (mobile medical apps) and that are intended to be used as an accessory to a regulated medical device or transform a mobile platform into a regulated medical device.

Other apps regulated by the FDA aim to help healthcare professionals improve and facilitate patient care. For example, the Radiation Emergency Medical Management (REMM) app gives healthcare providers guidance on diagnosing and treating radiation injuries. Some mobile medical apps can help physicians diagnose cancer or heart rhythm abnormalities, or function as the "central command" for a glucose meter used by an insulin-dependent diabetic patient.

12.5 Device Classifications and Controls

In the US, medical devices are classified as Class I, II, or III. This classification is based on (1) the level of control needed to provide reasonable assurance of safety and efficacy, (2) design complexity and potential risk, (3) intended use of the device, and (4) indications for use. Intended use and indications for use can be found in the device's labeling and are also conveyed orally during product sale.

The main pathways for FDA review of a new device include the 510(k), Premarket Approval (PMA) application, and others. These pathways and associated submissions for devices in each classification are discussed in detail in Section 12.8.

Descriptions, submission requirements, applicable regulatory controls, and examples of Class I, II, and III medical devices are presented in Table 12.2. All classifications require compliance with

Table 12.2 Descriptions, submission requirements, applicable regulatory controls, and examples of Class I, II, and III medical devices

Classification	Class I	Class II	Class III
Description	Not intended for use in supporting or sustaining life Not likely to present a potential unreasonable risk of illness or injury	General controls insufficient to ensure safety and effectiveness	Sustains or supports life Implanted Presents an unreasonable risk of illness or injury
Risk level	Low to moderate	Moderate to high	High
General controls	Yes	Yes	Yes
Special controls		Yes	Yes
Submission type	510(k)—most devices exempt	510(k)—few devices exempt	Premarket Approval (PMA)
Non-clinical data required?	Yes	Yes	Yes
Clinical data required?	Some	Some	Usually requires scientific evidence to prove safety and efficacy
Examples	Bandages, tongue depressors, hospital beds, surgical instruments, arm slings, nonpowered wheelchairs	Percutaneous transluminal coronary angioplasty (PTCA) catheters, cardiac devices, intravenous pumps, blood pressure cuffs, dialysis equipment, powered wheelchairs	Heart valves, breast implants, urological implants, pacemakers, orthopedic implants, intraocular lenses

Details on general and special controls are presented in Sections 12.5.1 and 12.5.2. Submission types are discussed in Section 12.8

general controls. Class I devices are exempt from the 510(k) process if they are similar to products sold prior to 1976, when the Medical Device Amendments were established. Devices sold prior to 1976 are pre-amendment devices and are "grandfathered" into the approval process (exempt from new requirements). Class II devices generally require compliance with special controls, a premarket notification (510[k]), and may require clinical study data to demonstrate substantial equivalence to a predicate device. Class III devices require PMA including clinical study data to prove safety and effectiveness.

12.5.1 General Controls

General controls are requirements that apply to all classifications of medical devices. They require manufacturers to:

- Register each manufacturing location with the FDA.
- List all marketed medical devices with the FDA.
- Manufacture devices in compliance with FDA Quality System Regulations (QSRs).
- Label devices in compliance with applicable regulations.
- Submit Premarket Notification, 510(k), unless device is exempt or subject to additional requirements (to be discussed in Section 12.8.1).

Table 12.3 Regulatory bodies responsible for medical device approvals for nine countries located in all inhabited continents

Country	Regulatory Body
United States	Food and Drug Administration Center for Devices and Radiological Health (CDRH) https://www.fda.gov/Medical-Devices
European Union	European Medicines Agency https://www.ema.europa.eu/en
Canada	Health Canada https://www.canada.ca/en/health-canada.html
Japan	Pharmaceuticals and Medical Devices Agency (PMDA) https://www.pmda.go.jp/english/index.html
Australia	Therapeutic Goods Administration https://www.tga.gov.au/
Brazil	Brazilian Health Surveillance Agency https://www.gov.br/anvisa/pt-br
Israel	Ministry of Health https://www.gov.il/en/departments/ministry_of_health/govil-landing-page
China	National Medical Products Administration (NMPA) http://english.nmpa.gov.cn/index.html
South Africa	South African Health Products Regulatory Authority https://www.sahpra.org.za/

12.5.2 Special Controls

Special controls apply to Class II medical devices for which General Controls alone are inadequate to ensure safety and efficacy. This is due to the potential health risk associated with the device. Special controls will vary among Class II devices but may include special labeling requirements, conformance with FDA guidance documents and mandatory performance standards, clinical trials, and post market surveillance.

12.6 Regulatory Requirements of Other Countries

Medical device manufacturers who wish to market their products in the United States must comply with the regulatory requirements of the FDA. Those who want to market their products in other countries must comply with the requirements established by each country's regulatory body. Many of these bodies require compliance with their own quality system regulations, various ISO standards, and, more increasingly, the Medical Device Single Audit Program (MDSAP) to market medical devices. The MDSAP is an international coalition whose goal is to jointly leverage the regulatory resources of member countries to manage an efficient, effective, and sustainable single audit program focused on the oversight of medical device manufacturers. It allows for a single audit of a medical device manufacturer that each member country would accept. Each of the countries listed in Table 12.3 requires their own registration for market access.

Different definitions of medical devices are used by the regulatory bodies in each of these countries. Some include devices used in humans or animals, whereas others are limited to use in humans. The examples presented here from the EU, Canada, and Japan were chosen to illustrate these differences. They are similar to the definition used by the FDA in that they address distinctions between drugs and devices.

12.6.1 European Union

According to EU Medical Devices Regulation and EU In-Vitro Diagnostic Medical Devices Regulation IVCR (as of May 2022), the term "medical device" refers to any:

instrument, apparatus, appliance, software, material or other article, whether used alone or in combination, including the software intended by its manufacturer to be used specifically for diagnostic and/or therapeutic purposes and necessary for its proper application, intended by the manufacturer to be used for human beings for the purpose of:
- diagnosis, prevention, monitoring, treatment or alleviation of disease,
- diagnosis, monitoring, treatment, alleviation of or compensation for an injury or handicap,
- investigation, replacement or modification of the anatomy or of a physiological process,
- control of conception,

and which does not achieve its principal intended action in or on the human body by pharmacological, immunological or metabolic means, but which may be assisted in its function by such means.

In contrast to the definition used in the United States, the European definition does not include devices used for animals.

12.6.2 Japan

In Japan, the Pharmaceutical Affairs Law (PAL) defines medical devices as:

instruments and apparatus intended for use in diagnosis, cure or prevention of diseases in humans or other animals; intended to affect the structure or functions of the body of man or other animals.

This definition does not directly address the differences between drugs and devices, and does not address software, but does include devices used for animals.

12.6.3 Canada

In Canada, the term "medical device" is defined in the Food and Drugs Act as:

any article, instrument, apparatus or contrivance, including any component, part or accessory thereof, manufactured, sold or represented for use in: the diagnosis, treatment, mitigation or prevention of a disease, disorder or abnormal physical state, or its symptoms, in a human being; the restoration, correction or modification of a body function or the body structure of a human being; the diagnosis of pregnancy in a human being; or the care of a human being during pregnancy and after the birth of a child, including the care of the child. It also includes a contraceptive device but does not include a drug.

This definition does not include devices used for animals, nor does it include drugs. It specifically mentions pregnancy (diagnosis, care during pregnancy, and care after birth for the woman and the child) and the child born thereafter.

In summary, many countries have their own regulatory bodies, regulatory requirements, and definitions of medical devices. To market their medical devices in the US, medical device manufacturers must comply with the regulatory requirements of the FDA. To market their products in other countries, they must comply with the requirements established by each country's regulatory body. In some countries, receiving FDA approval or clearance for a device to be marketed in the US often makes it easier to receive similar approval in those countries.

12.7 Quality Requirements

FDA requirements are based on a quality systems approach that affects design and development activities. For this reason, design engineers need to understand quality systems and quality-related regulatory requirements. The term "quality," as it relates to medical device design, refers to how well a design meets customer needs and performs as expected. Quality does not refer to high-end products with many features and "bells and whistles," nor is a simpler design considered to be of low quality. For example, a Foley catheter with multiple lumens, check valves, and irrigation ports may be considered a low-quality product if it does not meet customer needs. A simple rubber intermittent catheter with a single lumen and no additional features may be considered a high-quality product if it meets customer needs and expectations. Additional quality-related terms defined in ISO 9000 and used in ISO 13485 are described below:

Quality management refers to those aspects of a company's overall management function (including upper management) that determine and implement quality policy, which outlines a company's goals as they relate to product and process excellence.

Quality system (or **quality management system**) refers to a company's structure, procedures, processes, and resources for implementing quality management, with the goal of ensuring that devices remain safe and efficacious. Medical device companies create and maintain a **quality manual** that describes their quality system. This system also standardizes activities and should provide efficiency in how a business is run.

Quality assurance is achieved through the systematic and planned actions that are needed to provide an acceptable level of confidence that a product will meet the requirements for quality (will meet customer needs).

Quality control is a specific component of quality assurance and refers to the operational methods, techniques, and activities used by an organization to meet quality requirements. This includes incoming inspection procedures, vendor quality audits, and other activities.

12.7.1 Good Manufacturing Practice

Medical device manufacturers are required to design and build quality into their products. This means that they must have processes and procedures in place to ensure that products meet customer needs and performance requirements. Companies must document their quality processes and procedures and prove that these are being followed during product development and after market introduction. Without this documentation, devices will not be approved or cleared in the US and thus will not be allowed for sale in major international markets.

Federal regulations require drugs and devices to be manufactured in accordance with current good manufacturing practice (cGMP, sometimes referred to as GMP). These GMP's provide minimum requirements for what companies must do regarding methods, facilities, and controls used in the manufacturing, processing, packaging, and storage of products, and provides significant flexibility as to how the company accomplishes this. Current GMP requirements include an extensive quality systems approach that includes design and development activities. For medical devices, these are contained in the *Quality Systems Regulation* (QSR). The focus on quality has led to ongoing efforts to harmonize FDA requirements with those of international standards such as ISO 13485 (presented in Section 11.4), which is required by the EU and other countries. The FDA recognizes that requirements contained in the QSR are like those that have been part of the regulations of other countries, confirming that these are necessary to ensure that products are designed and manufactured to meet customer needs. For this reason, the FDA announced in October 2020 its intention to harmonize and modernize the QSR for medical devices. Revisions to the QSR will update the existing requirements with those of ISO 13485:2016. These revisions are intended to promote the use of more modern risk management principles and reduce regulatory burdens on device manufacturers and importers by harmonizing domestic and international requirements.

Currently, medical device manufacturers are required by the FDA to comply with the QSR. The proposed change will require them to meet the requirements of ISO 13485 as part of the QSR. Included in these are document control procedures and requirements for maintaining a DHF.

12.7.2 Design History File

As emphasized in almost every chapter, the DHF is a critical component of the medical device design process. It is required and contains or references the records needed to prove that the device was developed per the approved design and development plan and design control requirements of the QSR. The DHF documents the team's progress and the evolution of the design and provides useful information for future academic or industry design teams working on projects started by previous design teams. It contains information that is part of the DMR, which is needed to begin production of a medical device.

In an academic design project, the DHF may be in the form of a project notebook that includes meeting minutes, design input and output documents, design changes, results of design reviews, drawings, sketches, calculations, test procedures, assembly procedures, design verification and validation results, and evidence that design input = design output. Project definition documents, project schedules, customer interview questions and responses, customer needs, and target product specifications are often also included.

In industry, the DHF includes additional items such as manufacturing work instructions, labeling, packaging and shipping information, sterilization procedures, shelf life test results, maintenance and disposal instructions, and information on accessories and components. This information will also be a part of the DMR.

12.7.3 Registration of Medical Device Manufacturers

In the United States, there are more than 1700 categories of medical devices among 16 medical specialties produced by more than 10,000 medical device manufacturers. These companies, along with new startup medical device manufacturers, must register with the FDA to notify them of their intent to manufacture medical devices. This makes the FDA aware of the new manufacturer and allows them to audit the production facilities of the company to ensure that cGMPs are being followed and that requirements of the QSRs are being met. This is often accomplished through careful review of the documentation contained in the manufacturer's DHF. If violations are found during an audit, the FDA has the legal authority to force manufacturers to correct the violation or shut down a manufacturing facility until violations have been corrected.

12.8 Pathways to Market in the United States

The design of a new product can impact its regulatory pathway to market. Use of new materials not previously used in medical applications—or new features not part of existing medical devices—can present concerns to the FDA regarding device performance independent of device class and can trigger requirements for costly additional material safety tests and clinical studies, respectively. It is important for engineers to understand how their design can influence the required regulatory pathway. If speed to market is a high priority, then the design may need to be revised so that no unplanned additional testing will be required.

Most medical device companies employ experts in regulatory affairs to develop the regulatory strategy for each new product. It is their responsibility to identify the shortest path to allow the company to begin sales of a new product and determine the quickest appropriate pathway to market. Companies initially determine the regulatory pathway that they feel is appropriate; however, the FDA makes the ultimate decision regarding which path will be required. Regulatory personnel from medical device companies often meet with FDA representatives to get a better idea of which regulatory path the FDA will require. In some situations, it may be possible for academic design teams to contact FDA personnel for advice on the appropriate path for their devices.

In this section we discuss the main pathways for FDA review of a new device which include the 510(k), PMA, De Novo classification request, Breakthrough Device Designation, and Humanitarian Device Exemption (HDE). Each has a different impact on the project schedule, total project cost, and required testing, and has different requirements to qualify for the pathway.

12.8.1 510(k)

The 510(k) pathway gets its name from section 510, paragraph (k), of the FD&C Act that directs the FDA to clear medical devices for marketing. It allows manufacturers to market their devices if they can demonstrate *substantial equivalence* to a *predicate device* that is currently sold through interstate commerce through submission of a 510(k) premarket notification. This pathway is generally applicable to Class I and II devices where the specific classification regulation states that a 510(k) is required. To demonstrate substantial equivalence, manufacturers must show that the new device performs the same function(s) as the predicate device, is intended for the same use as the predicate device, and that any differences between the two devices do not raise any new safety or efficacy concerns. Once a manufacturer submits a 510(k) notification of its intent to market the device, the FDA has 90 calendar days to review the premarket notification (depending on the 510[k] used), and the company has 6 months to resolve any questions asked by FDA. A 510(k) submission includes:

- a description of the device,
- the indications for use,
- a discussion of the predicate device and how the new device is similar in design and function, and
- labels and labeling (including full operator manuals).

It also includes, as applicable:

- protocols and reports for shelf life, biocompatibility, reprocessing validation, and performance testing,
- protocols and reports for design verification and validation testing (with a specific focus on software documentation),
- cybersecurity information,
- results of clinical trials (when necessary),
- evidence of compliance with applicable standards, and
- other information.

There are three types of 510(k) submissions. First, the *Traditional 510(k)* is the most common, and is used for more significant changes or when a company does not own the predicate 510(k). It includes descriptive information regarding indications for use, the device, and results of performance testing to demonstrate substantial equivalence to a predicate device (if not identical to the predicate device). Second, manufacturers can use the *Special 510(k)* to modify their own 510(k) cleared device. If the modification does not change the intended use or the device significantly, then information from the design control process (DHFs) and other required 510(k) components can be used to obtain clearance. Third, the *Abbreviated 510(k)* allows manufacturers to submit summary reports on the use of FDA guidance documents, special controls, or declarations of conformity to FDA-recognized consensus standards to request clearance to market their devices.

The difference between "clearance" and "approval" is notable. If the data included in a 510(k) submission supports claims of substantial equivalence, then the FDA will "clear" the device for marketing. FDA approval is only granted to drugs through the New Drug Application process and devices through the PMA process, and possibly the HDE pathway presented in the next two sections.

12.8.2 Premarket Approval

As stated in Table 12.2, Class III devices are generally subject to PMA. This process involves a submission by the manufacturer and a scientific review of safety and efficacy data by an FDA Advisory Committee. If a device is not yet classified (completely new/novel device or significant change to an existing device), the device is automatically considered Class III and may be subject to PMA unless it is down classified via the *De Novo* process (based on risk). A PMA submission usually requires the submission of clinical study protocols and reports (discussed in Chapter 10). If they present a significant risk (which most PMA devices would), the clinical study requires a separate FDA review prior to study initiation. An Investigational Device Exemption (IDE) is required before the study can begin and allows an investigational device to be used in a clinical trial to support a

PMA application. It also allows manufacturers to ship their unapproved devices across state lines for purposes of conducting a clinical study. The PMA process takes an average of 250 days to receive approval. The PMA is a license granted to market the device, which prevents competitors from selling their similar devices without obtaining their own PMA license. This serves as a barrier to entry for competitors.

The PMA application contains scientific and regulatory documentation that demonstrates safety and efficacy of a Class III device. It includes indications for use, a device description, nonclinical study data (bench testing, animal studies, biocompatibility tests, shelf life studies, sterilization validations, software validations), clinical studies, manufacturing processes and validations, and labeling. PMAs have the strictest post approval requirements, meaning the level of required reporting and FDA scrutiny after approval is higher than that of other device classifications.

12.8.3 Humanitarian Device Exemption

The Humanitarian Use Device (HUD) classification was created in 1990 as part of the Safe Medical Devices Act to establish an alternate, expedited pathway to market in order to benefit patients with a disease or condition affecting fewer than 8000 people per year in the United States. It provides a financial incentive to manufacturers to develop devices for small populations in which, due to limited sales, the expected returns on investment would not exceed the required research and development costs. Examples of devices granted an HDE include a patient-specific talus spacer, pediatric esophageal atresia device, and an aneurism neck reconstruction device. Many academic design projects involve devices for people with disabilities. These projects often impact a small number of users and might qualify for the HDE.

To obtain approval for a HUD, an HDE application is submitted to the FDA. The application includes similar information to what is included in a PMA application but is not required to include results of scientifically valid clinical studies that prove efficacy for the device's intended use. The application must contain information to show that, in comparison to available devices or existing alternate forms of treatment, the device does not present an unreasonable or significant risk of illness or injury, and that the health benefits outweigh the risk of injury or illness due to use of the device. The case must also be made that there are no comparable devices available to diagnose or treat the disease or condition. Once approved by the FDA, a HUD is limited to use in facilities with an established local IRB (discussed in Section 10.10) that has granted approval to use the device to treat or diagnose the specific disease or condition.

12.8.4 Breakthrough Device Designation

A breakthrough device is one that provides for more effective treatment or diagnosis of life-threatening or irreversibly debilitating diseases or conditions compared to what is available. The goal of the Breakthrough Devices Program is to provide patients and healthcare providers with timely access to these medical devices by speeding up their development, assessment, and review, while still using the PMA, 510(k) clearance, and other established pathways to market, consistent with the FDA's mission to protect and promote public health. Breakthrough devices (or combination products) qualify for expedited review and a faster approval path if they meet all of the following criteria:

- intended to treat or diagnose a life-threatening or irreversibly debilitating disease or condition,
- represents a breakthrough technology that provides a clinically meaningful advantage over existing technologies; leads to clinical improvement in diagnosis or treatment of a life-threatening or irreversibly debilitating disease or condition,
- no approved alternative treatment or means of diagnosis exists,
- offers significant, clinically meaningful advantages over existing approved alternative treatments (if they exist), and
- device availability is in the patient's best interest; provides a specific public health benefit or meets the needs of a well-defined patient population.

12.8.5 De Novo Classification

The De Novo classification applies to novel devices that have not been previously classified. These devices are automatically classified as class III (high risk) when first reviewed by the FDA. These devices did not exist prior to the 1976 Medical Device Amendments and no predicate devices exist to allow the 510(k) pathway to be pursued. The De Novo classification process provides a pathway for manufacturers to request that a device which was automatically classified as class III to be down classified to class I or class II, based on risk.

If general controls alone, or a combination of general and special controls, can provide a reasonable assurance of safety and efficacy for the intended use of the device, but no legally marketed predicate device exists, then a manufacturer can request a De Novo classification. Devices that are classified as class I or class II as the result of a De Novo classification request can be marketed and cited as predicate devices for future 510(k) submissions.

There are two situations for which a manufacturer can submit a De Novo request for a risk-based evaluation to re-classify the device as class I or II. First, after receiving a determination of "not substantially equivalent" (NSE) in response to a 510(k) submission, a request can be submitted if there is no predicate device, no new intended use, or no different technological characteristics that raise different questions of safety and effectiveness. Second, upon the manufacturer's determination that there is no legally marketed device upon which to base a determination of substantial equivalence, a request can be submitted without first submitting a 510(k) and receiving an NSE determination.

De Novo requests include information such as:

- the device's intended use designation, prescription or over-the-counter use designation,
- device description, including technology, proposed conditions of use, accessories, and components,
- discussion of why general controls or general and special controls provide reasonable assurance of the safety and effectiveness of the device, and, if proposing a class II designation, what special controls would allow the FDA to conclude there is reasonable assurance the device is safe and effective for its intended use,
- clinical data (if applicable) to support reasonable assurance of the safety and effectiveness of the device,
- results of animal studies (if applicable),
- nonclinical data including bench performance testing,
- information on reprocessing and sterilization, shelf life, biocompatibility, software, electrical safety, and electromagnetic compatibility, and

- a description of the probable benefits of the device when compared to the probable or anticipated risks when the device is used as intended.

12.9 Post Market Surveillance and Medical Device Recalls

Post market surveillance activities are required by ISO 13485 and the QSRs. Healthcare systems or providers are required to report to the FDA and the device manufacturer any device-related incidents resulting in injury, damage to health, or death within a specified time period after learning of the incident. Reports are submitted to the FDA through the Manufacturer and User Facility Device Experience (MAUDE) Database. This database tracks a specific product's failure rate which can influence the FDA's decision to recall a medical device.

Manufacturers are also required to report these incidents to the FDA within a specified period and investigate the cause of the incident. The process through which a product recall is recommended (by the FDA) or initiated (by the manufacturer) will depend on the severity and frequency of the reported incidents. The engineer's potential role in the recall process, including conducting investigations into the cause of the incident, is presented in Section 14.5.

A product recall is the removal or correction of a marketed product, including its labeling and/or promotional materials. Recalls may affect a specific lot (batch of product made at one time) or all products in the field. This will depend on the cause of the recall. If the cause can be traced to a particular lot of material or specific components, then all lots of final product containing material from the faulty lot can be recalled, leaving properly functioning product in the field. If the cause is traced to a poor design, then the entire lot may be recalled. Corrections may involve actions taken to modify products in the field (repair or retrofitting with a new component) that would be difficult (and costly) to ship back to the manufacturer (e.g., CT scanners, MRI systems, PET scanners). In some situations, letters to hospitals and physicians notifying them of the problem with instructions on how to prevent it from occurring, software updates, or shipment of new parts or maintenance kits, can be used to correct the problem without removing product from the field.

12.9.1 Reasons for Medical Device Recalls

The approval/clearance process is not a guarantee of acceptable clinical performance. Medical devices that performed well in bench tests, animal tests, and human clinical studies may at times experience problems after market introduction. If serious enough, these problems can result in a product recall.

In a study of medical device recalls from 2003 to 2012, the FDA reported the most common reason for medical device recalls was device design, as shown in Figure 12.2. More than 60% of recalls were due to design or material defects. Other causes included process control and manufacturing errors and change control issues. Another common cause for recalls is misbranding, which includes problems with labels (e.g., nonsterile product labeled as sterile).

A more recent study of medical device recalls from 2013 to 2018 showed that recalls due to software problems increased due to the increasing complexity and sophistication of the technology used in medical devices. However, device design issues continued to be the most common reason for device recalls, followed by process control, component design, and software design issues. These issues often

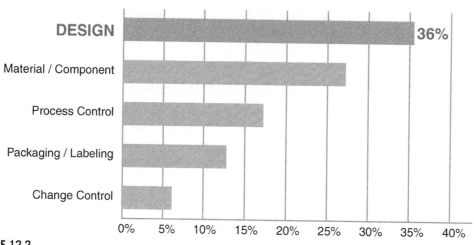

FIGURE 12.2

Common reasons for medical device recalls. According to the FDA, design-related issues were the most common reasons for medical device recall during the period of 2002 to 2013.

relate to how well a company's Quality Management System (a required "general control" for all medical devices) is implemented.

Problems not observed in clinical studies can occur for several reasons. First, in animal and human clinical studies, the sample size for device studies is typically fewer than 400 patients. Once released to the market, the device will typically be used in many more patients. Problems with small probabilities of occurrence that did not surface with the relatively small sample sizes used in testing may begin to surface when used in the much larger population available after market introduction.

Second, if prototype devices were used in the human clinical study instead of production units as discussed in Section 14.3.3, then slight differences between the two can result in differences in performance. These differences can lead to problems not observed in the clinical study but observed after market introduction. One lesson many recalls can teach us is that minor design changes or product or process improvements can often lead to unanticipated changes in product performance. It is important to ensure that all changes, no matter how minor they appear, are tested to confirm that new problems have not been created.

Third, if a product is used for an application not studied as part of a human clinical study, problems may surface after market introduction. This type of use is referred to as *off-label use*. Companies must provide clinical evidence that supports the claims they want to make about what their product can do (label claims). The FDA will approve use and label claims based on the population and application studied. Manufacturers are not allowed to promote off-label use of a device. However, physicians are not regulated by the FDA, which allows them to do what they feel is medically necessary and in the best interest of their patient. Off-label use as a label claim will not be approved by the FDA without clinical data to support the safety and efficacy of the device. Breakout Box 12.1 presents unexpected results of the off-label use of a drug.

Fourth, if a device is used with another device that was not used in the clinical study, interactions between the devices will not have been anticipated and thus not tested. This can result in unexpected outcomes and problems after market release.

Breakout Box 12.1 History of Minoxidil and its Off-Label Use to Treat Baldness

Minoxidil is an example of a drug intended for one use, but with an off-label use that led to the discovery of a new application for the drug. It was developed in the late 1950s by the Upjohn Company to treat ulcers. Animal studies indicated that it did not eliminate ulcers but was a strong vasodilator. In 1979, these studies resulted in FDA approval of minoxidil in the form of oral medication (pills) to treat hypertension (high blood pressure).

When Upjohn received FDA permission to test the new drug to treat hypertension, they found a physician to serve as principal investigator to conduct clinical studies to further study the drug. Results showed unexpected hair growth. The clinical investigator consulted a dermatologist and discussed the possibility of using minoxidil for treating baldness. They obtained the drug and conducted their own research without notifying Upjohn. The two doctors experimented with a 1% solution of minoxidil mixed with alcohol-based liquids.

Physicians began prescribing minoxidil to their patients in the 1980s to treat baldness (an off-label use). In August 1988, the FDA approved the drug for treating baldness in men under the trade name "Rogaine." Study results indicated that 39% of the men studied had "moderate to dense hair growth on the crown of the head." In 1991, Upjohn made the product available for women. In 1996, the FDA approved both the over-the-counter sale of the drug and the production of generic formulations of minoxidil.

Fifth, clinical investigators are typically well-trained in and very familiar with the proper use of the device being tested in the clinical study. After market introduction, the device will be used by medical providers who are not as familiar with the device or are not as well trained in its use. This can lead to improper use of the device and problems not observed during the clinical study.

12.9.2 Types and Classifications of Product Recalls

There are three types of recalls each with different levels of FDA involvement in the decision to recall the product: voluntary, semi-voluntary, and statutory. *Voluntary*, or company-initiated recalls, are those that the manufacturer or distributor voluntarily initiates at any time. The company is required to notify the FDA of the recall, which prompts the FDA to classify the recall according to the level of risk associated with the recalled product. The FDA will then monitor the recall process. *Semi-voluntary*, or FDA-requested recalls, are typically initiated by the FDA in urgent situations. If the company complies with the request to recall the product, it would be a voluntary recall. If the company does not agree to comply, then the FDA, using its mandated recall authority, will obtain a court order to authorize seizure of the product, resulting in a *statutory*, FDA-ordered recall.

There are three classifications of recalls, each representing a different level of risk to health posed by the recalled product. These recall classifications are not the same as the device classifications discussed earlier in this chapter. A Class I recall is used when there is a strong likelihood that the product will seriously affect health or cause death. A Class II recall applies when the use of or exposure to the product may cause temporary or reversible adverse health consequences, or when the probability of serious adverse health consequences is low. A Class III recall is used when the use of or exposure to the product is not likely to cause adverse health consequences. This would apply to a product that is recalled due to misbranding only.

When a Class I or II recall occurs, the FDA recommends that companies alert the public by issuing a press release. Recalls can tarnish a company's reputation. A 2017 McKinsey Report estimates manufacturers can experience as much as a 10% decrease in share prices after a major recall. To prevent this, medical device companies take steps to prevent recalls from occurring and have internal systems and procedures in place to quickly deal with recalls if they occur. Complying with the FDA QSRs and Design Control requirements presented in Section 11.4 can help prevent recalls.

12.9.3 Examples of Product Recalls

Despite the best efforts of medical device manufacturers to prevent product recalls, they do occur. In addition, recalls can result from lack of product testing, when testing was not conducted due to assumptions that minor design changes did not require testing or skipped to save time (reasons presented in Sections 12.9.1 and 14.3.3). Some of the companies involved notified the public immediately, and other companies chose to keep information from patients and physicians to protect the company's reputation and stock prices. Some of the recalls resulted in fines and/or settlements. Funds from settlements have been used to cover medical expenses of patients who wish to have their recalled implantable devices removed and replaced with newer, safer devices. Breakout Boxes 12.2, 12.3, and 12.4 present three examples of medical device recalls.

Breakout Box 12.2 Bjork-Shiley Heart Valve Recall, Pfizer Inc. (1986)

One model of an implantable heart valve in the 1970s, the Bjork-Shiley Convexo-Concave Heart Valve, consisted of several parts, as shown in Figure 12.3. The inlet and outlet struts worked together to keep the disc in place. If one of these struts failed, the disc could migrate into a chamber of the heart, and the heart valve would no longer function, resulting in death. Worldwide, 82,000 valves were sold and 31,230 were implanted in patients between 1979 and 1986. Among these, 619 strut fractures were reported (incidence rate of approximately 2%), of which roughly two-thirds resulted in death. The problem was traced to polishing of improperly welded struts that contained cracks instead of rewelding the struts or discarding the entire valve assembly. An investigation revealed that manufacturing documents were falsified and some test data regarding strut fractures observed during testing was not shared with the FDA (discussed further in Chapter 13). Lawsuits resulted in fines of $10.75 million and the company agreed to pay an additional $10 million to cover future medical costs for patients electing to replace their recalled heart valves. In 1991, a class action lawsuit resulted in the company agreeing to pay between $155 million and $205 million to create a fund to pay patients for cardiac consultation and research to identify which patients had significant risks of strut fractures.

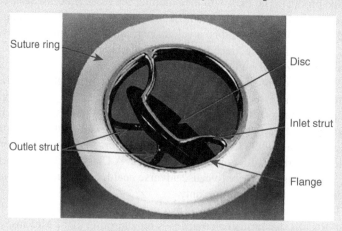

FIGURE 12.3

Design of the Bjork-Shiley heart valve. Note the location of the inlet and outlet struts.

Breakout Box 12.3 Sulzer Inter-Op Acetabular Shell Recall, Sulzer Orthopedics (2000)

Sulzer Orthopedics manufactured the Inter-Op Acetabular shell. These titanium alloy shells contained an acetabular bearing that articulated with a femoral head. The surface of the shell contacting bone included a porous coating to allow bony ingrowth to anchor the device into the bone without the need for bone cement.

Figure 12.4A shows an example of a femoral component of a hip implant with a porous coating on the upper half of the femoral stem, below the neck. (It is included here for illustration purposes only; it is not an image of the recalled Sulzer product.) Note the speckled appearance due to the rougher surface of the porous coating. Figure 12.4B shows a different femoral stem after removal (explantation) of the femoral component. Note the presence of ingrown bone tissue remaining in the porous coating.

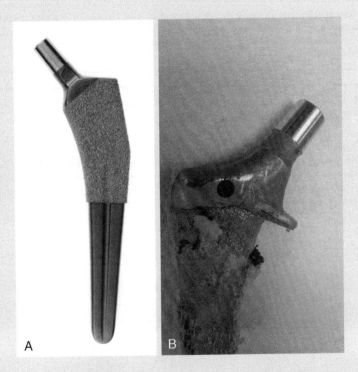

FIGURE 12.4

Porous coatings on femoral stems. (A) Shows porous coating on upper half of femoral stem prior to implantation. (B) Shows bone tissue remaining in an explanted device.

In 2000, the company noticed a higher than normal number of revision surgeries associated with the use of their shells. Although the shells were fixed to the pelvis with bone screws to prevent movement and allow bone to grow into the surface, they began to loosen. Patients experienced persistent, sharp groin pain due to this loosening and were unable to bear their weight on the leg with the implant. After completing an investigation of patient records, surgical techniques, and the product itself, the company ordered a voluntary recall. There were approximately 25,000 affected shells, of which approximately 17,500 had been implanted into patients. A letter was sent to orthopedic surgeons with the affected lot numbers, a list of physical symptoms associated with an affected shell (including an estimate of when these symptoms would normally occur), and instructions for using radiographs to verify loosening at the shell/bone interface.

Continued

Breakout Box 12.3 Sulzer Inter-Op Acetabular Shell Recall, Sulzer Orthopedics (2000)—cont'd

It is common practice to use lubricants during the machining of orthopedic implant components and remove the lubricants through a cleaning process. Passivation (per ASTM F86-13—*Standard Practice for Surface Preparation and Marking of Metallic Surgical Implants*) is a treatment used to clean the surface of a metallic implant and form a passive surface oxide layer that provides corrosion resistance. When the company transferred production from an outside manufacturer to its own production facility, it decided to eliminate the passivation operation used by the previous contract manufacturer. This operation was considered unnecessary due to the corrosion resistance of titanium alloy. However, unknown to the company, elimination of the passivation operation allowed a small residue of a lubricant to remain on the exterior porous surface of the shell. When implanted, the lubricant inhibited bone ingrowth into the surface of the shell, resulting in loosening and pain.

This is an example of a recall caused by a process change that was not adequately tested before being approved and implemented. It illustrates how seemingly minor and innocuous process changes can impact the clinical performance of a device. Legal action resulted in a $1 billion settlement with patients who were affected by the recalled devices.

Breakout Box 12.4 Guidant Ventak Prizm 2 Automatic Implantable Cardioverter Defibrillator Recall, Guidant LLC (2005)

A cardioverter converts a life threatening, abnormal cardiac rhythm (arrhythmia) into a normal rhythm. It uses electrical energy to shock the heart into a normal rhythm in the same way that a defibrillator is used to restart a stopped heart. Guidant manufactured an implantable cardioverter that was implanted in the patient's upper left chest as shown in Figure 12.5. If it detected an abnormal rhythm, it automatically delivered electrical energy to the heart to restore a normal rhythm. Patients would feel a significant shock when this occurred and would know that the cardioverter was functioning. Physicians and patients relied on this device to quickly detect and convert an abnormal heart rhythm.

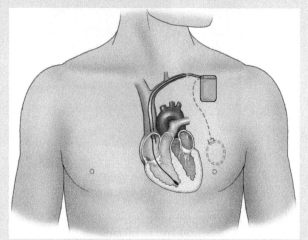

FIGURE 12.5

Implantation site of the Guidant implantable cardioverter defibrillator.

(From LaFleur Brooks M. Exploring Medical Language, *ed 9. St. Louis, Elsevier, 2013.)*

In 2005, the company initiated a voluntary Class I recall of these devices after a report of one death resulting from use of a similar Guidant device. At that time, there were 38,362 devices in the field, approximately 24,000 of which had been implanted. According to the FDA, "laboratory analysis of returned devices revealed that deterioration in a wire insulator within the lead connector block, in conjunction with other factors, resulted in an electrical short. The short circuit caused

---

Breakout Box 12.4 Guidant Ventak Prizm 2 Automatic Implantable Cardioverter Defibrillator Recall, Guidant LLC (2005)—cont'd

diversion of shock therapy energy away from the heart and into device circuitry." When this occurred, the device was unable to deliver a shock when needed, leaving the patient unprotected from an arrhythmia, which could result in death. Guidant changed the design of the Prizm 2 in November 2002 to correct the problem. In August 2003, Guidant falsely told the FDA that the design changes did not affect the device's safety or effectiveness.

When the design defect was discovered, the company failed to quickly notify physicians and patients of the risk and did not immediately remove affected devices from the field. In 2010, the company pled guilty to withholding information from the FDA regarding catastrophic failures and was fined $296 million.

12.10 Medical Device Labeling

Labeling (presented in Section 8.8) refers to product labels, package inserts, user manuals, websites, sales literature and advertising, and communications with customers. It is typically the responsibility of quality and/or regulatory personnel with input from marketing product managers regarding aesthetics. Labeling cannot contain unsupported claims about the product or inaccurate information. Warning labels and warnings appearing in user manuals help reduce risk and potential liability. Understanding how labeling is defined and its role in regulatory compliance is important to engineers working in medical device companies. In this section, we define what constitutes labeling, discuss how it can be used to reduce errors and risk, and discuss its role in communicating product claims. Section 14.4 discusses the role of engineers in developing medical device labeling.

12.10.1 Product Labels

Packaging components such as product labels and package inserts are forms of labeling. A typical product label for a medical device is shown in Figure 12.6. Product labels provide primary identification of the packaged product (product name, catalog number, size, etc.), quantity, and useful

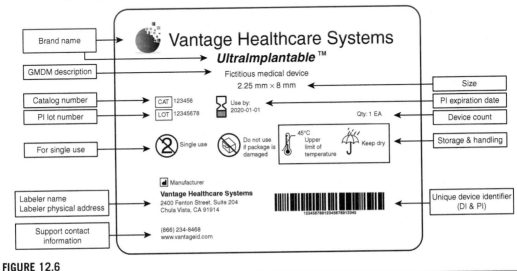

FIGURE 12.6

Typical product label for a medical device.

Intended Use:

This product is intended for limited physical assessment and other general purpose uses such as blood pressure assessment. It can be used for auscultation of heart, lung and other body sounds. This product is not designed, sold or intended for use except as indicated.

Precautions:

• Avoid extreme heat, cold, solvents, and oils.
• The entire stethoscope can be wiped clean with alcohol or soapy water.
• Eartips can be removed from the ear tubes for thorough cleaning.
NOTE: Do not immerse your stethoscope in any liquid or subject it to steam sterilization. If disinfection is required, the stethoscope may be wiped with a 70% isopropyl alcohol solution.

Explanation of Symbols:

 • Attention, see instructions for use.

 • This product and package do not contain natural rubber latex.

Instructions for Use:

Chestpiece:

On Littmann stethoscopes with traditional combination chestpiece, you select either the diaphragm or open bell side by holding the chestpiece stem in one hand and rotating the chestpiece with the other hand until a click is felt.

Changing Frequencies Using the Tunable Diaphragm:

Your Littmann Lightweight II S.E. Stethoscope is equipped with a 3M tunable style diaphragm that enables you to listen to both low and high frequency sounds without turning over the chestpiece.

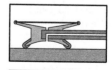

Low frequencies:
To listen to low frequency sounds, (traditional bell mode) you can use either the open bell or tunable diaphragm using very light skin contact.

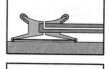

High frequencies:
To listen to higher frequency sounds, using the tunable diaphragm, press firmly on the chestpiece. To listen to low and high frequency sounds without removing and turning over the chestpiece, simply alternate between light and firm pressure on the tunable diaphragm.

Eartip replacement:

For maximum acoustic performance, comfortable patented 3M™ Littmann® Soft-sealing Eartips are provided with your stethoscope. This

FIGURE 12.7

Page from a package insert for a Littman® stethoscope showing the typical information included in a package insert for a medical device.

information such as directions for use and warnings. They also contain lot numbers, expiration dates, date of manufacture, method of sterilization, storage and handling information, and manufacturer/distributor contact information. For products sold outside of the United States, product labels may be required to be written in multiple languages. This is a requirement for products sold in the EU. Figure 12.6 includes a Unique Device Identifier (UDI) in the form of a bar code, which includes a Product Identifier (PI) and Device Identifier (DI). UDIs are discussed later in this section.

12.10.2 Package Inserts

An example of a package insert for a stethoscope is shown in Figure 12.7. It contains information regarding intended use, precautions, instructions for use, and an explanation. Package inserts for other

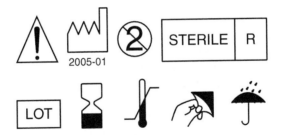

FIGURE 12.8

Commonly appearing graphical symbols specified in ISO 15223-1:2016. The symbols in the top row (*left to right*) mean: (1) warning—read instructions before using, (2) manufacture date, (3) single use only (do not reuse), and (4) method of sterilization (EO—ethylene oxide, R—radiation). The symbols in the bottom row (*left to right*) mean: (5) lot number, (6) expiration date, (7) store between temperature range shown (two numbers, not shown, indicating minimum and maximum temperatures), (8) open by pulling corner tab, and (9) keep dry.

types of products often include additional information such as indications for use (what the product can be used for), contraindications (when the product should not be used), potential adverse events, how to handle adverse events, and special handling instructions (if applicable). In some situations, instructions and warnings are also printed or attached onto the device for easier reference.

12.10.3 Standard Symbols for Medical Device Labels and Package Inserts

The current standard for symbols used in medical device labels and package inserts is ISO 15223-1:2016 *Medical devices—symbols to be used with medical device labels, labeling, and information to be supplied*. The symbols included in the standard are required for medical device labels on products sold internationally. Figure 12.8 shows nine examples of graphical symbols that commonly appear on medical device labels. Can you determine the meaning of each symbol without first reading the figure caption?

12.10.4 User Manuals

User manuals, either in print, video, or digital form, or available online, are examples of medical device labeling. As discussed previously in this chapter, companies cannot make label claims that are not substantiated by clinical evidence. Thus, user manuals cannot contain unsupported label claims. Warning labels and warnings included in user manuals are also forms of labeling and are used to reduce risk, as discussed in Chapter 8.

12.10.5 Sales Literature and Technical Bulletins

Sales brochures and advertisements (in print, online, or through broadcast media), are forms of labeling. Technical bulletins are used to provide updates to customers or users of a product regarding new features, trouble-shooting tips, ways to improve outcomes, how to handle adverse events, and other helpful information. As indicated earlier, these forms of labeling may not contain unsubstantiated claims about the product.

12.10.6 Trade Show Exhibits and Personal Conversations

Trade show exhibits are communication tools that inform potential customers and users of a product of new products, features, and enhanced performance. Companies need to make sure that they are not making unproven label claims through their displays. Companies seeking approval for a device through the PMA process cannot advertise the product for sale prior to receiving approval to market the device. Companies seeking clearance for a device through a 510(k) process may take orders for it but may not advertise or sell it prior to receiving clearance to market the device.

Personal conversations with customers are another form of labeling. This means that if a salesperson or engineer discusses a product during a private conversation with a customer, they must be careful not to make unproven claims of device performance or offer a new device for sale prior to it receiving FDA clearance or approval to be marketed.

12.10.7 Unique Device Identifiers

A UDI is a unique numeric or alphanumeric identification code assigned to medical devices by the labeler (manufacturer, reprocessor, or repackager) of the device. In 2013, the US FDA UDI rule was passed, mandating that single devices are to be identified through a unique identifier that appears on the label and package of the device. A UDI is required to appear on device labels in plain text and in a format that can be read using technology (similar to that used to read bar codes). The goal of the UDI rule was to enable identification of medical devices through the distribution and use of the device. The ability to identify devices and include device information (e.g., lot numbers, date of manufacture) in a patient's electronic medical record can be helpful if it becomes necessary to notify physicians and patients of device updates, replacements, and recalls. UDIs are a required component of device labels.

In summary, this chapter illustrates the importance of understanding regulatory requirements and their role in the design process. The examples of medical device recalls presented illustrate the impact that your work as an engineer can have on public health and on the lives of the patients you are trying to help. It is important for you to recognize and take seriously the responsibility that you have for designing and testing medical products that will perform as needed and not cause harm to patients.

To conclude this chapter, we present the perspective of a regulatory affairs consultant in Breakout Box 12.5.

Breakout Box 12.5 Designing Medical Devices in a Regulated Industry

What does it mean to design a product that has direct impact on a patient? Your design will enter an environment where someone's health can be directly impacted by the medical devices and medicines with which they interact. Engineers are involved with:

Communicating With Regulators

Over the course of your career, you may be involved in audits, responding to questions from regulators, and meetings with regulatory bodies such as the FDA. Communicating honestly in a concise manner is appropriate in these instances. Regulatory bodies do not have the same understanding of the product being reviewed that you have as a design engineer—you will need to communicate accordingly. A regulator's primary concern is whether patients are going to be safe. Questions and concerns arise when they cannot determine patient safety from documentation that has been submitted to them.

Considering Ethical Standards

There may be times when a company's management or your manager asks you to get a product out faster. As a leader and responsible engineer, you will need to make decisions about what testing to conduct and document. It is important to always use sound engineering practices and remember that your product is going to be used by real patients.

Robin Martin, RAC
Chief Regulatory Strategist
(Kinetic Compliance Solutions, LLC)

Key Points

- Medical device regulations exist to ensure safety and efficacy of medical devices and to protect the public. A medical device cannot be sold without regulatory approval.
- FDA regulations determine the requirements for obtaining regulatory approval to market medical devices in the US.
- Devices are classified as Class I, II, or III depending on the level of risk associated with use of the device.
- Different countries have their own regulatory bodies, with unique definitions of medical devices and regulatory submission and approval processes.
- In the US, medical device manufacturers cannot make claims about their products that have not been evaluated in clinical studies and demonstrated to the FDA.
- Mobile apps may be regulated by the FDA depending on their function and use.
- Devices are *cleared* for market through the 510(k) process and *approved* for market through the PMA process. Additional pathways include the Humanitarian Device Exemption, Breakthrough Device Designation, and the De Novo classification.
- Device recalls are classified according to the device failure's potential to result in injury or death, and are classified as voluntary, semi-voluntary, or statutory. They can trigger actions ranging from sending additional information to providers to removal of the product from the market (including removal of implants).
- Medical device labeling includes product labels, package inserts, sales literature and advertisements, and personal communications with physicians, customers, or users of the device.

Exercises to Help Advance Your Design Project

1. Does the device you are working on qualify as a medical device per the FDA definition presented in this chapter? If so, what class of device is it?
2. Attempt to determine the appropriate regulatory pathway to market for your device. Consider level of risk, predicate devices, and other criteria presented in this chapter.
3. Would your device qualify as a medical device in the EU, Japan, or Canada?
4. Create a user manual (in written or video form) for your product. Consult online FDA resources for guidance on writing user manuals.
5. What are the purposes and benefits of Unique Device Identifiers (UDIs) on medical devices? How might you incorporate a UDI on the device you are developing?
6. Develop search terms to yield FDA recall results related to the device you are developing. Search the FDA recall database and determine common causes of design-related device recalls. If any one of them is something you can address at the current stage of your project, what steps would you take to lower the risk for a potential recall of your medical device?
7. Find a recalled device that is like the device that you and your team are developing. What problem(s) led to its recall and how did the recalled devices affect patients? Is there anything that you can learn from the recall to improve your device?

General Exercises

8. What qualifies a mobile app for regulation by the FDA?

9. What items are forms of medical device labeling?

10. Find an example of a device that was classified as a humanitarian device and approved through a Humanitarian Device Exemption. What allowed it to be approved through this pathway and how many patients would benefit from this device?

11. The 2% failure rate of the Bjork-Shiley heart valve warranted a product recall. Discuss whether the same action should be taken if the incidence rate was 0.2% or 20%. Considering that all medical devices experience some failures, what would you consider to be an acceptable failure rate for a medical device?

References and Resources

Conrad, P. (2008). *The medicalization of society: On the transformation of human conditions into treatable disorders.* JHU Press.

Council Directive 93/42/EEC of 14 June 1993 concerning medical devices. (1993). Eur-lex.europa.eu. https://eur-lex.europa.eu/legal-content/EN/TXT/?uri=CELEX:01993L0042-20071011.

Definition of the term "biological product," 21 CFR § 600. (2018). https://www.federalregister.gov/documents/2018/12/12/2018-26840/definition-of-the-term-biological-product.

Fernandez, A. (2019, August 1). *Capturing unique device identifier data on non-sterile orthopedic implants.* Medical Design Briefs. https://www.medicaldesignbriefs.com/component/content/article/mdb/features/articles/34970.

Food and Drugs Act, R. S. C. (1985). *c.F-27.* https://laws-lois.justice.gc.ca/PDF/F-27.pdf.

Fuhr, T., George, K., & Pai, J. (2013, October). *The business case for medical device quality.* McKinsey Center for Government. https://www.mckinsey.com/~/media/mckinsey/dotcom/client_service/public%20sector/regulatory%20excellence/the_business_case_for_medical_device_quality.ashx.

Fundamentals of Medical Device Regulations. (2017). *Regulatory Affairs Professional Society.*

Goldfarb, N. M. (March 1996). When patents became interesting in clinical research. *The Journal of Clinical Research Best Practices*, 2(3).

https://www.lawyersandsettlements.com/lawsuit/bjork_shiley_heart_valves.html. (Accessed 24 November 2021).

https://yourlawyer.com. (Accessed 24 November 2021).

https://yourlawyer.com. (Accessed 24 November 2021).

Joshi, A., Zhu, B. Z., & Xu, L. (2019, May 21). *Trends in medical device recalls.* Medtechintelligence.com; MedTech Intelligence. https://www.medtechintelligence.com/feature_article/trends-in-medical-device-recalls/.

King, P. H., Fries, R. C., & Johnson, A. T. (2015). *Design of biomedical devices and systems.* CRC Press.

Lawsuit settled over heart valve implicated in about 300 deaths. (1992, January 25). *The New York Times.*

Lester, W. (1996, May 13). Hair-raising tale: no fame for men who discovered Rogaine. *The Daily Gazette.*

List of authority websites. (n.d.). (2021, October 19). Tarius.com. http://www.tarius.com/?page_id=612.

Martin, R. (2021, February 22). *Regulatory issues in medical device design.* Milwaukee, WI, United States: Marquette University. [class lecture].

Medical Device Overview. (2018). *US Food and Drug Administration.* https://www.fda.gov/industry/regulated-products/medical-device-overview#What%20is%20a%20medical%20device.

Medical Devices – Sector. (n.d.) European Commission. https://www.Ec.europa.eu. https://ec.europa.eu/health/md_sector/overview_en.

Office of the Commissioner. (2021). *U.S. Food and Drug Administration.* http://www.fda.gov.

Recalled Sulzer. (n.d.). (2021, October 19). Tripod.com. http://totallyhip1.tripod.com/Recall/sulzer_recall.htm.

The Sulzer Recall. (n.d.). (2021, October 19). Allthingsbiomaterials.org. http://allthingsbiomaterials.org/archives/227.

U.S. Department of Health and Human Services. (2018). *Harmonizing and modernizing regulation of medical device quality systems*. https://www.reginfo.gov/public/do/eAgendaViewRule?pubId=201810&RIN=0910-AH99.

U.S. Department of Justice, Office of Public Affairs. (2014, September 15). *Medical device manufacturer guidant Charged in failure to report defibrillator safety problems to FDA*. [Press release] https://www.justice.gov/opa/pr/medical-device-manufacturer-guidant-charged-failure-report-defibrillator-safety-problems-fda.

U.S. Food and Drug Administration. (n.d.). *Drugs@FDA Glossary*. https://www.accessdata.fda.gov/scripts/cder/daf/index.cfm?event=glossary.page.

U.S. Food and Drug Administration (n.d.). *FDA Learning Portal for Students, Academia, and Industry*. https://www.fda.gov/training-and-continuing-education.

U.S. Food and Drug Administration. (n.d.). *MAUDE—Manufacturer and User Facility Device Experience*. https://www.accessdata.fda.gov/scripts/cdrh/cfdocs/cfmaude/detail.cfm?mdrfoid=1578622.

U.S. Food and Drug Administration, Center for Devices and Radiological Health. (2014). *Medical Device Recall Report FY 2003 to FY 2012*. https://www.fdanews.com/ext/resources/files/03/03-31-14-Recalls.pdf.

U.S. Food and Drug Administration. (2007). *Class 1 device recall Ventak PRIZM 2 DR ICD*. https://www.accessdata.fda.gov/scripts/cdrh/cfdocs/cfres/res.cfm?id=39930.

U.S. Food and Drug Administration (FDA). (2014, January). *Medical device single audit program*. https://www.fda.gov/media/87043/download.

U.S. Food and Drug Administration. (2018). *Combination product definition, combination product types*. https://www.fda.gov/combination-products/about-combination-products/combination-product-definition-combination-product-types.

U.S. Food and Drug Administration. (2019). *Device software functions including mobile medical applications*. https://www.fda.gov/medical-devices/digital-health/mobile-medical-applications#a.

US. Food and Drug Administration. (2019). *Is This Product A Medical Device?* https://www.fda.gov/medical-devices/classify-your-medical-device/how-determine-if-your-product-medical-device.

U.S. Food and Drug Administration. (2021). *Breakthrough devices program*. https://www.fda.gov/medical-devices/how-study-and-market-your-device/breakthrough-devices-program#s3.

U.S. Food and Drug Administration. (2021a). *CDRH learn*. http://www.fda.gov/Training/CDRHLearn/.

U.S. Food and Drug Administration. (2021, July 21). *2020 agenda includes ISO 13485 harmonization, De Novo classification scheme*. Emergobyul.com. https://www.emergobyul.com/blog/2020/07/us-fda-2020-agenda-includes-iso-13485-harmonization-de-novo-classification-scheme.

Whitmore, E. (2012). *Development of FDA-regulated medical products: A Translational approach* (2nd ed.). Quality Press.

Wong, J., & Kaiyu, R. T. (2013). *Handbook of medical device regulatory affairs in Asia*. Pan Stanford Publishing.

Ethics in Medical Device Design

13

Your scientists were so preoccupied with whether or not they could, they didn't stop to think if they should.
– Character of Ian Malcolm in the movie *Jurassic Park*

That which is hateful unto you do not do to your neighbor.
– Hillel, Talmud Tractate Shabbat 31a

Chapter outline

Biomedical Engineering Design. https://doi.org/10.1016/B978-0-12-816444-0.00013-4

419

13.1 Introduction

A code of ethics is a set of rules that helps make decisions that impact others. You likely have developed your own internal code of ethics that guides how you behave in the real world. These deeply personal principles are intertwined with your background and experience. Professional fields and organizations also have ethical principles to ensure that the field and its members act responsibly. In some cases, global ethical declarations have been written about how all humans should treat one another.

As the designer of a medical device, you are making decisions that could impact others as you navigate the design process. Just as practicing and encouraging preventive medicine is an important approach to health care, so is developing your own personal and professional ethical compass before you encounter an ethical dilemma. The purpose of this chapter is to provide you with a way of thinking that can help you navigate the ethical issues that arise in developing a new medical device.

13.2 Applied Ethics

Ethics is one of the oldest academic disciplines, has origins in philosophy, and dates back to be beginning of recorded history. Our main purpose, however, is to focus on **applied ethics** as a rational way to resolve conflicts that arises when one or more "rules" conflict with one another. In this section we discuss the origins of applied ethics as well as an ethical reasoning framework.

13.2.1 An Overview of Philosophical Perspectives

It is helpful to have some grounding in the major philosophical ways of thinking about ethics. Ethics is derived from moral rules that govern how people treat each other. One of the more well-known collections of moral rules is from the moral philosopher Bernard Gert:

Not killing others	Not causing pain	Not disabling
Not depriving freedom	Not depriving pleasure	Being truthful
Keeping promises	Being honest	Obeying the law

Doing Your Duty

On the surface these all seem to be reasonable principles. However, the real world presents us with situations where these principles may conflict with one another, and we are forced to choose between them. The anthropologist and systems scientist Gregory Bateson calls these situations a **double bind**. For example, what should you do if your boss (an authority figure) asked you to break a minor corporate rule so that you can get a project back on track? His boss told him to break the rule because a late product release would violate the business principle of maximizing profits for shareholders. Here you have conflicting rules at multiple levels. In the real world these kinds of conflicts can be intertwined and span many levels from the individual to international law.

Ethics is a set of moral principles and how to resolve conflicts when two or more moral obligations are in conflict. One major divide is what matters most: right thinking or right outcomes? In other words, is ethics about what you intend or is it about what you do? Below are a few ethical theories that are often explained in the context of individuals but can also be applied to almost any discipline, organization, or societal group:

Utilitarianism is about doing the greatest good for the greatest number. The ethical measure is impact (as the product of significance and scale) on others. Engineering draws a great deal from utilitarianism and its focus on the consequences of actions. However, strict utilitarianism can also result in great harm to a small number of people if the overall benefit to society is high enough.

The **Categorical Imperative** is about unconditional duties and has two parts. The first is to obey rules that should be followed by everyone. These types of rules would be a moral obligation and would not be dependent upon context or potential benefits. The second is to treat all persons as ends in themselves and not means to some other purpose. In other words, people are not objects to be used to achieve a goal, even if that goal is desirable.

A **Social Contract** is a cooperative and voluntary agreement that an individual (or group) makes with society that they shall abide by the rules and norms of that society. In return the individual gains the rights (e.g., protection, welfare, membership) held by members of that society. Social contracts form the basis of almost all laws and professional codes of ethics. As the case studies in this chapter demonstrate, many of these laws come into being due to a real or expected major problem or concern that impacts society.

13.2.2 An Ethical Reasoning Framework

Applied ethics emerges when philosophical perspectives meet the real world. In the most difficult situations, there is no "right thing to do" in an absolute sense. Rather a decision must be made that has considered as many perspectives and dimensions as possible. The underlying idea of ethical reasoning is that a process has been followed such that a difficult decision is justified.

Aarne Vesilind has proposed the following eight steps for ethical decision making in his book *The Right Thing to Do: An Ethics Guide for Engineering Students*. They are:

1. What are the relevant facts?
2. What are the moral issues?
3. Who is affected by the decision you have to make?
4. What are your options?
5. What are the expected outcomes of each possible action?
6. What are the personal costs associated with each possible action?
7. Given the issues, alternatives, and costs, where can you get help in thinking through the problem?
8. Considering the moral issues, practical constraints, possible costs and expected outcomes, what is the best right action to take?

This framework, and every other ethical decision-making framework, seems simple and reasonable; however, answering these questions in the context of a real situation is not easy. To help you, the questions are ordered in such a way that the most straightforward come first and are designed to help you gather as much information and insight as possible before making a decision.

Questions three through five in the Vesilind framework ask you to consider the direct and indirect impacts on all stakeholders. To help you think more deeply about impacts, consider the STEEPLE framework:

Social
Technological
Economical
Environmental
Political
Legal
Ethical

STEEPLE is used by businesses to consider how external dynamics may be incorporated into decisions. For example, it may become socially unacceptable, although not illegal, to use certain plastics (e.g., polychlorinated biphenyl [PCB] is a known carcinogen) in packaging materials, prompting a company to change its packaging materials. This example highlights the interdependence of the STEEPLE forces; a societal force may have emerged from an environmental concern that leads to political and legal regulatory changes. This is the common way to use STEEPLE and focuses on how external factors influence internal decisions.

For the purpose of ethical reasoning, you can use STEEPLE in reverse; what external impacts may result from your internal decisions? Note that both ethics and technology are a part of STEEPLE, so in this sense you are asking about the ethical implications of your device on the other factors. For example, you could ask, "What economic changes might result from the introduction of this device?" To frame this as an ethical issue, you should narrow the question to: "Will this device economically disadvantage certain populations, directly or indirectly, by creating a wider disparity in wealth or access to health care?" Similar questions could be asked for the other factors. Passing your device through each of the elements of STEEPLE provides you with a systematic way to explore the ethical implications of bringing a new device to the world.

13.2.3 Exercising Your Ethical Reasoning

No one is ever fully prepared to navigate an ethical dilemma. Like the design process, however, you can develop your ethical reasoning through deliberate practice. A common way to practice ethical reasoning is through real or hypothetical dilemmas. Throughout this chapter we present a small sample of the types of ethical scenarios you may encounter as an engineer working on a healthcare-related project. As you read these scenarios, you are encouraged to practice using the frameworks presented above. To begin exercising your ethical reasoning, consider the dilemma in Breakout Box 13.1.

Breakout Box 13.1 Participating in a Clinical Procedure

Your team is visiting the clinic and you have scrubbed in to observe a surgery. During a particularly chaotic moment, the lead surgeon sees that the OR nurse is not in position and asks you to hand her hemostatic forceps that are on the table. What should you do?

This simple case may seem unusual but consider that in the medical profession it is often the case that a physician asks a medical student or resident to help in the context of a surgery. On the one hand, you would be helping out and you have been given a task by someone in a position of authority. You are after all "just a student" and you would not be interacting with a patient. Even if something does go wrong, you perhaps would be protected under the Good Samaritan Law (that allows a bystander to give reasonable assistance to those who are injured, ill, in peril or incapacitated, without fear of legal action). On the other hand, you likely do not know if you are violating any hospital policies. Furthermore, you may be violating an agreement that you have with your school to only act as an observer. What would you do in this situation? Why?

You may be thinking that complex ethical issues do not arise in an educational setting. It is true that it is unlikely in an academic design project that you would assume legal or ethical responsibility for the device you create. However, taking ethics seriously is critical as it is an opportunity to practice ethical reasoning. As in almost everything else, the more you practice, the better you become; your ethical reasoning and personal integrity should deepen and become more nuanced over time. As you advance in your career, integrity will become more and more important. Early in your career you may only see conflicts between your own personal code of ethics and that of your discipline or company. Individual ethical lapses may not even be noticed by others. But as your career develops, the conflicts and dilemmas you face will become complex. You will have less information upon which to make decisions, your responsibility will grow, and your potential to impact others will increase. Practicing now in simpler professional settings is preparation for making more difficult decisions later.

13.3 Engineering Ethics

Almost all professional disciplines abide by self-imposed ethical guidelines, which are social contracts as applied to the relationship between a field and its members. As engineers, we should be proud of our technological accomplishments but must keep in mind that medical technologies are meant to make human life better. Ethical dilemmas that can arise in biomedical engineering range from beginning and end of life issues to the balance between quality of life and extended life to changes in healthcare coverage and policy. Although not always at the forefront, ethical principles are incorporated into almost all decisions made during a design process. In this section, we discuss applied engineering ethics along with two professional codes of engineering ethics.

13.3.1 Applied Engineering Ethics

Ethical decisions are encountered in every design process, from the sources of information you use to how you test devices to ensure that they are safe and effective. There are many competing demands to consider. Hospitals and insurance companies need to make enough money to stay in business. Technologies cost money and often add cost to the healthcare system. What is best for an individual (or some select group of individuals) may not be best for a company, which also may not be best for the population at large. For example, a product that is only available to a small segment of the population can lead to inequities and institutionalized discrimination against already marginalized populations. The most obvious example is when the cost of a product means that it is only available to upper socioeconomic classes. Through your current project and your career, you should revisit the quote at the start of this chapter from the movie *Jurassic Park*, substituting "engineers" for "scientists."

To make this more personal, would you consider allowing a device you have created to be used for a loved one? If not, then it would stand to reason that it should not be allowed to be used for anyone. The scenario in Breakout Box 13.2 is one example of where ethical issues may arise in the context of a design project. This hypothetical scenario is followed by two codes of ethics that are most applicable to medical device design.

Breakout Box 13.2 Ethics and Testing

Imagine you are working on your design project and to stay on track you need the results of several tests (each taking over a day to run) in only two days. As a team you decide to divide the many tests between your members. After running the tests, you notice that two team members have remarkably similar data. What are you obligated to do? What do you feel is the right thing to do? Would you use the results as is?

This seemingly simple example highlights several ethical considerations. As student you may not be legally obligated to report the incident, but you may feel morally obligated to do so. If your school has an honor code, however, you may be bound by your social contract with the school to report the potential problem. There may even be a penalty for not reporting it, as is required by many businesses. At the root of this dilemma is your suspicion that test data has been corrupted. You do not know, however, if your classmates did this unknowingly (due to carelessness) or on purpose (due to intentional deception) and you do not know if it was a single individual or if there was cooperation. In the real world, professional negligence would not be an excuse, as a professional engineer is expected to only perform tests in which they are skilled. Intentional deception in this case would be considered falsification of data. An individual copying the work of another is also taking credit for something that was not their own work (a form of plagiarism). What other ethical issues are in play? How would you reason through your response? What other information would you need to help you decide how to proceed?

13.3.2 National Society of Professional Engineers Code of Ethics

The National Society of Professional Engineers (NSPE) was founded in the United States in 1934 as an association representing licensed professional engineers. All members must abide by the NSPE code of ethics, which outlines six professional duties engineers must fulfill:

1. Hold paramount the safety, health, and welfare of the public.
2. Perform services only in areas of their competence.
3. Issue public statements only in an objective and truthful manner.
4. Act for each employer or client as faithful agents or trustees.
5. Avoid deceptive acts.
6. Conduct themselves honorably, responsibly, ethically, and lawfully so as to enhance the honor, reputation, and usefulness of the profession.

The full code expands upon these fundamental tenants to include statements on confidentiality, conflicts of interest, giving credit where it is due, and accepting personal responsibility for professional actions.

13.3.3 Biomedical Engineering Society (BMES) Code of Ethics

The NSPE code of ethics applies to all engineering professionals, but many subspecialties have developed additional ethical guidelines. The BMES code of ethics outlines four areas for its members: Professional, Health Care, Research, and Training. The professional tenants largely mirror those from the NSPE code. Likewise, the healthcare obligations mirror many of the tenets of medical ethics and the Research obligations follow the tenets of research ethics, both outlined below. A somewhat unique aspect of the code is the responsibility to train "biomedical engineering students in proper professional conduct" as well as to "model such conduct" and to keep this training free from "inappropriate influence from special interests."

13.3.4 Legal and Ethical Duties of Medical Device Designers

The two codes of engineering ethics in the preceding sections are professional guidelines. In some engineering disciplines, a professional license can be obtained; through this license, the engineer is bound by the NSPE Code of Ethics. In the United States, becoming a licensed professional engineer (PE) requires passing two exams that demonstrate an ability to apply the fundamentals of engineering as well as specific disciplinary expertise. A PE can approve designs in their area of licensure. For example, a PE in civil engineering can sign off on plans for new construction. Ethical violations, which include negligence or lack of oversight, can result in the revocation of a professional license. In some cases, the PE assumes legal liability should a design fail. Breakout Box 13.3 explores a specific case of the ethical obligations of a PE.

In the United States, there is no professional license exam specific to medical device design. Though the industry exemption allowed by many states, engineers designing medical devices for medical device companies are not required to be licensed. As discussed in Chapter 12, all medical devices must demonstrate safety and effectiveness to pass through the proper regulatory pathway of regulatory bodies (e.g., The Food and Drug Administration (FDA) in the United States). Medical device companies typically assume liability for design failures.

13.4 Medical Ethics

The medical profession is not only one of the oldest professions, but also likely the first to have a written ethical code. The Hippocratic Oath of "do no harm" is the primary rule for all who heal. In many ways, this means to not make an existing problem worse through your actions. A secondary rule is to "do all that is possible." While medical device designers must consider the general societal-level ethical implications of healthcare products and decisions, it is medical professionals who are often directly confronted with individual cases. In this section we focus on two codes of ethics that apply to medical professionals and are present, in one form or another, in the design of a medical device.

Breakout Box 13.3 Professional Ethics

You are an engineer working for a company and your team wrote a specification for the material to be used in an implant that is about to be tested in clinical trials. When the implants come back from manufacturing, your team notices that the material used does not meet specifications for purity. Your immediate supervisor, however, says that the specifications will be changed because the functional tests performed by another group were acceptable. Devices made using the lower purity material have already been used in clinical trials but the material specification changes were never documented. After demonstrating safety and efficacy in clinical trials, devices were sold and implanted in patients.

Deconstructing this case illustrates a conflict between following established processes and accelerating the product development process. The utilitarian view would be that the risks posed by manufacturing implants from a material that did not meet specifications prior to documenting these changes are outweighed by the benefits of getting the device to market sooner so that it can help more patients. On the other hand, the proper procedures—put in place to protect the public and the company—were not followed. This is much closer to the categorical imperative that there are certain ways of behaving that are absolutes and should not be violated even when the outcome might be beneficial. You should also note that there is a power hierarchy involved in who would be considered responsible (both blame and credit) for any unwanted outcomes. How would you apply the ethical reasoning frameworks to further dissect this scenario?

13.4.1 Medical Code of Ethics

The American Medical Association professional code of ethics was first adopted in 1847. It is composed of several documents that span thousands of pages; however, a one-page summary can be found online. In general, the code outlines how physicians must: (1) respect the dignity and rights of their patients, other healthcare professionals, and the organizations under which they practice; (2) acknowledge the need to remain up to date with current practices; and (3) help train future physicians. Two unique aspects are also included. First, physicians can choose for whom they provide care, except in emergency situations. In many places (and sometimes enforced by law) physicians are obligated to help during an emergency if they are present and able to help. The Good Samaritan Law does not apply to physicians in that they can be prosecuted if they do not provide proper medical care during an emergency. Second, while physicians must follow the law, it is also stated that it is their responsibility to advocate for change that is in the best interest of patients. It is not uncommon for a medical device to be a driver of change in the healthcare ecosystem. For example, laparoscopic surgical techniques changed not only the standard of care but also led to innovations in processes, devices, and instrumentation, and played a primary role in the adoption of robotic surgery.

13.4.2 Technology and Medical Ethics

The Advanced Medical Technology Association (AdvaMed) is a professional organization of companies that develop, produce, manufacture, and market medical products and technologies. They have their own code of ethics that emphasizes how to do what is right for patients and, at a systems level, how companies and healthcare professionals should interact. This includes such topics as marketing, training, sales, direct and indirect compensation, transfer of privileged information, and trial samples.

13.4.3 The Health Insurance Portability and Accountability Act (HIPAA)

HIPAA is discussed in Chapter 3 in the context of clinical observations and in Chapter 10 in the context of clinical studies in the United States. One primary purpose of passing HIPAA in 1996 was to ensure that workers could carry their health insurance and medical records from one company or healthcare institution to another. A related section of the act, largely anticipating the move to electronic records, expanded upon the rights for patients to protect the privacy of their data. All institutions or people who may be in a position to gain access to medical data are included. The act provides guidelines which are then implemented by a local HIPAA Compliance Office. This officer is in charge of training, putting in place policy and process safeguards, and reporting violations. Generally, anyone who has access to patients or patient data is required to go through some training (described in Section 10.3.1) and become certified.

13.5 Research Ethics

You have likely encountered research ethics in the form of proper citations (giving credit for an idea), avoiding plagiarism (giving credit for original source material), and truthfully gathering, analyzing, and reporting data. Ethical considerations extend much farther in the context of medical research. For example, Chapters 3 and 10 introduce the requirements for Institution Review Board (IRB) approval for human subject research. In this section we expand upon these ideas by exploring the historical origins of these regulations and codes.

Two international codes have been adopted almost universally and directly intersect healthcare research and technology. First is the Nuremberg Code, which arose from the trial of Nazi doctors, shown in Figure 13.1. The Nuremberg Code (from which the terms "war crimes" and "crimes against humanity" originate) outlines 10 principles. Many ideas from this document are embedded in other ethical documents, such as voluntary consent, voluntary withdrawal at any time, animal experimentation when possible before human experimentation, the avoidance of undo mental or physical pain whenever possible, and the consideration of other means to test the same hypothesis. The principles were expanded upon in the 1964 Declaration of Helsinki, created by the World Medical Association, which applies largely to investigators involved in human subject research. Breakout Box 13.4 contains a scenario that considers using data gained through unethical means.

Second is the Universal Declaration of Human Rights of 1948, a set of 30 principles adopted by United Nations members. The document outlines rights that anyone living in any country can expect. In particular, Article 25 discusses access to adequate health care regardless of age, gender, race, ability, or circumstances. It also notes that mothers and children "are entitled to special care and assistance."

Many countries have their own legal and ethical laws that align with the Nuremberg Code and Universal Declaration of Human Rights. Policies and laws often come into being because of some ethical failure or event. Below are several cases and laws that have shaped research ethics laws in the United States.

FIGURE 13.1

The Nuremberg Code resulted from the international trial of Nazi doctors for their experiments on humans. The code defines several terms and outlines principles that were later built upon by declarations and codes that govern research using human subjects.

From Public Relations Photo Section, Office Chief of Counsel for War Crimes, Nuremberg, Germany, APO 696-A, US Army. Photo No. OMT-I-D-144.

Breakout Box 13.4 Using Data Obtained Through Unethical Means

A large body of data exist that may have been gained through unethical means. For example, some of the most robust data on the limits of human abilities and tolerances were gained through extraordinarily detailed and unethical experiments in Nazi camps. Given that these data exist, should they be used? Or does using these data go against the dignity of the subjects of these experiments?

Some scientists and clinicians face ethical dilemmas when it is discovered that data from their lab has been falsified. Such is the case of Francis Collins, a noted geneticist and driving force behind the Human Genome Project. Collins discovered that data contained in several publications had been falsified by a PhD student in his lab. Collins reported the problem immediately, but not all well-known researchers have reacted the same way in similar situations. What would you do if you discovered that much of your design of a new device is based upon research conducted by a lab that has recently been accused of falsifying data? How would you reason through the ethics of whether or not to use data for your project that may have been falsified?

FIGURE 13.2

A doctor treating one of the Tuskegee patients. The Tuskegee case led to the National Research Act and the Belmont Report that outline three ethical principles for conducting research using human subjects.

From the National Archives, Atlanta, GA.

13.5.1 The Syphilis Study at Tuskegee and The Belmont Report

From 1932 to 1972, a study was run by the US Public Health Service in collaboration with Tuskegee University to study "untreated syphilis in the negro male." The African-American men in the study, as shown in Figure 13.2, were given free medical care, meals, and burial, and were told the study would only last 6 months. In reality it spanned 40 years. The men also were not told that they had syphilis (being told only that they were being treated for "bad blood"). Furthermore, none were given penicillin, even after 1947 when it was known to be an effective treatment for syphilis. This case almost single handedly led to the passing of the National Research Act (1974) that established the National Commission for the Protection of Human Subjects of Biomedical and Behavioral Research, the Office for Human Research Protections, and federal laws mandating all research be conducted under an

---

Breakout Box 13.5 Performing Experiments Beyond Protocol

One biomedical engineering introductory class experiment involves measuring the effect of acceleration (produced by amusement park rides) on heart rate, measured with a Holter monitor. This is an IRB-approved study and, as a class exercise, is optional. One year, a design team developed a device to measure heart rate and acceleration. This device continued to be used in the course. Your team elects to also test other effects on the results (e.g., sleep deprivation). You propose to have one team member not sleep the night before being tested. It seems to be a straightforward and simple test. Are there safety or ethical concerns the team should consider?

This experiment raises several concerns. Aside from the science, sleep deprivation was not included in the IRB-approved protocol. Any change to the protocol would require further IRB review and approval. Sleep deprivation has many detrimental effects, including the risks of increased blood pressure, depression, forgetfulness, fuzzy thinking, and seizures in epileptic subjects. Most importantly, the effect of sleep deprivation would not be able to be measured separately from other factors; would performing this test with one individual yield statistically significant results? Under what situations would such a test be ethically and legally justified? You may wish to reach out to knowledgeable members of your community to help you reason through this scenario.

Institutional Review Board. The case also led to the Belmont Report which provides three ethical principles that should be considered in designing and executing human subject research. They are:

- *Respect for Persons* concerns protecting the autonomy, dignity, and value of subjects through informed consent by describing the research in simple terms.
- *Beneficence* goes beyond the maxim of "do no harm" by minimizing risks to the individual while maximizing benefits of the research study to human subjects.
- *Justice* ensures the fair and equal distribution of costs and benefits, equitable access to potential subjects, and special protections for vulnerable persons (e.g., minors, prisoners, persons with mental or other disabilities).

These three principles are often used to assess the ethical dimensions of a proposed research or clinical study. Updates occurred in 1981 and 1991 when several federal agencies created what is known as the "Common Rule," which covers government-funded researchers and institutions and clarifies the IRB requirements for biomedical and behavioral research. Breakout Box 13.5 contains a scenario to help you think through the three principles outlined in the Belmont Report.

13.5.2 The Stanford Prison Experiment

The Stanford Prison Experiment was based upon a 1963 study by Stanley Milgram to test the hypothesis that humans were more inclined to obey an authority figure even if that meant harming another person. The original Milgram experiment, conducted within months of some of the high-profile Nazi war trials, involved subjects being directed to deliver painful shocks to others. The receivers of the shocks were in fact actors who were faking various levels of pain and discomfort, up to and including becoming unresponsive. The Stanford Prison Experiment took the next step by placing college males into a simulated prison and randomly assigned them roles as either prisoners or guards. Over a short amount of time the "guards" began to subject the "prisoners" to psychological abuse. The experiment was terminated after only 6 days. The topic is brought up here because the various ethical and methodological problems in the study (e.g., fuzzy informed consent, researchers preventing subjects from leaving the study, researchers instructing guards to behave in a certain way) prompted revisions to the guidelines for institutional review boards.

13.5.3 Origins of the Food and Drug Administration

The FDA has a history extending back to the Agriculture Division of the Patent Office in 1848. Over the next fifty years, several high-profile cases—specifically, the unsanitary conditions of the Chicago meatpacking industry described in Upton Sinclair's book *The Jungle*—led to the establishment of the Pure Food and Drug Act of 1906. Although the modern FDA does not oversee ethical violations, it is important to know that their origins were to correct what many considered to be unethical practices through federal oversight.

13.5.4 Clinical Studies

In some cases, academic projects may be tested clinically. The details of this process are covered in Chapter 10. Ethical issues, however, sometimes arise in conducting the trials and interpreting the results. For example, some clinical trials have been shut down early because early results had already reached statistical significance. The results were sometimes very clearly negative, often due to extreme side-effects, including death. For example, Jesse Geisinger was the first person known to have died in a gene therapy clinical trial in 1999. On the other hand, there are rare cases in which the results are so positive that it cannot be ethically justified to deny the treatment to a placebo (control) group. The HOPE (Heart Outcomes Prevention Evaluation) clinical study of Ramipril (the first ACE inhibitor) ended one year early because of such clear positive outcomes with minimal side effects. Most clinical results are not as clear, given that they inherently involve biological organisms and often involve studies by multiple researchers at multiple sites. Breakout Boxes 13.6 and 13.7 focus on additional ethical considerations that may arise in clinical testing of medical devices.

Breakout Box 13.6 Clinical Trials in Developing Countries

Imagine that after several successful product designs, you have been promoted many times and now find yourself as the Vice President of Product Development. One of the divisions you oversee has a product that they feel is almost ready for clinical trials. The CEO of the company suggests that some significant portion of the planned trials be conducted in lower and middle income countries (LMIC). Many countries either do not have laws and processes in place to regulate clinical trials or do not enforce such laws. As a result, costs are lower, recruitment and consent are simpler, and logistical barriers are decreased. Also, some countries welcome the additional income of resources and medical expertise that comes with a clinical trial. The business argument is that the time-to-market may be reduced by several years, saving the company resources and accelerate the widespread use of what is expected to be an impactful product. Furthermore, your competitors are conducting clinical trials in developing countries for a similar product. Not said at the meeting, however, is that US rules and protections will not be extended to the subjects, and the final product cost will likely prevent it from ever benefiting the population of the countries in which clinical trials are conducted. From the meeting, it is also not clear if policies put forward by the World Medical Association (WMA) or International Organization for Medical Sciences (CIOMS) standards will be followed.

Deconstructing this case could fill many pages, or even an entire chapter, but it is realistic; these types of decisions are discussed and often made by companies. A utilitarian may argue that the clinical trial will need to be conducted on some subjects eventually and that there are potential secondary benefits to subjects in the developing world. However, it seems as though the company is proposing to use subjects as a means to an end, and it is unclear why someone would deserve less protection simply because of their geographic location. As the Vice President of Product Development, how would you lead your team in a discussion of the ethical dimensions of this case? What recommendations would you hope to hear from your Product Development managers?

Breakout Box 13.7 Interpreting Data

A unique aspect of medical device design is that data is often collected from living systems, ranging from organic molecules and cells to humans and human interaction. In some arenas of science, data is expected to be very clean, meaning that the effect of an intervention or independent variable can be easily distinguished from a control case. Consider the more usual case in biological and clinical situations in Figure 13.3.

In this experiment, a line is proposed to separate the impact of some intervention, where dots represent measurements from treated (solid dots) and control (open dots) cases. You should note that, although one line is presented, there is no single line that can completely separate the treated and control data. There are, however, statistical measures (not to be discussed here) that can quantify the degree of certainty that this line represents. List and discuss the ethical implications to consider when using statistical analysis techniques to make decisions toward the later stages of a medical device design project.

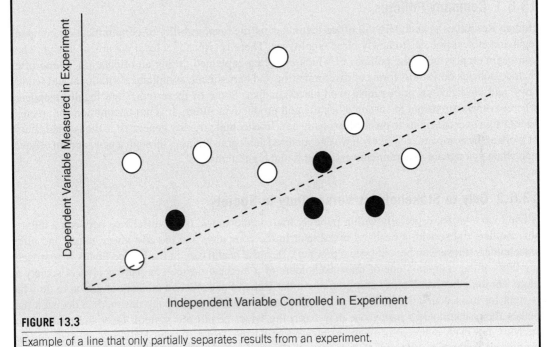

FIGURE 13.3

Example of a line that only partially separates results from an experiment.

13.5.5 Animal Rights

In the United States, the body that oversees the rights of animals is a local Institutional Animal Care and Use Committee (ACUC), as described in Section 10.4. This is parallel to the IRB for human research. Animals require a degree of protection that is graded based upon the species, especially because they cannot consent to an experimental protocol. The origin of ACUC came from a 1989 update to the Animal Welfare Act of 1966. Many of the same underlying ethical principles for humans are extended to animals; for example, minimizing pain and discomfort, ensuring that any benefits outweigh the risks of the research, receiving adequate care before, during, and after an experiment, and training for personnel involved in the study.

13.6 Organizational Policies and Corporate Ethics

When working for or with any organization, it is important to recognize that the organization and professionals who work for that organization may have their own codes of ethics. For example, coworkers in marketing, finance, and other divisions follow ethical and legal codes from their own professional organizations. Companies also have internal policies that supplement the various professional codes of ethics. In this section we cover some basic ideas in corporate ethics of which you should be aware before joining the workforce.

13.6.1 Company Policies

Human Resources is generally the office that takes on the responsibility of communicating company legal and ethical policies to newly hired employees. There is typically a legal document to sign when you begin employment, the policies of which are often explained during an orientation. These often include policies on various forms of discrimination and harassment, compliance with local and federal laws, and guidance on professional workplace conduct. Some of these topics are legally mandated whereas others are meant to sustain a healthy and productive culture. It is not uncommon for an engineer to sign over the rights to ownership of any new intellectual property generated in the normal course of work at the company. You likely have encountered analogous training through a new student orientation when you agreed to certain university rules and regulations.

13.6.2 Duty to Stakeholders Versus Duty to Society

A common conflict, especially within publicly traded companies, is the difference between a duty to shareholders (those with a business investment in the company) and the customers, users, and other stakeholders (those who benefit from a product). Because health care in the United States is approaching 20% of the economy, one of the stakeholders of a medical device company is always society at large. On the other hand, many corporations in the United States feel an ethical obligation to do what is right for their shareholders. There are several cases in which a company has made a decision that makes their shareholders happy but may harm the wider healthcare system. Over the past several decades, however, many companies are now adopting the Triple Bottom Line (e.g., planet, people, profit) and Triple Aim of Health Care (e.g., reducing cost, improving health, improving patient experience) (both explored in Breakout Box 3.9) as a way to balance competing demands.

13.6.3 Confidentiality and Conflicts of Interest

Information and power within organizations often follow hierarchies, a practice which can lead to problems of confidentiality and conflicts of interest. Using privileged information or power to influence an outcome for your own direct or indirect benefit is known as **conflict of interest (COI).** From a healthcare perspective, this may be due to the nature of privileged clinical or research data. It may also be because companies carefully guard certain information from competitors. Legal means used to protect critical information include not filing patents and other forms of intellectual property. Many companies also require their employees to sign confidentiality agreements which state that they cannot share information outside of the company. As employees often move from one company to another, often to a competitor, maintaining confidentiality can be very important. Many companies require employees to sign a "non-compete" agreement that extends past the period of employment. This agreement prevents past employees from working for competitors, with certain restrictions.

A related idea is disclosing any potential COI with the intent of making the nature of the conflict known. For example, you may make money or gain a promotion as an inventor. As the inventor of a medical device, should you be an investigator involved in clinical testing? In this case, there is a conflict between objective evaluation and potential financial gain. For this reason, some companies do not allow an inventor to be directly involved in the decisions regarding whether or not a new product is ready to be introduced to the market. COIs also extend to external partners or consultants who would stand to gain from particular decisions.

13.6.4 Corporate Wrongdoing

Corporations worry about the legal, political, and brand risks that could result should a major ethical violation become widely known. For that reason, many larger corporations have internal policies and procedures for detecting ethical and legal violations as early as possible. Some companies have a Department of Ethical Compliance or Internal Affairs Department to monitor employees for compliance with company policies. If there is suspicion, an internal investigation would be launched to determine if there was any wrongdoing. It would then be up to management to decide what to do if problems are found. An issue arises, however, if the wrongdoing involves management. There is then an internal COI because anyone who points the finger at management could be in danger of retaliation and loss of employment. It is for these types of cases that the Federal Whistleblower Protection Act of 1989 was passed. The law protects any employee who reports company wrongdoing.

13.7 Conclusions and Final Scenarios

We present a variety of ethical scenarios in this chapter and throughout the text, such as the recalls of the Bjork-Shiley Heart Valve (Breakout Box 12.2), Sulzer Hip Implant (Breakout Box 12.3), and Guidant Ventak Prizm 2 Defibrillator (Breakout Box 12.4). The purpose of sharing them is to help you think about how to react when you encounter an ethical dilemma. By arming you with real examples, our aim is to help you build a well-developed ethical sense that can aid you in detecting when something is not right. Although we constructed these scenarios to help you, it is important to note that it is very rare in the medical device industry for failures to be due to malintent. Rather, most ethical issues and failures in the medical device industry are due to negligence (the failure to catch a mistake before it has an impact on others). However, you should note that negligence is considered a violation of the ethical codes of professional organizations. For example, the NSPE may revoke a PE's license for negligence. We conclude this chapter with two final scenarios, in Breakout Boxes 13.8 and 13.9.

Breakout Box 13.8 Johnson & Johnson and the Chicago Tylenol Murders

Some measure of a company's ethical principles can be seen *after* a problem occurs and the company responds to the situation. As mentioned in Section 2.3.4, seven people were found poisoned to death in 1982 due to someone slipping potassium cyanide into Johnson & Johnson bottles of Tylenol. In response, the company quickly (almost overnight) voluntarily retrieved all Tylenol bottles from the shelves of stores, pharmacies, and hospitals. This voluntary recall (discussed in Chapter 12), was implemented not only in Chicago but throughout the entire country at a cost of more than $100M. The company also immediately alerted the public. Johnson & Johnson was praised for their quick action and the case has become a model for how corporations should handle problems with their products. In addition, the case led to reforms in over-the-counter tamper-proof packaging and federal anti-tampering laws. Find Johnson & Johnson's core values and mission statement and discuss how they put them in action in this particular case. Can you find other examples of companies that made a difficult ethical decision based upon their values and mission?

Breakout Box 13.9 The Dalkon Shield Intrauterine Device

Intrauterine devices (IUDs) help prevent women from becoming pregnant. An IUD is placed by a healthcare provider and held in place by spurs on the device. When an IUD is removed, the healthcare provider pulls upon a string to release the IUD. In 1969, the Dalkon Shield IUD replaced the single string with a multifilament string that was expected to be stronger, as shown in Figure 13.4. The co-inventor, physician Hugh Davis, was the founder and 35% owner of the Dalkon Corporation. He also conducted the clinical trials for the Dalkon Shield IUD. At the study conclusion, Davis claimed that the pregnancy rate was 1.1%, which was far below the current rates of competing products. Within days, Davis published an article on the findings and never disclosed that he was the inventor or that he had significant financial interest in the product. Only later was it found that the tests were only conducted for 5 months and women were asked to use a spermicide with the device. Corrected studies showed that the pregnancy rate was in fact much higher. The device was sold in 1970 to A.H. Robins Co. which mass produced the device and sold over 4 million devices in over 80 countries. It was then that reports of severe cramping due to an infection called pelvic inflammatory disease began to be reported. In many cases, the infection impaired or destroyed a woman's ability to bear children, and at least 18 women died. Later studies showed that the cause was the multifilament string ends. Because they were not sealed, the ends frayed, allowing bacteria to travel up the vagina to the uterus. What ethical principles do you see at play in this example? How could the ethical reasoning framework have been applied to prevent this outcome?

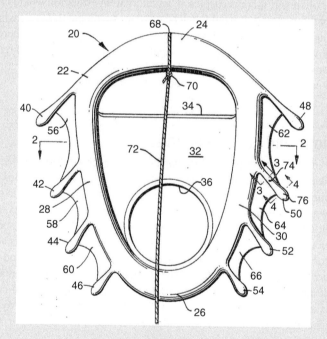

FIGURE 13.4

Sketch of the Dalkon Shield IUD, from US Patent 3782376, showing the anchoring string.

Key Points

- Ethics is a set of moral principles and how to make decisions when two or more moral obligations are in conflict.
- Applied ethics applies broadly to all medical technologies and professional conduct.
- The public failure of a medical device or study, or misconduct by an individual or organization, can sometimes lead to establishing laws and rules to guide future ethical decisions.
- The ethics of medical device design lies at the intersection of professional codes from engineering, medicine, research, and business, as well as federal and international law.
- Scenarios are a powerful way to internalize and clarify one's own ethical principles.
- Practicing ethical reasoning now will help you navigate difficult ethical dilemmas later.

Exercises to Help Advance Your Design Project

1. List the top three worst ethical violations you can imagine for the device you are developing. These can be violations of the rights of any stakeholder and may involve misuse, user error, or corporate wrongdoing. What could you do as an engineer to mitigate these ethical risks?
2. Evaluate your device through the lens of the Triple Bottom Line discussed in Breakout Box 3.9.
3. Evaluate your device through the lens of the Triple Aim of Health Care discussed in Breakout Box 3.9
4. For your device, create three scenarios that would involve an ethical dilemma. Discuss the nature of the ethical tension. Are there ways to change your design to mitigate the dilemma?
5. Navigate to the Collaborative Institutions Training Initiative (CITI) course on Conflict of Interest at https://about.citiprogram.org/en/series/conflicts-of-interest-coi/. If your university has access, take the course and save the certificate to show that you have passed the course.
6. Navigate to the Collaborative Institutions Training Initiative (CITI) course on Bioethics at https://about.citiprogram.org/en/series/bioethics/. If your university has access, take the course and save the certificate to show that you have passed the course.

General Exercises

7. Choose one of the hypothetical scenarios above and have a debate within your team (or perhaps with another team). Be sure to agree upon which "side" you expect to take before the debate so that you can organize the points you wish to make. You are encouraged to use the various professional and philosophical ideas from this chapter to support your viewpoint.
8. Choose one of the real scenarios above. Perform as much research as you can and then present the ethical perspective of at least three critical stakeholders that were involved in the case.
9. Find an example of how a medical device company reacted to a problem with one of their products and dissect their decision. You may wish to search for recent FDA recalls on the fda.gov website.
10. Explain why authors need to disclose financial interests in publications and when conducting human studies.

11. Choose one of the historically significant events in Figure 12.1 and research both the origin and impact of the event.

12. Create your own hypothetical scenario using one of the ideas below. Then write up your own discussion of the case.

    a. A mentor or sponsor asks you to do something you suspect is a violation of a university rule.

    b. You suspect that one of your teammates has falsified the results of a verification test.

    c. A nurse demonstrates on a team member how a medical device is used.

    d. Your team discovers after the fact that some medical data was collected from home caregivers before you were approved by your Institutional Review Board.

    e. You are asked in an interview to take part in a timing experiment (e.g., how quickly you can change the batteries) for a device being developed by the company. You were never asked to sign anything—the experiment was presented as part of the interview.

13. On a routine prenatal visit, a healthcare worker prescribes Amoxicillin, an antibiotic safe in pregnancy, to treat an unspecified urinary tract infection (UTI). When the financial office submits the CPT and ICD numbers for reimbursement, the insurer denies payment, claiming the treatment was part of the overall pregnancy treatment. If you were part of the financial office, what would you do and why?

14. In 1996, a healthy 19-year-old sophomore was paid $150 to volunteer to take part in an IRB-approved study of how the lungs defend against infections and pollutants. The study involved sampling lung cells from healthy patients. To collect lung cells, a bronchoscopy is needed; a tube was thus inserted through the volunteer's mouth into her lungs to collect cells. The procedure was performed by a researcher using local anesthetic Lidocaine. Because the volunteer was coughing more than usual, additional Lidocaine was administered. Sadly, the volunteer died two days later due to an overdose of Lidocaine. What information could have been included in the IRB application that may have averted this tragic outcome?

15. Some IRB-approved clinical trials involve inducing a medical condition, such as taking a drug that triggers a mild form of a disease. One example is inhaling hexamethonium to trigger a mild asthma attack. What are some ethical concerns about such a clinical trial?

16. In a class exam, calculators are allowed but first need to have the memory cleared by the proctor before the exam. A number of students have deviously brought a second calculator with memory intact to the exam, and switched the cleared calculator for the other one while going back to their seat. You find out about this in advance of the test. Your options are to join them, report them, or say nothing. Which would you choose and why?

References and Resources

Advanced Medical Technology Association (AdvaMed). (2021). *AdvaMed code of ethics 2021*. https://www.advamed.org/member-center/resource-library/advamed-code-of-ethics-2021/.

American Medical Association. (n.d.). *Medical code of ethics*. https://www.ama-assn.org/delivering-care/ethics/code-medical-ethics-overview.

Bateson, G. (1972). *Steps to an ecology of mind: Collected essays in anthropology, psychiatry, evolution, and epistemology*. University of Chicago Press.

Biomedical Engineering Society. (2014). *Biomedical engineering society code of ethics*. https://www.bmes. org/files/CodeEthics04.pdf.

Blot, W. J., Ibrahim, M. A., Ivey, T. D., et al. (2005). Twenty-five-year experience with the Björk-Shiley convexoconcave heart valve: A continuing clinical concern. *Circulation, 111*(21), 2850–2857. https://doi. org/10.1161/CIRCULATIONAHA.104.511659

Centers for Disease Control and Prevention. (n.d.). *Public Health Professionals Gateway: Health Insurance Portability and Accountability Act of 1996 (HIPAA)*. https://www.cdc.gov/phlp/publications/topic/hipaa.html.

Centers for Disease Control and Prevention. (2021). *The U.S. Public Health Service syphilis study at Tuskegee*. https://www.cdc.gov/tuskegee/.

Cyranoski, D. (2019). The CRISPR-baby scandal: What's next for human gene-editing. *Nature, 566*(7745), 440–442. https://www.nature.com/articles/d41586-019-00673-1.

Davis, H. J. (1970). The shield intrauterine device: A superior modern contraceptive. *American Journal of Obstetrics and Gynecology, 106*(3), 455–456.

Fletcher, D. (2009). A brief history of the Tylenol poisoning. *Time*. http://content.time.com/time/nation/article/ 0,8599,1878063,00.html.

Friend, C. (n.d.). Social contract theory. *Internet Encyclopedia of Philosophy*. https://iep.utm.edu/soc-cont/.

Gert, B. (2004). *Common morality: Deciding what to do*. Oxford University Press.

Horowitz, R. (2018). The Dalkon Shield. *The Embryo Project Encyclopedia*. https://embryo.asu.edu/pages/ dalkon-shield.

Jarmusik, N. (2019). The Nuremberg Code and its impact on clinical research. *Compliance in Focus*. https://www. imarcresearch.com/blog/bid/359393/nuremberg-code-1947.

Johnson, R., & Cureton, A. (2016). Kant's moral philosophy. In E. N. Zalta (Ed.), *The Stanford encyclopedia of philosophy* (Fall 2016 ed.). Stanford University. https://plato.stanford.edu/entries/kant-moral/.

Markel, H. (2014). How the Tylenol murders of 1982 changed the way we consume medication. *PBS News Hour*. https://www.pbs.org/newshour/health/tylenol-murders-1982.

McLeod, S. (2020). The Stanford prison experiment. *SimplyPsychology*. https://www.simplypsychology.org/ zimbardo.html.

Milgram, S. (1963). Behavioral study of obedience. *Journal of Abnormal and Social Psychology, 67*(4), 371–378.

Nathanson, S. (n.d.). Act and rule utilitarianism. *Internet Encyclopedia of Philosophy*. https://iep.utm.edu/util-a-r/.

National Institutes of Health Office of Laboratory Animal Welfare. (n.d.). *Tutorial: PHS Policy on Humane Care and Use of Laboratory Animals*. https://olaw.nih.gov/resources/tutorial/.

National Institutes of Health Office of NIH History & Stetten Museum. (n.d.) *The Nuremberg Code*. https://history. nih.gov/download/attachments/1016866/nuremberg.pdf?version=1&modificationDate=1589152811742&api=v2.

Normile, D. (2018). CRISPR bombshell: Chinese researcher claims to have created gene-edited twins. *ScienceInsider*. https://www.sciencemag.org/news/2018/11/crispr-bombshell-chinese-researcher-claims-have-created-gene-edited-twins.

Panko, B. (2017). Where did the FDA come from, and what does it do? *Smithsonian Magazine*. https://www. smithsonianmag.com/science-nature/origins-FDA-what-does-it-do-180962054/.

Rinde, M. (2019). The death of Jesse Gelsinger, 20 years later. *Science History Institute: Distillations*. https://www. sciencehistory.org/distillations/the-death-of-jesse-gelsinger-20-years-later.

Sleight, P. (2000). The HOPE study (heart outcomes prevention evaluation). *Journal of the Renin-Angiotensin-Aldosterone System, 1*(1), 18–20. https://doi.org/10.3317/jraas.2000.002.

United Nations. (1948). *Universal declaration of human rights*. https://www.un.org/en/about-us/ universal-declaration-of-human-rights.

U.S. Department of Agriculture National Agricultural Library. (n.d.). *Animal Welfare Act*. https://www.nal.usda. gov/awic/animal-welfare-act.

U.S. Department of Health and Human Services. (n.d.). *Health Information Privacy*. https://www.hhs.gov/hipaa/index.html.

U.S. Department of Health and Human Services. (2016). *The Belmont Report: Ethical principles and guidelines for the protection of human subjects of research*. https://www.hhs.gov/ohrp/regulations-and-policy/belmont-report/index.html.

U.S. Food and Drug Administration. (2018). *FDA history*. https://www.fda.gov/about-fda/fda-history.

Vesilind, P. A. (2006). *The right thing to do: An ethics guide for engineering students*. Lakeshore Press.

World Medical Association. (2013). *Declaration of Helsinki*. https://www.wma.net/what-we-do/medical-ethics/declaration-of-helsinki/.

Zimbardo, P. G. (n.d.). Stanford prison experiment: A simulation study on the psychology of imprisonment. https://www.prisonexp.org/.

COMMERCIALIZATION AND POST MARKET SURVEILLANCE

7

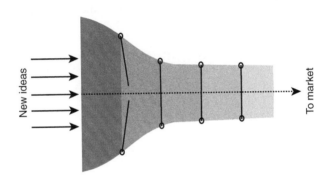

In the concluding two chapters, you will learn more about aspects of the stage-gate model of product development that have only been mentioned in the preceding chapters. As this text was intended to support an academic design process, these final chapters will build your awareness of the many additional considerations and tasks that are required to commercialize a medical device.

Chapter 14 considers additional industry roles engineering designers may play. These include design transfer and post market surveillance as well as a supporting role in regulatory, reimbursement, and patent submissions. Design transfer is the formal hand-off of a finalized design to manufacturing and the last step of the engineering design process in industry. After commercialization, companies often ask engineers to assist with certain post market activities such as troubleshooting customer problems, investigating customer complaints, product liability and patent litigation, and exploring future versions of the product. Some industry engineers will spend a significant amount of time and effort on these tasks.

Chapter 15 provides you with a holistic view of the complex actions and decisions that move a medical device through the stage-gate process. These perspectives include, but are not limited to, finance, sales, marketing, corporate law, regulatory and reimbursement, supply and distribution chains, and operations. In most companies, dedicated professionals will be considering each of these areas throughout the stage-gate process, making decisions that will impact all other areas, including how you navigate the engineering design process. The goal of Chapter 15 is for you to understand and respect the terminology, frameworks, perspectives, and processes of these other professional disciplines. This broad understanding will allow you to be a more effective engineer and team member throughout your career.

By the end of Chapters 14 and 15 you will be able to:

- Prepare documentation and other materials necessary for design transfer.
- Describe the functional roles of professionals with whom engineers may interact during the development and commercialization of a medical device.
- Explain how the return on investment is determined for a medical product, used in making business decisions, and tracked after a product is launched.
- Recognize how intellectual property, brand awareness, and other assets are obtained and deployed when commercializing a medical device.

Beyond Design: The Engineer's Role in Design Transfer, Commercialization, and Post Market Surveillance

14

When we launch a product, we're already working on the next one.
And possibly even the next, next one.
— Tim Cook, CEO, Apple

Chapter outline

14.1 Introduction

The goal of a medical device company is to commercialize new products. Most academic design projects conclude after verification and validation testing and do not involve design transfer activities. In industry design projects, several stages beyond design verification and validation are required, including design transfer (per Design Controls) prior to market introduction. Although activities included in these stages are typically the responsibility of personnel in other functional groups within the organization (e.g., manufacturing, purchasing, regulatory affairs, quality assurance, marketing, sales), most of these activities require input from engineers. As the project leader and expert on the technical aspects of the product, you may be asked to assist with tasks that are not part of the formal design process. For example, after market introduction, you may be asked to participate in complaint investigations, recalls, or patent litigation.

In this chapter, we introduce you to the tasks of design engineers working on industry design projects that are not typically encountered in academic design projects. We discuss these activities and describe how you as an engineer working for a medical device company may become involved in moving products out of the lab and into the market. Understanding your potential role in this process will help prepare you for a career involving medical device design.

As discussed in Chapter 12, the US represents the largest market for medical devices and many other regulatory agencies have modeled their policies and approval processes on those used by the Food and Drug Administration (FDA). For this reason, our discussion of your potential role in supporting regulatory activities will focus on activities to support the FDA regulatory review process.

14.2 Regulatory Support

In industry, regulatory activities begin once a project is initiated, in parallel with design activities. Regulatory personnel are part of the initial project team and monitor the progress and evolution of the design. Their responsibilities include determining regulatory strategy to obtain clearance to market or approval of the new product from regulatory bodies such as the FDA, and considering if other pathways (presented in Chapter 12) such as De Novo and Humanitarian Device Exemptions, would be more appropriate. They also compile documents and information required to demonstrate the safety and effectiveness of the new product, submit this information to the FDA, and answer questions the FDA needs answered prior to allowing the company to market the new product. The documentation contained in your Design History File (DHF) will be needed by regulatory personnel as they prepare submissions to the FDA and international regulatory bodies.

As an expert in the design of the new product with an understanding of customer needs, how the product will be used, and the history of the project, you may be asked to assist regulatory personnel (in-house team members or outside consultants) with a variety of tasks at any time during the project. These tasks are described in the following sections.

14.2.1 Participation in Meetings With the FDA

At some point in the project, you may be asked to attend a pre-submission meeting with FDA personnel to discuss and determine the probable regulatory pathway (510[k] vs. Premarket Approval (PMA)), expected device classification, and testing required to prove safety and efficacy of the new product. Although your team may have completed a preliminary assessment of potential regulatory pathways for your new product, FDA personnel can provide insights into the nuances of the regulatory process that may affect your device classification and ultimate FDA pathway. This information helps set regulatory strategy for the company to determine how long the regulatory approval process will take. Knowing the required testing and pathway early in your project allows you (if you are the project leader) to create a more accurate project schedule that includes the tasks needed to obtain regulatory approval. A large percentage of the time needed to complete a medical device project will include tasks involving the project team to support regulatory submissions. These tasks are often a major factor in determining the overall project timeline.

14.2.2 Assisting in Training of Clinical Investigators

You may be asked to assist in the training of clinical investigators either before a clinical study of your device begins, or during the study, as new investigators participate. Depending on the complexity of the new product and how new it is to the target market, your technical familiarity may be needed to help explain how the product functions, how it is to be used by various end users such as physicians, therapists, nurses, caretakers, and patients, and what to do if a problem occurs during use of the product. Regulatory personnel from your company would typically be responsible for designing the study, writing the study protocol, obtaining Institutional Review Board (IRB) approvals, and establishing the training methods and content. They will also determine your role in the training process.

14.2.3 Writing Technical Sections of Regulatory Documents

As the technical expert on issues relating to the new product, you may be asked to write, revise, and approve specific technical sections of a 510(k) or PMA submission. These may involve technical descriptions of the product (including product specifications and functional requirements), how it is to be used, how it is manufactured and tested, and what to do in the case of an adverse event caused by the product.

14.3 Design Transfer

Design Transfer is the last formal stage in the design process as shown in Figure 1.3 and described in Section 11.4.9. To avoid introducing a product that is no longer needed, companies often confirm its expected return on investment, market demand, and its fit with company goals prior to transferring it to production. Once this is confirmed, the project will move to the design transfer phase, during which all design output is provided to manufacturing personnel. Changes after this point become very costly, as discussed in Chapters 8 and 9. Design transfer involves handing off all documentation needed to manufacture the new product to manufacturing personnel, including product and component drawings, bills of materials, purchasing specifications, inspection requirements, assembly drawings, and work instructions. These are part of the Device Master Record (DMR) and are discussed in detail in this section.

14.3.1 Documentation to Manufacture a New Product

Several forms of design output are provided to production personnel during design transfer in the form of the DHF and DMR. The DHF contains information regarding the evolution of the new product's design and documents compliance with Design Control requirements. The DMR contains information needed to manufacture, inspect, package, install, maintain, and service the new product.

Product and component drawings communicate information on geometry, dimensions, tolerances, materials, part and component numbers, and other design characteristics of all components and the final product. A **Bill of Materials** (BOM) lists all parts, part numbers, quantities of each part, and other related information. An example of a BOM for a syringe is shown in Breakout Box 14.1. Purchasing specifications document all requirements for purchased parts and materials. They typically specify material composition, required compliance with standards, required vendor test and inspection procedures, levels of purity, and other quality requirements. These are legal documents that are referred to in contracts and

Breakout Box 14.1 Example of a Bill of Material

A Bill of Materials (BOM) is a list of all parts needed to produce a finished product. It often indicates the hierarchy of components by listing top level assemblies, subassemblies, and individual components. Table 14.1 shows an example of a BOM for a polypropylene syringe, similar to that shown in Figure 11.1. Lower-level components are combined to create subassemblies, which are combined to create assemblies. As the lower-level components are assembled together, they create higher level subassemblies, until the process creates the final assembly (highest level).

In this example, polypropylene resin is injection-molded to create the syringe barrels and plungers. Silicone resin (level 5) is injection-molded to create the plunger tips (level 4), which are attached to the plungers to create the plunger subassemblies (level 3). Next, these plunger subassemblies are inserted into the barrels to create the syringe assembly (level 2). Finally, one syringe assembly is placed into a Tyvek®/Mylar pouch (level 2), and a preprinted label (level 2) is attached to complete the final packaged and labeled syringe assembly (level 1).

Bills of Materials (BOM) typically contain information regarding the level of assembly or part within the hierarchy, part numbers (for raw materials, purchased components, assemblies, and subassemblies), including the most current version, and a part name with descriptive information regarding the part or subassembly, as indicated in Table 14.1. Some BOMs may also contain cost information for each part as well as supplier information (not shown in Table 14.1). Supplier information is often found in purchasing documents; for this reason, it may not be appropriate to list a single supplier in a BOM in cases where multiple suppliers are used to source components and raw materials. Component costs can change frequently, requiring constant revisions of cost information in BOMs, and can be found in other purchasing and manufacturing documents. In some companies, project engineers work with manufacturing personnel to create a BOM for a new product.

Table 14.1 Example of an Indented Bill of Materials for the Assembly of a Polypropylene Syringe.

Assembly Level					Part Name	Part No./ Rev	Quantity	Type
1	2	3	4	5				
X					Syringe, packaged, labeled	SYR-01-B	1	Final Assy
	X				Syringe Assembly	SYR-10-C	1	Assy
		X			Barrel, injection molded	BAR-20-B	1	Sub
			X		Polypropylene resin	PPR-40-C	20 gm	Pur
		X			Plunger	PLU-20-B	1	Sub
			X		Plunger, injection molded	PLU-30-B	1	Sub
				X	Polypropylene resin	PPR-40-C	15 gm	Pur
			X		Plunger tip, injection molded	TIP-30-C		Sub
				X	Silicone resin	SIL-40-B	3 gm	Pur
	X				Pouch, Tyvek/Mylar	PKG-50-A	1	Pur
	X				Label, preprinted	PKG-51-A	1	Pur

Assy, Assembly; Pur, purchased component; Sub, subassembly.

purchase orders that define acceptable components or materials and are used to determine if a shipment of parts from a vendor can be accepted or should be rejected. Inspection requirements specify how the product will be sampled and inspected during incoming quality inspections (purchased items), in process testing (at various stages of production), and final inspection (finished product). They also specify the equipment and fixtures to be used and the dimensions to be checked.

Assembly drawings show how parts are assembled and work instructions include detailed step-by-step lists of how to produce, assemble, and test the new product. Work instructions are typically used by production workers and must be carefully written to include all information needed to complete each manufacturing operation. Transfer of these documents to manufacturing and quality assurance personnel is often the responsibility of design engineers.

14.3.2 Concurrent Engineering

In a well-managed project, some design transfer activities begin before the design is finalized and are completed concurrently (or in parallel) with other project tasks to save time. Some tasks (such as designing, constructing, and validating multi-cavity injection molds) have long lead times (4 to 6 months); beginning the process as early as possible helps shorten the time-to-market. To accomplish this, you will work with a multidisciplinary team of manufacturing, manufacturing engineering, quality assurance, and purchasing personnel (concurrently with design activities) once the design solution has been determined but well before it is frozen. This **concurrent engineering** approach ensures a smooth transition of the new product from the design laboratory to the manufacturing floor. These team members will typically be responsible for training production workers, designing assembly fixtures and tooling, designing production processes, qualifying suppliers, establishing contract manufacturing requirements, determining inspection procedures, and other tasks.

During design transfer, you will share your knowledge of the product's functional requirements with production personnel to ensure that manufacturing processes result in a product that eventually meets all design requirements and performance specifications. Keep in mind that you are the expert on functionality and how it relates to customer needs and will know which critical dimensions and performance characteristics will impact function. You will need to communicate this to quality assurance personnel who are responsible for ensuring that critical dimensions are inspected.

For example, to assist in placement of catheters, stents, and other tubular devices inside the body, guide wires are often placed into the internal lumens of these devices. Guide wires provide stiffness to the flexible catheters and assist in steering and advancing the catheter through intact natural channels such as blood vessels and ureters. To function properly, adequate clearance is needed to allow the guide wire to easily slide up through the lumen of the catheter. If the inside diameter of the catheter is too small, or the outside diameter of the guide wire is too large, then a press fit condition will occur, preventing placement and movement of the guide wire through the catheter. If this occurs during a procedure, the two components will not work together, and serious consequences for the patient can result. The design engineer would be well aware of the importance of maintaining the proper tolerances for the various inside and outside diameters of the mating parts to ensure proper function of the catheter and guide wire. Breakout Box 14.2 illustrates how design engineers use their expertise and familiarity with the design of a medical device to prevent efforts to reduce costs from impacting the function of the device.

Breakout Box 14.2 Cost Savings Versus Functionality

In industry, purchasing departments are responsible for ordering purchased items that meet the material, dimensional, and functional requirements determined by design engineers. Purchasing specifications are created by design engineers and purchasing agents to ensure that the purchased component will perform acceptably without compromising the function or safety of the medical device in which they are used. Other responsibilities include identifying reliable vendors, negotiating prices, expediting vendor deliveries, and managing relationships with vendors. Their goal is to prevent disruption of product flow on the manufacturing floor by ensuring that purchased parts and components are acceptable (meet the requirements established by design engineers), delivered on time, and available for production operations when they are needed. They also play an important role in reducing production costs, which is often accomplished by searching for new vendors who can provide components at lower costs.

Allowing the use of lower cost, alternate materials, and parts made with relaxed tolerances are two common ways to reduce the costs of purchased parts. Purchasing agents often request changes to specifications to allow purchasing of these lower cost parts. The design engineer is the most knowledgeable person regarding whether such a change will affect product performance or safety.

For example, ureteral stent manufacturers often purchase guide wires from vendors to package with their ureteral stent products. By providing a guide wire with the stent, companies have more control over the guide wire that urologists use during a stent placement procedure, ensuring that urologists use the proper diameter guide wire with their stents. This helps prevent (and reduces the risk of) the potential press fit condition described earlier.

The purchasing manager of one of these companies determined that by changing the purchasing specification to allow for slightly larger diameter guide wires (an additional 0.002 in [0.05 mm]), significant cost savings could be realized from the vendor. She requested that the design engineer approve a slight increase in the tolerance of the outside diameter of the guide wires, allowing acceptance of guide wires with a slightly larger diameter.

This request constituted a design change that had to be approved by the design engineer (the person most familiar with the design and potential sources of failure). In this case, for the reasons previously discussed, the design engineer did not approve the change and was able to defend this decision based on his knowledge of product design, use, and function.

As manufacturing engineers, industrial engineers, and other personnel suggest design and process changes that will allow the product to be manufactured more efficiently and at lower cost, you have an ethical and legal responsibility to ensure that any design or process changes do not compromise function or safety, as discussed in Chapter 13. This is important because many device recalls, such as those presented in Section 12.7, can be traced to product and/or process design changes or improvements that, due to a lack of proper testing, unknowingly affected product performance and function. Good documentation as part of the DHF helps ensure (and prove) that proposed design changes were carefully considered using good technical judgment and data to support decisions, and that these changes were approved using the company's established change control procedures.

14.3.3 Testing of Production Units

The design transfer phase requires verification that the production (not prototype) version of the new medical device meets specifications and is safe and effective prior to market introduction. This must be done through documented testing of the product under simulated or actual use conditions and is ideally done with sterilized (if the product is sold as sterile) production units (not prototypes). As discussed in Section 12.9.1, product recalls can be due to failure to test production units. In this section, we present the significant differences between prototype and production units that can lead to these recalls.

First, prototypes are typically made from prototype molds (or other tooling) that may be made of less expensive, easier to machine, and less wear-resistant materials (such as aluminum) compared to production tooling made of more durable steel. This difference can produce parts slightly dimensionally different than those made from production tooling.

Second, prototypes used for clinical studies or bench testing are often made by more skilled production workers who have been better trained in assembling prototype units than typical production personnel who make production units. These more experienced assemblers often have more time to produce prototypes and only produce a small quantity of units for testing. These differences can result in subtle dimensional and functional differences between prototype and production units that can lead to differences in device performance.

Third, production units are made using production quality tooling and production processes designed for producing larger quantities of product. The differences in tooling and processes can introduce subtle dimensional and other changes that can result in differences in performance. For example, release agents may be used on multicavity production molds to prevent parts from adhering to the molds, thus making part removal easier. Small amounts of these release agents can be absorbed by the molded parts and, when in contact with body tissues, can leach out and cause adverse reactions in patients. If tests are not conducted with these parts, then the effect of the leached release agent will not be known.

Fourth, the effects of sterilization and packaging design can impact product functionality. If non sterilized product is used for testing, then the effect of sterilization on the finished product will not be tested and could result in an undiscovered problem occurring after market introduction, leading to a potential recall. For example, the biodegradable ureteral stent discussed in Chapter 4 was sterilized using gamma radiation, which is known to produce chain scission in polymers over time, with a resulting decrease in molecular weight of the polymer. This chemical change resulted in a reduction in the polymer's tensile strength over time. This could have had a significant impact on stent function, but shelf-life tests (as described in Section 9.6.6) indicated that the polymer maintained an acceptable tensile strength for a 2-year period after sterilization. For this reason, product labeling indicated a 2-year shelf life. Packaging can also affect device performance if damage to the device occurs during shipping due to a poor package design that failed to protect the product. Packaging design and testing are presented in Sections 8.8 and 9.6.4, respectively.

Testing the final product, subject to all actual production processes and distribution conditions, is necessary to validate processes and verify product performance. When done properly, it will ensure that production units that are sterilized, shipped, and stored when used by the customer will meet all performance specifications and functional requirements. This protects the user, patient, and company.

14.4 Commercialization

Once a design has been transferred to production and all testing of production units has been completed with acceptable results, product specifications are then frozen. Production can then begin, and the product is ready to be commercialized (introduced to the market). Commercialization refers to the activities resulting in placing a product into the stream of commerce to get new, innovative medical devices to the people that can benefit from them. Commercialization is necessary for companies to sell their products and generate profits, but it also allows not-for-profit organizations to support their mission by selling or donating their products to the people who need them. A single organization can contain both for-profit and nonprofit entities. For example, Laerdal Medical Corporation, a Norwegian-based international for-profit company, is a global leader in the design and manufacture of medical simulation training equipment and educational materials. Laerdal Global Health (LGH), on the other hand, is a nonprofit corporation that is focused on improving maternal and child health in the developing world. LGH does not need to recover development costs because these are funded through the profits from Laerdal Medical Corporation.

In most companies, tasks completed in preparation for commercialization are the responsibility of sales, marketing, purchasing, supply chain, and legal personnel. These include development of sales literature, user training materials, sales training presentations, and other marketing and sales support materials. Supply chain and distribution plans and strategies for protecting intellectual property are finalized. Your participation may decrease during this phase as the remaining tasks become the responsibility of others in the company, and you may be assigned to your next project. However, you may be asked to continue to support the original project through the following activities:

- Provide support in developing sales and marketing materials related to technical details of the new product (e.g., performance characteristics, capabilities). This may include support for sales literature, advertisements, and sales training sessions.
- Participate in the training of company sales representatives.
- Provide technical support through additional testing and evaluation of competitive products (competitive benchmarking) and comparing performance results to those of the new product.
- Provide technical sales support through field visits to customers to answer technical questions about the new product. This may include attendance and support at trade show events and medical conferences.
- Provide customer support through product troubleshooting and technical problem solving.
- Review sales literature and product labeling to ensure use of appropriate label claims regarding safety and efficacy of the product. This is to ensure that labeling (sales literature, package labels, package inserts, advertisements, user manuals, etc.) do not make claims that have not been verified through testing.
- Provide technical support to patent attorneys in their preparation of patent applications.
- Provide continued technical support to regulatory personnel.

Chapter 15 provides more information on the tasks, processes, terminology, and mindset of the personnel who are responsible for the tasks you will be supporting.

14.5 Post Market Surveillance

According to the Code of Federal Regulations, Title 21, Part 822, section (i), *post market surveillance* is the active, systematic, scientifically valid collection, analysis, and interpretation of data or other information about a marketed device. It is a tool for providing continuous feedback about a marketed medical device to ensure that the device remains safe and effective after market introduction.

Post market surveillance occurs after a new medical device is introduced and is typically the responsibility of regulatory or quality assurance personnel in a medical device company. However, due to your expertise and familiarity with the design and function of the new product, you may be asked to help with post market surveillance activities such as complaint and failure investigations, corrective and preventive actions (CAPA), device recalls, and providing expertise to assist attorneys in defending product liability and patent infringement lawsuits.

14.5.1 Complaint Investigations

Companies receive complaints about their products from hospitals, physicians, and patients, which are reported through the Medical Device Reporting (MDR) process (presented in Section 12.7). The FDA defines a complaint as "any written, electronic, or oral communication that alleges deficiencies related to the identity, quality, durability, reliability, safety, effectiveness, or performance of a device after it is released for distribution." Per ISO 13485 (presented in Chapter 11), companies are required to investigate all product complaints and respond back to the complainant (customer) to close the feedback loop, as shown in Figure 14.1.

Complaint investigations often involve inspection (and sometimes testing) of the returned product by someone who is familiar with the design and function of the product. As the design engineer, you are an ideal candidate to conduct these investigations to confirm if the returned product (including the original packaging if available) is damaged and/or does not function as intended and determine the source of the problem. Sources include design defects, mishandling, misuse, manufacturing defects, inadequate packaging, shipping issues, and loss of functionality during aging, any of which may result in a defective or damaged product.

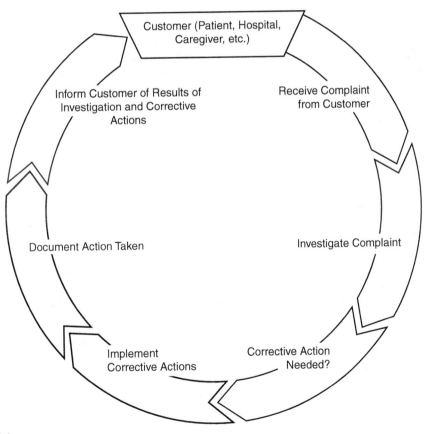

FIGURE 14.1

Process for complaint investigations.

If it is determined that all product currently in process, stored in inventory, or in the field is defective (due to a design or manufacturing defect), then a decision regarding how to correct the defect must be made. This may require a field action or product recall, as presented in Section 12.7. A field action consists of notifying customers of the problem and taking corrective actions where the product is physically located, instead of requesting that all units in the field be returned to the manufacturer. Examples of possible field actions include sending technical or repair personnel to the customer's location to replace a defective part or make adjustments to the product, providing customers software updates, or sending instructions on how to properly maintain or sterilize the product between uses. The appropriate use of field actions depends on the type of defect and its potential to cause death or serious injury.

There are situations when a returned device is found to function properly and claims of device failure or inadequate performance cannot be confirmed through testing. This can occur when the customer does not use the device properly or has unreasonable expectations of device performance.

Once the complaint investigation is concluded, you will determine if any corrective or preventive actions are required to prevent the problem from occurring again in future production units. These could involve changes to the design of the product, manufacturing process, instructions for use, or warnings, and are similar to actions taken to reduce risk as part of a risk management process, as presented in Sections 8.9 and 9.11. As shown in Figure 14.1, corrective actions are identified, implemented, and documented. Finally, to close the feedback loop, the customer is informed by quality assurance personnel of the results of the investigation and any corrective actions taken.

Breakout Box 14.3 presents an example involving a complaint for a medical device. It includes a discussion of the problem, results of an investigation, and potential corrective and preventive actions.

Breakout Box 14.3 Complaint Investigation and Resulting Corrective and Preventive Actions

Prior to injecting a patient using the medication pump device described in Breakout Box 9.2, the user needs to aspirate fluid from the patient's tissue to ensure proper needle placement. When functioning properly, seconds after the needle is inserted, the device creates a negative pressure by retracting the piston to create a slight negative pressure that aspirates fluid from the tissue into the clear tubing. If this fluid is clear, then proper placement of the needle into tissue is confirmed. If the fluid is dark red (blood), incorrect placement is confirmed and indicates that the needle is in a vein and must be repositioned. This aspiration phase is repeated until clear fluid is observed in the clear tubing. Once correct placement is confirmed, the injection can safely continue until the entire dose of medication is delivered.

Suppose that 1) accidental injections of medication into veins when using conventional hand syringes resulting in temporary adverse reactions had been reported, and 2) a few months after this new product was introduced the company received a report of an accidental injection into a vein. This event resulted in an adverse reaction and was attributed to excessive wear of an O-ring used to seal the piston that forced the medication up into the tubing, through the needle, and out into the tissue. As wear of this O-ring increased, the seal degraded, and reversing of the piston failed to create sufficient negative pressure needed to aspirate fluid from the tissue. This was a critical step necessary to ensure that the needle had not been placed into a vein prior to injection of the medication. Due to difficulty in confirming the correct needle placement due to wear of the O-ring, the medication was accidentally injected directly into a vein, causing the adverse reaction reported by the patient.

To correct this problem, the design engineer could create a maintenance kit consisting of a set of new O-rings and silicone lubricant along with written instructions for customers explaining how to replace their existing O-rings and apply the lubricant. Customers would be warned not to use their devices for more than two weeks without replacing the potentially worn O-rings. This combination of design change (addition of lubricant), procedure change (replacement of potentially worn O-rings with silicone lubricant every two weeks), and warnings constitute corrective and preventive actions, and could be implemented to reduce residual risk and increase safety. This field action could solve the problem without a recall of the product.

14.5.2 Product Recalls

You may become involved in device recalls for products that remain in the field but need to be monitored or modified. For example, the Bjork-Shiley heart valve recall, presented in Breakout Boxes 12.2 and 13.9, involved many patients for which the valve could not be removed due to risk of a replacement surgical procedure. The company determined that a simple chest X-ray could be used to predict the potential for a valve strut to fail, allowing patients with a recalled device to be monitored. In a situation such as this, you may be asked to help develop procedures for determining if a patient is at risk for device failure *in-vivo*, or field corrective actions that eliminate or reduce the risk of a product failure in the field.

14.5.3 Failure Investigations Involving Legal Action

As with complaint investigations, you may become involved in failure investigations regarding medical devices that are the subject of a lawsuit. The company's legal counsel will typically hire an outside expert witness to assist in defending the case, but as the design engineer, you may be asked to consult on the case due to your expertise regarding the function and design of the product in question.

The goal of a failure (and a complaint) investigation is not to simply comply with regulatory requirements, but to confirm that the failure occurred, determine the cause of the failure, and determine what can be done to prevent the failure from occurring again. Determining what can be done to reduce the probability of the failure from reoccurring is valuable information needed to reduce residual risk. This improves the design, reduces liability, and provides better and safer outcomes for patients and end users of the product.

The purpose of an investigation is to determine the cause of the failure, not to assign blame. However, product liability lawsuits typically allege that a failure is the fault of the manufacturer, due to either a design or manufacturing defect. Product failures are not always caused by design or manufacturing defects; they can also be caused by surgical error, misuse, and damage sustained during shipping. If the product is damaged during shipping, the shipping company or distributor may be liable. If a healthcare provider uses a product improperly, then a medical malpractice lawsuit may result. Medical negligence occurs when a provider performs below the standard of care, causing harm to the patient. In these cases, the health provider(s) can be held liable. Liability is not clear in the situation presented in Breakout Box 14.3. While the adverse outcome resulted from an injection of an anesthetic medication into the vein, the cause of this event was a clear-looking fluid aspiration caused by an improperly functioning product.

If an investigation indicates that the manufacturer's actions caused or contributed to the failure, then a product liability lawsuit will focus on the cause of failure. For this reason, the results of a failure investigation resulting from a lawsuit are kept confidential and shared with the manufacturer's legal counsel, who will determine how to best communicate these findings. Depending on the cause of the failure, the manufacturer may decide to either settle the lawsuit or defend it in court. If the case goes to trial in court, then you, as the design engineer conducting the investigation, may be asked to testify as an expert witness. For this reason, all notes and reports associated with the investigation need to include factual information, and not mislead, exaggerate, or reflect any bias toward the manufacturer. Information from the DHF is often used as evidence in product liability lawsuits.

One of the first steps that should be completed as part of a failure investigation is to confirm that the returned device in question was manufactured by your company. It is not uncommon for plaintiffs to name the incorrect manufacturer in a lawsuit. Confirming the manufacturer of the device early in the process can save time and lead to dismissal of the lawsuit if the incorrect manufacturer is identified. If your company did not manufacture the device, then it cannot be held liable for the failure.

To determine the cause of the failure (in complaint and failure investigations), the device itself needs to be inspected and possibly tested, and information relevant to the failure needs to be obtained and reviewed. Often, a failed device is sent to the manufacturer for testing to determine if it performs to the manufacturer's specifications and complies with existing applicable standards. Critical information needed to determine how and why the device failed should be gathered upon return of the device to be inspected and includes:

- name of product, model number, serial number
- photographs of the device and any visible damage to the device
- sketches and measurements describing visible damage to the device

14.5.4 Patent Litigation

Engineers are sometimes involved in patent litigation, both before and after a product has been introduced. This typically occurs when one company claims that another company (usually a competitor) is selling a product that infringes one of their patents and issues a cease and desist order demanding that the infringing company stop selling the product in question. If the infringing company does not comply, patent litigation may result.

As your company prepares its case (either as the plaintiff or the defendant), the legal team will need someone who is "skilled in the art" (as presented in Sections 5.8, 6.3.5, and 15.3) to provide an opinion regarding the claims of the patent and how the invention meets the criteria for patentability (novel, useful, and non obvious to one skilled in the art). As an engineer familiar with the design and related existing products, you are considered to be "skilled in the art" and therefore qualified to assist the legal team as an expert. You may be deposed (interviewed under oath) by the other company's legal team to answer questions about subpoenaed documents from the project's DHF.

Key Points

- For a commercial product, considerable work is required of design engineers after the initial design, prototyping, and testing phases.
- Design engineers are experts in how a design impacts their product's functional characteristics and performance.
- Design engineers know best if and how a design change might negatively impact the function of a medical device.
- Proposed design (and some process) changes must be approved by design engineers to prevent changes from affecting device performance.
- Due to their knowledge of product design and prior art, design engineers qualify as experts who are "skilled in the art."
- The Design History File contains information that can be helpful to a legal team that is defending a product liability lawsuit.

- Design engineers who are familiar with a specific product's design history often conduct complaint and failure investigations. Product failures may be due to design defects, manufacturing defects, packaging or distribution problems, or misuse of the product.
- Failure and complaint investigations are conducted to determine the root cause of the problem. They provide information that helps design engineers improve device performance and reduce risk resulting from design or manufacturing defects.

Exercises to Help Advance Your Design Project

1. Create a Bill of Materials for your product. Include all components and organize by subassemblies.
2. Write work instructions for one subassembly of your product.
   a. Share it with another person who is not familiar with the design of your product and ask them to follow the instructions to build the subassembly.
   b. Ask for feedback regarding the information provided in the work instructions.
   c. Was there any information missing, were there details that were left out, and was it easy to follow and not confusing to use?
3. Visit the websites of a few medical device manufacturers and find product literature for a medical device that is similar to your product. Read through the literature and note where claims have been made regarding product performance. Are claims substantiated by data presented in the product literature or cited in any references? What similar or different claims could you make regarding the device that you are designing?

References and Resources

Abuhav, I. (2012). *ISO 13485: A complete guide to quality management in the medical device industry.* CRC Press.

Association for the Advancement of Medical Instrumentation. (2015). *The quality system compendium: CGMP requirements and industry Practice.*

Code of Federal Regulations. (2002). *Part 822—Postmarket Surveillance.* https://www.ecfr.gov/current/title-21/chapter-I/subchapter-H/part-822.

DeMarco, C. (2011). *Medical device design and regulation.* American Society for Quality Press.

Geddes, L. A. (1998). *Medical device accidents.* CRC Press LLC, 216.

Teixeira, M. (2014). *Design controls for the medical device industry* (2nd ed.). CRC Press.

Yock, P. G., Zenios, S., Makower, J., et al. (2015). *Biodesign: The process of innovating medical technologies* (2nd ed.). Cambridge University Press.

Collaborating on Multifunctional Teams to Commercialize Medical Products

Education is not preparation for life. Education is life itself.
– John **Dewey**, philosopher and educational reformer

Persons appear to us according to the light we throw upon them from our own minds.
– Laura Ingalls **Wilder**, author of the *Little House on the Prairie* series

Chapter outline

Biomedical Engineering Design. https://doi.org/10.1016/B978-0-12-816444-0.00015-8

15.1 Introduction

In the healthcare ecosystem, moving an idea forward, whether it is a device, a process, a policy, or a new research idea, requires collaboration with others. These include experts in intellectual property, accounting, finance, sales, marketing, operations, manufacturing, and other areas. As it is not possible for one person to be an expert in all of these areas, commercializing a medical device requires a team effort. Whether you plan to attend graduate school, work at a large company, or start your own company, to be effective you must be able to work effectively with others.

As presented in Chapter 2, being a good team member and collaborator includes building and maintaining trust, a sense of mutual accountability, strong lines of communication, and a respect for viewpoints that are different from your own. While engineers navigate a medical device project by thinking about specifications, feasibility, risk, compliance with standards, and testing, other professionals view the same project through their own lenses. Their terminology, processes, and mindsets may seem strange to you. On the other hand, engineering design may be a mystery to them. Both groups should keep in mind that all team members are working toward the same goal.

The purpose of this chapter is to help you understand how team members representing these disciplines contribute to the commercialization of a medical device. As an engineer you may be asked to consider their terminology, frameworks, perspectives, and processes, recognizing that every company navigates the stage-gate process differently. The real goal, however, is for you to understand and respect their mindsets, opinions, and individual approaches so you can meet other professionals in the spaces between your respective disciplines. It is in these complex and often unexplored interstitial spaces that innovations can be developed and eventually have an impact on the world.

15.2 Engineering Economics and Finance

Moving a medical device through the stage-gate process requires resources. These resources include (but are not limited to) people, materials, equipment, and money. In the medical device industry, all of these resources are generally translated into a financial value so that they can be compared on equal footing. For example, a company might compare the cost of hiring two employees (e.g., salary, bonuses, vacation, sick pay, contributions to a retirement fund) to the cost of buying and maintaining a robot for an assembly line. As pointed out in Chapter 1, most product development decisions in a medical device company are based upon the relative long- and short-term Return on Investment (ROI) of competing solutions. For example, a company may decide that investing in a new manufacturing facility to scale up production for an existing product would provide a greater return than launching a new product. Although most companies employ financial and accounting professionals to evaluate these options, engineers should possess a basic level of economic and financial literacy. In this section, we introduce some of the terminology and mindsets of how finance is used to make business decisions.

15.2.1 Time Value of Money

When invested over time, money typically earns interest; a dollar in hand now is worth more than a dollar received in the future. The **time value of money** refers to how the value of money grows over time due to compound interest. The equations that govern this behavior are presented below along with examples. We also introduce a **cash flow diagram** as a graphical way to show revenues (flow of cash into the company) and expenses (flow of cash out of the company).

Given the following definitions:

Interest rate (i): rate of gain received from an investment,
Present value (P): value of a sum of money at time 0 (t = 0; present),
Future value (F): value of a sum of money at some time in the future, and
(n): number of interest periods;

in addition to the following assumptions:

1. The end of one year is the beginning of the next,
2. P is at the beginning of the current (present) year (t = 0), and
3. F is at the end of the nth year from the present;

consider the return on investing $100 at an annual interest rate of 10% per year two years from now. In this example, you invest $100 at t = 0 and want to know how much it will be worth at the end of two years. To calculate this future value (F_2), you are given that P = $100, i = (annual interest rate) = 0.10, and n = 2 years. To find F_2, we first draw the cash flow diagram shown in Figure 15.1 with expenses (negative cash flows) indicated with an arrow pointing downward and revenues (positive cash flows) pointing upward.

At the end of one year : $F_1 = \$100 + \$100\ (0.10) = \$100\ (1 + 0.10) = \$110$

At the end of two years: $F_2 = \$110\ (1 + 0.10) = \$121$

Or : $F_2 = \$100\ (1 + 0.10)\ (1 + 0.10)$

$= \$100\ (1 + 0.10)^2 = \$121$

FIGURE 15.1

Cash flow diagram indicating an initial expense of $100 (negative cash flow) and future value (F_2) at the end of two years.

We can generalize for any amount of money, present or future values, or time period (n) using the following equations:

Calculate F when P is known $F = P (1 + i)^n$

Calculate P when F is known $P = F / (1 + i)^n$

15.2.2 Using Cash Flow Diagrams to Make Business Decisions

Cash flow diagrams enable the outcomes of financial decisions to be directly compared to one another by estimating when positive and negative cash flows are expected (revenues and expenses, respectively). The simplified example below is meant to help you gain insight into how a financial professional would identify and evaluate cash flows for a new medical device project.

To start, a list of all known expenses should be generated. The list may include:

Product development cost : $315,000

Production costs : $43.25 / unit

Tooling cost : $230,000

Maintenance costs : $20,000 / year

This list reveals some terms that are often used by financial professionals. One-time costs are those that are incurred only once. Examples include product development costs and tooling costs, which are also sometimes called *start-up costs* or *up-front costs* because they are incurred before any product has been sold or manufactured. Product development costs include salaries, materials and supplies for prototypes and testing, prototype construction, testing (bench, animal, and clinical), sales and marketing materials, and any other expenses required to design and develop a product to the point where it is ready for commercialization. This could include costs associated with intellectual property protection, sales training, and regulatory and reimbursement approvals. High start-up costs can be a significant barrier to innovation in the medical device industry relative to other types of products.

Depending on the manufacturing costs and expected sales volumes, some companies will build an initial inventory to support a new product launch that is included in the start-up costs. For example, initial inventory for a new, low-cost, high-volume product such as a syringe would be produced prior to market introduction. Other companies build to order. For example, an MRI scanner is a high-cost, low-volume product and would be made as orders are received.

Recurring costs are those incurred at regular, anticipated intervals when operating a business. These include rent, electricity, insurance, salaries, and costs associated with manufacturing a product. In the example, production and maintenance costs can be considered recurring costs. These often include costs for raw materials and parts, assembly labor, maintenance of production equipment, packaging materials, sterilization, and distribution.

An additional distinction is made between *fixed costs* (which do not depend on production levels) and *variable costs* (which change based on production levels). These terms can be applied to both up-front and recurring costs. In the example, initial tooling costs are based on expected production levels and are incurred once before production begins. They do not change with higher production levels (unless sales exceed forecasted levels and additional tooling is needed to meet demand). On the other hand, maintenance costs will likely vary depending upon production levels. For example, higher production levels will result in increased use of production tooling, which can lead to increased wear that will require more frequent maintenance.

The next step would be to determine the expected *revenue*; all inflows of money into the company, over time. In our example, assume that market research has determined that a reasonable selling price for the product would be $100 per unit. Given that the production cost (including materials, labor, and overhead) for a single unit is $43.25, the *gross profit margin* would be ($100 – $43.25)/$100 = 56.75%, indicating that 56.75% of each dollar of sales revenue would be profit. A marketing product manager would generate a *sales forecast* that details the expected number of sales over time. The selling price and sales forecast could then be used to determine the expected revenue over time. The forecast in Table 15.1 considers sales revenue over time from United States (US) and European Union (EU) markets, but additional markets could easily be included.

The inflows and outflows can be used to estimate the *net cash flow* as shown in Table 15.2 for our example.

A visual summary of the net flow of cash over time is shown in the cash flow diagram in Figure 15.2.

15.2.3 Determining Return on Investment

To maximize overall ROI and ensure growth, businesses need to determine where to invest their available funds. These decisions are based to a large extent on the ROI of each investment alternative

Table 15.1 Forecast of the first five years of product sales revenue per year for US and EU markets

Year	US Sales (units)	Sales revenue @ $100/unit	EU Sales (units)	Sales revenue @ $100/unit	Total Units Sold	Total Revenue ($)
1	2000	200,000	1000	100,000	3000	300,000
2	2200	220,000	1100	110,000	3300	330,000
3	2500	250,000	1230	123,000	3730	373,000
4	3000	300,000	1500	150,000	4500	450,000
5	3500	350,000	1750	175,000	5250	525,000

Table 15.2 Summary of five-year forecast of cash flows for a new medical device

	Year 1	Year 2	Year 3	Year 4	Year 5
Expenses					
Product development	$315,000	0	0	0	0
Tooling	$230,000	0	0	0	0
Equipment maintenance	$20,000	$20,000	$20,000	$20,000	$20,000
Production @ $43.25/units X units sold	$129,750	$142,725	$161,323	$194,625	$227,066
Units sold	3000	3300	3730	4500	5250
Total expenses	$694,750	$162,725	$181,323	$214,625	$247,066
Total revenues @ $100/unit X units sold	$300,000	$330,000	$373,000	$450,000	$525,000
Net cash flow	-$394,750	$167,275	$191,677	$235,375	$277,934

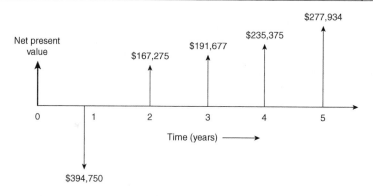

FIGURE 15.2

Cash flow diagram for the first five years of a new medical device.

which may include stocks and bonds, a new product development project, or expanding production facilities. The ROIs of many alternatives are calculated and compared to help determine which are to be funded.

Companies need to grow and maintain profitability. This is typically accomplished through increased sales, reduced costs, or both. One way to increase sales is to introduce new products to give a salesforce more products to sell. This is why new product development is extremely important to the growth of companies. However, companies have limited resources. Often, the resources needed to fund potential new product development projects are inadequate to fund all projects. In this situation, the ROIs of each project are calculated and compared to help determine which projects will be funded. Here we present several methods used to determine ROI. These include net present value (NPV), breakeven point, payback period, and internal rate of return.

15.2.3.1 Net Present Value

The cash flow diagram in Figure 15.2 does not consider the time value of money. This prevents us from directly comparing cash flows occurring in different years of the project. For example, a cash flow from year three might seem at first glance to be significantly higher than that from year two. However, due to the time value of money, this may not be true. To allow for "apples to apples" comparisons, future cash flows must be translated back (discounted) to their equivalent values at the same point in time (t = 0) based on the number of years of compounding and the interest rate. This adjustment results in a series of *discounted* cash flows that represent the present value of each cash flow (at t = 0), which can be directly compared. These are added to determine the *net present value* (NPV) which represents the sum of the series of discounted cash flows translated back to the start of the project (t = 0).

We can use equations from the previous two sections to compute present values for each cash flow in Figure 15.2 given an interest rate (i) of 5%, and the formula $P = F / (1 + i)^n$. We can then sum them to calculate the NPV for the project.

$$NPV = \sum PVn$$

$$= -394,750/(1.05)^1 + 167,275/(1.05)^2 + 191,677/(1.05)^3$$
$$+ 235,375/(1.05)^4 + 277,934/(1.05)^5$$

$$= -375,952 + 151,723 + 165,524 + 193,645 + 217,817$$

$$= +\$352,757$$

The NPV is an indication of a product's potential profitability. A positive NPV indicates that the product will be profitable (over time, the discounted revenues exceed the discounted expenses). A negative NPV indicates that the project will not be profitable.

15.2.3.2 Other Methods of Determining Return on Investment

Other common methods used to determine ROI include breakeven point, payback period, and internal rate of return. Each considers discounted cash flows in its calculations.

The *breakeven point* is the quantity of a product needed to be sold to generate enough positive cash flow to gain back the initial investment (negative cash flow) in a project. In other words, it is the quantity of product sold where the positive discounted cash flows equal the negative discounted cash flows (revenues = expenses).

Figure 15.3 illustrates how the breakeven point can be calculated. The dashed line shows the relationship between revenue and the number of units sold. The slope of this line is equal to the average selling price per unit. The more units sold, the higher the sales revenue. The solid line shows the relationship between costs and the number of units produced. The slope of this line is equal to the cost to produce a single unit. This line does not begin at the origin; it begins at a point on the y-axis equal to the fixed costs. As production begins, variable costs begin to increase. The more units produced, the higher the variable costs incurred, and the higher the total cost. When few units are sold, costs exceed revenues, resulting in a loss. As more units are produced and sold, revenue and costs increase at different rates. It is at the breakeven point that sales revenue exceeds the production costs. Sales beyond this breakeven point result in higher revenue than costs, resulting in a profit.

A related measure is the *payback period*, defined as the length of time required for positive discounted cash flows to equal negative discounted cash flows. This occurs when the cumulative discounted cash flows first become positive (+). In the simplified example, the projected net (discounted) cash flows are expected to recover the initial investment sometime during the fourth year of the project as shown in Table 15.3. This is not an atypical time frame to recover initial investments for a new medical device.

The payback period can also be determined from Figure 15.3. If the sales forecast for a particular product is 100 units per year, and the breakeven point is determined to be 250 units, then it will require (250 units/100 units/y) = 2.5 years to sell 250 units, resulting in the payback period of 2.5 years.

The *Internal rate of return (IRR)* is the interest rate at which discounted positive cash flows equal negative cash flows and NPV = 0 for a project. At interest rates above the IRR, the project becomes more profitable than simply investing funds at the currently available interest rate.

The examples presented here are simple and the calculations are straightforward. They can become more complicated if revenue from intellectual property, changes in regulatory requirements, or tax incentives (or increases) are included. The initial cash flow estimates are updated as a product is moved

Table 15.3 Cumulative cash flows for example					
	Year 1	**Year 2**	**Year 3**	**Year 4**	**Year 5**
Cumulative discounted net cash flow	- $375,952	- $224,229	- $58,505	$135,140	$352,757

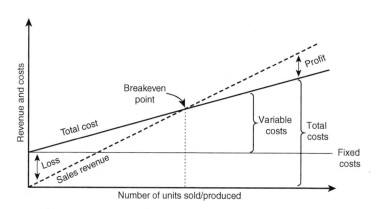

FIGURE 15.3

Calculation of Breakeven Point. The solid line represents total costs comprised of fixed and variable costs and has a slope equal to the cost to product a single unit. The dashed line represents revenue resulting from sales and has a slope equal to the average selling price. The breakeven point occurs at the point where these two lines intersect, where sales revenue begins to exceed production costs, resulting in a profit.

through the stage-gate model and more information becomes available. Updated calculations also occur after a product is launched to reflect changes in production costs, sales, and profitability. In many ways, this is similar to how an engineer considers conflicting and dynamic constraints, factors, and options when navigating the design process.

15.2.4 Comparison of Financial Alternatives and Project Selection

Just as an engineer evaluates designs, financial professionals evaluate financial projections and models. Financial projections are generated when a new project is considered whether the project is for a novel product or to improve an existing product.

Typically, several possible cash flow diagrams would be created for the same project, each exploring alternative financial models. For example, a company might consider whether to purchase a costly piece of manufacturing equipment from existing funds or from a loan to be repaid over time. There is also the possibility of leasing the equipment or hiring another company to make a custom part. Although each of these alternatives would result in different cash flow diagrams, the estimated ROI could be directly compared. In addition, many variables could be adjusted such as the selling price, expected sales, inflation, and interest rates. Adjusting these variables allows for upper (best case) and lower (worst case) scenarios to be generated that place bounds on the payback period and long-term ROI. Some sophisticated models may also estimate levels of financial risk, generally linked to the uncertainty of the estimates.

Management often establishes a *minimally acceptable return rate* (MARR) that all new projects must meet or exceed to be funded. A project that fails to meet the MARR at any point along the stage-gate process may be terminated. Any resources already expended would be considered unrecoverable *sunk costs* and thus would not impact any additional financial calculations. Each company updates its MARR based upon internal and external factors. External financial investments (e.g., stock market, mutual funds) may provide a greater rate of return than a proposed new product. The MARR is often chosen to be equal to the interest rate that can be earned by these external investments. Financial professionals help determine the best options for investing the company's money and recommend these to upper management.

15.2.5 Company Portfolios

Established medical device companies usually maintain a *portfolio* that includes commercialized products, new product development projects, financial investments, intellectual property, real estate, and equipment. These are all examples of *business assets* that can be leveraged to transform a new idea into a reality.

Sound financial decisions do not narrowly focus on the bottom line, but rather consider the holistic impact on the overall portfolio of the company. Other project selection criteria often include the fit of the new product with the company's existing research and development, production, sales and marketing capabilities, strategic plan and mission, and barriers to entry (regulatory hurdles, patent landscape, and competition). An individual project moves forward based upon its merits relative to other projects and its contribution to the portfolio. These decisions may vary over time, depending on the health of the company. When a company is doing well, management may invest in riskier, innovative projects, accepting that most may be terminated early and thus result in sunk costs. The investments and payoffs are therefore not dependent upon any single project. The overall strategy of funding many projects is that at least a few are successful and over time help strengthen the company portfolio. On the other hand, in hard times a company may terminate a project, or even existing products, to focus only on those that generate a stable profit.

Companies need to create a balanced, diversified portfolio of new products and new product development projects. To diversify risk in a new product development portfolio, different types of projects are funded. Companies may fund short- and long-term projects to ensure that new products are introduced (and new revenue streams are received) each year. These may include the development of some riskier new-to-the-world products, but also less risky line extensions, product enhancements, and cost reduction projects (see Section 15.5.4).

Management also considers intangible factors (those whose impact cannot be assigned a dollar amount) that indirectly impact the long-term financial health of the company. These factors include company reputation, brand loyalty, customer satisfaction, changes in regulations, social trends, competitive environment, and strategic investments of competitors. Furthermore, many medical device companies have a mission statement and core principles that aim to add value to the healthcare ecosystem. A company may therefore choose to pursue projects that may not return the greatest profit but are in alignment with its core principles. For example, Johnson & Johnson includes "Caring and Giving" as a core value and channels profits into the Johnson & Johnson Foundation that manages a wide range of humanitarian health projects around the world. An additional example is presented in Section 14.4 of the nonprofit work of the for-profit company Laerdal Medical Corporation.

15.2.6 Funding New Products in Established Companies

Within an established company, the assets of a portfolio can be directed toward the development of new products or the redesign of an existing product. An old heuristic was to invest 5% of sales from the previous year into research and product development. This number is often much higher today, sometimes reaching 30% of sales. Nearly all products become obsolete at some point in time. Therefore, there are many high-level business decisions to be made concerning how to dynamically channel the profits from an existing product to advance various projects that may yield value in the future. Boston Consulting Group created the Growth Share Matrix, shown in Figure 15.4, as a simple framework that describes how companies can reinvest profits.

The Growth Share Matrix divides products into four classes based upon their current market share and rate of market growth. For example, Cash Cows are products that have a high market share but are in a low-growth market and require little maintenance (low cost) to sustain high sales. Such products make a reliable profit for the company and often bolster the company brand. A good example is Band-Aid® produced by Johnson & Johnson Consumer Inc. Cash Cows can be contrasted with the Question Marks (upper left quadrant of the matrix). These products currently have a low market share but their growth potential is high. In other words, there is great potential, but the true ROI is difficult to calculate. When the potential for growth is realized, these question mark products become Stars—outstanding products that gain market share in a relatively short time. If the question mark products do not realize their market potential, the product becomes a Dog—products that do not make a sufficient profit and are generally cut from the company portfolio. Over time, some Stars become Cash Cows.

Using the matrix within a company means directing profits, usually from the Cash Cows, to test and grow products that might become Stars and eventually mature to become Cash Cows themselves. On the other hand, the profits from a Cash Cow could be reinvested in the Cash Cow to maintain market share. In other words, a company must make strategic decisions about how to reinvest profits to maintain the long-term sustainability of the company. For example, Zimmer Biomet started in 1927 manufacturing aluminum splints. They used this cash cow to continue to grow a diverse portfolio of

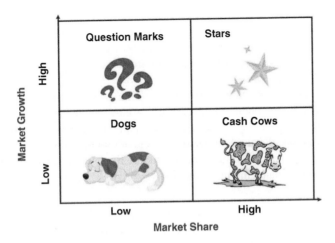

FIGURE 15.4

The Boston Consulting Group Growth Share Matrix.

orthopedic products that include artificial joints, dental replacements, prostheses, and devices used in implant surgeries. Section 15.5.5 explores more on how companies consider product life cycles.

15.2.7 Funding New Products in Start-Up Companies

For a new company, the dynamics of funding change. Large companies generally channel internal funding to new products. A new company, however, typically requires external funding to cover up-front costs. Entrepreneurs often piece together funding from many sources. Examples include the following:

- Federal, state, and local grants may be awarded through a competitive process. The SBIR (Small Business Innovation Research) and STTR (Small Business Technology Transfer) programs are two examples in the United States. These funds are made available to help stimulate the overall economy.
- Many types of philanthropic foundations (e.g., public, corporate, private family) have built the stimulation of business development in a particular market segment into their mission. A familiar example is the Bill and Melinda Gates Foundation, which has awarded several billion dollars to groups working on health care in the developing world.
- Competitions—whether regional, national, or international—can be a source of funding for start-ups competing against one another for monetary prizes. These usually are in the form of a pitch competition; some competition awards also come with peripheral benefits such as access to space, equipment, expertise, or other sources of funding.

One benefit to all three abovementioned funding sources is that funds do not need to be repaid. On the other hand, there may be other funding sources that do need to be repaid as the funders expect a return on their investment. These include:

- *Venture Capitalists (VCs)* generally work for firms that invest the money of other people in new companies. A VC does not expect a return on any individual company but rather hopes that the firm gains an overall return on its investment by having a diverse portfolio of start-up companies. A start-up company generally negotiates a percentage of its profits to be returned to the VC.
- *Angel Investors* are generally wealthy individuals willing to invest in an early-stage start-up. They often care deeply about the potential impact of the product and are therefore willing to be the first significant source of funding. Funding from an angel investor is typically used to keep the business growing and enables the founders to attract additional funding. In exchange, most angel investors own some percentage of the company but are likely to have an exit strategy, as they typically have no interest in managing the company.

In general, the key financial goal of a start-up company is to gain enough early funding to survive long enough to begin making a sustainable profit. Some academic design projects (e.g., Avitus, Ecovative, Keen Mobility, Tissue Analytics) have successfully transitioned to for-profit companies by mixing and matching the start-up funding opportunities above.

15.3 Protecting Intellectual Property

Intellectual property (IP) includes patents, trademarks, copyrights, and trade secrets. Governments grant IP "to promote the progress of science and useful arts by securing for limited times to authors and inventors the exclusive right to their respective writings and discoveries" (US Constitution, Article I, Section 8). Just as with physical property, IP is considered to be the property of the owner. Many companies maintain a portfolio of IP, a valuable and protected business asset. As such, companies pay significant fees to file legal documents to the appropriate governmental offices, detect infringement of their IP, and litigate when necessary. Similar to the landscape for medical regulations, each country has unique intellectual property laws. The nuances of intellectual property can become very complex and many medical device companies either employ a legal team or use an external legal firm. It is not unusual for a company to have a mixed strategy, employing a small internal legal team that then contracts out larger legal projects to a specialized IP firm. In this section, we introduce the basics of trademarks, copyrights, and trade secrets.

15.3.1 Trademarks, Copyrights, and Trade Secrets

Logos, names, processes, marketing materials, and instructions are all considered important parts of a product and are often legally protected. In this section, we discuss the means by which companies protect these valuable assets.

15.3.1.1 Trademarks

According to the USPTO website, a **trademark** is a word, phrase, symbol, and/or design that identifies and distinguishes the source of the goods of one party from those of others. A service mark is a word, phrase, symbol, and/or design that identifies and distinguishes the source of a service rather than goods.

Some examples include brand names, slogans, and logos. The term "trademark" is often used in a general sense to refer to both trademarks and service marks.

A trademark is valuable intellectual property because it allows customers to easily recognize a brand, a topic discussed in Section 15.4. Trademarks are ubiquitous and many are associated with common medical products. A quick perusal of products in your bathroom (e.g., dietary supplements, drugs, toothpastes) will reveal examples of trademarks indicated by the symbol®. Registered trademarks have been reviewed by a government office and deemed different enough from other registered trademarks for similar goods and/or services as to be unlikely to cause confusion in the marketplace. It is important to note that registered trademarks, like patents, are territorial; a registered trademark in one country does not carry over to other countries.

Trademarks are sometimes fiercely defended. For example, in 1977 the 3M corporation registered 3M® for rubber and plastic parts. In 2005 the Changzhou Huawei Advanced Material Co (CHAM) registered the trademark 3N in the United States for nonmetal building materials. CHAM argued that 3N is not the same as 3M® because it stands for "New Concept, New Technologies, New Products" and targets a separate market. 3M, however, filed a trademark infringement suit against CHAM claiming it attracted 3M's market share and customer base, potentially resulting in marketplace confusion and financial loss. The court decided that 3M® was well-known enough and that 3M and 3N were similar. The United States Supreme Court upheld the verdict in favor of 3M.

15.3.1.2 Copyrights

A **copyright**, indicated by the symbol ©, is a form of intellectual property that is most often reserved for the creative works of writers, artists, photographers, videographers, software developers, and musicians. It is similar to a patent in that it excludes others from copying the work without written consent or an agreement to provide royalties to the creator. Unlike patents and trademarks, however, the author by default owns the copyright; the work does not need to be registered with a government copyright office. This is the legal origin of plagiarism—passing off someone else's work as your own. The ethical and legal obligation is to credit the original author. In the world of written works, this is often achieved by crediting the author of a quote or making a citation to the original work, or both.

Although copyrights are the default in the United States when a work is created, there are still advantages to registering the copyright with the USPTO. Having a copyright on the public record can allow the author to have a much stronger case should there be a legal dispute. It is common for corporations to have their employees sign documents upon being hired that make any creative works generated during normal duties the property of the company. In this way, instruction manuals, labeling, and warning signs can become the intellectual property of the company.

15.3.1.3 Trade Secrets

Trade secrets refer to any information, such as technology, processes, formulations, or even customer lists, that has economic value, typically because it provides a commercial advantage to its owner, that is not generally known or ascertainable by others and that the owner takes reasonable measures to keep secret. Trade secrets are the original form of intellectual property in that protection is achieved simply by keeping it secret. Many trade secrets could be protected through other legal means (e.g., a patent). There are, however, some reasons why a company would prefer to maintain some information as a trade secret. First, the exclusive rights provided by patents have a defined time limit, after which the inventions disclosed can be freely used by others. Trade secrets have no time limit; for example, Coca-Cola's signature recipe is a trade secret that has allowed the company to maintain its market share for more than 125 years. Second, a trade secret can delay when the clock starts ticking on more formal legal protection. Such is the case with a design that has not been presented publicly or exploited commercially; it is essentially a trade secret known only to the members of the design team. It can remain a trade secret until the owners of the IP begin to commercialize it or decide to formally apply for legal protection. However, if an idea or product enters the public domain (even if by independent development by a third party or legitimate reverse engineering), a trade secret may lose the possibility of patent protection. For example, the Coca-Cola product has been available to the public for so long that it is no longer eligible for patent protection.

Although not protected in the same way as other IP, trade secrets are often protected in other ways. Trade secrets in the United States are protected by state laws guided by the 1985 Uniform Trade Secrets Act. Companies often require employees to sign legal documents upon hiring, forbidding them to reveal trade secrets. Penalties can be up to 10 years in federal prison and a $250,000 fine per secret stolen. For example, in 2019 a former employee used 33 trade secrets of a rideshare company to form his own startup. He is now serving time in federal prison.

15.3.2 Nondisclosure Agreements and Memoranda of Understanding

During the development of a product, parties outside the company often need to know about the product, including confidential and proprietary information. These parties include third-party manufacturers, experts and consultants, potential customers, suppliers of raw materials or parts, or distributors. Although there are situations when these parties can be included without revealing any sensitive intellectual property, that is sometimes not possible. For example, if a specialized part is being contracted out to another company, the contractor must have access to the design drawings of that part. Likewise, if a consultant is hired to help streamline a proprietary process, the company would want to prevent the consultant from disclosing this information to anyone outside of the company. In these situations, companies can draft legal a legal agreement that will protect their confidential information.

The most common agreements are the **Nondisclosure Agreement** (NDA) and the **Memorandum of Understand** (MOU). An NDA is the most common type of contract used when a company does not want proprietary information to be shared. Some unspoken nondisclosure agreements are culturally relevant, as discussed in Breakout Box 15.1. However, in the business world, an NDA is a legal contract signed by two or more parties who agree not to reveal proprietary information or knowledge to others. Common forms include unilateral (sometimes known as a one-way NDA), in which only one party agrees not to disclose information, or bilateral (sometimes known as a two-way NDA), in which each party agrees not to disclose proprietary information they may have learned from each other. Common components include the parties involved, the information to be kept confidential, the terms of the agreement, and the consequences for violating the agreement. Typically, an NDA would be written in legal terminology and be signed by all parties before disclosure of any sensitive information.

Some NDAs include a *noncompete clause*, which contains language that prevents someone with confidential information from using it to compete with the company. Typically, there is a time length given, such as one or five years. Such a clause may also be added to employee contracts, such that if an employee leaves the company, he is prevented from disclosing protected information. The employee of the ride-sharing company who used proprietary information to start his own company was violating a noncompete clause. Similar noncompete language is also used in other legal contracts with consultants or other companies.

Breakout Box 15.1 Magic and Unspoken Nondisclosure Agreements (NDAs)

Some NDAs are universal and do not require a contract. These include doctor–patient relationships, congregant–clergy relationships, and client–service relationships. Magicians are a special case in that proprietary information is generally not revealed—when the method is revealed, the trick no longer remains magical. An example where this was an issue occurred when Goldin Horace, an internationally known stage magician in the early 20th century, created a new illusion (at the time) of cutting a woman in half. He opted to patent it and described his method for doing so in great detail. Figure 15.5 is an image from the patent, which claims in part, "An illusion device comprising a long and deep box container specially designed for the herein described illusion of sawing the box and a person therein in half…" Other magicians started performing variations on the trick. Many legal battles ensued, but since the method to create the illusion was patented, arguments to prevent others from performing it were not honored by courts. The patent lawsuits did help spark creativity as Mr. Horace subsequently developed an advanced trick using a transparent box and an electric saw with an oversize blade as shown in Figure 15.5. This trick was not patented; it was kept a trade secret.

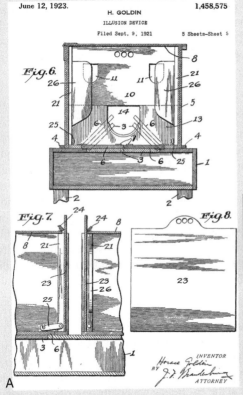

FIGURE 15.5

(A) An image for how the "sawing in half" illusion is performed. (B) An advanced version of sawing a woman-in-half.

(From US patent US1458575A, US Patent and Trademark Office.)

An MOU is not a legally binding document but rather a good-faith expression of a desire to work together. It is often written in plain language which more generally lays out the high-level goals of the relationship. In other words, legally binding promises are not made, and both parties can walk away from the relationship at any time for any reason. An MOU demonstrates a degree of loyalty (e.g., to a particular supplier or distributor).

NDAs and MOUs can become relevant to academic design projects in a variety of instances, as discussed in Chapter 2. Industry sponsors often require you to read and sign an NDA to protect their confidential information. Likewise, a university may also require you to sign an NDA to protect their own IP. Another instance in which an NDA may be relevant is if you present your work at a public event, such as a design exposition. Once a novel design is presented publicly or published, it becomes part of the public domain. In the United States, the inventors have one year from the date of the first public disclosure to file a patent application. However, such a grace period does not exist in the European Union, and all patent rights will be lost upon public disclosure. Sharing information about your academic design project with others (e.g., job or graduate school interview, a phone call with the supplier of a custom part or mold) may be prevented by the terms of an NDA. When in doubt, consult with your project mentor; there are always ways to talk about your project without revealing sensitive information.

15.3.3 Patents

Patents are the most common form of IP encountered by engineers. Patents are discussed in Chapter 5 in the context of idea generation, in Chapter 6 in the context of avoiding infringing existing IP, and in Chapter 14 in the context of providing design information in the event of litigation. In some companies, engineers might also engage in the dissection of competitive products (similar to the dissection discussed in Section 5.10) to detect any infringement of the company's patents. There is much more, however, to intellectual property that enters into how decisions are made within a medical device company. Large medical device companies may own thousands of patents, with new patents being filed often. For example, Medtronic owns over 10,000 patents. This section reviews some additional information regarding how patents create value for a company. If you wish to know more, many universities have a Technology Transfer Office that is staffed with legal professionals who may be able to answer your questions.

15.3.3.1 Patent Application Filing and Prosecution

Obtaining a patent is a process, much like the stage-gate process, where progressively more restrictive filters are applied. Very few ideas result in patents. The patent process typically includes a rigorous search for prior art, the preparation of a patent application, and a determination by the patent office that an idea is useful, novel, and nonobvious to one skilled in the art. Below we briefly describe how a legal professional would approach each of these steps.

A company must ensure that it is not infringing on any existing patents. A patent owner could prevent an infringing product from being commercialized or initiate legal action after the product is launched. A company would not want to advance a design into the stage-gate process without assessing its *Freedom to Operate* (FTO). Broadly, FTO is an assessment of the IP landscape to determine if the risk of potential infringement is low. A first step in determining FTO consists of assessing the patent landscape, often in multiple countries, to determine if the current design would infringe upon the claims of an existing patent. This would be a much more targeted search than described in Chapters 3 and 5 because at that phase of the design process the design was not known. In most companies, such a search would be conducted by a patent attorney employed by the company or a consultant from a patent law firm. It is not uncommon for existing IP to be found that closely relates to a design in development. A second step in determining FTO is to develop a plan for clearing the legal pathway to commercialization. This may include buying or licensing a patent from the owner (which allows the new owner or licensee to practice the invention without the threat of legal action) or filing a new patent application after modifying the design so as not to infringe the claims of the original patent.

FTO is an important legal milestone for a design to be deemed commercially viable, whether or not a company pursues patent protection. However, if a design is likely to be eligible for patent protection and pursuing a patent makes strategic sense for the company, some initial steps can be taken. A logical first step would be for the inventors to submit an invention disclosure to their patent attorney who will then draft an abstract detailing the unique features of the design. A draft of the patent claims will often also be included. These claims would then be used to conduct an even more targeted patent search. Based upon this search, a preliminary determination would be made as to whether or not the invention seems to be novel, useful, and nonobvious (criteria used by patent examiners). Patent attorneys representing the inventors will consult with patent examiners to gain additional guidance.

Some companies choose to file a *provisional patent application* if a design shows promise. This is an optional step but is relatively simple to file—sometimes only including information about the design—and can cost as little as a few hundred dollars. In the business world this is a fairly small investment and legally establishes the filing date for the invention. It is considered a public disclosure and notification of the inventors' intent to seek patent protection and therefore deters others from filing an application for the same idea. It also allows for public presentations and publications without jeopardizing future patent rights. After a provisional patent application is filed, a full patent application must be filed within one year in the United States or the owner loses legal rights to the IP. A provisional patent application allows time to prepare the full patent application. In effect, a provisional patent application establishes the filing date and holds the inventors' place in line.

A full patent application contains all necessary components described in Section 5.8, most importantly a full description of the claims. A patent is written in legal terminology so that it is clear to a patent examiner, and precise and defensible in the event of litigation. For this reason, a patent is almost always written by a patent agent or attorney, working with the inventors. These professionals are trained in the art of writing patents, specifically claims, so that they are as broad as possible, yet do not infringe on existing patents.

A patent application is submitted to the appropriate government agency and some companies simultaneously file in multiple countries. Filing fees can become significant and often depend upon the complexity (e.g., number of claims) of the patent. It is not unusual for a medical device patent filing in the US to cost $25,000. Filing in multiple countries can become costly. However, there are agreements between some countries such that only a single filing is necessary. In most instances, the application is confidential for a designated period of time, after which the application is published, typically on the patent office's website. In the United States, this period occurs approximately 1.5 years after the date of filing.

After the filing, a patent examiner prosecutes the patent, the legal term for the review. This is most often a two-step process. First, the examiner performs an initial review and indicates which claims will be allowed, if any. This is communicated to the patent attorney representing the inventors who, in the

United States, then have three months to respond (for more complex patents it may be 6 months). Claims can either be explained, amended, or removed. Second, the patent examiner performs a final review, makes a final determination, and provides the reason for the decision. In general, patent prosecution can take several years, especially for a complex patent. There are some options to accelerate the process that are used by some companies, but this must be requested at the time of the filing along with justification.

15.3.3.2 Inventorship and Ownership

Regarding patents, intellectual property is comprised of two parts. The intellectual component is assigned to the inventors, who are legally defined as those persons who had "intellectual domination" over the inventive process "or those who have made inventive contributions" to the invention. In practice, an inventor is anyone who contributed to at least one claim. In fact, as the claims are adjusted (both during the drafting of the application and in response to the feedback of the patent examiner), the list of inventors may be modified. It is important to point out that reduction to practice does not merit inventorship status. For example, someone who created or tested prototypes is not considered an inventor, unless he also contributed an idea that was included in the design and forms the basis for at least one claim. Listing someone as an inventor who did not contribute or excluding someone who made an inventive contribution can invalidate the patent in the event of litigation.

A patent is assigned to those who legally own the patent, typically those who paid the filing fees. It is considered the owner's responsibility (not that of the government patent office) to ensure that others are not infringing upon the patent. In this way, it is the owners who manage the financial and legal risks as well as receive the rewards from the patent.

It is possible for the inventors and the assignees who own the patent to overlap, for example, if an individual or a team files a patent to protect an idea that may form the basis of a startup company. In the corporate world, however, it is most common that the business would be the owner and the design engineers would be listed as inventors. It is not uncommon for the US-based corporations to require employees to sign over ownership rights to a patent in exchange for employment and salary. In Japan and Europe, however, the law requires that employee inventors be compensated. While ownership of a patent may change, inventorship never changes.

As an engineer, being listed as an inventor is a point of pride as well as an excellent professional demonstration of expertise in the inventive engineering design process. Many engineers frame and display their patents and list them prominently on their resumes. Justifying inventorship may sometimes become contentious. In these instances, the Design History File (DHF; including project notebooks) can be used to establish inventorship; it is therefore in your best interest to be sure that all of your contributions have been documented. In addition, should there be a dispute or litigation, the DHF may be submitted as part of a legal proceeding.

15.3.3.3 Patent Assignment, Licensing, and Royalties

Once a patent has been prosecuted and the intellectual property has been assigned, a company may use the patent to exclude others from manufacturing or selling the device. They can also exclude others from assigning or licensing it to another party. Both licensing agreements and assignments are complex legal contracts with many intricacies. As a patent is considered property it can be used in most of the same ways as physical property; it can be bought, sold, and traded between legal entities. For example, a company may buy a patent from the owner (perhaps another company) to obtain FTO. In some cases, no money changes hands, as companies may trade IP, or one company may purchase another. These instances require a change of *patent assignment* because the legal owner has changed. The USPTO website allows one to search by assignment, as shown in Figure 15.6.

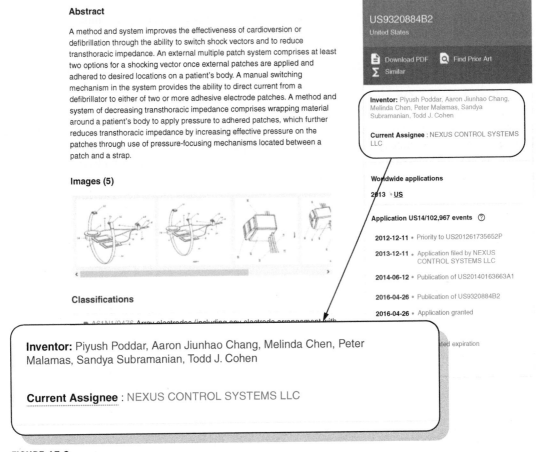

Method and system for switching shock vectors and decreasing transthoracic impedance for cardioversion and defibrillation

Abstract

A method and system improves the effectiveness of cardioversion or defibrillation through the ability to switch shock vectors and to reduce transthoracic impedance. An external multiple patch system comprises at least two options for a shocking vector once external patches are applied and adhered to desired locations on a patient's body. A manual switching mechanism in the system provides the ability to direct current from a defibrillator to either of two or more adhesive electrode patches. A method and system of decreasing transthoracic impedance comprises wrapping material around a patient's body to apply pressure to adhered patches, which further reduces transthoracic impedance by increasing effective pressure on the patches through use of pressure-focusing mechanisms located between a patch and a strap.

US9320884B2
United States

Download PDF   Find Prior Art
Similar

Inventor: Piyush Poddar, Aaron Jiunhao Chang, Melinda Chen, Peter Malamas, Sandya Subramanian, Todd J. Cohen

Current Assignee : NEXUS CONTROL SYSTEMS LLC

Images (5)

Worldwide applications

2013 · US

Application US14/102,967 events ⓘ

2012-12-11 · Priority to US201261735652P

2013-12-11 · Application filed by NEXUS CONTROL SYSTEMS LLC

2014-06-12 · Publication of US20140163663A1

2016-04-26 · Publication of US9320884B2

2016-04-26 · Application granted

Classifications

Inventor: Piyush Poddar, Aaron Jiunhao Chang, Melinda Chen, Peter Malamas, Sandya Subramanian, Todd J. Cohen

Current Assignee : NEXUS CONTROL SYSTEMS LLC

FIGURE 15.6

A patent originally assigned to an academic project team in March 2015 that was assigned to Nexus Control Systems LLC in February 2016.

Although a patent is not physical property, there are ways to share the intellectual component with outside parties. A patent can be *licensed*, meaning that the owner retains ownership but grants another party FTO. The party thereby has permission to manufacture, commercialize and sell the associated product, and ultimately profit from it. In some cases, the license might be exclusive, meaning that the inventors do not grant FTO to anyone else, i.e., a competitor. In return, the owner of the patent gets something of value. This may be a one-time financial sum, or periodic payments in the form of *royalties*, which are a percentage of net sales of the product. These two methods of compensation may also be mixed; a flat fee to be paid upfront in addition to royalties. The US Securities and Exchange Commission website (sec.gov) provides examples of licensing agreements.

Companies and individuals use assignments and licenses for a variety of strategic business reasons. First, allowing another the FTO can improve the probability of sales and commercial success. Second, it removes the work and expensive burden of commercialization from the inventor/owner, yet some financial benefits are still retained. Third, it is becoming increasingly common for commercial entities to either cross-license their technology or to create patent pools. Both mechanisms allow two or more companies in the same technical domain to use each other's IP to generate new technology. Amgen, for example, is one company that has many licensing agreements with companies such as Infinity, Biopharma, and Genentech. Fourth, the assignment or licensing of a patent may eventually yield a much higher return rate than commercializing the product. For example, Kyphon, LLC was a start-up company that received over 300 patents by 2002. One 1997 patent was for a unique method of ameliorating spinal compression laparoscopically using a balloon, a procedure now called kyphoplasty. The company was acquired by Medtronic for $3.9B in 2007, in large part due to that patent.

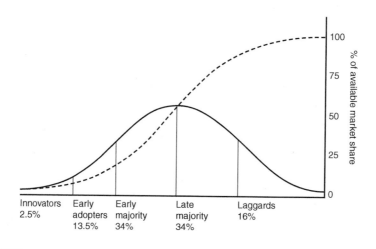

FIGURE 15.7

Everett Rogers' Diffusion of Innovation Curves. The solid curve shows an estimated percentage of each type of adopter archetype. The dotted curve shows the percentage of available market share over time as different adopter types begin using the innovation.

15.4 Marketing and Sales

Marketing and sales are two groups that work together and are intimately involved in the decision-making process as a product moves through the stage-gate process. Both groups also follow a product during and after product launch. As the most customer-facing part of a company, they identify market gaps, needs and opportunities, add value to products by increasing awareness and accessibility (e.g., branding, advertising, streamlining ordering), user education (e.g., online and in-person demonstrations, tutorials, customer support), and build and maintain trusted relationships with many stakeholders. They also collect and share important real-world perspectives and customer feedback that can be helpful in early and late validation of a product. For these reasons, it is important for engineers to have some understanding of how marketing and sales groups operate within a company. It is not usual for engineers to assist in some aspects of marketing and sales, as presented in Chapter 14. In this section, we introduce marketing and sales terminology and frameworks used in medical device companies.

15.4.1 Diffusion of Innovation

The underlying theory of much of modern marketing is built upon the work of Everett Rogers, which spanned the 1950s to 1980s and is detailed in the book *Diffusion of Innovations*. Rogers studied the factors that influence how new innovations (e.g., products, ideas, movements) spread throughout a population. Rogers' work has many dimensions, but one has become the cornerstone of marketing a new product; the diffusion of innovation curve, as shown in Figure 15.7. The blue bell-shaped curve is meant to represent the percentage of customer archetypes for a particular innovation. Rogers hypothesized that there would be a small number of *Innovators*, many of whom were involved in the creation of the innovation. There are a few more *Early Adopters* (a term coined by Rogers) who would readily adopt a new innovation. The bulk of a population, however, would be made up of an *Early Majority* and *Late Majority*. Finally, there are the *Laggards* who will only accept a new innovation after it becomes difficult to resist any longer. In the context of medical devices, these potential adopters might include patients, caretakers, healthcare providers, or purchasing agents.

Rogers documented the evolution of an innovation spreading from the Early Adopters to the Early Majority to Late Majority and Laggards until all capturable parts of the population had adopted. This progressive adoption is captured in the yellow S-shaped curve, as more and more people adopt the innovation. Although the bottom axis is not time, it does roughly mirror time as more and more people adopt the innovation. The maximum market share per cent (the y-axis) is 100%, but this does not mean 100% of the overall population has adopted; most new innovative medical devices are not intended to diffuse to an entire population. Rather, the goal is to capture 100% of the available market, which is usually a small subset of the overall population. In the healthcare ecosystem, the reachable market often includes a complex and varied range of healthcare providers.

Rogers used the diffusion curves to derive several additional theoretical frameworks that have greatly influenced modern marketing. We only cover one; how an individual moves throughout the decision to adopt. The model is shown in Figure 15.8 and includes five steps. First, an individual becomes aware of the existence of the innovation. Second, more is learned about the relative advantages of adoption. Third, a cognitive decision is made to adopt, but the adoption has not yet taken place. Fourth, the innovation is tested. Last, after a trial or demonstration, a decision is made whether to continue to use the innovation (to become an advocate) or discontinue use (to become a skeptic). Rogers imagined a potential adaptor to progress through these steps in stops and starts. Furthermore, a

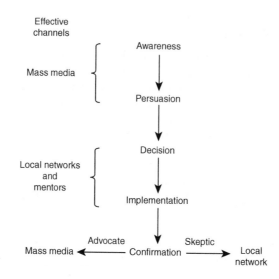

FIGURE 15.8

Diffusion of innovation framework for the individual decision-making process of adoption.

potential adopter can exit the adoption process at any point. As a potential adaptor progresses from stage to stage, she is reevaluating the value of adoption.

Where marketers can intervene is in the type and timing of information that is presented to individuals. Rogers hypothesized that certain *channels* for information, shown on the left side of Figure 15.8, would be most effective and efficient at particular points in the decision-making process. To build awareness and persuasion, mass media (e.g., radio, television, billboards) could be used. However, as an individual progresses, she requires more and more detailed and trusted information and therefore will turn to local networks, peers, and mentors.

Over the past few decades, the internet has changed the way some elements of marketing occur. For example, the distribution of helpful materials (e.g., user training, repair manuals), advertising (e.g., pop-up ads, targeted emails), customer service (e.g., chat lines, online FAQs), and customer feedback (e.g., unfiltered customer comments) all have a significant online component. These online elements have added several new channels that can help (or in some cases hinder) how a customer navigates the decision-making pathway toward adoption.

Several features of modern marketing can be derived from this decision-making framework. For example, when an individual adopts a product, it would be most helpful if he could help recruit others. One way to accomplish this is to imprint a catchy logo, name, or phrase (the essence of *branding*) on a product as an outward display of adoption. On a daily basis, we encounter hundreds (if not thousands) of branded products, which builds our awareness of the adoption of others.

15.4.2 Marketing and the Four P's

Marketing is a function of almost all businesses with the goal of helping potential customers navigate the adoption pathway in Figure 15.8. Practically, marketing is composed of four main components that are often referred to as the Four Ps: Product, Price, Place, and Promotion. In this section we briefly review this common marketing framework.

Product is about the functionality, features, ease of use, branding, packaging, and any service (e.g., repair, servicing, customizations) that will make the product attractive to potential customers.

Price is the sum of all financial commitments of the customer. This includes the list price, any discounts, credits, payment terms or contracts, warranties, maintenance and service fees, and payment periods. It could also involve bundling with other products such that the product price is reduced.

Place is about where a product is marketed, distributed, and sold. It focuses on how a company can make it easier for its customers to find products.

Promotion is how existing and potential customers learn more about a product, from awareness through adoption. The tools of promotion include advertising and sales materials (e.g., websites, product literature), mass media channels (e.g., radio, TV, billboards, internet ads), as well as in-person and virtual meetings and demonstrations. The combination of promotion techniques and the channels that are activated for a particular product is often referred to as the **marketing mix**. It is the task of the marketing department to determine the most effective marketing mix for a given product. A marketer carefully considers how to recruit early adaptors, often with special attention and incentives, so that they may help recruit others (e.g., early majority and late majority). Marketing a medical device often has some specialized marketing channels. Two such channels are displayed at professional conferences and meetings with hospital purchasing agents.

The four Ps have been a mainstay of marketing, but most modern marketing teams have reoriented to building customer relationships rather than simply selling products. This is especially true in the healthcare ecosystem where brand loyalty is critical. **Customer Lifetime Value** is the term used to describe returns that may be gained when a loyal customer continues to purchase products from the same company. Many companies estimate the financial value to the company of gaining a new customer versus losing a customer. A common adage is that it costs five times more to recruit a new customer than it does to retain an existing customer. To track customer analytics, many medical device companies use a *Customer Relationship Management* (CRM) system (often a software program) that tracks data (e.g., orders, visits, calls, emails, other contact points) for all of their customers.

15.4.3 Sales and Forecasting

Sales is often grouped with marketing, yet it performs a very different role within a business. At the most basic level, the sales department manages all of the logistics and paperwork to complete a sales transaction. In that role, however, they work closely with marketing and finance to track the current level and rate of sales as well as develop projections of future sales. This is a separate function from creating the financial projections in Section 15.1, which were developed prior to commercialization. Once a product is launched, the sales department uses real sales data to refine future sales projections.

Sales are most often summarized in a master spreadsheet. In most modern companies this spreadsheet is tied directly to the sales process such that as a sale is made it is automatically entered. The most basic spreadsheet would contain the elements shown in Table 15.4. The left-most column represents the important measures that are tracked. The top row represents the time period over which measurements are made, often by month or by business quarter.

Three critical dynamics can be extracted from Table 15.4. First, between January and February, there was an increased cost to manufacture and deliver the product (from $4.25 to $4.45). In this instance, the company may have invested capital to enhance the manufacturing process (e.g., new

Table 15.4 Example of a sales spreadsheet showing actual sales data. Forecasting (extending the table past the current time) makes estimates of how trends are expected to continue into the future.

	January 2022	February 2022	March 2022
Units sold	500	525	575
Unit price	$19.99	$19.99	$19.99
Revenue (Units sold * Price)	$9,995.00	$10,494.75	11,494.25
Sales Growth	-	5%	9%
Unit COGS (Cost of Goods Sold)	$4.25	$4.45	$4.05
Margin per unit (Unit price – Unit COGS)	$15.74	$15.54	$15.94
Gross profit (Margin per unit * Unit sold)	$7,870.00	$8,158.50	$9,165.50

machine, hiring a new employee) that resulted in a temporary increase in manufacturing costs. You should recognize this as an investment in assets that may pay off in the future. Second, over the same period, the gross profit increased (from $7,8700.00 to $8,1,58.50) because the company simultaneously sold more units, resulting in a greater profit. As described above, this increase in profit helps payoff initial investments and also verifies the long-term potential for the product. Third, the cost to manufacture product decreased between February and March. This is one example of **economies of scale**. In many industries, including the healthcare industry, the Cost of Goods Sold (COGS) often decreases as more units are manufactured and delivered. There are many factors in play that cause this decrease, including but not limited to, lower cost of materials from suppliers, more efficient manufacturing methods, and streamlining of processes. For example, setup time for a CNC mill to run a batch of parts is the same if for one part or for 100 parts. Furthermore, it is often the case that a company can purchase raw materials or parts in bulk at a reduced cost. In fact, when the ROI of a project is assessed during development, as in Section 15.1, these economies of scale are often anticipated.

The trends in Table 15.4 can be used to help track the financial success of a product. The growth rate of sales is a simple measure of how well a product is diffusing into a market. It can also be used to detect when a product has saturated the available market (e.g., when growth rates approach zero for a sustained period). Most companies create projections of all of the entries in the table as a way to forecast the expected timeline for reaching the breakeven point. Such data can also verify and update whether or not previous predictions were accurate. Likewise, it is possible to explore the impact of changes, such as decreased or increased selling price or the addition of an employee to help scale production, on the overall profitability of a product.

Table 15.4 is greatly simplified. Most sales spreadsheets break out elements in much more detail. The purpose of including these details is to better see how funds are flowing through the company. Such information can be used to target specific areas for improvement, as discussed in Section 15.8. Likewise, how many units have been sold by region (or another relevant category) helps marketing product managers detect where their efforts have been most (or least) effective. Finally, the spreadsheet can become much more complex when multiple products are included; consider that some products are bundled (e.g., surgical stapler, removal staples, and tool to remove staples) or share common aspects of the business (e.g., customer service, repair contracts, warranties).

15.5 Business Acumen

The most impactful engineers usually possess good business acumen, more specifically an understanding of the dynamics of how decisions are made and how internal processes are used to accomplish a goal. For example, Hewlett and Packard were two engineers who demonstrated superior business acumen. You might find yourself as the first technical person in a new start-up company. Or you might become an "intrapreneur"—someone who innovates from within a larger company to create a new product line, division, or business entity within the company. During design reviews, you may be questioned by members of the company who are analyzing your progress through a business lens. In all of these cases, it is important that you have a basic understanding of the concepts, processes, and terminology that influence business decisions. Many of these factors are discussed in other chapters and the preceding sections of this chapter. This section briefly covers some additional ideas and concepts that engineers often encounter in start-ups and in large companies.

15.5.1 Types and Classifications of Businesses and Companies

The medical device industry is composed of so many different business, company types, and classifications that it is not possible to cover each in enough depth. Although there is a difference between a business and a company, we use the terms interchangeably for the purposes of this section. A business may be defined by legal classification, size, ownership, flow of money, or several other attributes, as shown in Figure 15.9. There are even businesses composed of other businesses, which may all have their own classifications. A few points are essential to consider. First, a single business can often be classified in several ways. For example, a medical device company may be small, for-profit, and operate as a business-to-business entity. Second, the type or classification of business usually comes with unique incentives and restrictions that impact tax status, hiring practices, finance rules, liability, responsible parties, and the hierarchical structures of decision making. Third, companies can change their status over time, changing any of the dimensions in Figure 15.9. For example, a Limited by Guarantee company (known in the US as a Limited Liability Company or LLC) can become a Limited by Shares company if it becomes a publicly traded company.

Type of company based upon

Size	Ownership	Liability	Control	Customer	Financial
Sole proprietor	One-person	Limited by shares	Holding	Business to business	For profit
Start-up	Private	Limited by guarantee	Subsidiary	Business to customer	Non-profit
Small business	Public	Unlimited	Associate	Business to government	Governmental
Business					

FIGURE 15.9

Examples of business attributes that can be recombined to create many different types of companies.

15.5.2 Creating a Business Case

Every type of business has a value proposition and a business structure that aims to realize this value. As a new value proposition is created (or an existing value proposition is modified or updated), there are a variety of elements usually explored and communicated to others. In this section, we explore the business plan and the business model as mechanisms for clarifying and communicating the value proposition of a product. Many of the components would be prepared by personnel across several departments, compiled, and then shared with senior management.

15.5.2.1 Business Plans

A business plan is generally a formal (e.g., well-written and formatted) paper or electronic explanation of the value proposition. The plan usually contains the following:

- Executive Summary
- Company and/or Product Description
- Market Analysis
- Competitive Analysis
- Strategy and Implementation
- Organization and Management Team
- Details on Product and Services
- Marketing Plan
- Sales Strategy
- Budget or Request for Funding
- Financial Projections
- Exit Strategy (usually reserved for start-up companies)

A business plan aims to make a convincing argument to executives or potential investors that they can expect a return on their investment in a reasonable amount of time. The exact timeline of the ROI depends upon the type of business and product. For example, a software-only product may have a short development time and time-to-market with a comparatively short product life cycle, meaning that the ROI may be realized very quickly, but the product may also be replaced much more rapidly. On the other hand, a new medical device may take years (or even decades) to commercialize and require many years to realize an ROI. The trade-off is that a medical device often stays on the market, earning profits for a much more extended time period than other types of products.

Within a medical device company, the business case is often made several times throughout product development. The stage-gate model introduced in Chapter 1 has decision points when the business case is reevaluated. In this way, the business plan becomes an evolving document used to track a project's progress. As an engineer, you may be asked to help build a business case and defend it during stage-gate reviews.

15.5.2.2 Business Model Generation

Business plans often require a great deal of information, time to develop, and a relatively high level of certainty. In a relatively static environment, such long-term and careful planning may be possible. In a volatile space, however, a formal business plan may be incorrect before completion. In these situations, a solution is to create a *business model*. A business model is often developed by filling out a Business

Model Canvas, a one-page graphical tool that you can find online and is explained in detail in Osterwalder and Pigneur's book *Business Model Generation*. A business model is continuously updated by making a series of hypotheses about each element of a business plan and then performing simple experiments to verify or refute the hypothesis quickly. Just as in the scientific method and engineering design process, clarity of the business model emerges through iteration.

Three core entrepreneurial ideas stem from developing a business model. First is the concept of *failing fast*, sometimes also called failing forward. In the context of business model generation, refuting a hypothesis is a pathway to more quickly gaining clarity. The second is the concept of the *pivot*. A pivot is a change in direction, usually prompted by realizing the current business model has become untenable. Rather than give up, use the new understanding of the business landscape to generate a new business model. Often a pivot is associated with the development of a new value proposition or the identification of a new target customer. The third idea is the concept of rapidly getting a *Minimally Viable Product* (MVP) into the hands of potential customers. In Chapter 7, an MVP is defined as the most basic fully functional product needed to make a sale and allow a customer or user to understand the essence of the value proposition. The information gained from these early customers is then fed back into the redesign of the product.

In the medical device industry, the ideas of failing forward, pivoting and MVPs, seem to be in direct tension with the highly regulated healthcare ecosystem. However, these entrepreneurial ideas are being used in volatile and emerging areas such as medical smartphone applications, artificial intelligence applied to diagnosis, and deep learning algorithms that can search for genetic and epidemiological patterns in patient data. Likewise, many of the ideas of business model generation are thriving *inside* traditional medical device companies. A Minimally Viable Prototype (defined in Chapter 7) is often shown to trusted potential customers to gain immediate feedback during design reviews. The process of iterating on a prototype encourages failing forward. Furthermore, companies sometimes pivot when a product under development might benefit from a change in the value proposition; a surgical tool initially developed for liver transplantation could find a larger market by refocusing on kidney transplantation. An additional framework for considering a new value proposition is shown in Breakout Box 15.2.

15.5.3 The Pitch

Both a written report and a presentation are most often used to communicate a new business plan or model. In the start-up world, this presentation would be to potential venture capitalists or other investors, whereas in an established company the presentation would be to managers and those designated

---

Breakout Box 15.2 Will It Fly?

Thomas McKnight interviewed many venture capitalists (VCs) to understand what factors they consider when deciding whether or not to fund a new idea. Keep in mind that when a VC invests in a project, there is an expectation of return on that investment. McKnight summarized his findings in a book, *Will it Fly*, and created an Innovator's Scorecard shown in Figure 15.10. The scorecard has 44 categories for assessment where each area is weighted (1, 2, or 3) to reflect relative importance. Each area receives a score between 10 (perfect), 0 (neutral), and -10 (disaster). The greyed-out areas are for entrepreneurs only. To use the scorecard, make a rough guess as to the score (-10 to 10) in each area and then compute a total score using the weights. To raise the score, find entries with low scores and brainstorm ways to improve that part of the solution. Although the score card was created for VCs, established companies often consider similar areas.

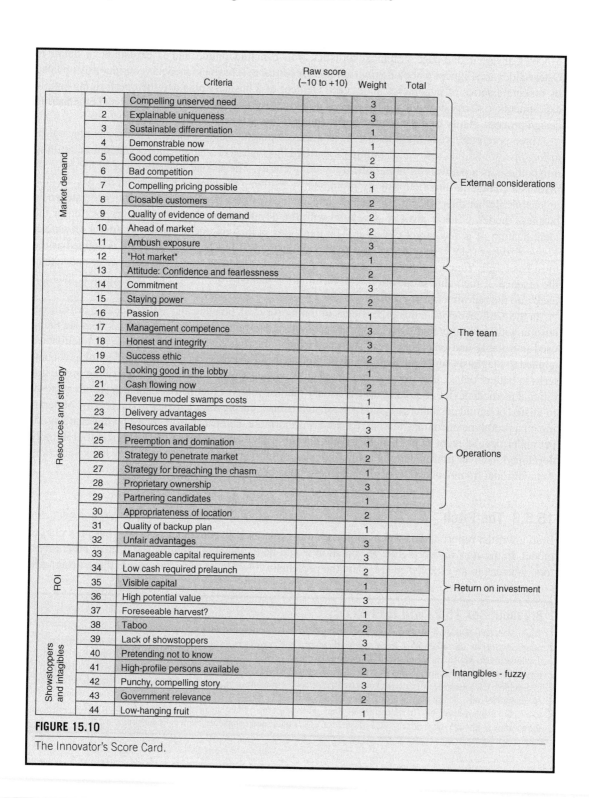

		Criteria	Raw score (−10 to +10)	Weight	Total	
Market demand	1	Compelling unserved need		3		External considerations
	2	Explainable uniqueness		3		
	3	Sustainable differentiation		1		
	4	Demonstrable now		1		
	5	Good competition		2		
	6	Bad competition		3		
	7	Compelling pricing possible		1		
	8	Closable customers		2		
	9	Quality of evidence of demand		2		
	10	Ahead of market		2		
	11	Ambush exposure		3		
	12	"Hot market"		1		
Resources and strategy	13	Attitude: Confidence and fearlessness		2		The team
	14	Commitment		3		
	15	Staying power		2		
	16	Passion		1		
	17	Management competence		3		
	18	Honest and integrity		3		
	19	Success ethic		2		
	20	Looking good in the lobby		1		
	21	Cash flowing now		2		
	22	Revenue model swamps costs		1		
	23	Delivery advantages		1		
	24	Resources available		3		
	25	Preemption and domination		1		Operations
	26	Strategy to penetrate market		2		
	27	Strategy for breaching the chasm		1		
	28	Proprietary ownership		3		
	29	Partnering candidates		1		
	30	Appropriateness of location		2		
	31	Quality of backup plan		1		
	32	Unfair advantages		3		
ROI	33	Manageable capital requirements		3		Return on investment
	34	Low cash required prelaunch		2		
	35	Visible capital		1		
	36	High potential value		3		
	37	Foreseeable harvest?		1		
Showstoppers and intagibles	38	Taboo		2		Intangibles - fuzzy
	39	Lack of showstoppers		3		
	40	Pretending not to know		1		
	41	High-profile persons available		2		
	42	Punchy, compelling story		3		
	43	Government relevance		2		
	44	Low-hanging fruit		1		

FIGURE 15.10

The Innovator's Score Card.

to make decisions. In either case, a *pitch* is the most common presentation format. A pitch intends to quickly communicate the most potent arguments for funding and keeping the project moving forward.

A pitch is usually driven by a *slide deck* (e.g., presentation slides) with approximately ten slides presented within ten minutes. Pitch guru Guy Kawasaki recommends the following slide flow, but keep in mind that his suggestions are for entrepreneurial start-ups:

- Title Slide
- Problem/Opportunity
- Value Proposition
- Underlying Novelty
- Business Model
- Marketing Plan
- Competitive Analysis
- Management Team
- Financial Projects and Key Metrics
- Current Status and Timeline Going Forward

Generally, there is a substantial amount of time allocated for questions once a pitch is concluded. Developing a robust business case allows for elaboration on the key points of your pitch.

Another pitch type is the *elevator pitch*, an oral articulation of the business case lasting between 30 seconds and 2 minutes. Even if speaking quickly, 30 seconds allows for about 75 to 100 words in which to state the medical problem, articulate the solution's novelty, and explain how this is an exciting commercial opportunity. One effective opening line is something like: "Prostate cancer patients have two treatment options where the safe choice results in life with impotence." The clinical problem and target patient population are defined, and the sentence invites more to be explained. There is a great deal more information online about how to deliver pitches, and you may find that there are pitch competitions and other resources on your campus.

15.5.4 Types of Industry Design Projects

Most industry design projects will fall into one of seven types, each of which offers unique benefits to the company, as well as a level of risk. The first four involve incremental changes to existing products, the fifth and sixth include current products, and the seventh involves the creation of a new product that did not exist before.

Line Extension—this type of project involves changing the size, color, or other attributes of an existing product, typically to expand the offer to the customer. No new features or functionality are added. An example of a line extension would be the addition of a 22 cm long ureteral stent to expand the stent product line offering; previously, the longest stent available was 20 cm. Line extensions typically involve established, validated designs and manufacturing processes, with a low risk of failure in the marketplace.

Product Enhancement—this type of project involves the addition of a new feature or function to an existing product. An example of a product enhancement would be the addition of a hydrophilic coating to a ureteral stent to reduce friction during insertion of the stent through a cystoscope. Such a functional change would make stent insertion easier for the urologist and more comfortable for the patient.

Product enhancements contain new features that may represent newer, unproven technologies or manufacturing processes compared to earlier product versions. Although it is expected that a product with new features will gain market support, there is no guarantee. Therefore, product enhancements generally involve a higher level of risk than a line extension.

Cost Reduction—this type of project focuses on implementing changes to lower the total manufacturing cost of an existing product. Reduced manufacturing costs either allow companies to reduce the selling price to better compete in the market or maintain the selling price to increase profit margins. Cost reduction strategies often include (1) finding lower-cost materials and parts, (2) combining multiple parts into a single component, (3) optimizing designs to reduce material requirements, and (4) finding more efficient production and assembly methods. These projects may present certain design risks (e.g., use of lower-cost materials, unfamiliar new production processes); design engineers are therefore often involved in generating and assessing cost reduction opportunities.

Regulatory Compliance—this type of project involves changes to existing products to ensure compliance with regulatory requirements. Changes may be made to the device's design or to the processes used to manufacture the product. These projects often result from changes in regulatory requirements from agencies such as the FDA, Environmental Protection Agency (EPA), or Occupational Safety and Health Administration (OSHA). Redesigning a medical device to be sterilized using a method other than ethylene oxide (EtO) to reduce exposure of production workers to EtO would be an example of this type of project.

Line Addition—this involves the addition of a product (or product line) that exists in the market but is not currently offered by the company. These products are already sold by competitors but would be new to the company. Depending on the complexity of the design, the company's experience with required manufacturing processes, and the competitive landscape, line additions can represent a medium to high level of risk.

Product Repositioning—this involves selling an existing product in a market in which the company does not currently sell the product. It is not a new product for the company, but it would be a new product for the new target market or market segment. An example would be selling the dental anesthesia pump (described in Breakout Box 9.2) to help podiatrists with painful heel injections. The device was initially designed for and sold to dentists. Still, when the company became aware of a new application for the same product in the podiatry market, the product was repositioned to be sold to a new market for the company. A product repositioning may require minor design changes, a new product name, or cosmetic changes to adapt it to the specific needs of the new market. These minor changes present a low technical risk, but uncertainty regarding market acceptance (in a new market) presents a moderate to high risk of failure.

New-to-the World Product—this type of project involves creating a new product that does not exist and thus is not currently in the company portfolio. It provides new features and new functions and addresses a problem in a new way. Some new-to-the-world products are disruptive in that they replace the currently accepted method of solving a problem. An example is Extracorporeal Shockwave Lithotripsy shown in Figure 4.3. When introduced, it was the only product available to treat kidney stones noninvasively, and it challenged the currently accepted method of removing kidney stones via open surgery. Due to unproven market acceptance, new, unproven technology, and possibly new manufacturing processes, new-to-the-world projects have the highest risk of all types of projects.

15.5.5 Product Life Cycles

When a product is already on the market, releasing an updated version is a critical decision. In deciding to update a product, many companies consider a typical **product life cycle**. The Diffusion of Innovation S-shaped curve in Figure 15.7 shows the adoption of a new innovation. Not pictured, however, is what happens once the maximum market diffusion has occurred; if a product is not refreshed (e.g., unique design, rebranded, etc.), it usually begins to lose market share. Figure 15.11 shows this trend as a downward turn that includes retirement (loss of market growth) and death (the product being discontinued). Companies can usually detect when a product is beginning to lose market share through sales data analysis, as shown in Table 15.4. This detection triggers a decision to retire the product, refresh the product with a new updated version, or reallocate funding (as an up-front investment) to an entirely new type of product. This high-level decision is usually informed by making an initial business case (e.g., the first stage of the stage-gate model). To maintain or grow market share, it is common to develop the next product version long before the previous version has been retired. The goal is to introduce Product 2.0 (growth phase) as Product 1.0 sales decline (retirement phase).

Start-up companies consider the product life cycle somewhat differently. With only one product initially, the goal is usually to get through conception, birth, and growth as quickly as possible to the point where profits can sustain the business. If this goal is achieved, the company may look for other markets for that same product (product repositioning) or diversify by adding new products that are similar along some dimension (e.g., used for the same disease, user, or clinical environment). An alternative used by some entrepreneurs is an exit plan in which the company is sold along with its brand, intellectual property, customer lists, and sometimes even employees to a larger company. The story of Kyphon, LLC in section 15.3 is an excellent example.

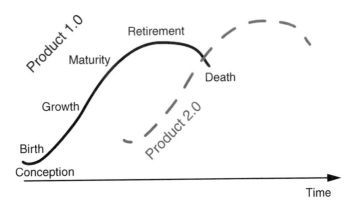

FIGURE 15.11

The Product Life Cycle. In this example, Product 1.0 grows, matures, and begins to decline. The detection of this decline (often measured by sales growth rates) triggers efforts to develop a similar, but refreshed product, called Product 2.0.

15.5.6 Start-up Companies and Local Resources

Many locations, both urban and rural, have support systems in place for new business ventures. The logic is that small local start-up companies help the local economy. Support may take many forms, from private incubators and accelerators to Small Business Development Centers (SBDC) funded through state and local governments. Each of these entities often sponsors free clinics, talks, and workshops that you can attend to learn more. In addition, many academic institutions have internal resources that you can access. For example, many business schools have free courses, clinics, and consultancies to help develop your business plan elements. Schools of Design can often help with naming, graphic design of marketing materials, logos, and websites. Law schools also have workshops or clinics that can help determine the appropriate legal structures of a business and strategies for protecting intellectual property (e.g., patents, trademarks). These services are often free or of minimal cost and provide excellent advice. Even if these services are available at your institution, neighboring institutions may be willing to help.

15.6 Supply and Distribution Chains

All businesses rely on other businesses in one way or another. Even software-only companies rely on the infrastructure that has been created and maintained by other institutions or companies. For this reason, many companies develop relationships with other businesses. Unlike in business-to-customer relationships, which are often one-off transactions, a business-to-business relationship may be sustained and backed by legal contracts. Over a long period, the relationship can become a trusted partnership. The most common partnerships are those of manufacturers and their suppliers and distributors. Many companies employ dedicated professionals who manage relationships with other businesses and track incoming raw materials and products shipped.

Suppliers (sometimes also called vendors) are the source of raw materials and parts that will be transformed and assembled into a product. Price, quality, service, reliability, delivery time, customization and flexibility, and capacity may all be considerations in choosing a supplier. In the medical device industry, suppliers must be audited and validated for compliance with Quality Management Systems requirements. This is typically done by purchasing and quality assurance personnel.

For a complex medical product, there often is a **supply chain**, whereby one supplier is supplying another company that acts as a supplier to another company. Consider a device such as a prosthetic hand. The company that sells the hand negotiated a relationship with companies that supply materials, electronics, motors, and sensors. However, those suppliers have their own suppliers and so on, tracing back to the raw materials extracted from some natural source. Problems (or practical innovations) at any point in the supply chain can ripple upward to impact all companies in the supply chain.

Distributors are responsible for moving a product to the customer. Generally, a distributor buys a product in bulk and then resells it in smaller quantities, often adding additional services along the way. Just as with suppliers, there are **distribution chains**, with the potential for several distributors being involved in moving a product from the manufacturer to the customer. At each step, the distributor requires some form of financial compensation in return for which it adds value, either to the customer or to the company making the product. This can take many forms (depending upon the distributor), ranging from warehousing and tracking of inventory and sales to packaging, sterilization, and quality control. When a distribution chain is working correctly, there is seamless communication between all parties, on-time delivery, maintenance of product quality, and effective communication of the product's value to the customer.

15.7 Health Insurance and Reimbursement

In most businesses, a customer directly pays the company that makes the product. This is the model used for some medical products sold directly to hospitals (e.g., medical simulation equipment or surgical tools), a medical office (e.g., tongue depressors), or an over-the-counter solution (e.g., bandages). However, payments for many medical devices do not flow through this pathway. Instead, health insurers (both public and private) manage the flow of money. Whether or not health insurers reimburse for the use of a device is, therefore, a significant factor in determining if a product will be viable in the marketplace. For example, if a healthcare provider is considering two similar products, (e.g., cervical collars for a fractured neck) and a patient's health insurance covers one but not the other, the provider will most likely choose the covered product.

In this section, we briefly outline some of the reimbursement factors considered during medical device design and commercialization. The information presented below is a greatly simplified view of reimbursement; the dynamics are far more nuanced. We begin with some preliminary information and clarification and then proceed to how two complementary sets of codes (CPT and ICD, discussed in Section 6.3.2) help determine whether or not an insurer reimburses for a new product.

15.7.1 Reimbursement Considerations in Medical Device Development

Reimbursement is generally considered early in the stage-gate process, perhaps even helping to determine if a new project is initiated or not. Later in the process, company representatives often approach health insurance agencies to determine the medical situations in which a device is likely to be reimbursed. Complicating this process, as is discussed below, is that each country has different reimbursement systems which are constantly changing. Furthermore, whether the device is used as part of a procedure (e.g., surgical tool) or as a part of the treatment (e.g., implant), it can profoundly impact the reimbursement strategy. As a result, larger companies often have an entire department that manages the company's reimbursement strategy. Industry engineers are typically not involved in developing a reimbursement strategy.

Most modern healthcare systems are single-payor systems where government health services pay for up to 75% of healthcare expenses. For example, patients in France generally are reimbursed for approximately 70% of their healthcare costs through a national health insurance plan. On the other hand, reimbursement in the United States involves the most complex system in the world. Payors include federal and state programs (e.g., Medicare, Medicaid, and CHIP), private for-profit and non-profit insurance companies (e.g., Anthem, UnitedHealth, Kaiser Permanente), and patients themselves. Some large employers provide health insurance for their employees. Each of these payors may reimburse for portions of a medical or surgical procedure, often basing coverage on pre-existing conditions, age, ability to pay, and other factors. Furthermore, the charges for the same medical procedure can vary significantly from one hospital system to another.

15.7.2 The Flow of Reimbursement Payments

Every medical service incurs some cost to the healthcare provider. The provider expects to be paid back either by an insurer or the individual receiving the service. After a service is rendered, the provider submits a *claim* to the insurer. This is essentially a request, on behalf of the patient, to pay for the service. This claim includes ICD and CPT codes, which are explained in more detail in the following sections. When an insurer receives a claim, it determines whether or not the patient's disease classification (summarized in the ICD code) justifies the service (outlined in the CPT code). If the claim is accepted, the insurer reimburses the healthcare provider for their service. The amount paid for the service is often negotiated with the healthcare provider or the medical facility. If full payment is expected at the time of service, the patient may be directly reimbursed. If an insurance claim is denied, it can be resubmitted. However, negotiations can become complicated if multiple parties are involved.

There are many nuances to the process of reimbursement. First, each insurer has unique databases and heuristics to help determine when there is alignment between ICD and CPT codes (resulting in a claim being accepted or denied). For example, when the suspected diagnosis is a broken bone, the cost of an X-ray is justified, and the claim is accepted. However, an insurer would not reimburse for an X-ray if the patient's diagnosis was a rash. Second, most healthcare providers have enough experience to know whether or not a patient's insurer will accept a claim. In other words, before performing a service, the healthcare provider generally informs the patient if it expects insurance to cover the service. Third, some insurers only approve partial coverage, leaving the remainder to be paid by the patient.

Reimbursement is a significant factor in how money flows through the healthcare ecosystem and is therefore a critical factor in whether or not healthcare providers adopt a medical device. Hospital systems generally purchase devices in bulk. An exception is a one-time large purchase, such as an imaging system. However, the procedures involving an imaging system may be reimbursed by insurers, allowing a healthcare provider to recover its initial investment in the imaging system.

Hospitals typically have purchasing committees that make decisions about which products to buy. These committee members know that healthcare providers would not usually use a device if the costs are transferred to the patient (e.g., the patient's insurance will not reimburse for the device). If a hospital is not reimbursed for a new product, the market for that product may be limited.

15.7.3 Current Procedural Terminology Codes

Current Procedural Terminology (CPT) codes are five-digit numerical codes that healthcare providers and insurance companies use to document services and prescriptions. A registry of thousands of codes is maintained and regularly updated by the American Medical Association (www.ama-assn.org) and helps provide some uniformity across healthcare providers. New codes are always being added; however, new codes typically require one to three years to be published.

CPT codes are divided into the following categories:

- Category I: Devices and drugs, including vaccines
- Category II: Performance measures and quality of care
- Category III: Services and procedures using emerging technology
- Proprietary Laboratory Analyses (PLA) codes: alpha-numeric CPT codes used for lab testing

ICD codes are applied to nearly all medical procedures and are used by hospitals to document all services and medical products used carefully. Healthcare systems often employ professional coders to ensure that treatment is appropriately coded. As a simple example, 99213 and 99214 are CPT codes for general check-ups. However, during a check-up, an individual might have received additional services such as a flu shot (CPT code 90658) and a shingles vaccine (CPT code 90750). In this way, each interaction with a healthcare provider generates a record of that interaction. Some CPT codes have also been created that represent bundled services—for example, a kidney transplant has a single CPT code (50365) that references several other critical services, each with a unique CPT code.

CPT codes are used beyond simply tracking services. For example, healthcare systems may use CPT codes to track the actions of individual healthcare providers, track inventory and aid in bookkeeping. They are also used to clearly communicate with insurers what services have been provided.

15.7.4 International Classification of Diseases Codes

When a healthcare provider meets with a patient, she may make a diagnosis. This diagnosis is entered into a database (often in an electronic patient record) and assigned one or more International Classification of Disease (ICD) codes. ICD codes are a library of international diagnoses and symptoms. The library, published and maintained by the World Health Organization, is one of the few uniform elements of healthcare systems worldwide. At the time of the writing of this text, the most current version is ICD-10-CM (Tenth Revision) that contains over 70,000 different codes. These codes are used for various purposes, ranging from tracking epidemics and worldwide mortality rates to processing health insurance claims. New codes are being created almost daily as new diseases are discovered, and diagnoses are refined.

Each ICD code is a unique identifier where the first three characters broadly indicate the disease, disorder, infection, or symptom. For example, codes starting with M00-M99 are for musculoskeletal system diseases and connective tissue (e.g., rheumatoid arthritis), while codes starting with J00-J99 are for diseases of the respiratory system. The remaining characters indicate body site, the severity of the problem, cause of the injury or illness, and other clinical details. Using this hierarchical classification, additional characters are added to make more specific diagnoses. For example, M05.7 is rheumatoid arthritis without organ or system involvement. M05.73 is more precise when indicating the bodily location of the wrist, whereas M05.732 is for the left wrist.

15.7.5 Reimbursement of Medical Devices

Convincing insurers that a new medical device should be reimbursed is a critical part of the pathway to the adoption of a new product. Three major steps are typically followed. First, a new medical device must have an associated CPT code to clarify the intervention or diagnostic procedure to insurers and healthcare providers. In many cases, an existing code describes the device and can be found online at sites such as www.findacode.com. Novel devices may require a new CPT code, which is obtained through an application and review process. For a device that might be used in various settings or considered part of a bundled service, multiple CPT codes would apply.

Second, the ICD codes would be found that correspond to the disease or symptom that would prompt the use of the device. There are several websites where ICD codes can be found. As with other searches, keywords are critical and should be based upon your knowledge of the medical situation. For example, if developing improved delivery forceps, keywords should include diagnoses such as "obstructed delivery" (ICD-10 code O66.9), "protracted labor" (ICD-10 code 0.63.9), "nonreassuring fetal heart tracing" (ICD-10 code O76), and the treatment "forceps delivery" (ICD-10 code O81). Each of these codes has more specificity to them (e.g., low-forceps, mid-cavity forceps, mid-cavity forceps with rotation). The more ICD and CPT codes are used, the more potential value is added to the hospital. A newly recognized disease or diagnosis may not have an ICD code yet. New codes are added through a submission and approval process administered by the World Health Organization.

Third, there must be a clear, data-driven alignment between the specific patient situation (ICD code) and the service provided (CPT code). These data are often obtained through clinical trials. However, as noted above, each country's regulatory and reimbursement pathways are different, and each insurer makes its own determination as to how ICD and CPT codes are aligned. It is entirely possible for one insurer to accept claims, while another does not. This puts a medical device company in the position of needing to negotiate with many insurers. This is especially true in the United States. A common strategy used by many medical companies is to start by gaining the approval of a few large insurers, with the assumption that other smaller insurance companies will follow. As Medicare and Medicaid are the largest insurers in the US accounting for more than 40% of all claims, many companies use the guidelines posted by the Center for Medicare & Medicaid Service (CMS) as a starting point for their reimbursement strategy. In general, CMS covers claims that are deemed to be both "reasonable and necessary." More information can be found at www.cms.gov.

15.8 Operations Mangagement

There are many jokes made about the tension associated with internal company operations, exemplified in cartoons like Dilbert, television shows like *The Office,* and movies like *Office Space.* Many of the jokes are about the seemingly absurd policies and procedures that no one understands or follows. The complex logistics of transforming raw materials and parts into a finished product, however, are critical to delivering a quality product, satisfying customer demands, and ensuring sustainable profit. This is especially true when no single person knows all of the details of how a product is made from start to finish. Furthermore, unlike in an academic design project where the goal is typically to create one unique final prototype, industry processes must scale to hundreds, thousands, or even millions of units. Maintaining an efficient, robust, high-quality manufacturing process requires companies to carefully control how materials, people, information, equipment, spaces, and supplies come together in the right way at the right time. Despite the jokes in popular culture, policies and procedures (sometimes called Standard Operating Procedures or SOPs) are a vital part of every modern medical device company. Furthermore, after a product is on the market, operations management experts often add new value by looking for ways to optimize production through process innovations.

The formal study of processes within a company is known as *operations management*, sometimes also called decision science or operations research in academia. The study of processes may go by many names within a company, including process engineering, systems engineering, or industrial engineering. A study of processes typically begins with equations that describe flows, which are measured as a quantity per unit of time. In this context, the quantity might include but is not limited to the number of parts delivered from a supplier, the percentage of work hours of a given employee dedicated to a particular project, the space taken up in a warehouse, waste generated, or the amount of product shipped. Time is typically measured by hours, days, weeks, or months. Dividing a quantity by a time period yields a flow rate. An operations management professional would attempt to determine all of the critical flows necessary to keep production moving forward. Each of these flows would have one or more equations assigned, many of them containing other dependent flows. Then, through a process of minimizing variables (e.g., waste generated, warehouse floor spaced used) and maximizing others (e.g., product shipped) an overall optimum production rate and process) would be determined.

Many factors and tradeoffs are quantified in the flow equations considered when determining an optimum production rate. First, production can only proceed as fast as the slowest step, often called a *time bottleneck*. If the capacitors on a circuit board can only be mounted and soldered at a specific rate, other downstream processes will be slowed. In this situation, an operations management person might suggest investing in a higher throughput device to allow other machines to run at a faster rate. The higher throughput may lower production costs, resulting in higher profits, and the new machine may soon pay for itself.

Second, production can only progress as fast as the inputs (e.g., materials, labor) can be replenished. This is a *material bottleneck*. If a supplier can only deliver a limited quantity of a critical component each week (a flow rate) or is inconsistent in its supply, the production rate will be limited. An operations management person might suggest renegotiating the contract with the supplier or switching to another supplier. A similar argument can be made for the outputs; there is no reason to make a product faster than needed to meet customer demand.

Third, delays in flow at any point in the process (either known or unexpected) can either halt or decrease the production rate. This might be due to the breakdown of a machine, a delay in the delivery of a component, or problems with shipping. These are all tests of the *robustness* of the system. Operations personnel might propose ways to maintain a degree of robustness of production even when other flows vary. For example, allocating floor space for critical supplies in a warehouse for a week (although at a cost to maintain) would allow production to continue even if a shipment is delayed.

Fourth, there are materials needed to keep production moving forward, and operations personnel try to strike a balance between maintaining efficient production operations and low production costs. For example, 100lb of a mold release agent might be ordered each week from a supplier to allow molded parts to be removed quickly and to keep molding operations running efficiently. This material is an example of a consumable that is not recovered or reprocessed and must be replaced after use. It is not part of the final product and does not appear in a Bill of Materials (BOM) but represents a cost to the company. Without it, all molding operations would be slowed or come to a halt. In this situation, using more release agents may accelerate production but would increase production costs. A careful analysis would be required to determine if the lower part costs resulting from increased production rates would justify the increased cost of using more release agents during molding operations.

Medical device companies, especially those that produce life-sustaining products, must consider the quality and integrity of their products. This consideration is summed up in the term *quality control,* described in Section 12.5. In this context, quality does not refer to how a customer might judge a product but rather to the process for creating the product and ensuring that all specifications have been met. The goal of quality control is to minimize the number of defective products that reach customers. As pointed out in Chapter 12, many regulatory bodies will consider the company's compliance with its internal quality system requirements and processes when assessing whether or not to approve a product for use.

The same type of "thinking in flows" described above can help in reducing defects. First, companies inspect or test parts and materials that come from suppliers. Defective parts can be returned to the supplier if the parts do not meet purchasing specifications. Second, it is not uncommon for testing and inspection to occur during the production process. For example, each circuit board might be functionally tested before being installed. Defective circuit boards would either be sent back to the production line to be corrected or discarded. Testing might also reveal a systematic problem with the line (e.g., defective diode placement because a machine is out of calibration and requires maintenance). Catching such a problem early, and shutting down the manufacturing line, halts further production of defective circuit boards. It is not uncommon for companies to also inspect and test finished devices. Those that do not pass inspection might be returned to the line for correction or discarded. Third, packaging of individual products (discussed in Chapters 8 and 12) and bundling of many products on a pallet for shipment are carefully monitored to reduce the number of products that may be damaged during shipment.

The above factors only consider the flow of the physical aspects of a product. An operations professional would also consider how the flow of information, people, space and other process components impact production. For example, models may include variables such as employee sick days and vacation, employee retention, or injury rate and severity. Likewise, an optimization method may determine that maintaining a small warehouse would be the most efficient way to maintain production even if the supply of parts is variable. Similar kinds of optimizations would be performed and used to find how to best deploy company resources.

As in all optimizations, trade-offs often mean that one internal group or department is asked to change so that another group can realize efficiency. These types of internal tensions and misunderstandings are captured in the popular portrayal of company dynamics. As an engineer, it is important to recognize that even though internal company procedures and policies may seem unnecessary or misplaced, they originate from a holistic view of how to best satisfy demand, make a profit, and deliver a safe and efficacious product.

15.9 Documentation of the Stage-Gate Process

We have encouraged you to document all of your design work. However, all of the functional areas mentioned in this chapter have their documentation processes and standards. There are financial regulations and accounting standards that govern bookkeeping. Marketing must document compliance with ethical and legal requirements affecting medical device marketing. Supply chain and distribution experts keep detailed records of numbers of incoming parts and outgoing devices and the quality, reliability, performance, and service of their suppliers and distributors.

In addition to each department keeping its own records, many companies holistically document all activities for various internal reasons. Such documentation could be used to track the progress of a project as it moves through the stage-gate process, aid in communication across functional groups, or facilitate the transfer of a project to another internal team or another company. Documentation might be used in performance reviews or as evidence to support the promotion of an employee.

Key Points

- Commercializing a medical device is a complex process and requires a diverse team of professionals who each support and respect each other's contributions.
- Understanding other disciplines' mindsets, frameworks, and terminology will help in effective communication with other professionals. Learning to develop respectful and trusting relationships with these professionals is a critical step in becoming an engineer who can thrive throughout a career.
- Finance and accounting professionals compare and update financial forecasts for a new project or product by considering the variations of how present assets may be deployed to earn a return on investment. They continue to monitor sales and costs after a product is launched.
- Intellectual property professionals consider how to protect the nontangible assets of the company through a mix of patents, trademarks, copyrights, and trade secrets, as well as nondisclosure agreements.
- Marketing professionals increase awareness and accessibility (e.g., branding, advertising, streamlined ordering) and user education (e.g., online and in-person demonstrations, tutorials, customer support), and build and maintain trusted relationships with many stakeholders.
- Sales professionals work with the logistics of completing the financial transactions of a sale and keeping records of sales that can then be used to make future sales projections.
- Supply and distribution chain professionals manage the complex flow of raw materials into the company and how the product reaches customers.
- Senior management often makes decisions based upon a holistic view of the company. These decisions are often based upon a business plan or business model.
- Documentation of commercialization is an important component of almost all professional disciplines. This is especially true in the medical device industry.

Exercises to Help Advance Your Design Project

1. Where would you source parts and raw materials to manufacture production quantities of your final product? Find (or estimate) the unit cost of these parts and materials.
2. Find suppliers or manufacturers who might be able to supply your entire product or critical components. Use online resources, or contact these companies, to determine estimates for contract manufacturing. Ask them to quote prices for quantities of 100, 1000, and 10,000 units or other relevant quantities. List any economies of scale in these estimates.
3. Based upon what you know of your product (e.g., selling prices of competitive or similar products, manufacturing costs), estimate your expected selling price. Project expected sales numbers 5 years into the future and use them to determine the anticipated total revenue for your product.
4. Generate a draft of Table 15.1 for your product. To do so, you should first determine all one-time and reoccurring costs that you anticipate before the product begins generating revenue.
5. Assume that claims for your current design could be written so that it does not infringe an existing patent. List the pros and cons of using this patent as an asset. These should include

reassigning the patent to a company, licensing, retaining the patent to commercialize the product yourself, and any liabilities. Remember to include both the benefits and risks.

6. Outline a marketing plan for your product, using the four Ps and the Diffusion of Innovation framework as a guide.

7. Imagine building a new company with your design as your first product. What would you name your company? What would you name your product? Design a logo for your company and product. You may wish to consult the many online tutorials for best practices in creating good company and product names and logos.

8. Determine alternative and/or secondary markets for your device. Create a marketing plan to reach these markets that includes the four Ps.

9. Create a product brochure that a salesperson might leave with a potential customer. You may want to consider what regulations are in place for the marketing of your product. An online search for "Healthcare Marketing Laws" reveals the nuances of this field.

10. Assume that you are creating a start-up company around your product. What mix of funding sources might you use? What is required to successfully obtain this funding (e.g., a government funding agency requires a proposal, a venture capitalist would require a pitch).

11. Use the Scorecard in Figure 15.10 to score as many aspects of your product as you can. Where are the scores low (or negative)? Are there ways to mitigate these barriers?

12. Find a company with an extensive portfolio of medical products that might be interested in buying the rights to commercialize your product. What would the value proposition be for this company in adding your product to their portfolio?

13. Look up the CPT codes you expect to be associated with your product. You may start by considering similar products and services.

14. Look up the ICD-10 codes you expect to be associated with the diagnoses and diseases that would lead to the use of your device. You may start by considering similar diagnoses and disease states.

15. What data do you think would convince health insurers that your device warrants coverage? Include specifics on the connection between CPT and ICD-10 codes.

16. How might you distribute your product? Would value be added by bundling it with other products? Would it require any special shipping requirements?

General Exercises

17. Find a medical product that is already on the market. Create a table, like Table 15.1, that outlines what you hypothesize were the start-up costs for the company. What investments (e.g., patents, new employees, new tooling) were also likely required before the product could enter the marketplace?

18. Perform research on the marketing materials for a medical device that is on the market. Dissect the approaches used by the company in the context of the four Ps and the Diffusion of Innovation framework. If you have a clinical mentor, you may ask them for materials that have been shared with them.

19. To cover the wide range of medical problems that healthcare providers encounter, some strange and very specific ICD codes have become a part of the database. Try searching online for "Strange ICD Codes." Report some of the most striking that you find.

References and Resources

Boardman, B. (2020). *Introduction to industrial engineering*. Mavs Open Press.

Boston Consulting Group. (2021). *What is the growth share matrix?* https://www.bcg.com/about/our-history/shiftentergrowth-share-matrix.aspx.

Brown, G. R. (1994). In *Sawing a Woman in Half*. American Heritage.

Duening, T. N., Hisrich, R. A., & Lechter, M. A. (2010). *Technology entrepreneurship: Creating, capturing and protecting value*. Academic Press.

Grissom, F. E., & Pressman, D. (2008). In *The Inventor's notebook: A "patent it yourself" companion*. NOLO.

Heizer, J., & Render, B. (2013). In *Operations management: Sustainability and supply chain management*. Pearson.

Kawasaki, G. (2015). In *The art of the start 2.0: The time-tested, battle-hardened guide for anyone starting Anything*. Penguin.

Kawasaki, G. (2015). *The only 10 slides you Need in your pitch*. https://guykawasaki.com/the-only-shiftenter10-slides-you-need-in-your-pitch/.

Kotler, P. T., & Armstrong, G. (2012). In *Principles of marketing*. Pearson Prentice Hall.

Ma, M. (2015). In *Fundamentals of patenting and licensing for scientists and engineers*. World Scientific Publishing Co.

McKnight, T. (2004). *Will it fly? How to know if your new business idea has Wings*. Pearson FT Press.

Osterwalder, A., & Pigneur, Y. (2010). *Business model generation: A Handbook for visionaries, game changers and challengers*. John Wiley & Sons.

Park, C. S. (2012). In *Fundamentals of engineering economics*. Pearson.

Rogers, E. M. (2003). In *Diffusion of innovations*. Free Press.

Thuesen, G., & Fabrycky, W. (1984). In *Engineering economy*. Prentice Hall.

Yock, P. G., Zenios, S., Makower, J., et al. (2015). In *Biodesign: The process of innovating medical technologies*. Cambridge University Press.

Index